T0327333

AC Circuits and Power Systems in Practice

AC Circuits and Power Systems in Practice

Graeme Vertigan

Registered Offices
John Wiley & Sons, Inc., 111 River Street, Hoboken, NJ 07030, USA
John Wiley & Sons Ltd, The Atrium, Southern Gate, Chichester, West Sussex, PO19 8SQ, UK

Editorial Office
The Atrium, Southern Gate, Chichester, West Sussex, PO19 8SQ, UK

For details of our global editorial offices, customer services, and more information about Wiley products visit us at www.wiley.com.

Wiley also publishes its books in a variety of electronic formats and by print-on-demand. Some content that appears in standard print versions of this book may not be available in other formats.

Library of Congress Cataloging-in-Publication Data

Names: Vertigan, Graeme, 1956– author.
Title: AC circuits and power systems in practice / by Mr. Graeme Vertigan.
Description: Hoboken, NJ, USA : Wiley, [2017] | Includes bibliographical references and index. |
Identifiers: LCCN 2017026963 (print) | LCCN 2017028362 (ebook) | ISBN 9781118924600 (pdf) | ISBN 9781118924617 (epub) | ISBN 9781118924594 (cloth)
Subjects: LCSH: Electric circuits–Alternating current. | Electric power systems.
Classification: LCC TK1141 (ebook) | LCC TK1141 .V47 2017 (print) | DDC 621.31/33–dc23
LC record available at https://lccn.loc.gov/2017026963

Cover design: Wiley
Cover image: Graeme Vertigan

Set in 10/12pt Warnock by SPi Global, Pondicherry, India

For Louise

Contents

Preface

This book is written as a practical power engineering text for engineering students and recent graduates. It contains more than 400 illustrations and is designed to provide the reader with a broad introduction to the subject and to facilitate further study. Many of the examples included come from industry and are not normally covered in under-graduate syllabi. They are provided to assist in bridging the gap between tertiary study and industrial practice, and to assist the professional development of recent graduates. The material presented is easy to follow and includes both mathematical and visual representations using phasor diagrams. Problems included at the end of most chapters are designed to walk the reader through practical applications of the associated theory.

The text is divided into two parts. The first (Chapters 1 – 6) is primarily intended for undergraduate students. It includes a general overview of the power system, AC circuit theory, network theorems and phasor analysis, in addition to a discussion of active and reactive power, magnetic circuits and an introduction to current and voltage trans-former operation. Part 1 concludes with a discussion of symmetrical component theory and the parameters affecting the flow of power in AC networks, including the phenom-enon of voltage collapse.

Chapter 1 provides a general overview of low, medium and high voltage power sys-tems, including the changes to the generation profile presently occurring and their implications for future network development.

Chapter 2 introduces RMS quantities and phasor representation of alternating volt-ages and currents. Elementary relationships between the voltage and currents in resis-tors, capacitors and inductors are derived and represented as phasor quantities. This chapter demonstrates the use of phasor diagrams as a tool for analysing complex cir-cuits and for gaining a visual insight into their operation. It also reviews voltage and current sign conventions, Kirchhoff's current and voltage laws and their application to the principle of superposition, as well as Thévenin and Norton's Theorems. Series and parallel resonant circuits are also introduced, together with the concept of the quality factor, Q and its application to resonant circuits.

Chapter 2 also contains several phasor analysis examples, including balancing the load of a single-phase induction furnace across three phases, the operation of a phase sequence indicator, power factor correction and capacitive voltage support for an inductive load.

The concepts of active power, reactive power and power factor are explained in Chapter 3, together with a discussion of the electrical characteristics of large and small commercial loads. The need for power factor correction (PFC) is considered together with a practical method of sizing PFC equipment for a given load. The chapter concludes with a general introduction to energy retailing, including transmission and distribution loss factors and maximum demand limits and charges.

Chapter 4 introduces the idea of a magnetic circuit and its application to voltage and current transformers. The properties of magnetic materials are considered, including air gaps necessary in reactor design. The constant flux model of a two-winding transformer and its equivalent circuit are developed, and per-unit quantities are introduced through numerical examples. Finally, the apparent difference between current and voltage transformers is explained. Examples include the reactor design and the determination of the Q factor, the magnetic analysis of an electromagnetic sheet-metal folding machine, and transformer operation from the point of view of mutual coupling.

Symmetrical component theory is introduced in Chapter 5, with the concept of positive, negative and zero-sequence components, sequence networks and sequence impedances. Sequence network connections are analysed for common system faults, and their use in determining the primary phase currents of a transformer with a faulted secondary is described. Examples include a method of locating faults in MV feeders as well as the design of an electronic negative sequence filter.

Chapter 6 considers the flow of active and reactive power in AC networks, including a discussion of the degree of coupling between them and the network parameters influencing each. The phenomenon of voltage collapse in networks is also discussed as well as steps generally taken by authorities to prevent one. Examples include voltage drops in transmission and distribution networks as functions of the system X/R ratio, and the practical application of phase shifting transformers to control active and reactive power flows in transmission networks.

The second part of the book (Chapters 7–14) contains material appropriate to final year students and recent engineering graduates and is written to assist a rapid integration into the engineering profession. It introduces the practical application of engineering standards and compares IEEE standards published in the USA with those published by the IEC in Europe.

Part 2 begins in Chapter 7 with a detailed discussion of three-phase transformers, including impedance calculations and the influence of core architectures and winding arrangements on positive, negative and zero-sequence impedances. Vector grouping, transformer voltage regulation, magnetising characteristics and zero-sequence impedances are examined, as are tap-changing techniques and the parallel operation of transformers. Examples include a detailed operational analysis of step voltage regulators and phase shifting transformers.

Chapter 8 examines the characteristics of both inductive and capacitive voltage transformers. It begins with a detailed examination of the inherent phase and magnitude errors and presents equations relating them to elements within the transformer equivalent circuit and the applied burden. IEEE and IEC voltage transformer standards are compared, with particular reference to ratings and accuracy classes. A simple method for error conversion between different burdens is presented, together with a discussion of the use of voltage transformers in protection and metering applications. The definitions of earth fault factor, effective earthing and the phenomenon of ferroresonance are discussed. Finally, the operating principles of non-conventional voltage transformers are briefly examined.

Chapter 9 analyses the operating principles and limitations of magnetic current transformers in metering and protection applications. The relevant IEEE and IEC standards are again compared. Magnitude and phase errors as well as ratio and transformer correction factors are defined and evaluated from elements within the CT equivalent circuit including the connected burden. Magnetising admittances and saturation effects are discussed and the concept of composite error and methods of measuring it in protection cores are described. The significance of the knee point voltage and accuracy limit factors in protection CTs are explained and the various protection classes defined in each standard are also considered. The derivation of the over-current ratio curve from the magnetising characteristic is described together with the series and parallel CT connections used in both protection and metering applications. Finally, non-conventional current transformers are introduced together with a discussion of their operating principles. Examples include the design of a simple current transformer test set, and the evaluation of CT errors from magnetising admittance data.

Three-phase energy metering circuits are described in Chapter 10. The concept of a metering interval is introduced and the advantages offered by static meters as compared to accumulation meters are explained. Both the three-element and two-element approaches to three-phase metering are analysed, including Blondel's theorem. Several non-Blondel compliant metering topologies are also described. This chapter considers the response of these circuits to negative and zero-sequence components and the degree of error they introduce. It also evaluates the overall metering error as a result of the inherent errors in current and voltage transformers. The final correction factor as defined in the IEEE standard is described and is related to the transformer correction factors for the associated voltage and current transformers. Examples include a comparison of MV and LV metering across a transformer, the recovery of the correct metering data from a faulted two-element metering installation and the analysis of a non-Blondel compliant metering topology.

Chapter 11 provides an introduction to the various earthing systems used in MV and LV electricity networks. It begins with an examination of the effects of electric current on the human body, which determine to a large extent the operational requirements of earthing systems. Chapter 11 continues with a discussion of the TT, TN and IT low voltage earthing systems, as well as impedance earthed and un-earthed neutrals used in medium voltage systems, including resonant earthing. An example of an earth grid design is presented according to the process outlined in the American standard IEEE 80.

Power system protection is introduced in Chapter 12, beginning with a general discussion of protection principles including primary and backup protection, check relays, zones of protection, discrimination and protection reliability. It concludes with a discussion of overcurrent, restricted earth fault, differential transformer protection and busbar protection systems as well as impedance protection schemes used throughout the transmission network. The class of current transformers required for each scheme is explained. Examples include the detection of a cable fault on a resonant earthed system, the interpretation of inverse time-current characteristics for establishing grading margins and the operation of a high impedance bus protection scheme in the presence of CT saturation.

Chapter 13 considers the issue of harmonics in power systems. It describes the resolution of a periodic waveform into its harmonic components using Fourier series and the simplifications that can be made due to waveform symmetry. Common harmonic measures including total harmonic distortion, total demand distortion, crest factor and transformer K ratings are defined, as are positive, negative and zero-sequence (triplen) harmonics. The adverse effects of triplen harmonic currents are discussed, as are methods used to contain them. The effects of harmonic losses in transformers are described and the harmonic loss factor is used to calculate a transformer de-rating factor appropriate to the load harmonic spectrum. The power factor definition is amended in the presence of harmonics to allow for the resulting harmonic VAr flow. Harmonic filters and harmonic cancellation techniques are discussed and the approaches taken under IEC and IEEE standards to harmonic management are described, together with the assessment of a distorting load prior to network connection.

Finally, Chapter 14 brings together several operational topics of interest to graduate engineers. These include a discussion of the one line diagram and device numbers used to designate primary and secondary items of equipment, followed by a discussion of common switchgear and busbar topologies including suggestions for optimal busbar arrangements. Switching plans, equipment isolation and permit to work procedures are discussed as well as workplace safety and the observation of limits of approach relative to live equipment. Arc flash injury and the selection of arc rated personal protective equipment are also discussed.

Graeme Vertigan
January 2017

Acknowledgements

Writing this book would not have been possible without the assistance, encouragement and advice of many colleagues, and the support of numerous organisations. I would like to acknowledge the assistance provided by Andrew Halley, Matthew and Chris Simmons, Jimmy Chong, Dominik Ziomek, Dan Sauer, Anasthasie Sainvilus, Daniela Chiaramonte and John Thierfelder in reviewing various chapters and assisting with copyright permissions.

In particular, I would like to thank Dr David Lewis and Mr Tim Sutton for their attention to detail, both technical and grammatical, and their patience in reviewing each chapter, usually more than once, and finally to David Vertigan for his skill and dedication in preparing the illustrations.

I would also like to thank the International Electrotechnical Commission (IEC) for permission to reproduce information from its International Standards. All such extracts are copyright of IEC, Geneva, Switzerland. All rights reserved. Further information on the IEC is available from www.iec.ch. IEC has no responsibility for the placement and context in which the extracts and contents are reproduced by the author, nor is the IEC in any way responsible for the other content or accuracy herein.

The author is also grateful to the Institute of Electrical and Electronics Engineers (IEEE) for permission to reproduce material from IEEE Standards. All such extracts are copyright of the IEEE, New Jersey, USA. The IEEE accepts no responsibility for the use of its material nor is it in any way responsible for the other content.

The Eaton Corporation of the USA has kindly provided illustrations and ongoing assistance with the preparation of material relating to its Cooper step regulators, for which I would like to express my appreciation.

Schneider Electric has made available material from its *Cahiers Techniques* series of industrial publications for use in this book, for which I would like to express my sincere thanks.

Finally, I would also like to extend my appreciation to the following companies and individuals for their assistance and use of materials:

- Lynda O'Meara of *SAI Global* on behalf of *Standards Australia*, Sydney
- Denis Berry of the *National Fire Prevention Association*, Quincy, Massachusetts
- Hughes Zhang of the *Guizhou Changzheng Electric Co., Ltd*, Peoples Republic of China
- Mark Ridgway of *AEM Cores*, Gillman, South Australia

- Romano Sartori of *Mitre Core Technologies*, Germiston South, Republic of South Africa
- Alan Bottomley the Tasmanian *Inventor of the Magnabend Electromagnetic Sheet Metal Bending Machine*, Hobart, Tasmania
- *The Beckwith Electric Company*, Largo, Florida
- *The International Energy Agency*, Paris, France

G. Vertigan

Part I

1

Power Systems: A General Overview

In this chapter we present an overview of the structure of a modern power system, from the low voltage distribution networks with which we are partly familiar, to the high voltage transmission system bringing energy from remote electrical generators.

Before we begin, we need to define the different potentials which we will encounter throughout the power system. We use abbreviations to denote different voltage levels, as outlined in Table 1.1. However, as is often the case, there appears to be no universally accepted definition, so you may see slightly different definitions used elsewhere.

1.1 Three-phase System of AC Voltages

The alternating voltage distributed to our homes has (ideally) a sinusoidal form. Sinusoidal waveforms are chosen because, when pure, they contain only one frequency; that is, they should contain no *supply frequency harmonics* (i.e. multiples of the fundamental frequency). Unfortunately, due to the increasing number of non-linear loads connected to the power system and the non-sinusoidal currents they consume, harmonic voltage distortion is becoming an increasing problem. This is particularly so in the LV network and, as a result, the AC voltage we receive often contains some *harmonic distortion*. However, despite this it is still approximately sinusoidal.

Power distribution systems universally use a system of *three-phase* sinusoidal voltages, with each phase displaced from the next by 120°, as shown in Figure 1.1. These potentials are generated by rotating a magnetic field having a sinusoidal *spatial flux distribution* inside a machine having three sets of fixed *stator* windings. The sinusoidal magnetic flux cutting each stator conductor induces a time-varying sinusoidal potential within it and the addition of the induced potential in each conductor in the winding produces the associated *phase voltage*.

In mainland Europe, the UK, China, India and the Middle East, the supply frequency is 50 Hz, whereas in the USA, Canada and parts of South America and Asia, 60 Hz is used.

Phase voltages are measured with respect to the *neutral* terminal, located at the *star point* of the winding, and are often represented by *colours*, in order to distinguish one phase from another. *Red, white and blue (R, W, B)* are frequently used, but in some locations *red, yellow and blue (R, Y, B)* are preferred. Alternatively, the letters *A, B and C, or U, V and W* are often used instead. For convenience throughout this text, we will use *A, B, C, or a, b, c.*

AC Circuits and Power Systems in Practice, First Edition. Graeme Vertigan.
© 2018 John Wiley & Sons Ltd. Published 2018 by John Wiley & Sons Ltd.

Table 1.1 Voltage definitions.

LV (low voltage)	<1000 V AC
MV (medium voltage)	1–35 kV AC
HV (high voltage)	35–230 kV
EHV (extra high voltage)	>230 kV

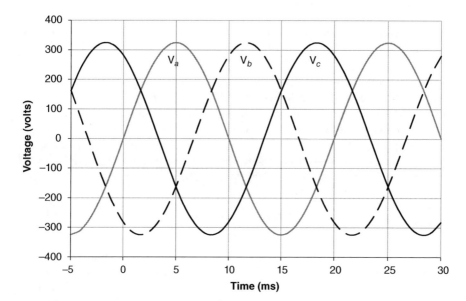

Figure 1.1 Three-phase alternating voltages.

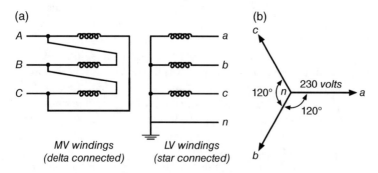

Figure 1.2 (a) Typical distribution transformer winding arrangement (b) Low voltage representation.

Low voltage customers are generally supplied from a three-phase *distribution transformer*, although single-phase transformers are often used for smaller loads. Figure 1.2a shows the transformer winding arrangement most commonly used. The primary windings or the *medium voltage (MV)* windings are *delta* connected (Δ) and are supplied from the three-conductor MV bus. The secondary or the *low voltage (LV)* windings are connected in a *wye (Y) or star* configuration, within which the LV potentials are induced.

Each low voltage phase is referenced to a common *neutral* terminal (n), and while the three-phase voltages each have the same magnitude, they differ from one another in phase by 120°. The magnitude of the phase voltages varies throughout the world, but in many countries a phase potential around 230 V is used.

The *sequence* in which the phase voltages reach their maximum value is important. Just as there are only two possible rotational directions for the magnetic field within a machine, (clockwise and anticlockwise), there are also only two possible *phase sequences*; ABC and ACB. The phase sequence of the supply determines the direction of rotation of polyphase motors. Reversing the phase sequence of a three-phase supply, by swapping any two phases, will thus reverse the direction of rotation of any machine connected to it.

Some loads, such as three-phase motors, do not always require a neutral connection; instead they operate from the potential difference between the phase voltages. These potentials are known as *line voltages* (or *line-to-line voltages*) and are illustrated in Figure 1.3. The line voltage amplitudes are greater by a factor of √3 than the phase voltages from which they are derived. Thus if the phase voltage is 230 V then the associated line voltage is about 400 V (often expressed as 230/400 V), and this is used to designate the system potential.

It is also worth noting that there is a 30° phase shift between phase voltages and the associated line voltage. For example, the *a* phase voltage *leads* the *ac* line voltage by 30°, i.e. it passes through its maximum 30° before the AC line voltage peaks. Similarly, the *b* phase voltage *leads* the *ba* line voltage by 30°, and finally the *c* phase voltage *leads* the *cb* line voltage by a further 30°, as shown in Figure 1.3.

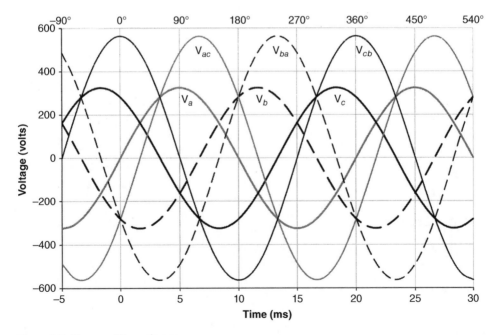

Figure 1.3 Phase and line voltages.

1.2 Low Voltage Distribution

Most European countries use 230/400 V for LV distribution, or in some cases 240/415 V instead. The majority of residential customers receive a *single-phase supply* (i.e. one phase conductor and the neutral) which terminate at the customer's premises via a *service drop* (or *service connection*). These conductors frequently run from the overhead LV network, to a roof or wall mount on the customer's dwelling, and from there to a *service fuse*, which can be used for isolation purposes. The supply cables or *customer mains* then run via the electricity meter to the customer's switchboard. Customers who have an underground LV supply are likely to have mains running from a *distribution kiosk* on the street to their premises.

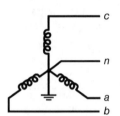

Figure 1.4 Typical 230/400 V three-phase supply.

Commercial and industrial low voltage customers in Europe, Australia and New Zealand are usually provided with a 230/400 V *three-phase supply* (three-phase conductors and the neutral) from an LV transformer arrangement like that in Figure 1.4. These customers may therefore run single-phase loads between any phase conductor and neutral, as well as line connected three-phase loads. The four-wire wye (or star) 230/400 V connection is probably the most common LV distribution arrangement, and it is used in many countries worldwide.

In some countries, large commercial or industrial LV loads can also be supplied with a three-wire 400/690 V supply. In the USA and Canada, however, residential customers are often supplied with two voltages: 120 V for lighting and low-current loads and 240 V for larger single-phase loads, such as water heaters or power tools. These voltages are often generated within the same single-phase transformer using a centre-tapped winding as shown in Figure 1.5. The centre tap (or *neutral terminal*) is earthed (connected to ground potential) and therefore the potentials of the two active conductors are 180° apart, that is, they are symmetrical with respect to the earthed neutral conductor. Such single-phase transformers are usually relatively small, (<100 kVA) and generally supply fewer than ten residences.

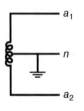

Figure 1.5 Single-phase transformer using a centre-tapped winding.

In older parts of North America other customers are supplied with a *three-phase four-wire* supply from a *delta connected* secondary winding (Figure 1.6), where one winding has been fitted with a grounded centre tap. This arrangement provides two 120 V supplies for lighting and low-current loads, while also providing a three-phase 240 V supply for high-current loads such as water and space heating. Because one phase lies 208 V away from ground while the others are only 120 V from it, this arrangement is known as a *high leg delta connection*.

One advantage of the high leg delta arrangement is that, if necessary, it can be provided in an *open delta* configuration from *two* single-phase transformers rather than three, as shown

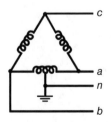

Figure 1.6 Three-phase four-wire supply from a delta connected secondary winding.

in Figure 1.7; however, the maximum three-phase load that can be applied in this situation is only 1/√3 or 57.7% of the rating with three transformers present. In other parts of the USA, residential customers are provided with a three-phase dual-voltage supply derived from a delta/wye transformer (Δ/Y) with a phase voltage of 120 V and a line voltage of 208 V.

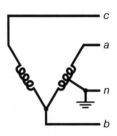

Figure 1.7 Open delta configuration from two single-phase transformers.

Commercial and industrial customers in the USA frequently require a three-phase supply with line voltages greater than either 208 or 240 V; for these a conventional delta/wye (Δ/Y) connected transformer is used, having phase/line voltages of either 265/460 V or 277/480 V.

Finally, there is one LV winding arrangement that has historically been used in the USA for mainly three-phase loads, which deserves special comment. It is known as a *corner grounded delta*, or a *grounded b phase system* (Figure 1.8). This configuration is no longer used in new installations but it was used for many years in irrigation and oil pumping applications and is occasionally still in use today. One advantage of the corner grounded delta is its ability to continue to supply load when operating as an open delta, as described above. This feature provided the possibility of doing maintenance on each *single-phase* transformer in turn, from a three-phase bank, without having to disconnect the entire load. Today, three-phase transformers are more reliable and are not assembled from single-phase devices, and therefore the corner grounded delta has lost its popularity.

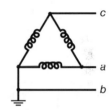

Figure 1.8 Corner grounded delta, or a grounded *b* phase system.

There are some disadvantages associated with this configuration that should also be mentioned. Firstly, for reasons of safety it is necessary to identify the grounded phase at all points along the line. Secondly, a *phase-to-ground fault* in a corner grounded delta network is essentially a *phase-to-phase fault*, and in 480 V systems quite large fault currents can flow through *one pole* of the circuit breaker. Interrupting such single-phase fault currents can be challenging, especially if the circuit breaker chosen does not have a sufficiently high *single pole interruption rating*.

Many moulded case circuit breakers used in LV applications have a considerably higher *three-phase* fault interruption capacity than a *single-phase* one. A *bolted* phase-to-phase fault (i.e. one with zero impedance) can generate as much as 87% of the current available in a three-phase fault, and the breaker chosen must be capable of clearing this level of fault current. Not all moulded case circuit breakers have the ability to interrupt single-phase currents of this magnitude. Circuit breakers used in corner grounded delta networks must therefore be carefully chosen with the necessary single pole interruption capacity in mind.

1.2.1 Voltage Tolerance

The LV distribution voltage levels mentioned above are all subject to a prescribed *tolerance*. This is partly because the MV system voltage fluctuates as a function of changes in the reflected LV load. Although the MV system potential is regulated by the HV/MV transformer (through the use of an *on-load tap-changer*), voltage drops that

Table 1.2 Typical distribution voltage ranges.

System frequency (Hz)	Nominal voltage	Highest supply or utilisation Voltage	Lowest supply voltage	Lowest utilisation voltage for lighting	Lowest utilisation voltage for other loads
50	230/400	253/440	207/360	200/348	196/340
60	120/208	127/220	109/190	106/184	104/180
60	230/400	253/440	219/380	212/368	207/360

(IEC 60038, Copyright © 2009 IEC Geneva, Switzerland. www.iec.ch)

occur throughout the MV network, as well as within the MV/LV transformer itself, are directly reflected into the LV network since the MV/LV transformer usually does not have on-load tap-change facilities, that is, it operates at a *fixed transformation ratio*.

The *International Electrotechnical Commission* (IEC) standard IEC 60038, entitled 'Standard Voltages' defines a ±10% tolerance[1] on the *nominal supply voltage* for 50 Hz LV systems, and +10% −5% tolerance[1] for 60 Hz systems. It also allows a 3% voltage drop for lighting circuits and 5% for all other circuits *within a customer's premises*. Collectively this means that for 50 Hz systems the *Utilisation Voltage* seen at outlets within a customer's premises may vary from the nominal voltage by +10% −13% for lighting circuits, and +10% −15% for all other circuits. For 60 Hz systems, the overall tolerance is +10% −8% for lighting circuits and ±10% for all other circuits. Table 1.2 shows the typical ranges of common LV distribution voltages.

1.2.2 Load Balance

It is important that the total connected load is shared almost equally across all three phases, that is, it is *balanced*. This is generally achieved in a residential distribution system by ensuring that the number of customers connected to each phase is roughly equal, given that residential loads are generally similar and they also have comparable daily load profiles. Whether a small commercial or industrial customer's load is balanced usually depends largely on how well the single-phase load is distributed, since three-phase loads, (motors for example) tend to be inherently balanced. On the other hand, very large industrial customers (*base load* customers) are required to present a balanced load to the network as part of their connection agreements.

1.3 Examples of Distribution Transformers

An engineer can learn much by observing equipment in the field. In particular, reading equipment nameplates can reveal a lot about an installation. We will consider two examples of distribution transformers: Figure 1.9 shows a public *pole-mounted* LV

---

1 These tolerances may vary slightly from one country to another if national standards differ from IEC 60038.

Figure 1.9 A 500 kVA pole-mounted distribution transformer (China).

Figure 1.10 A 500 kVA ground-mounted transformer (Australia).

distribution transformer in China, and Figure 1.10 shows an Australian ground-mounted distribution transformer enclosure (also known as a *pad-mounted* transformer). The pole-mounted device is fed from *three-wire* 10 kV *droppers* on the left-hand pole, while the *four-wire* 400 V LV supply leaves the transformer as an aerial bus via the right-hand pole. MV *drop-out* fuses in series with the incoming droppers protect the high voltage side, and exposed LV fuse elements mounted on the LV isolator on the right protect the low-voltage side. The LV conductors also pass through a set of *metering* current transformers (in the metal enclosure), so that the energy supplied by this transformer can be measured.

The pad-mounted transformer enclosure shows little from the outside; internally, however, it is actually a miniature 11 kV:415 V substation. It contains both MV and LV switchgear, located in opposite ends of the enclosure, with the distribution transformer mounted in between. This particular unit is dedicated to one customer and is fed from an underground 11 kV ring main; it supplies a large building on an industrial site. Both the MV and LV cabling is underground, and therefore this installation is aesthetically more pleasing than the pole-mounted one.

1.4   Practical Magnitude Limits for LV Loads

There are significant advantages in moving to higher potentials for larger loads, particularly the associated reduction in both supply current and electrical losses. This change also permits smaller section conductors to be used throughout the distribution network, which can provide considerable savings. It is for these reasons that commercial and industrial LV customers operate on 400 or 460 V rather than 230 or 120 V.

However, as industrial loads increase in size, eventually a limit is reached where it is no longer economic to provide the service from an LV supply. This limit generally occurs in industries where the site load is considerably larger than that usually found in the public network, for one of the following reasons. Firstly, very large phase currents demand very large supply conductors, which can become expensive, especially if they have to run a considerable distance from the distributor to the customer. Secondly, depending on the available transformer capacity, it may be necessary for the distribution company to install a *dedicated* transformer to cater for the requirements of particular customers. Thirdly, for particularly large loads and low impedance LV supplies, the prospective fault currents may become too large for conventional LV switchgear (circuit breakers) to handle.

The third point requires further clarification. Moulded case circuit breakers, frequently used in LV applications have several current ratings. For the purpose of this discussion we will consider only two: the *continuous current rating* and the *service fault interruption capacity*. The continuous current rating is the current that the breaker can continuously carry and interrupt safely; it is the *nominal rating of the device* (although it may be higher than the current at which the breaker is actually set to trip). The service fault interruption capacity is the value of the (balanced) three-phase fault current that the breaker is capable of interrupting. This is the practical limit of the ability of the device to clear a downstream three-phase fault.

The continuous current ratings of LV circuit breakers can be as large as 6000 A, but breakers of this size are generally for special applications. The current rating of LV circuit breakers used in general customer applications seldom exceed around 3200 A, and for these devices the service fault interruption capacity is frequently of the order of 50–70 kA.

A practical limit to the size of an LV load arises when the prospective three-phase fault current cannot be interrupted by the selected LV circuit breaker. For example, consider a 2.2 MVA Δ/Y distribution transformer delivering 400 V to an LV bus. The *rated secondary current* for such a transformer is 3175 A (=2.2 MVA/(400 V√3)), so a 3200 A breaker would *just* suffice in this application (although it is not good engineering practice to operate either the transformer or the circuit breaker continuously at the limit of their ratings).

Assuming that the MV bus impedance is small with respect to that of the transformer, then the latter will limit any three-phase fault current. As we shall see in a later chapter, the *impedance of a transformer* (sometimes called the *voltage impedance)* is usually expressed in *per cent*, with typical values being 4–20%. The percentage impedance is defined *as that fraction of the nominal supply voltage which, when applied to a transformer with short circuits on all phases, results in the rated current flowing in each phase.* Therefore the three-phase fault current that would flow with the transformer *fully excited*, can be calculated by dividing the *rated current* by the transformer's impedance.

For example, if our transformer has an impedance of 5% (0.05) then the three-phase fault current will be approximately 3175/0.05 = 63.5 kA, and this is the current that the LV breaker must be able to successfully clear in the event of such a fault.

Choosing a 3200 A breaker with a 70 kA service fault interruption capacity would therefore be acceptable in this instance, but a circuit breaker with only a 50 kA interruption capacity clearly would not. If it were deemed necessary to use two such transformers on the same LV bus, then neither of these circuit breakers would be acceptable, unless the transformer impedances were increased substantially.

Because of the limitations imposed by switchgear capacities and the size of the LV busbars that would be required, industrial customers generally segregate their load into smaller portions, fed individually from smaller transformers. This is often dictated by the location of equipment across a customer's site and by a desire to minimise the load lost in the event of a transformer outage.

As an alternative to using one large transformer, a customer may take several independent LV supplies from the distribution network, perhaps at different locations across a site. Or instead, energy may be taken directly from the MV bus, and distributed to one or more private MV-LV substations. While this demands a substantial investment in electrical infrastructure, and the ability for maintenance staff to operate both LV and MV equipment, the MV tariffs available may make this approach attractive.

Often the distribution company will provide a direct MV connection to a customer's own MV-LV transformer, either pole or ground mounted (Figures 1.9 and 1.10). This arrangement can be used to provide the customer with a dedicated LV supply, while obviating the need for MV distribution across their site. In both these cases the MV supply will pass through a *metering unit*, containing the necessary voltage and current transformers so that the energy used can be appropriately metered.

The VA threshold at which a customer will be directed to the MV network varies from country to country, in France this can be as little as 250 kVA, in some parts of the USA it can be as high as 2 MVA, while in Australia it is of the order of 1 MVA.

1.5   Medium Voltage Network

The purpose of the medium voltage distribution network is to transport electricity from the transmission or sub-transmission substations to public customers via the LV network, or to industrial customers supplied directly at MV potentials. It represents a major infrastructure investment on the part of the distributor, and its design has a significant influence on the level of service and cost experienced by customers.

The IEC standard IEC 60038 '*Standard Voltages*' specifies the MV potentials shown in Table 1.3 for both 50 Hz and 60 Hz systems. It defines two voltage levels, the *nominal system voltage* which is used to identify an MV system, and the *highest voltage for equipment* which is the highest voltage which may occur under normal operating conditions. Some or all of the voltages shown in Table 1.3 are used by MV distributors throughout their networks. It should be noted that although the nominal system voltage is usually used to identify a system voltage, equipment must be designed to operate continuously at the highest voltage for equipment. For example, a device designed for 11 kV must be able to tolerate a supply potential as high as 12 kV.

Table 1.3 Medium voltage levels.

Australia, NZ & Europe (50 Hz)		North America (60 Hz)	
Nominal system line voltage	Highest voltage for equipment	Nominal system line voltage	Highest voltage for equipment
3.3 kV	3.6 kV	4.16 kV	4.4 kV
6.6 kV	7.2 kV	12.47 kV	13.2 kV
11 kV	12 kV	13.2 kV	13.97 kV
22 kV	24 kV	13.8 kV	14.52 kV
33 kV	36 kV	24.94 kV	26.4 kV
—	—	34.5 kV	36.5 kV

(IEC 60038, Copyright © 2009 IEC Geneva, Switzerland. www.iec.ch)

While networks may adopt potentials within the range outlined in Table 1.3, some jurisdictions provide a tighter tolerance in the case of large MV customers, typically the nominal system voltage ±5%.

The voltage chosen depends upon the length of the feeder concerned and the total load (MW) that it must supply. Since losses are proportional to the square of the currents flowing, they can be minimised by choosing as high a system voltage as possible; however, equipment costs rise steeply with system voltages above about 33 kV, and therefore an economic compromise must be made as to the choice of voltage and the acceptable network efficiency.

In addition, the voltage drop along an MV feeder must be maintained within prescribed limits. IEC60038 specifies that the highest MV voltage and the lowest MV Voltage should not differ from the nominal voltage by more than ±10% for 50 Hz systems and +5% −10% for 60 Hz systems. Given these tolerances the practical limit to voltage drop along an MV feeder will generally be less than 10%, although this may often be relaxed under contingency conditions.

The voltage drop along a feeder is a function of both the line resistance and its reactance, and therefore for a given power transfer, the line voltage drop will be influenced by the average *power factor* along the line. As the line power factor gets smaller (worse), the line current will become larger and with it the voltage drop and also the line losses.

The line's *reactance to resistance, (X/R) ratio* also has a significant influence on the voltage drop. Large X/R ratios mean that a transmission line's impedance is largely inductive, and therefore lagging currents will contribute significantly to the cumulative voltage drop along the line. For this reason the transmission of energy at high power factors improves the apparent capacity of a line.

For a given power flow, it can easily be shown that the line loss is inversely proportional to the square of the system voltage. Figure 1.11a shows the maximum power that can be transmitted along a 33 kV MV feeder, subject to a 7.5% voltage drop, as a function of line length. Four curves are shown: the upper two apply for 120 mm² conductors with 0.9 and 0.8 power factors, while the lower two apply for

(a)

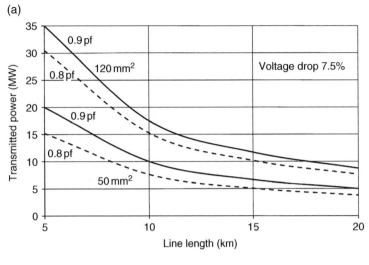

(b)

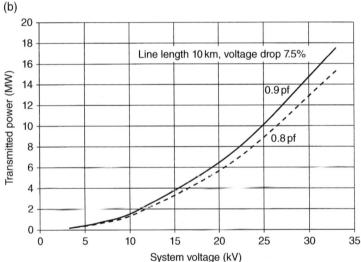

Figure 1.11 (a) Transmitted power vs line length (b) Transmitted power vs system voltage.

$50\,\text{mm}^2$ conductors with the same power factors. Clearly, as the line length increases, the power transmitted must be reduced if the voltage drop constraint is to be met. Further, as the average power factor becomes worse, the apparent line capacity diminishes and the addition of shunt capacitance at regular intervals can be used to improve MV voltage regulation.

Figure 1.11b shows the quadratic nature of the relationship between transmitted power on a 10 km line and system voltage, for the same voltage drop requirement. The choice of the transmission voltage clearly depends on both the *power to be transmitted* and the *distance over which it must be sent*. In reality, the problem is a little more complex since feeders rarely deliver all their power to an end-connected load.

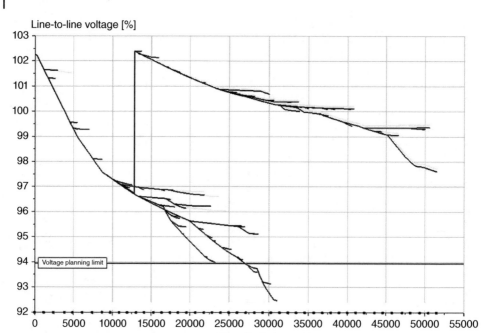

Figure 1.12 Voltage profile of MV feeder trunk and spurs.

The sending end power tends instead to be absorbed progressively by loads distributed along the line's length, as shown in Figure 1.12. In order to extend the transmission distances available from MV feeders, autotransformer based regulators or *step regulators* are frequently inserted at intervals along a feeder to restore the voltage drop.

Figure 1.12 shows an example of voltage profiles of an MV feeder trunk and spurs, as functions of distance from the source substation. Close to the substation the trunk feeder potential falls quite steeply, since the load carried is heavy. A regulator located about 13 km from the substation produces a sharp rise in the voltage downstream, but the potential of the remaining spurs continues to fall, with the most heavily loaded decaying fastest. The lower voltage *planning limit* in this case is 94% (0.94 per unit), and the potential on one heavily loaded spur falls below this limit at about 27 km from the substation. Assuming that this voltage profile represents the worst case, and depending on how frequently it arises, a new regulator may be required at a suitable upstream location. Voltage profiles like this are useful tools for managing feeder performance and planning network upgrades.

1.5.1 Three-Wire and Four-Wire MV Circuits

The circuit arrangements used for MV distribution vary considerably from one country to another, the main differences being the number of conductors and the earthing arrangements used on the MV side of the HV/MV transformer.

In North America and associated countries, a *four-wire* system is used. The HV/MV transformer is generally a delta/wye configuration (Δ/Y), generally with the MV star point directly (solidly) grounded at the transformer. The MV network consists of four wires: the three-phase conductors and the neutral. The neutral is grounded at regular intervals of about 250 m, thereby ensuring that the neutral conductor is close to ground potential along the route of the feeder. This permits the use of single-phase MV/LV distribution transformers energised from a phase potential for residential loads, in addition to three-phase transformers for commercial and industrial loads.

In the USA, single-phase MV transformers are available in 10, 15, 25, 37.5, 50 and 100 kVA capacities, and often have dual primary voltage ratings so that they can also operate from the local line voltage if required. The use of single-phase transformers like these provides a significant advantage where light loads are supplied from branch or spur lines running off a main trunk feeder. In this situation, the branch line can consist of only one phase conductor and the neutral, resulting in a considerable saving in material costs. (Similar techniques are also used in other countries where *two-phase* spur lines supply single-phase MV/LV transformers.)

The solidly earthed neutral means that phase-to-earth faults can generate large currents, limited only by the network and local ground impedances. These currents therefore tend to diminish with the distance from the parent substation. Unfortunately, the single-phase loads on the MV network also result in substantial currents in the neutral, to the point where it can be difficult to discriminate between the two. As a result, it is necessary to break the line up into *protection zones* and to use a distributed protection system for each one, capable of discriminating between earth fault currents and residual load current, within each zone.

This arrangement means that the protection system is highly dependent on the configuration of the network, making it difficult to temporarily rearrange during fault recovery or indeed during major upgrades. In North America, the MV network is mainly an aerial one, emanating radially from a parent MV substation.

By contrast the MV network in most of Europe, Australia and New Zealand is based on *three-wire distribution*. As before, HV/MV transformers are frequently delta/Wye (Δ/Y) connected, although Y/Y connected transformers with the HV star point open, are used as well. The MV star point is usually grounded, sometimes directly and sometimes through an impedance designed to limit the magnitude of earth fault currents. In some countries, the star point is not grounded at all, and the MV system is constrained around ground potential by the effects of stray phase-to-earth capacitances.

An advantage of three-wire MV distribution lies in the fact that there are no phase-to-neutral loads and therefore no neutral current flows, except for a small *residual capacitive current*, flowing back to the transformer's star point. (This is the result of unbalanced stray capacitances that exist between the phases and ground.) MV/LV transformers are either three-phase, for larger LV loads or line connected single-phase units, for small residential loads.

The lack of neutral current makes the detection of phase-to-ground fault current relatively easy, either by directly measuring the neutral current, or by a residual over-current measurement on the phase conductors themselves. Both these measurements can be made at the parent substation, which considerably simplifies the protection system. In particular, this protection approach to three-wire MV networks is tolerant of changes to the network, thus facilitating network reconfiguration during fault recovery. In this respect, three-wire MV distribution networks can provide a higher level of service.

While impedance earthing is effective in reducing the magnitude of earth fault currents, it can also lead to significant *over-voltages* throughout the network. This is because the neutral potential is not tightly constrained and is therefore able to rise in potential during fault conditions.

1.5.2   MV Network Topologies

There are several topologies used in MV networks, and the use of each depends upon the density of the connected load and its relative importance. In rural areas where there are relatively few customers connected to each distribution transformer, the failure of part of the local MV network or of a distribution transformer itself, may not inconvenience many customers, and if the distribution company is prompt in carrying out repairs, the inconvenience will be brief. (Many distribution companies take advantage of this and allow LV distribution transformers *to run to failure*, thus avoiding the expense of a transformer monitoring and replacement programme.)

By contrast, in more densely populated areas, the number of customers supplied from an entire MV feeder can be substantial, and system faults may cause serious interruptions to supply, including critical loads such as hospitals and health centres. Because of this, parts of the MV network are often designed to provide a high level of redundancy that can be exploited during fault recovery so as to minimise the number of customers ultimately affected. The cost of this approach must be weighed against the benefits to the community, and in general the more redundancy provided, the more expensive the network becomes.

There are many different network arrangements in use worldwide and while we can't discuss them all, we will consider three fundamental topologies; radial feed, open loop feed, (or ring main feed) and mesh networks.

Radial Feed Lines

In low population density areas, MV distribution is generally achieved using aerial radial networks. These are relatively inexpensive but tend to suffer from a lack of redundancy and are more prone to faults through contact with lightning, foliage or animals. Each customer fed from a radially connected feeder has only one source of supply, and the failure of part of the upstream network means an outage until the fault can be repaired. Radial feed arrangements tend to be tree-like in structure, with many *branch* feeders emanating from a major *trunk* feeder. Although there is no redundancy in terms of source of supply, these networks generally have the ability to automatically detect and isolate faulty network segments. Thus a minimum number of customers will experience power loss until repairs can be made.

Faults to aerial networks arise for several reasons: lightning strikes or wildlife and foliage impinging on the line, are but a few. Many line faults are cleared by the fault current during the fault or during a subsequent re-energisation of the line. A modern MV feeder is equipped with numerous devices that collectively are able to detect, clear or isolate faults and then restore supply, while causing a minimum of disruption to customers.

Recloser Circuit Breakers

Reclosers (*or automatic circuit reclosers, ACRs*) are intelligent electronic devices (IEDs), containing *current and voltage transformers*; they are capable of measuring and storing both operational and fault record data and can communicate with the MV network control centre, from where they can be operated remotely (Figure 1.13). The prime function of a recloser is to *segment* the line in the event of a 'permanent' fault that cannot be cleared by several *reclosing attempts*. In the event of a fault, the nearest upstream recloser will initially trip on the fault current and after a brief period will *reclose* for a short time, in an effort to clear whatever may have caused the fault. This process occurs a maximum of four times before the recloser permanently locks itself open. (The number of reclose attempts may be remotely altered as required, if necessary. It is usual to reduce this number to zero in times of severe fire danger, or when live-line work is being carried out.)

For example, a downstream transformer fault can usually be cleared by the operation of its local protection fuse(s). An animal carcass or foliage on the line can often be blown clear by subsequent applications of the fault current itself, but if the recloser attempts fail and the fault current remains, the recloser will *lock itself out* in the open position, thereby isolating the faulty portion of the network.

It is common to use several reclosers in series at intervals along a line and in this event it is necessary for downstream devices to be *graded* with those upstream. They must be

Figure 1.13 Recloser, bypass switches and isolator switches.

programmed so that the downstream devices trip earlier and possibly on a lower current threshold than those upstream, so that the number of customers disrupted is
minimised.

Reclosers can also be used to reconfigure a network under remote control from the
network control centre. This is frequently done during fault recovery, where normally
open points in the network are temporarily closed, so that power can be restored to
sections which would otherwise suffer an outage.

As an example, Figure 1.14 shows a typical radial network, consisting of two trunk feeders originating from an HV/MV substation. This drawing is an example of a *one-line diagram*, where one line is used to represent all three phases, and the neutral conductor if
required. The outgoing feeders supply many MV/LV distribution transformers, each of
which is protected by a set of fuses. At the fork in the left-hand feeder are two reclosers, R1

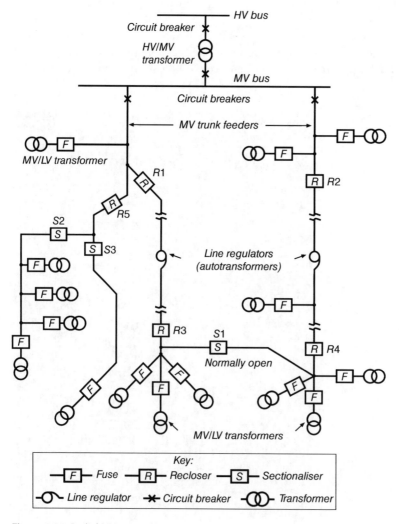

Figure 1.14 Radial MV network.

and R5. Should a 'permanent' fault occur downstream of R5, this device will lock itself open, isolating the fault, while R1 remains closed, having seen no fault current. This situation can be further improved through the uses of *sectionalisers* or *load break switches*.

Sectionalisers or Load Break Switches

Sectionalisers are also IED devices and are thus capable of being remotely controlled and providing operational data to a network control centre (Figure 1.15). They are used in conjunction with reclosers to isolate a faulty section of line. While they can interrupt *load* currents, they have limited *fault* current making and breaking abilities. In particular, they do not have the fault current interruption capabilities of reclosers.

Figure 1.15 Sectionaliser or load break switch.

Returning to our example in Figure 1.14, a permanent fault on the line downstream of S3, will ultimately see R5 lock out, resulting in the loss of all the associated load. In order to prevent this, sectionaliser S3 measures fault current pulses created by the reclose attempts of R5, counting the number of reclose cycles. During the open interval of the penultimate reclose cycle of R5, S3 opens, thus isolating the faulty circuit so that when R5 recloses for its final time no fault current is seen. R5 consequently remains closed and thus supplies the S2 load. Therefore by working together, recloser R5 and sectionaliser S3 collectively act to isolate only the faulty portion of the network, while maintaining supply to the majority of customers *without* the need for inter-device communication.

Consider now the feeder sections downstream of R3 and R4 in Figure 1.14. Because there is significant load downstream of these reclosers, provision has been made for a *backup supply* from the adjacent feeder in the event of a fault. For example, in the case of a permanent fault on the line between R2 and R4, recloser R2 will ultimately lock out, isolating the faulty line segment, but this will see the loss of a considerable amount of load, normally supplied through R4. If recloser R4 is remotely opened and sectionaliser S1 is closed, the network can be reconfigured so that these customers are maintained via R3 and R1 until the fault can be repaired. Therefore by temporarily reconfiguring the network, much of the load can be restored immediately. In many instances it is possible to program a normally open sectionaliser with a set of rules which will be followed in the event of a disruption to the supply, thereby permitting automatic reconfiguration of the network, without the need for remote intervention.

Voltage Regulators

The voltage drop constraints mentioned earlier can be overcome to a large extent by inserting *step voltage regulators* in the line. Two such regulators are included in Figure 1.13 on the long line sections running from reclosers R1 and R2. These are generally *autotransformers* equipped with on-load tap-changers and can automatically regulate the downstream voltage over a range of about ±10%, thus virtually removing the effects of line voltage drop.

Figure 1.16 Three-phase 22 kV autotransformer regulator.

On long lines with end-connected loads and a mid-line regulator, it is also possible to configure these devices to regulate the voltage *at the receiving end* of the line, a technique known as *line drop compensation*. This is achieved by measuring both the current flowing and the voltage at the output terminals of the regulator, and from knowledge of the line's impedance it is possible to calculate and therefore regulate the voltage at the receiving end.

Figure 1.16 shows an example of a three-phase wye/wye (Y/Y) connected 22 kV autotransformer, fitted with a tertiary delta winding and a 16-position on-load tap-changer. It is capable of regulating line voltages from 18,700 V to 22,550 V, and has a line current rating of 131 A and a VA rating of 5 MVA. Although the nameplate rating suggests that the transformer will deliver 5 MVA, this is really just *the nominal rating of the line* at 131 A. The maximum VA contribution delivered by this transformer is:

$$VA\ Contribution = \sqrt{3} \times Max\ Boost\ Voltage \times Rated\ Line\ Current \approx 750\text{kVA}$$

Thus a relatively small autotransformer can regulate a much higher capacity feeder.

Figure 1.17 shows a pair of *single-phase*, 330 kVA, 150 A *step voltage regulators*, arranged in an open delta configuration on a 22 kV line. In this configuration these are able to regulate a 3.8 MVA feeder; they are also based on an autotransformer design, complete with an on-load tap-changer. Single-phase regulators of this type can be easily replaced or maintained, without the need to remove the line from service through the

Figure 1.17 Single-phase 22 kV step regulators (open delta connection).

inclusion of bypass and isolation switches as part of the installation. They can also be installed *singly* on two-wire feeders, in *pairs* in open delta configurations and in *threes* in either a three-wire closed delta or a four-wire grounded wye arrangement. It is common to *series connect* line regulators on long MV feeders.

Open Loop Feed

The open loop feed arrangement, also known as a *ring main* arrangement, is used in more highly populated areas where many customers are supplied from each MV transformer, and consequently where supply reliability is important. This arrangement is shown in Figure 1.18 and it includes two possible MV supply routes to each downstream substation. In each loop there is one *open point* which defines the path of supply to each distribution transformer. Ideally, each loop will originate from a different busbar within the parent MV substation, or possibly from a different MV substation, ensuring a further level of supply redundancy.

Should a fault develop in a section of one loop, the open point can be relocated and the fault isolated by opening the loop in a second location as well. Therefore most if not all of the customers can have their supply rapidly restored. Open loop feeder arrangements are frequently implemented as part of public underground cable distribution networks. Private MV networks in large buildings, educational campuses, hospitals and other public facilities also frequently use this distribution system, enabling multiple MV/LV transformers to be supplied from within a customer's premises. Each node in the loop requires a small suite of MV switchgear, sufficient to provide for at least one

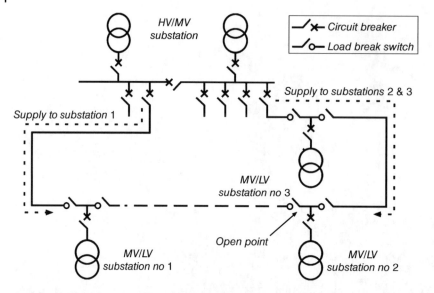

Figure 1.18 Open loop feed. (Extracted from Cahier Technique No. 203, 'Basic selection of MV public distribution networks' with kind permission from Schneider Electric).

Figure 1.19 11 kV ring main switchboard.

incoming switch *from* the loop, one outgoing switch *to* the loop, and an MV circuit breaker through which the local transformer is supplied.

A switchboard of this kind, used in a hospital's private 11 kV network, is shown in Figure 1.19. The incoming feeder enters via the extreme left-hand cubicle and the

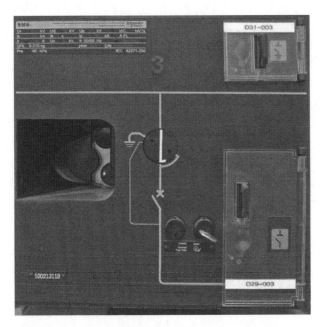

Figure 1.20 Circuit breaker and mimic diagram.

outgoing one departs from the cubicle adjacent to it. These cubicles are provided for *isolation purposes* only; they contain a *load break switch* rated at 630 A and insulated with *sulphur hexafluoride gas* (SF_6), as well as an interlocked *earth switch*, capable of earthing the incoming cable. The right-hand cubicles provide feeds to two local distribution transformers via SF_6 insulated circuit breakers. The operating mechanism for these can be seen protruding from the lower left of each cubicle. Incorporated into each circuit breaker is a *protection current transformer*, and the associated protection relay can be seen above the mimic panel.

On the front of each cubicle is a *mimic diagram* showing the function of the switchgear within. Figure 1.20 shows the mimic diagram on the transformer circuit breaker panel. The horizontal line represents the 11 kV bus, while the vertical one running from it represents the feed to the transformer, which passes through an *isolator* before entering the circuit breaker. This is operated by inserting a lever into the receptacle behind the clear panel on the lower right of Figure 1.20; an interlock mechanism prevents the isolator from being operated unless the circuit breaker is *open*.

The earth switch is used to ground the outgoing bus connection to the transformer during maintenance activities. It is operated using the same lever; this time it is inserted into the receptacle on the upper part of the mimic panel, shown in Figure 1.20. The interlock mechanism also prevents the earth switch from being closed unless the circuit breaker is open.

Also visible on the left-hand side of Figure 1.20 is a *viewing window*, which permits the operator to view the state of the isolator. Viewing windows are also provided for the earth switch, and these can been seen on the breaker cubicle in Figure 1.19. Verification of the position of both these switches is a critical safety requirement when maintenance is to be carried out.

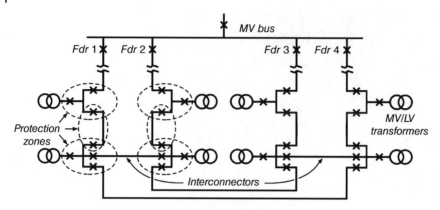

Figure 1.21 Medium voltage mesh network.

Mesh Networks

Mesh networks are used in cities where the population density is high and there are many MV/LV distribution transformers. Supply security is therefore important, and a mesh arrangement similar to that shown in Figure 1.21 is often used. This is really a high density application of the open loop arrangement discussed above. Cables are often run beneath footpaths or streets and service nearby buildings, either with an LV supply from a local distribution transformer, or with a private MV supply for the transformers within the building.

Figure 1.21 includes numerous ring main switchboards, supplying an MV/LV distribution transformer, each with a supply from both sides of the ring. Interconnectors may also run between strategic feeders providing a further level of redundancy. Protection zones are defined within sections of each feeder as shown, and should a fault occur in any one it will automatically be isolated from the network with the minimum loss of load.

1.6 Transmission and Sub-Transmission Networks

While there is an increasing amount of *embedded generation* within today's distribution networks, the bulk of the electrical energy is still generated by large coal, hydro, gas or nuclear-powered machines. These are generally located near their respective fuel source, since it is generally more economical to transport electricity than either coal or gas in the quantities required. Likewise hydro generation must, by definition, occur where the water power is available, while the location of nuclear-powered plant tends to be a compromise between safety concerns and the proximity to the electrical load.

It is therefore the job of the transmission system to transport the electrical energy from where it is generated to the load centres where it will be used. This can be further broken into *transmission* and *sub-transmission* networks. The transmission network often extends over considerable distances and can carry very large quantities of energy. In most applications the transmission system consists of three-phase open wire lines, supported by large transmission towers and operating at *extra high voltages* (EHV), typically 220 kV–1 MV. EHV transmission cables are used within cities where it is impractical to erect transmission towers, but their lengths are generally limited due to their expense and the large capacitive charging currents that must be supplied. With the advent of high power semiconductor devices in the 1970s and 1980s, high voltage DC links (HVDC) began to appear. Initially these tended to be between different AC networks, spanning borders, in the form of open wire lines, or as DC cables crossing waterways, where it was impractical to operate AC cables. More recently, however, HVDC links have begun to compete with AC transmission lines within AC networks. This is due to the absence of the capacitive charging effects, as well as the lack of an inductive voltage drop along the length of a DC line.

The sub-transmission system generally operates at slightly lower voltages (66–132 kV) and supplies somewhat smaller loads. It is generally used to interlink major distribution substations within or near city boundaries, and provides redundant supply paths, so that the demand for electricity can be met in the event that part of the network is out of service, either through *planned* maintenance or through the *unplanned* operation of the protection system, in response to a fault.

The choice of supply voltage (either AC or DC) is an economic compromise between acceptable transmission losses and system cost. Generally, the higher the chosen transmission voltage the lower will be the total loss, but the more costly will be the terminal equipment which must operate at that potential. Open wire transmission lines being air insulated, do not substantially add to this cost, but the transmission towers they require certainly do. As the transmission potential rises, so do the electrical clearances required, both between the phase conductors themselves and over objects on the ground. These requirements lead to bigger and therefore more costly transmission towers. The requirement to carry two, three or four conductors per phase and the desire to increase the span between towers, further adds to this cost. Finally, although each transmission tower is quite large, it does not provide a span of more than a few hundred meters (see Figure 1.24), so long transmission lines requiring many towers are costly to build.

Transmission Redundancy and Network Planning

In the case of LV networks – and to a lesser extent with MV networks – the loss of a single feeder usually results in the loss of supply to relatively few customers. Therefore occasional outages can be tolerated, so long as their duration is kept short. This is definitely not the case in the EHV network, where transmission lines carry substantial amounts of energy and the unexpected loss of a line may result in severe disruption to the overall network. For this reason, transmission networks must be designed to provide a considerable degree of *redundancy*, so that the sudden loss of part of the network can be automatically accommodated without a loss of supply to customers.

Redundancy in this context effectively means the provision of *excess network capacity* together with sufficient *network flexibility*, in order that the entire load can be carried at times when parts of the network are out of service, either due to a fault or perhaps as a result of a planned maintenance outage. The decision as to how much redundancy should be provided is essentially an economic one. Since the provision of redundancy in the transmission network is expensive, the most likely contingencies should be provided for first, while those considered less likely may be deferred until money becomes available. Sometimes a particular contingency can remain unforeseen until it actually occurs, the recovery from which can involve a widespread loss of supply which can be very expensive, particularly if a *black start* is required. A black start is required following the total collapse of a power system. Generators must be restarted a few at a time and base load must be progressively restored in such a way that it matches the generation capacity available. The time required to fully restore power to all customers can be considerable, so it is far better to shed load early in order to avoid this possibility.

Often a degree of redundancy can be provided at a discounted cost, if it is carefully planned prior to construction. For example, transmission lines like those shown in Figures 1.22–1.24 can be built as either a *single circuit line* or a *dual circuit line*, without a proportional increase in cost. While the dual circuit option provides twice the transmission capacity and therefore appears an attractive option, both circuits share the same route and therefore failures due to environmental factors (such as ground fires) are equally likely to affect both circuits. Truly redundant circuits require route diversity in addition to spare capacity. Dual circuits sharing the same transmission towers are often used none the less, and provide the ability for one line to be taken out of service for maintenance, while the other carries the load.

Redundancy provision also includes the terminal equipment within each transmission substation. For example, the bus arrangements provided have a major impact on the substation's flexibility; the provision of autotransformers linking one transmission potential with another also affects the ease with which the loss of one transmission line can be mitigated by the others in the substation.

Figure 1.22 275 kV single circuit transmission tower (France).

Figure 1.23 400 kV dual circuit transmission tower (France).

Figure 1.24 Each tower provides a relatively short span, so many towers are required for a long line.

As a result of their importance to network security, individual EHV transmission lines are also equipped with a far higher degree of electrical protection than their MV counterparts. Frequently, independent protection schemes are used simultaneously on the same line, achieved through the use of *duplicate protection relays*. Even the substation batteries powering these relays must be duplicated, so that there is no single point of failure in common to the two schemes.

Network planning is thus a complex task, which not only includes the provision of adequate redundancy, but must also anticipates future load growth so that the network can be adapted for changes in population density and generation profile. Transmission networks are therefore not only expensive to build, but also to maintain, since they require ongoing upgrading in order to meet community expectations.

Load Classifications and Demand Response

The connected load on an electrical system falls roughly into two classifications: *base load* and *variable load*. Base load is provided by large industrial customers, many of whom are connected to the network at either EHV or HV potentials. Aluminium, zinc and nickel refineries are typical base load customers, since they consume large amounts of electricity, usually at high potentials, and operate 24 hours per day. Paper and automotive manufacturers also fall into this class. Base load customers are usually required to maintain power factors, in excess of 0.95, and collectively they provide a stabilising effect on a power network. The power they consume generally has a flat profile, and usually a substantial proportion of the available generation must run continuously in order to meet this demand, thus providing a *spinning reserve* of deliverable power, which can quickly be made available should the load suddenly increase.

The *variable load* component consists of residential, commercial and small industrial loads, most of which are supplied at LV potentials. These loads are not 'flat', but instead exhibit a seasonal and cyclic daily profile, in addition to which they usually also consume significant reactive power. They are also largely responsible for creating the daily peaks in demand, which the transmission and distribution networks must have sufficient capacity to deliver. These peaks in daily demand generally correspond with meal times, particularly in the morning and evening. Meeting this demand is one of the main aims of the network planners, and is made worse during extreme weather events, either hot or cold, when the additional cooling or heating load substantially increases the maximum demand imposed on the system. The transmission, distribution and generation system must cope with such extreme loads, and it is during these events that the system may become overloaded, in which case the load must be artificially reduced.

Some distribution companies reduce the potential they supply to residential and commercial customers slightly during times of peak demand, in the expectation that the demand will fall sufficiently to avoid overloading the network. This works well in the case of resistive heating, where the power demanded is proportional to the square of the voltage, but it does not have the same effect with other load types. Others companies dynamically reduce the residential hot water heating load, using a control signal imposed on the network itself. A third load control mechanism is to reduce the speed at which inverter-driven air conditioners operate, and hence the load they impose on the network. This is achieved by communicating with the air conditioner via a smart meter installed on the premises or through a direct internet connection. These mechanisms are all examples of *demand response*, whereby the demand on the electrical system is reduced during extreme load events.

However, perhaps the best way to control the load at times of extreme demand is via a *price signal*. It is during extreme weather events like these, that the market price for electricity is greatest. Base load customers (and many smaller customers too) are usually quite price sensitive, and where electricity contracts include an exposure to the market price, they will usually curtail their consumption at times when the price is high.

Building owners and industrial customers often take advantage of the services offered by *demand response aggregators*. These companies manage blocks of electrical load offered by their clients for load shedding purposes. In an extreme weather event or when network security is threatened the aggregator in conjunction with the network operator, notifies its clients of the need to curtail load, and manages the load reduction throughout the event. This may be achieved by temporarily altering heating or cooling set point temperatures in building management systems, or by the shedding of industrial load for intervals as short as half an hour. Through the load shed by its many clients, often on a rolling basis, the aggregator can offer the network operator substantial blocks of distributed load for shedding when needed. Payments are usually made by the network operator to the aggregator for the *provision* of this service, as well as for the actual load shed when an event occurs. Clients who can provide larger blocks of load or who can respond rapidly to a load shedding event are paid more than those who can supply less or require advanced notice. Since many extreme weather events can be predicted well in advance, the load can generally be progressively shed as an event unfolds, consistent with network requirements.

Demand response represents one of the major benefits of an electricity market; it flattens the overall load profile and reduces periods of peak demand, thereby deferring the need for network upgrades and capital expenditure.

1.6.1 Transmission System Operation

Unlike MV transmission lines, which must supply the active as well as any reactive power demanded by the loads they service, the transmission system is used primarily for the transmission of *active* energy from a source of generation to a load centre, or between transmission substations. There are two reasons for this.

Firstly the voltage drop along a transmission line increases substantially as the power factor becomes smaller and for very long lines this can be substantial. This effect is illustrated in Figure 1.25, which shows the effect of a worsening power factor on the fractional voltage drop, for a constant VA flow as a function of the X/R ratio of the line. As load power factor decreases, the line voltage drop increases, as it rapidly becomes dominated by the larger inductive impedance. This effect is particularly evident at higher X/R ratios.

While this drop can be corrected using series connected line regulators in MV networks, voltage regulation of this kind is much more expensive to achieve at transmission potentials.

Secondly, the *current capacity* of the transmission network represents an expensive asset, one that should not normally be used for the transmission of reactive power,

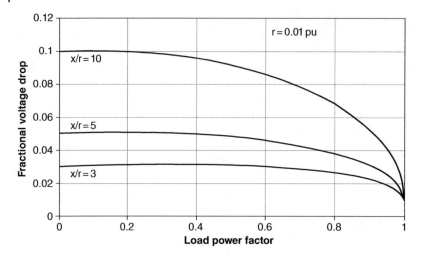

Figure 1.25 Fractional voltage drop as a function of the load power factor, for a constant VA load.

which can easily be generated near the loads requiring it, through either the provision of network capacitors or the installation of power factor correction equipment within customers' premises.

Reactive power flows along a transmission line as a result of a *voltage difference* between its ends, flowing towards the *lower potential*. In contrast, *active* power flows along a line when there is a *phase difference* between its ends, and flows towards the more *lagging potential*. Therefore depending on the conditions at each end of a line, it is quite possible to have these quantities simultaneously flowing in opposite directions.

EHV transmission lines, unlike their MV counterparts, run between substations at which the bus voltages can usually be independently controlled by adjusting the taps on network transformers or by adjusting the excitation of nearby generators. Therefore it is a relatively simple matter to reduce the reactive power flowing along a line to an acceptable level, by adjusting the voltage difference between its ends. This can generally be achieved almost independent of the active power flow required. For these reasons, most transmission systems are operated in such a way that they mainly transport active power.

At transmission potentials the power quality also tends to be high, meaning that, in addition to power factors close to unity, there will be relatively few harmonics, and the line currents will also be well balanced. For this reason the negative or zero-sequence voltages present on a transmission bus should also be very small.

Frequency Control

It is almost taken for granted that the frequency in large power systems is particularly stable. However, the maintenance of a stable operating frequency requires a precise match between the power being generated and that consumed. Should this balance be upset, for example through a sudden increase in connected load, the system frequency will begin to fall. The rate at which it falls is proportional to the power mismatch and the energy stored in the inertia of the generators themselves. The imbalance will be absorbed from the rotational energy of the machine rotors at the expense of a fall in the speed of each machine until the governor reacts and increases the energy delivered by the prime mover. The energy collectively stored in the system's rotating inertia is given by the equation:

$$Stored\ Energy = \frac{1}{2} J\omega(t)^2 \tag{1.1}$$

where J is the collective inertia of the system, and $\omega(t)$ is the system frequency, which must now be considered to be a function of time. Differentiating this equation and putting $t = 0$ yields the following expression for the initial rate of change of frequency:

$$\left.\frac{d\omega}{dt}\right|_{t=0} = \frac{\Delta P}{J\omega_0} \tag{1.2}$$

where ΔP is the initial mismatch between the power generated and that demanded. As shown, the initial rate of change of frequency is proportional to ΔP and inversely proportional to the collective machine inertia, J and the *nominal* system frequency ω_0. Should an *excess* of generation capacity become available (as the result of a loss of load), then this energy will be absorbed by the machine inertia and the frequency will begin to *rise*.

The frequency is regulated by assigning a *frequency control function* to many of the generators operating within the system, so that when a fall in frequency is detected, the governor associated with each machine proportionally increases its machine's output to restore the balance. Since the load on the network is rarely static, the frequency tends to oscillate slowly about its target level. Because each governor is particularly sensitive to small changes in frequency, the variations about the target are generally very small. Large networks with many generators collectively have a proportionally large rotating inertia and therefore are able to achieve a very tight operational range; typically within ±0.2% of the target frequency. On the other hand, smaller networks with fewer generators will see larger frequency swings as predicted by Equation (1.2). Therefore each governor must work harder to make up for any deficit (or excess) of generation, and as a result a wider frequency tolerance must be accepted.

Under-Frequency Load Shedding

All networks can be vulnerable to a large and sudden loss of generation, particularly when one or more transmission lines are unexpectedly removed from service due to a fault or perhaps a protection mal-operation. Depending on how much generation is lost, the remaining machines may not have sufficient capacity to make up the shortfall, or they may not be able to restore the power balance while maintaining the system frequency within prescribed limits. When this situation occurs, it is critical that the generation–load imbalance is rectified as quickly as possible, in order to avoid a complete network collapse and the need for a black start.

At such times *under-frequency load shedding* is often used to help rapidly restore the balance. This is a process whereby large predefined blocks of MV or HV load are automatically tripped, in response either to the frequency falling below a predefined threshold, or when the *rate of change of frequency* becomes excessive. Equation (1.2) shows that the rate of change of frequency is proportional to the power deficit ΔP and therefore this derivative is an ideal control parameter.

Base load is ideal for this, since the inconvenience is felt by relatively few customers, and large amounts of load can be shed rapidly, through the action of local under-frequency relays. For this reason, most large industries will be required to provide a portion of their load for an under-frequency load shedding scheme, and acceptance of this will usually be a condition of the network connection agreement. If the frequency continues to fall, then residential and commercial load may also need to be shed, so as to avoid a system collapse.

Restoration of the load tripped usually occurs progressively and reasonably quickly. As soon as the generation shortfall has been rectified or the network is reconfigured to account for the equipment lost and the frequency can again be maintained within its prescribed limits, the load can be progressively restored. In this way, industrial production is disturbed minimally and importantly the security of the electrical system is preserved. Power networks generally employ several stages of under-frequency load shedding, triggered as the frequency progressively falls, until such time as it finally begins to rise, and the power balance begins to recover. This enables critical loads, such as hospitals, to be shed last.

Voltage Control

The control of bus potentials within the transmission system is usually the responsibility of the associated transmission company, although where a *regional electricity market* is in operation, this task may also be overseen by the *market operating company*.

Voltage control is generally achieved by adjusting taps on EHV transmission transformers coupling MV generators to the transmission network as well as on autotransformers operating between different transmission and sub-transmission potentials, or by adjusting the generator potentials directly. EHV to MV distribution transformers also are equipped with on load tap-changers, to regulate the MV load voltage. Generally each transformer is equipped with a *voltage regulator relay*, controlling its tap-changer so that the local bus voltage is automatically maintained within the prescribed limits, independent of the applied load.

The transmission, sub-transmission and MV distribution networks are generally monitored remotely by *network control operators*, who can override the automatic voltage control function should the need arise. They generally also have the ability to remotely reconfigure the parts of the network in emergency situations, in order that supply may be maintained.

Both the HV and MV networks usually contain *voltage support capacitors* which are either automatically or manually switched into the network prior to periods of peak load. The effect of this is twofold: firstly, they cause the local bus voltage to rise slightly, typically by about 2–3%; secondly, they inject a source of reactive power into the local network sufficient to satisfy the reactive demand of the *variable load*, the increase in which is creating the peak load condition. This reactive injection also acts to preserve the *reactive margin* of the system, which may be regarded as the excess in reactive capacity required throughout a network to avoid the possibility of a *voltage collapse*.

1.7   Generation of Electrical Energy

The worldwide demand for electricity has increased substantially in the past two decades, most of which has come from developing nations. The People's Republic of China, for example, has seen an almost exponential growth in electricity production

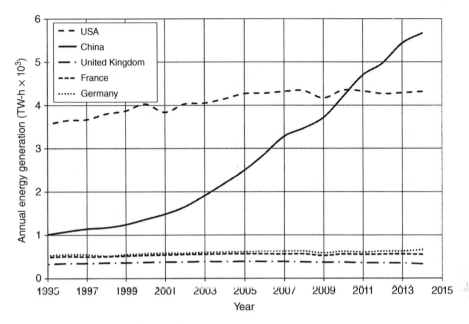

Figure 1.26 Annual electricity production by country. (Based on IEA data from © OECD/IEA 2016, IEA Energy Atlas, www.iea.org/statistics, Licence: www.iea.org/t&c; modified by the author).

this century, accompanied by a parallel increase in coal and wind generation. It is now the largest electricity producer in the world. In contrast, developed nations such as the USA, the UK, France and Germany have seen an almost constant electrical demand during the same period, as shown in Figure 1.26.

Electricity has traditionally been generated thermally using steam produced through the burning of fossil fuels such as coal, oil or gas, or through the use of nuclear energy, to drive turbines. These fuels have a high calorific value which has permitted the construction of sizeable machines capable of producing electricity on a large scale and relatively cheaply. Thermal machines, however, require long start-up and shut-down intervals making them impossible to start up and shut down in response to daily cyclic load variations. They are most efficient when run continuously, supplying base load. In locations where these machines must also cater for the peaks in demand, sufficient *spinning reserve*[2] must be maintained in order to meet the next peak load event. In between, these machines will be only partially loaded and will operate at slightly lower efficiencies.

Renewable Energy
For much of the twentieth century hydro-powered generators represented the main source of renewable energy. The power available from these machines depends on the topography of the generation site, specifically the volume of water and the head available. Hydro generation schemes were among the first to be built and have complemented thermal generators for many years.

---

2 Spinning reserve or spare generation capacity, must also be provided as a contingency against unexpected machine outages or loss of transmission assets, so that such events will be transparent to consumers.

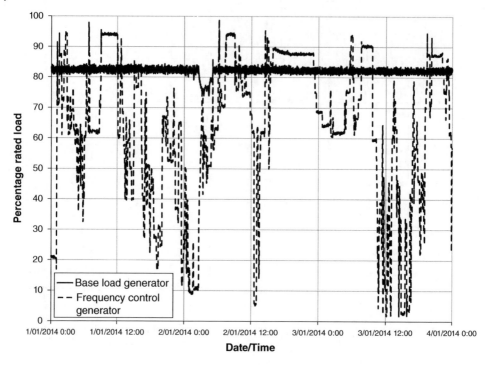

Figure 1.27 Operational comparison between a base load machine and a frequency control machine.

In recent years natural gas driven turbines operating on a similar principle to aircraft jet engines have also become popular. Both gas and hydro turbines offer the advantage of rapid response to changes in system load. They can be quickly shut down or partially unloaded during times of low demand, and therefore they often find application as *peaking generators*, operating largely to meet the peaks in the daily load cycle. They can also supply base load when required to do so, or undertake frequency regulation, by accepting or rejecting load in order to maintain the system frequency within its prescribed limits. Figure 1.27 shows an operational comparison between a machine supplying base load and a hydro machine used for frequency control.

In recent decades advances in technology have enabled various sources of renewable energy including solar, wind, wave and tidal energy to also become part of generation mix. While these technologies all offer the ability to offset the CO_2 emissions characteristic of thermal generators, they do so at the expense of relatively low energy dense 'fuels'. Wind turbines, for example, must present a large swept blade area to the wind in order to extract a moderate amount of energy. As a consequence, physically large turbines have relatively small ratings, typically between 2 and 4 MW, and many turbines are required to produce the same quantity of electrical energy available from a single thermal machine. Offshore wind farms are not limited by the same construction, environmental and noise constraints as land-based installations and are therefore able to use considerably larger turbines to harvest energy from more reliable winds. The largest currently installed offshore machines have capacities of the order of 7 MW.

Similarly, the solar energy incident at the surface of the earth on a clear day is about $1\,\text{kW/m}^2$, and given that the conversion efficiency presently available from commercial photovoltaic cells is only about 18%, the area required to collect $1\,\text{kW}$ is a large $5.6\,\text{m}^2$. While this limitation is acceptable in the case of small roof-top installations in residential applications it represents one reason why large-scale solar farms are relatively few. Despite this, there are several very large solar farms throughout Europe and many in the USA, with nameplate ratings as high as 600 MW on sites as large as 2500 hectares.

A significant improvement in the solar collector area required can be had by using a large number of mirrors (known as *heliostats*) to track and focus sunlight onto a receiver which uses solar energy to heat molten salts to produce steam for the generation of electricity. This technology has the advantage that sufficient thermal energy can be stored in molten salt to permit electricity to be generated both day and night. *Concentrated solar power systems* as they are known are becoming popular in locations where land is relatively inexpensive and the daily solar radiation is high.

Tidal generation systems use tidal flows to power turbines in a similar way to wind turbines. However, because the density of water is about 850 times greater than that of air, considerably more energy is recoverable from a tidal generator than can be had from a wind turbine of the same size, permitting useful quantities of energy to be harvested from relatively low flow velocities. The performance of tidal generators can be further improved by fitting a duct or shroud around the machine. This increases the water velocity through the turbine and since the power output is proportional to the *cube* of velocity, a shrouded design can provide a useful return from small tidal or river flows. While the harvesting of tidal energy is not a new idea, recent advances in technology have made the construction of large tidal machines possible, many more of which are currently planned.

Renewable sources such as wind, solar and tidal all share the characteristic of being *variable*. The wind speed is rarely constant for more than a few hours, and since the energy recovered from a wind turbine is also proportional to the cube of the wind speed, small speed variations can result in large changes in machine output, as shown in Figure 1.28.

Solar intensity varies with the passage of clouds and with the diurnal movement of the sun. Figure 1.29 contrasts the power generated by a large PV array on a clear day with its performance on a cloudy one. Both these graphs highlight the uncontrollable nature of renewable energy sources and the fact that they cannot be relied upon during times of peak demand.

1.7.1 Synchronous and Asynchronous Generation

The problem of global warming, and in particular the greenhouse effects of CO_2, mean that the world's reliance on traditional fossil fuel generation must be reduced. While in some developing countries coal-fired generation is still being installed, in others older, less efficient and more polluting machines are progressively being withdrawn from service. This, together with the simultaneous increase in renewable energy systems, is creating a change in the generation mix, from one characterised by a few large thermal and hydro machines, towards one in which many smaller renewable generators are playing an increasingly significant part. The former may be generally classified as *synchronous generators*, while the latter are broadly *asynchronous*. The electrical differences

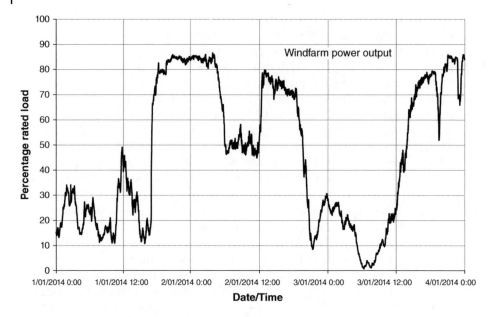

Figure 1.28 Daily variations in a wind farm generation.

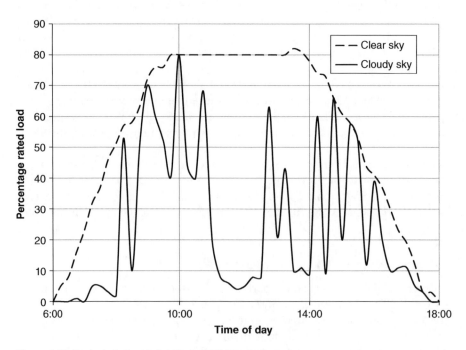

Figure 1.29 Typical photovoltaic solar installation performance.

between the two are significant, and the networks to which they are connected must adapt to the change in generation if they are to maintain their current levels of security and reliability, particularly under fault conditions.

Both fossil-fuelled and hydro turbines are capable of delivering reliable and controllable power. These are connected to the local network through large *synchronous* generators which rotate at speeds synchronised to the network in both frequency and phase. These machines frequently produce hundreds of megawatts and collectively possess characteristics that are particularly suited to ensuring stable and reliable electricity networks. For example, their large rotors provide considerable inertia, and since their governors are sensitive to tiny variations in shaft speed, they can effectively respond to sudden load variations while maintaining a near constant system frequency.

Thermal and hydro turbines also have the capability of providing ample reserve torque (spinning reserve), and the synchronous generators they drive can provide the necessary active and reactive power for peak load events, network contingencies and voltage regulation purposes. The *power quality* provided by synchronous machines is also high. They generate sinusoidal waveforms of high purity and can provide significant voltage support during network faults.

Asynchronous Generation

Asynchronous generation is a common characteristic of renewable energy sources. The variable nature of these sources means that it may not be possible for renewable rotating machines to achieve synchronous speed while delivering useful power to the grid. For this reason, wind-powered turbines generally use induction generators that operate over a range of speeds dependent on the wind energy available; they operate *asynchronously*. For example *doubly fed induction generators* (DFIG) employed on many wind turbines use a back-to-back voltage converter fed from the grid to supply the rotor windings with the necessary low frequency excitation. This allows the machine to appear to have synchronous characteristics and to deliver both active and reactive power to the network under normal conditions. Other wind-powered machines also use induction generators rotating at asynchronous speeds and deliver energy to the grid via back-to-back AC-DC and DC-AC converters. Both these topologies are limited by the current capacity of their electronics, and each has the propensity to suffer considerable electronic damage in the event of a nearby fault. Their inbuilt protection systems have therefore been primarily designed to preserve the integrity of the machine rather than to provide fault support for the network.

Network Fault Level and Fault-Ride-Through Ability

When a fault occurs, the local network voltage becomes depressed and, depending on the proximity and impedance of the fault, it may even fall to zero. A strong network will inject a large current into a fault, in an effort to support the network voltage. Further from the fault the voltage depression is less pronounced, supported by the fault current flowing towards the fault from nearby sources.

The *strength* of a network determines the extent of the resulting voltage depression as well as the degree to which its effects propagate throughout the remainder of the system. It is generally measured by the magnitude of the three-phase fault current at a particular location, known as the *fault level* and is usually expressed in amps (or kA), or sometimes in volt-amps (or MVA).

Fault levels vary considerably throughout a network and depend on both the network voltage and the network impedance at a given location. They tend to be highest near large synchronous generators where the network impedance is very low, and generally become lower as the distance from the generation sources increases. Synchronous machines are capable of generating very large fault currents, many times their continuous current ratings, whereas asynchronous generators and inverter connected machines generally contribute far less fault current, often equal to or less than their nameplate rating.

The ability of a generator to assist in supporting the network during a fault depends upon its ability to ride through the fault. Generators without *fault-ride-through* (FRT) ability are likely to trip in response to significant voltage depressions and a strong demand for reactive current. A significant loss of generation capacity during fault conditions can lead to a frequency collapse, and if load shedding is either inadequate or too slow, a system blackout may result. Therefore FRT is an important requirement in maintaining a network's transient stability.

Synchronous machines have the ability to ride through faults, and their high fault current contribution and the large quantities of reactive power they provide can help to support the system voltage during faults. Further, they can rapidly re-establish normal operation once a fault is cleared.

Asynchronous machines powered by renewable sources, on the other hand, have traditionally had limited FRT capacity, disconnecting themselves from the network during moderate voltage depressions. This upsets the post-fault load–generation balance, and removes a useful current contribution that would have helped support the network voltage during the fault.

1.7.2 Effects of Renewable Energy Sources on Network Performance

When there were relatively few asynchronous sources in the power network, the loss of some under fault conditions was of little consequence. However, the increasing proportion of asynchronous sources at a time when synchronous generators are being retired has increased the requirements for FRT and fault current capacity from those machines remaining. This may even necessitate the provision of synchronous spinning reserve solely on this account.

Limited FRT ability is but one of the drawbacks associated with asynchronous machines. The variable nature of renewable energy and the low rotational inertia of asynchronous machines make them generally unsuitable as peaking generators. Further, as their penetration increases, the resulting reduction in system inertia makes the frequency control task more difficult. In addition, large converter installations used in wind or solar farms frequently inject considerable harmonic current into the local MV or HV bus, thereby generating a degree of harmonic voltage distortion, necessitating the installation of harmonic filters near the injection point.

In 2016, the European Commission published a code for the grid connection for both synchronous and asynchronous generators. This document prescribes the characteristics required of new generation equipment installed throughout the European Union. Chief among these is the requirement for *all* generation sources to possess FRT ability under a prescribed voltage profile, both during and post-fault, and to provide reactive power during a fault, up to a specified fraction of the machine's rated capacity.

In addition, the code requires that generators shall remain connected over a defined range of network frequencies and shall provide a degree of frequency support. This may require a *reduction* in power delivery during over-frequency events, an *increase* during under-frequency events, or the capacity for both, depending on the size of the machine. It also requires a voltage support characteristic for machines connected to the HV or

EHV networks, in the form of a defined reactive power exchange with the network commensurate with the local network voltage. This means that the machine must *consume* reactive power from the network when the voltage *exceeds* the machine set point, and *export* it when the voltage *falls below* the set point.

Residential Photovoltaic Installations

Low-voltage residential photovoltaic (PV) solar arrays with capacities up to 5 kW have become very popular in countries such as Australia, where government incentives and high feed-in tariffs initially encouraged a rapid uptake of this technology. These arrays are interfaced to the LV network using small DC-AC inverters. They require the network voltage as a synchronising signal, and shut down when this falls below a defined threshold, so they cannot support the network during fault conditions or periods of depressed voltage. The inverters also inject harmonic rich currents into the LV network which contribute to LV harmonic distortion. Distributed LV generation tends to be visible only in so far as it offsets the consumer demand; it is transparent to network operators and cannot be controlled.

These PV systems were often initially installed with little consideration of the problems that unplanned distributed generation might create. In suburbs where the uptake was dramatic, the high density of solar generation created a significant voltage regulation problem. This has been caused by the *reverse* flow of power from the LV to the MV network, via local distribution transformers. The LV network is designed for a power flow in the opposite direction, and the fixed tap setting on distribution transformers allows for a potential drop within the transformer as well as in the LV network and the customer's premises. In the middle of the day when residential demand is low and solar radiation is high, the net power flow may be *towards* the MV network, causing these voltage drops to reverse. The LV network potential therefore rises and it may exceed the maximum level permitted. In many cases, this rise is sufficient to cause inverters to shut down, reducing the solar energy collected at a time when it is most abundant. Since most distribution transformers have off-load tap-changers, there is no easy way to rectify the situation. Further, the voltage experienced by customer equipment may vary considerably because the solar radiation can change dynamically throughout the day.

Regulators have begun to react to these issues, and new standards for low voltage inverters now demand a degree of both voltage and frequency support. For example, in Australia the 2015 edition of AS 4777, 'Grid connection of energy systems via inverters Part 2: inverter requirements', now requires both voltage and frequency support modes of inverter operation, in addition to a total harmonic current distortion of less than 5%. The voltage support mode is designed to permit an increase in the penetration of solar generation throughout the LV network, without creating the voltage regulation issue described above. This requires that the ratio of active to reactive power be modulated by the local voltage. Inverters must be capable of *exporting* up to 60% of nameplate rating as *reactive power* when the voltage is low and *import* up to this fraction when it is high.

1.7.3 Transmission Performance Improvement Through FACTS Devices

System strength has generally not been improved by the introduction of multiple asynchronous generators and inverter connected sources, but in recent decades various power electronic devices aimed at improving transmission network performance have become available. Collectively these fall into a category known as *flexible AC transmission systems* (FACTS). They can be used to increase or control the power flow in transmission lines and assist in the integration of renewable sources into a network by improving transient stability and providing voltage support during faults. FACTS devices can be broadly classified according to their method of connection as *shunt* compensators, *series* compensators or a combination of the two.

Shunt Compensators

SVCs – Static VAr compensators are devices that consist of switched capacitors in parallel with thyristor controlled reactors. Unlike a simple shunt capacitor, they can either consume or deliver reactive power to a bus in response to rapid changes in its voltage, and are therefore useful in improving network transient stability. Although the principle of operation is quite different, their characteristics are not dissimilar to traditional synchronous capacitors, but because they have no moving parts, they cannot provide inertial frequency support.

STATCOMS – Static compensators are based on a voltage source converter and are also able to exchange reactive power with a network in response to changes in network voltage. They are often used to maintain a desired power factor or regulate a bus voltage. When powered from a separate DC source, a STATCOM can also dynamically exchange active power with the bus, thereby damping power oscillations and improving transient stability.

Series Compensators

TCSC – The thyristor controlled series capacitor is connected in series with a transmission line and consists of a capacitor in parallel with a thyristor controlled reactor. By varying the thyristor firing angle the impedance of the combination can be progressively varied between that of an inductor and a capacitor. They are useful for controlling the power flow in transmission lines, damping power oscillations, limiting short circuit currents and providing voltage support.

Shunt-Series Compensation

UPFC – The unified power flow controller is the most complex and versatile member of the FACTS family. It is applied to a transmission line, and consists of *shunt* and *series* connected DC-AC converters, both supplied from a common capacitively supported DC bus. The series connected converter injects a voltage into the line that may be varied in both magnitude and phase, while the shunt connected converter absorbs real power from the line and delivers it to the DC bus supplying the series converter. Reactive power can flow bidirectionally from both converters, each independent of the other.

Depending on the phase of the injected potential, the UPFC can be used to regulate a line's receiving end voltage, to insert series (capacitive) compensation into the line, to adjust the phase shift across the line, or a combination of all three. As a result, it can independently control the active and reactive power flows as well as damping power oscillations and improving the transient stability.

1.7.4 Future Transmission Network Planning and Augmentation

Traditionally transmission networks have been designed around the need to transport large quantities of energy from fossil-fuelled power stations to load centres, or from one transmission substation to another. However, the location and capacity of new energy sources will be different from those presently in service; this will require transmission companies to augment their networks accordingly. In addition, it is likely that the proliferation of more residential PV systems, the introduction of storage batteries and smart metering, together with an improved understanding of time-based electricity pricing, will see changes in future consumer demand profiles. Collectively these changes may mean that future network planning will no longer be largely driven by a need to provide sufficient infrastructure to meet peak demand; instead it may be targeted at accommodating new sources of generation into the network together with the voltage and frequency services necessary to support it.

Two things are certain: considerable expenditure will be required to move the world towards more sustainable generation, and the transmission and distribution networks of the future will look quite different to those of today.

1.8   Sources

1 IEC 60038 '*Standard Voltages*' Edition 7, 2009–06, International Electrotechnical Commission, Geneva.
2 Schneider Electric, Cahier Technique No. 203. '*Basic selection of MV public distribution networks*'. 2001.
3 Schneider Electric/Square D '*Corner-grounded delta (Grounded B Phase) systems*'. 2012.
4 Zhang X, Cao X, Wang W, Yun C, '*Fault ride through study of wind turbines*', School of Electrical Engineering, Xinjiang University, Urumqi China, Journal of Power and Energy Engineering, 2013.
5 Australian Standard AS 4777.2, 2015, '*Grid connection of energy systems via inverters Part 2: Inverter requirements*', Standards Australia, Sydney.
6 AEMO '*National transmission network development plan 2016*', Australian Energy market Operator, Sydney.

Further Reading

1 Mohanty AK, Barik AK, '*Power system stability improvement using FACTS devices*' International Journal of Modern Engineering Research, Vol. 1, Issue 2, pp. 666–672.
2 Gyugyi L, Schauder CD, Williams SL, Rietman TR, Torgerson DR, Edris A, '*The unified power flow controller: A new approach to power transmission control*' IEEE Transactions on Power Delivery, Vol. 10, No. 2, April 1995.
3 Varma, Rajiv K '*Elements of FACTS controllers*', IEEE Power Engineering Society.
4 Abdulrazzaq AA, '*Improving the power system performance using FACTS devices*', Journal of Electrical and Electronics Engineering (IOSR-JEEE), Volume 10, Issue 2, pp. 41–49, March-April 2015.
5 Beck G, Breuer W, Povh D, Retzmann D, Telsch E, '*Use of FACTS and HVDC for power system interconnection and grid enhancement*', Siemens Power Transmission and Distribution, Power–Gen Conference, Abu Dhabi, United Arab Emirates, 2006.

2

Review of AC Circuit Theory and Application of Phasor Diagrams

In this chapter we review the network theorems fundamental to the operation of AC circuits and the use of *phasor diagrams* as a tool for circuit analysis. A firm understanding of the theory and application of phasors is essential in analysing the operation of AC circuits. This chapter provides an explanation of the concepts that underpin phasor analysis as well as numerous examples of their use in analysing AC circuits. We will see that phasor analysis frequently avoids the need for complicated calculations while providing an insight into how AC circuits actually work.

We will also study the resonant behaviour of LC series and parallel circuits, and the effect that resistive losses have near resonance. We begin by considering in more detail the three-phase system of voltage and currents.

2.1 Representation of AC Voltages and Currents

We saw in Chapter 1 that voltages in AC systems are ideally sinusoidal, and therefore they only contain energy at the system frequency ω. We also saw that three-phase voltages are displaced from one another by 120°, like those shown in Figure 2.1.

When $t = 0$ we see that $v_a(0) = 0$, so we can express the 'a' phase voltage, $v_a(t)$ as:

$$v_a(t) = \hat{v}\sin(\omega t) = \hat{v}\sin(2\pi ft) = \hat{v}\sin\left(\frac{2\pi}{T}t\right)$$

where $\hat{v}$ is the amplitude, ω is the angular frequency, expressed in radians per second, and T is the period of the waveform.

From Figure 2.1 V_b can be seen to lag V_a by $T/3$ or 120° electrical *(2π/3 radians)*. This is because the 'b' phase voltage passes through its positive going zero crossing 6.7 ms after the corresponding 'a' phase zero crossing.

Similarly, V_c lags V_b by 120° electrical, or alternatively we can say that V_c leads V_a by the same amount. These phase voltages have an *abc* phase sequence, since this is the order in which they pass through zero.

We can therefore express $v_b(t)$ and $v_c(t)$ as follows:

$$v_b(t) = \hat{v}\sin(\omega t - 2\pi/3)$$
$$v_c(t) = \hat{v}\sin(\omega t - 4\pi/3) = \hat{v}\sin(\omega t + 2\pi/3)$$

AC Circuits and Power Systems in Practice, First Edition. Graeme Vertigan.
© 2018 John Wiley & Sons Ltd. Published 2018 by John Wiley & Sons Ltd.

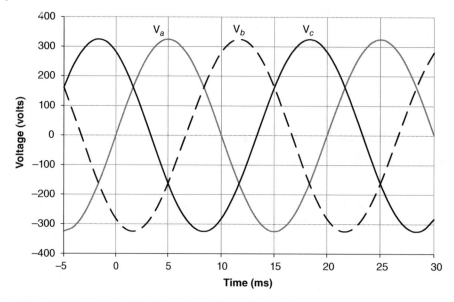

Figure 2.1 Three-phase 50 Hz alternating voltages ($T = 20$ ms).

These waveforms also are periodic, with period T, therefore we can write:

$$v(t+T) = v(t)$$

For a system frequency of 50 Hz, $T = 20$ ms, while for 60 Hz systems $T = 16.67$ ms.

We will confine our analysis to the *steady state* behaviour of AC circuits, where any transients arising at start-up have long since died away. Since AC voltages are sinusoidal, so will be the AC currents flowing in circuits comprising *linear components* such as resistors, capacitors and inductors.

Non-linear devices, (variable frequency drives for example), consume current at the system frequency, as well as integral multiples of it, known as *harmonics*. Supply authorities limit the distorting effects of harmonic currents so that system voltages generally do not contain more than about 5% *total harmonic distortion*. In the analysis that follows, both the voltages and the currents are assumed to be sinusoidal.

2.2 RMS Measurement of Time Varying AC Quantities

While it is relatively easy to measure the *amplitude* of a sinusoidal waveform (i.e. its *peak value*), this is not the most useful representation. Instead, engineers use a *root mean square (RMS)* calculation in the measurement of AC currents and voltages. The *RMS* concept is not restricted to sinusoids; it can be applied to *any periodic time varying waveform*.

The RMS value of an AC current or voltage is that which creates *the same average heating effect in a resistor as a DC voltage or current of the same magnitude*. Numerically, the power dissipated in an R ohm resistance by a source of DC potential V_{DC} $(= V_{RMS})$ is constant and is given by:

$$P_{DC} = \frac{V_{DC}^2}{R} = \frac{V_{RMS}^2}{R} (\text{watts})$$

The *average power* delivered by a sinusoidal AC source to the same resistor, can be expressed as:

$$P_{AC} = \frac{Average\{\hat{v}\sin(wt)\}^2}{R} = Average\left\{\left[\frac{\hat{v}\sin(wt)}{R}\right]^2\right\}$$

If these quantities are equated, we find:

$$V_{RMS}^2 = Average\{\hat{v}\sin(wt)\}^2$$

The root mean square voltage can therefore be expressed as:

$$V_{RMS} = \sqrt{Average\{\hat{v}\sin(wt)\}^2} = \sqrt{\frac{1}{T}\int_0^T \{\hat{v}\sin(wt)\}^2 \, dt} = \frac{\hat{v}}{\sqrt{2}}$$

Thus the power delivered to an R ohm resistor by this voltage is $\dfrac{\hat{v}^2}{2R} = \dfrac{(V_{RMS})^2}{R}$ watts.

Therefore for a sinusoidal waveform we may write: $V_{RMS} = \dfrac{\hat{v}}{\sqrt{2}}$.

Similarly, RMS *currents* can be defined by:

$$I_{RMS} = \sqrt{\frac{1}{T}\int_0^T \{\hat{i}\sin(wt)\}^2 \, dt} = \frac{\hat{i}}{\sqrt{2}}$$

The RMS value of *any* periodic function $v(t)$, can be calculated using the equation:

$$V_{RMS} = \sqrt{\frac{1}{T}\int_0^T \{v(t)\}^2 \, dt}$$

RMS quantities are usually written in upper case, e.g. V or V_{RMS}; they are not time varying quantities, but are constants. So as to distinguish them from the upper case notation that we will use to designate phasors, RMS values in this text will be displayed within a modulus symbol, i.e. $V_{RMS} = |V|$.

2.3   Phasor Notation (Phasor Diagram Analysis)

All the analyses that follow are aimed at establishing the *steady state response* of AC circuits to sinusoidal excitation at the system frequency – that remaining after any short-term transients have died away. Manipulating and keeping track of *time varying waveforms* of the kind shown in Figure 2.1 is very tedious, especially when three-phase voltages and currents are involved. Fortunately, a shorthand notation has been developed that enables engineers to easily analyse quite complex multi-phase circuits, known as *phasor notation*, or alternatively *phasor diagrams*. Phasor diagrams are based on the sine and cosine functions that describe alternating currents and voltages. They provide a simple and elegant visual representation of the voltages and currents that exist in AC circuits, and are applicable to Kirchhoff's voltage and current laws, making them a very powerful tool for analysing the steady state behaviour of quite complex AC circuits.

2.3.1 Sine and Cosine as Circular Functions

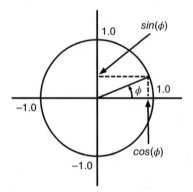

Figure 2.2 Definitions of sine and cosine functions.

Sinusoidal waveforms are strongly related to circular motion; indeed, the sine and cosine functions are often referred to as *circular functions*. The *unit circle* in Figure 2.2 can be used to *define* the sine and cosine functions of an arbitrary angle ϕ. If we use the positive x axis as the angular origin, then $\sin(\phi)$ is defined as the projection of the radius on the *vertical* axis while $\cos(\phi)$ is its projection onto the *horizontal* axis.

If we allow the radius to rotate *at a constant angular frequency* ω, in an anticlockwise direction, commencing at $t=0$ from a position on the positive x axis, then its projections onto the x and y axes describe the time varying functions $\cos(\omega t)$ and $\sin(\omega t)$ respectively, as illustrated in Figure 2.3.

The cosine function can be seen to *lead* the sine, since it passes through its maximum value 90° *before* the sine function does. Alternatively, we can say that the sine function *lags* the cosine, by 90°. The shape of these functions is representative of the voltages and currents that exist in a power system.

If we consider the expressions presented above for our set of phase voltages, we see that only *two* quantities are required to define each one. The first is its *magnitude*, expressed in volts RMS, and the second is the *phase angle* ϕ that it takes when $t=0$. For example, the '*b*' phase voltage in the example above has an *RMS* magnitude of $\hat{v}/\sqrt{2}$ and a phase angle of $\phi = -2\pi/3$ radians, while the '*c*' phase voltage has the same magnitude and a phase angle of $\phi = -4\pi/3$ radians.

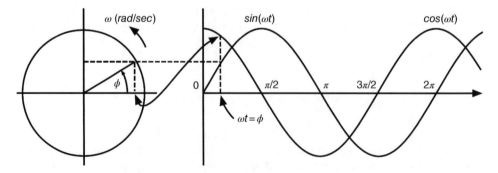

Figure 2.3 Generation of time varying sinusoidal functions from rotary motion.

Let us now consider our circle to be drawn on the *complex plane*. Euler's formula states that:

$$e^{j\omega t} = \cos(\omega t) + j\sin(\omega t)$$

This expression represents a *unit* vector, rotating anticlockwise at ω radians per second, whose real coordinate is $\cos(\omega t)$ and whose imaginary coordinate is $\sin(\omega t)$. A rotating vector with an *RMS* amplitude of $|V|$ can therefore be written as:

$$\sqrt{2}|V|e^{j(\omega t+\phi)} = \sqrt{2}|V|\cos(\omega t+\phi) + j\sqrt{2}|V|\sin(\omega t+\phi)$$

The left-hand side of this equation is the *polar form* of the vector's coordinates, while that on the right is their *Cartesian form*. Alternatively, we can represent an arbitrary *cosine* function by the **Real** part of $\sqrt{2}|V|e^{j(\omega t+\phi)}$, where:

$$\text{Re}\left[\sqrt{2}|V|e^{j(\omega t+\phi)}\right] = \sqrt{2}|V|\cos(\omega t+\phi)$$

and a *sine* function by the **Imaginary** part:

$$\text{Im}\left[\sqrt{2}|V|e^{j(\omega t+\phi)}\right] = \sqrt{2}|V|\sin(\omega t+\phi)$$

2.3.2 Phasor Representation

The *phasor representation* of an AC quantity is obtained by taking a *snapshot* of its rotating vector, *thereby freezing its motion in time*. The frozen vector is called a *phasor*. Its length represents the phasor's magnitude in volts or amps RMS, and the exact point in time when its rotation is frozen can be chosen arbitrarily. This is usually determined with regard to a *reference phasor*, generally either a system voltage or current, which is assigned a phase angle of *zero degrees*.

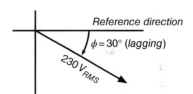

Figure 2.4 Phasor representation of an AC voltage.

The phasor in Figure 2.4 represents a voltage whose magnitude is $230V_{RMS}$ and whose angle *lags* the reference position by 30°.

If we describe a rotating voltage vector by $\sqrt{2}|V|e^{j(\omega t+\phi)}$ and *freeze* its rotation in time, by putting $t = 0$, we obtain the polar form of the phasor V, representing it. Thus:

$$V = |V|e^{j\phi} = |V|\angle\phi \tag{2.1}$$

Since in the steady state, the frequency of all voltages and currents in a network is the same, the only parameters required to uniquely define any phasor are its magnitude and its phase, both of which are included in the phasor expression in Equation (2.1).

As an example, consider Figure 2.5, which shows a single-phase induction motor. Because motors are partially *inductive*, as we shall see later in this chapter, the current they consume *lags* the applied voltage. In the associated phasor diagram, the applied voltage has been chosen as the *reference phasor*, and therefore its phase has been

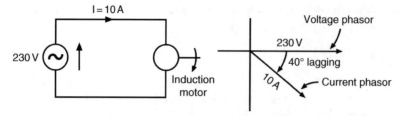

Figure 2.5 Single phase circuit and its phasor representation.

assigned a value of zero degrees. The motor current *lags* this voltage, in this case by an angle of 40°, and therefore it can be described in polar form by the phasor $10 \angle -40°$ amps.

2.3.3 Three-Phase Systems and Phasor Diagrams

Figure 2.6a shows the phasor representation of a three-phase system of voltages. Here the 'a' phase has been chosen as the reference phasor, with an angle of zero degrees. Therefore its phase to neutral voltage lies along the positive real axis. If we assume a conventional phase sequence of *abc*, then 'b' phase voltage lags the 'a' phase by 120°, and the 'c' phase lags the 'a' by 240°. The neutral (or star) point lies at the origin. The phase sequence is *abc* because, with an *anticlockwise* rotation, this is the sequence in which the phasors pass through the reference position. It is also useful to note that *abc*, *bca* and *cab* all represent this phase sequence.

There are only two possible phase sequences, the other being *acb*. Thus, if the sequence *abc* corresponds to an anticlockwise rotation of a machine, then the sequence *acb* will correspond to a clockwise shaft rotation.

Line voltages can also be represented as phasors, as shown in Figure 2.6b. Note that, in this case, there is no connection to the origin, and the angle between adjacent line voltages on the diagram is 60°, not 120°. The line voltages in Figure 2.6b together with the two associated phase voltages form an isosceles triangle. Since the included angle is 120°, the magnitude of each line voltage is √3 times that of a phase voltage.

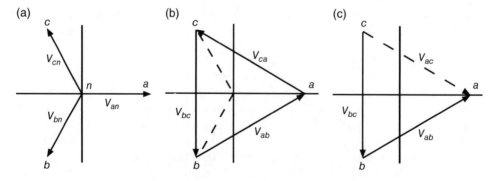

Figure 2.6 (a) Phase voltages (b) Line voltages (c) Summation of voltages $V_{ab} + V_{bc}$.

Any voltage phasor represents a difference in potential between two points in a network. These points are often included as subscripts in the name of the phasor. For example, Figure 2.6b shows the line voltages derived from Figure 2.6a; these are labelled V_{ab}, V_{ca} and V_{bc}. In each case, the subscripts represent the phases between which each potential is measured. Thus V_{ab} is the voltage of phase 'a' with respect to phase 'b' and therefore the arrowhead corresponds to phase 'a'. Similarly phasor V_{ca} represents the voltage of phase 'c' with respect to phase 'a'. Therefore $V_{ac} = -V_{ca}$. Where the reference potential is the *neutral terminal* (0 volts), the second subscript is often omitted, thus $V_{an} = V_a$.

The addition (or subtraction) of phasors is similar to that of vectors, since each has a particular magnitude and phase. Therefore a phasor can be moved about on the complex plane, so long as its magnitude and phase are preserved. Two phasors can thus be added by simply placing them end to end, while preserving the magnitude and phase of each one. For example, in Figure 2.6c we see that $V_{ab} + V_{bc} = V_{ac}$. The voltage V_{ac} results from the vector addition of V_{ab} and V_{bc}. Note also that the common subscript 'b' vanishes in the summation.

We can see from Figure 2.6b that when laid end to end the line voltages sum to zero, i.e. $V_{ab} + V_{bc} + V_{ca} = 0$. We can also demonstrate this as follows:

$$\left(V_{ab} + V_{bc}\right) = V_{ac} = -V_{ca}$$

Therefore $(V_{ab} + V_{bc}) + V_{ca} = -V_{ca} + V_{ca} = 0$.

We shall see that the ability to add or subtract phasors is particularly useful when applying Kirchhoff's voltage and current laws.

Figure 2.7 shows phase currents included on a phasor diagram as well as phase voltages. In this example, each phase current lags its respective phase voltage by a small angle, typical of an industrial load. Three-phase loads whose currents have the same magnitude and the same relative phase are said to be *balanced*. Sometimes, as in this case, the neutral reference is dropped from the voltage phasor nomenclature, especially in the case of phase voltages, since it is understood that these are always measured with respect to the neutral terminal.

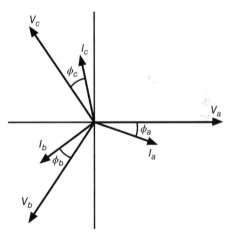

Figure 2.7 Phasor representation of three-phase balanced voltages and currents.

2.4 Passive Circuit Components: Resistors, Capacitors and Inductors

We will now review three linear components – resistors, capacitors and inductors – and from the equations describing their fundamental behaviour, we derive expressions for the impedance presented by each one.

2.4.1 Resistors

The simplest passive component is the resistive element, R. The voltage developed across a resistance is in direct proportion to the current flowing through it, as defined by Ohm's law, i.e. $v(t) = i(t)R$, or in terms of RMS quantities we can write: $|V| = |I|R$.

The current flowing in a resistor lies exactly in phase with the voltage developed across it, therefore the resistance R is a real number. Thus the voltage and current phasors V_R and I_R lie in the same direction when represented on a phasor diagram. The resistance of a conductor of constant cross-section is proportional to its length and inversely proportional to its cross-sectional area; the proportionality constant is called the *resistivity* and is given the symbol ρ, measured in *ohm-meters*. Therefore we may write:

$$R = \frac{\rho l}{A}$$

where R is the resistance in ohms, l is the length of the conductor in metres and A is its cross-sectional area in m^2.

The resistivity is equal to the resistance between opposite faces of a 1 metre cube of the material in question, and therefore for metals it takes on very small values. For example, the resistivity of copper is 1.68×10^{-8} ohm-m at 20°C, increasing approximately linearly with temperature to 2.04×10^{-8} ohm-m at 75°C.

The resistivity is the reciprocal of the *conductivity* (σ) of the material, which is related to the *electric field strength E* within the conductor and the resulting current density, J (A/m^2) according to:

$$\sigma = \frac{J}{E}$$

Sometimes it is more convenient to use *conductances* rather than resistances, where the conductance G, is the *reciprocal of resistance*. Conductances are measured in *siemens*, and are particularly convenient when resistances are connected in *parallel*, where they can be added, just as resistances connected in series are additive.

Current flows through a metallic conductor under the influence of a small internal electric field. For example, a copper conductor with a cross-sectional area of 1 mm^2, carrying a current of 1 amp has a current density $J = 10^6$ amps/m^2. According to the equation above, the electric field strength E within this conductor is:

$$E = J/\sigma = 1.68 \times 10^{-2} \text{ V/m}.$$

Thus an electric field of *less than* 20 mV/m is sufficient to establish a current density as high as of one million amps per square metre! This is because copper is a particularly good conductor and it is for this reason that it is widely used in the construction of transformers, busbars and electrical cables.

Transmission and distribution losses should be kept as low as practical, and despite having resistivity values slightly higher than that of copper, *steel reinforced aluminium conductors* are increasingly used in transmission and distribution networks. This is due to the low density of aluminium, coupled with the relatively low coefficient of thermal expansion of the steel reinforcing, enabling greater span distances to be achieved between transmission towers. The minimum clearance below a transmission line is a critical parameter, and it varies with temperature. At high ambient temperatures, or high line loadings, thermal expansion causes an increase in the conductor length, increasing the sag, and reducing the clearance beneath the line.

2.4.2 Capacitors

A capacitor typically consists of two metallic plates, separated by an insulating medium known as a *dielectric*, as shown in Figure 2.8. If a DC voltage V is applied to a discharged capacitor, a transient current briefly flows, establishing a positive charge on one plate and a negative charge on the other, retarding further charge accumulation. Once the potential between the plates equals the applied voltage, no further current will flow. As a result of the accumulation of this charge, an *electric field E* is established between the plates, equal to V/d volts per metre, where V is the voltage applied and d is the distance between the plates, illustrated in Figure 2.9.

For a given plate area A and separation d, the accumulated charge is proportional to the voltage applied, according to the equation:

$$q = \frac{A\varepsilon}{d}V$$

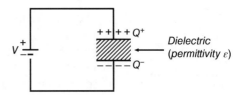

Figure 2.8 Capacitor construction.

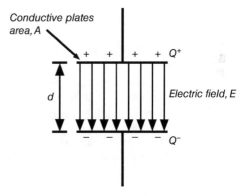

Figure 2.9 Electric field distribution.

Where ε is the *permittivity* of the dielectric material between the capacitor's plates. We define the term $\dfrac{A\varepsilon}{d}$ as the *capacitance C*, measured in *farads* (and named in honour of Michael Faraday). Therefore we can rewrite this equation in the form:

$$q = \frac{A\varepsilon}{d}V = CV \qquad (2.2)$$

The permittivity ε is a measure of the degree to which a dielectric material is *polarised* by the presence of an electric field. If we consider a capacitor holding a *constant charge*, we can show that the inclusion of a dielectric material between its plates acts to *reduce* the magnitude of the electric field between them. This is because the dielectric materials become *polarised* by the external electric field. They contain *polar molecules* that form tiny electric dipoles when the positive and negative charges they contain are slightly separated by the action of the external field. This effect is shown in Figure 2.10.

These electric dipoles create an electric field of their own, which as shown *opposes* the external field that created them. This can be demonstrated numerically by rewriting Equation (2.2) in terms of the electric field strength E, between the plates:

$$q = A\varepsilon_r \varepsilon_o E \qquad (2.3)$$

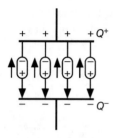

Figure 2.10 Polarisation of the dielectric by the applied electric field.

Since q is assumed to be constant, then increasing the relative permittivity ε_r will result in a corresponding decrease in the electric field strength. This in turn leads to an increase in capacitance value, compared to one with an air dielectric (for which $\varepsilon_r = 1$), since $C = \dfrac{A\varepsilon_o\varepsilon_r}{d}$.

When the plates are separated by an air space, or by a vacuum, the permittivity takes on a value given the symbol $\varepsilon_o = 8.85 \times 10^{-12}$ farads per metre (F/m). When dielectric materials such as polypropylene or polycarbonate are used, the permittivity *increases*. The permittivity of dielectric materials is usually expressed *relative* to the permittivity of free space, to which we give the symbol ε_o. We define the *relative permittivity* ε_r as:

$$\varepsilon_r = \frac{\varepsilon}{\varepsilon_o}$$

where ε is the *absolute permittivity* of the dielectric material.

The relative permittivity ε_r is always greater than or equal to unity, and therefore the absolute permittivity is always greater than or equal to ε_o. For example, the relative permittivity of polycarbonate is 2.9 while that of polypropylene is 11.9. Both of these are used as dielectric materials in capacitor construction.

Capacitive Reactance and Impedance

Equation (2.2) can be rewritten in terms of the capacitance C, the accumulated charge $q(t)$ and the applied voltage $v(t)$, both of which may be considered as functions of time:

$$q(t) = Cv(t) \tag{2.4}$$

Unlike the case of DC excitation, which only results in the establishment of a *static* electric field between the plates, an alternating potential will result in a continuously changing electric field. This is accompanied by an alternating current required to deliver the charge necessary to establish the field. This current can be obtained by differentiating Equation (2.4):

$$i(t) = \frac{dq(t)}{dt} = C\frac{dv(t)}{dt} \tag{2.5}$$

Equation (2.5) shows that the current flowing in a capacitor is proportional to the *rate of change* of the voltage applied to it. Therefore we might reasonably expect larger currents to flow at higher frequencies. If we let $v(t) = \sqrt{2}|V|\cos(\omega t)$ then Equation (2.5) provides an expression for $i(t)$:

$$i(t) = -\omega c\sqrt{2}|V|\sin(\omega t) = \omega c\sqrt{2}|V|\cos(\omega t + \pi/2) \tag{2.6}$$

Alternatively, the RMS current can be expressed in terms of the RMS voltage according to $|I| = \omega C|V|$. The *capacitive reactance*, to which we give the symbol X_c, is therefore given by:

$$X_c = \frac{|V|}{|I|} = \frac{1}{\omega C} \text{ ohms}$$

Thus the capacitive reactance does indeed fall as the frequency rises, consistent with our previous expectation. Examination of Equation (2.6) also shows that the current is phase shifted with respect to the capacitor voltage, leading it by 90°. If we define the capacitor's *complex impedance* Z_c as the ratio of its voltage and current phasors V_c and I_c, we find that:

$$Z_c = \frac{V_c}{I_c} = \frac{|V|\angle 0°}{\omega C|V|\angle 90°} = \frac{|V|\angle 0°}{j\omega C|V|\angle 0°} = \frac{1}{j\omega C} = -jX_c \quad \left(\text{Recall that } 1\angle 90° = j\right)$$

The complex impedance Z_c includes the phase information necessary to generate the observed phase shift in the capacitor current:

$$I_c = \frac{V_c}{Z_c} = j\omega C V_c \tag{2.7}$$

It is worth pointing out that although the imped-ance is generally a complex number, it is *not* a phasor, rather it is the *ratio* of two phasors and does not rep-resent a sinusoidally varying quantity. Equation (2.7) suggests that if we choose the voltage as the reference phasor, then the current flowing in our capacitor lies along the positive j axis, as shown in Figure 2.11. Alternatively, we can say that the *current leads the capacitor voltage by 90°.*

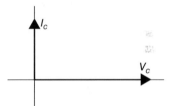

Figure 2.11 Capacitor voltage and current phasor diagram.

Capacitive Susceptance and Admittance

When capacitors are parallel connected it is often more convenient to work in terms of *capacitive susceptance* rather than the capacitive reactance, since susceptances in parallel may be added. The susceptance B_c, is equal to the reciprocal of the reactance, and is measured in siemens, thus $B_c = 1/X_c = \omega C$. The complex *capacitive admittance* to which we give the symbol Y_c, contains phase information in the same way as the capaci-tive impedance does, since:

$$Y_c = \frac{1}{Z_c} = j\omega C = jB_c \text{ siemens}$$

2.4.3 Inductors (Reactors)

An inductor essentially consists of a coil, within which a magnetic field is created by the action of a current flowing within it. Whereas the current flowing in a capacitor is related to the rate of change of the electric field that exists between its plates, the volt-age dropped across an inductor is related to the rate of change of the magnetic field linking its windings. Inductors may be wound on an easily magnetised iron core, or they may be air-cored. We will discuss the theory of magnetic circuits in a later chap-ter; for now we will simply consider the electrical properties of inductors, also known as *reactors*.

Faraday's law of induction states that a time varying magnetic flux will induce an electric potential within a winding linking that flux. Numerically, this is expressed in Equation (2.8).

$$v(t) = N \frac{d\Phi(t)}{dt} \tag{2.8}$$

where $v(t)$ is the induced potential, N is the number of turns on the winding and $\Phi(t)$ is the time varying magnetic flux linking the winding. In a *linear* inductor where the flux is proportional to the current flowing within its winding, we may write:

$$\Phi(t) = N \frac{i(t)}{\Re} \tag{2.9}$$

where $i(t)$ is the current flowing within the winding and $\Re$ is the *reluctance* of the magnetic circuit. Reluctance in magnetic circuits is an analogous quantity to resistance in electric circuits. It relates to the physical construction of the magnetic circuit and the material from which it is made. We will discuss reluctance in more detail in Chapter 4.

By substituting Equation (2.9) into Equation (2.8) we find:

$$v(t) = N \frac{d\Phi(t)}{dt} = \frac{N^2}{\Re} \frac{di(t)}{dt} = L \frac{di(t)}{dt} \tag{2.10}$$

where L is the inductance of the winding, and is equal to $\dfrac{N^2}{\Re}$ *henries*.

Inductive Reactance and Impedance

If we represent the inductive voltage by $v(t) = \sqrt{2}|V|\cos(\omega t)$ and substitute this into Equation (2.10), we find that in the steady state the current is given by:

$$i(t) = \frac{\sqrt{2}|V|}{\omega L} \sin(\omega t) \tag{2.11}$$

By analogy with the capacitive case, the *inductive reactance* X_L is given by the ratio of its RMS voltage to its RMS current, thus:

$$X_L = \frac{|V|}{|V|/\omega L} = \omega L \text{ ohms} \tag{2.12}$$

Equation (2.12) suggests that the inductive reactance *increases* with frequency and therefore, for a given excitation voltage, the current will decrease as the frequency increases. Examination of Equation (2.11) also shows that the inductor current is *phase shifted* with respect to its voltage, in this case lagging it by 90°. As before, we define the inductor's *complex impedance* Z_L as the ratio of its voltage and current phasors, V_L and I_L, thus we find that:

$$Z_L = \frac{V_L}{I_L} = \frac{\omega L|V|\angle 0°}{|V|\angle -90°} = \frac{j\omega L|V|\angle 0°}{|V|\angle 0°} = j\omega L = jX_L \text{ ohms}$$

The phasor diagram for an inductor appears in Figure 2.12.

Inductive Admittances

As in the capacitive case, when inductive elements are *parallel* connected, working in terms of admittances is usually more convenient. The inductive admittance Y_L is the reciprocal of the inductive impedance Z_L, thus $Y_L = 1/Z_L = 1/j\omega L = -jB_L$, where B_L is the *inductive susceptance*: $B_L = \dfrac{1}{\omega L}$ and is measured in siemens.

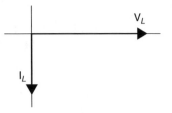

Figure 2.12 Inductor voltage and current phasor diagram.

2.4.4 General Impedance and Admittance Expressions

When resistive, inductive and capacitive elements all appear in series, their collective impedance Z can be expressed in *rectangular form* (Cartesian) as:

$$Z = R + j(X_L - X_C) = R + jX \text{ ohms}$$

Similarly, when these elements appear in *parallel*, their collective *admittance Y*, is given by:

$$Y = G - jB_L + jB_C = G + jB \text{ siemens}$$

Since $Y = 1/Z$, in general we may write:

$$Y = G + jB = \frac{R}{R^2 + X^2} - j\frac{X}{R^2 + X^2} \text{ siemens}$$

Therefore it is only when $R = 0$ that $B = 1/X$ and similarly only when $X = 0$ does $G = 1/R$. Impedances and admittances may also be expressed in *polar form*:

$$Z = |Z|\angle\phi \text{ or } Y = |Y|\angle\theta$$

Where

$$|Z| = \sqrt{R^2 + X^2}, \phi = ar\tan(X/R), |Y| = \sqrt{G^2 + B^2} \text{ and } \theta = ar\tan(B/G).$$

2.5 Review of Sign Conventions and Network Theorems

We now briefly review the major network theorems used in the analysis of AC circuits. We begin with a review of the *sign convention* used in defining AC voltages and currents.

2.5.1 Sign Convention

The application of the network theorems that follow requires that the voltage dropped across circuit elements be defined with the correct polarity in relation to the current flowing through them. Although these polarities reverse every half cycle in an AC circuit, the following sign convention defines the instantaneous polarities throughout the circuit at one point in time, in much the same way as they would be defined in a DC circuit.

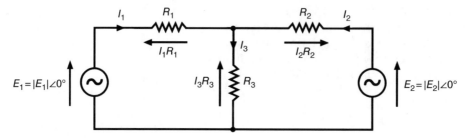

Figure 2.13 Sign conventions.

The instantaneous potentials dropped across elements in an AC circuit arise as a consequence of the current flowing through them. We indicate this polarity by arrows, pointing in the opposite direction to the flow of conventional current which created them. *Conventional current* is defined as the flow of *positive charge* (as opposed to *actual* current which relates to the movement of negatively charged electrons within a conductor).

Generally, we can define the direction of currents within a circuit however we choose, so long as we are consistent in our application of the voltage sign convention. Should the value of a particular current be found to be negative, then in reality it flows in the opposite direction to that in which it was defined.

The polarity of voltage (or current) sources is also shown by use of arrows. The magnitude of a source is the RMS potential *between* or the RMS current flowing through its terminals. The arrowhead indicates the instantaneous positive terminal and therefore the direction in which conventional current will flow from the source.

This sign convention, when applied correctly, defines the polarity of currents and voltages as they are used in loop and mesh analyses. It is depicted in the simple circuit of Figure 2.13, where phasor E_1 establishes current I_1 and phasor E_2 establishes current I_2. Both these currents return to their respective sources through R_3.

We review the major network theorems that find application in AC circuit analysis. Each of these applies to the instantaneous voltages and currents existing throughout a network, and therefore they also apply to the phasors representing them as well. In the analyses that follow we will use the phasor representation of currents and voltages.

2.5.2 Kirchhoff's Voltage Law

Kirchhoff's voltage law (KVL) states that *the sum of all voltages dropped around any closed loop is equal to the sum of voltage sources contained within that loop*. From the circuit in Figure 2.13a, using KVL we may write:

$$E_1 = I_1R_1 + I_3R_3 \quad \text{and} \quad E_2 = I_2R_2 + I_3R_3$$

Although in this simple example we appear to add voltage drops algebraically, in general AC analysis KVL requires that voltage drops be added *vectorially*, as complex numbers, thereby taking both their magnitude and phase into account.

Figure 2.13a Example circuit.

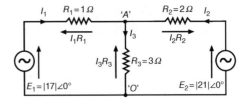

2.5.3 Kirchhoff's Current Law

Kirchhoff's current law (KCL) states that *the sum of the currents entering a junction is equal to the sum of those leaving it.* Therefore from Figure 2.13a we may write:

$$I_1 + I_2 = I_3$$

Kirchhoff's current law generally requires that currents be added *vectorially,* as complex numbers. In this simple example, however, since the voltage sources are in phase with each other and the impedances are resistive, then the currents will also lie in phase.

Suppose, for example, we wish to solve for I_1 and I_2 in Figure 2.13 using KVL and KCL then we might proceed as follows:

We can combine the equations obtained above using KVL and KCL to yield:

$$E_1 = I_1 R_1 + (I_1 + I_2) R_3$$

and $\quad E_2 = I_2 R_2 + (I_1 + I_2) R_3.$

Using the values given in Figure 2.13, these simultaneous equations are satisfied if $I_1 = 2\angle 0°$ amps and $I_2 = 3\angle 0°$ amps.

2.5.4 Principle of Superposition

The principle of superposition applies to circuits containing linear components only. It states that *any voltage or current in a linear circuit, containing multiple energy sources, can be evaluated from the sum of the contributions from each source when acting on its own (with all other sources set to zero).*

The contribution from each source acting alone deserves further comment. Consider the circuit of Figure 2.13a where there are two voltage sources. When we consider the source E_1 acting alone, we must remove the effects of the source E_2, *without* limiting the current that may flow through this branch. Accordingly we must force E_2 to zero, and replace this source by its *equivalent internal impedance*, which for a voltage source is zero ohms. Thus a voltage source is replaced by a *short circuit*. Similarly, if the network contains a *current source*, then when removed, *no current* must be permitted to flow in the branch, regardless of the potential developed across it. It must therefore be replaced with *infinite* impedance, i.e. an *open circuit*.

If we wish to solve for I_1 and I_2 in Figure 2.13a using the superposition theorem, we must add the contributions to I_1 and I_2 from both E_1 and E_2, thus:

$$I_1 = \frac{E_1}{(R_1 + R_2\|R_3)} - \frac{E_2}{(R_2 + R_1\|R_3)(R_1 + R_3)} \cdot R_3 = 7.727\angle 0° - 5.727\angle 0° = 2\angle 0° \text{amps}$$

$$I_2 = \frac{E_2}{(R_2 + R_1\|R_3)} - \frac{E_1}{(R_1 + R_2\|R_3)(R_2 + R_3)} \cdot R_3 = 7.636\angle 0° - 4.636\angle 0° = 3\angle 0° \text{amps}$$

where $R_2\|R_3$ represents the *parallel combination* of R_2 and R_3, i.e. $R_2\|R_3 = R_2 R_3/(R_2 + R_3)$.

2.5.5 Thévenin's Theorem

The equivalent circuit of a linear network as seen from any pair of terminals, can be represented by a *Thévenin equivalent voltage source* V_{th} in series with the *Thévenin equivalent impedance* Z_{Th} shown in Figure 2.14. The Thévenin equivalent voltage V_{th} is the open circuit potential between the terminals in question, i.e. that which appears with no external impedance connected.

The Thévenin equivalent impedance Z_{th} is the impedance seen looking into the network terminals, with all voltage sources in the network replaced by short circuits, and all current sources replaced by open circuits.

If we wish to solve for I_1 and I_2 in Figure 2.13a using *Thévenin's theorem*, we can consider the Thévenin equivalent of the circuit between nodes 'A' and 'O', with R_3 removed. The Thévenin equivalent voltage source can be found using the superposition theorem, and is equal to the open circuit voltage, V_{AO}:

$$V_{th} = V_{AO} = \frac{E_1 R_2}{(R_1 + R_2)} + \frac{E_2 R_1}{(R_1 + R_2)} = 18.33\angle 0° \text{ volts}$$

and the Thévenin equivalent resistance R_{th} is equal to $R_1\|R_2 = 0.666$ ohms.

If we now connect R_3 to the Thévenin equivalent circuit, we find that the current flowing through it is given by:

$$I_3 = \frac{V_{th}}{(Z_{th} + Z_{load})} = \frac{18.33\angle 0° \text{ volts}}{(3 + 0.666) \text{ ohms}} = 5\angle 0° \text{ amps}$$

And therefore the voltage across R_3 is equal to the voltage at point 'A' in Figure 2.13a with respect to that at point 'O', i.e. V_{AO}.

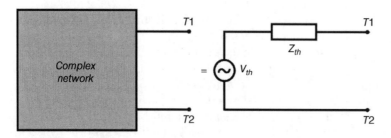

Figure 2.14 Thévenin equivalent network.

So $V_{AO} = 5\angle0° A \times 3\Omega = 15\angle0°$ volts

$$I_1 = \frac{(E_1 - V_{AO})}{R_1} = 2\angle0° \text{ amps}$$

and $I_2 = \dfrac{(E_2 - V_{AO})}{R_2} = 3\angle0°$ amps.

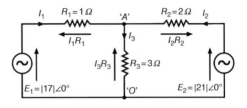

Figure 2.13a Example circuit.

2.5.6 Norton's Theorem

Norton's theorem is an alternative approach to Thévenin's theorem. It states that the equivalent circuit of a linear network referenced to any pair of terminals in that network can be represented by a *current source* I_N in parallel with an *admittance* Y_N, as shown in Figure 2.15.

The Norton equivalent current source is the *short circuit current* that would flow from the network terminals T1 and T2. The Norton equivalent impedance $(1/Y_n)$ is the same as the Thévenin equivalent impedance and it is found in the same way. From these definitions, we can see that the Thévenin equivalent voltage is equal to the Norton equivalent current divided by the Norton admittance Y_n:

$$V_{th}(t) = I_n(t) / Y_n$$

N.B. It should be stressed that both Thévenin and Norton's theorems do not imply anything about the internal operation of the network to which they refer. They merely provide information relating to the combinations of voltages and currents that can be obtained from the terminals T1 and T2, under various loading conditions.

We can also use Norton's theorem to find I_1 and I_2 in Figure 2.13a. As before, we proceed by determining the Norton equivalent circuit across R_3, with R_3 removed. The *short circuit currents* flowing from E_1 and E_2 are $17\angle0°$ amps and $10.5\angle0°$ amps respectively. Thus the total short circuit current available from node 'A', is $27.5\angle0°$ amps. As in the previous example, the Norton impedance is 0.666 ohms. From the Norton equivalent circuit the potential across R_3 is given by:

$$V_x = 27.5 \frac{R_3 Z_N}{(R_3 + Z_N)} = 27.5 \frac{3 \times 0.666}{(3 + 0.666)} = 15\angle0° \text{ volts}$$

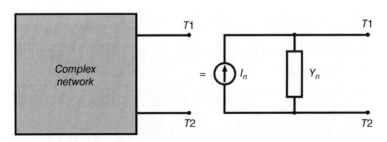

Figure 2.15 Norton equivalent network.

And therefore we again find that:

$$I_1 = \frac{(E_1 - V_x)}{R_1} = 2\angle 0° \text{ amps}$$

and: $I_2 = \frac{(E_2 - V_x)}{R_2} = 3\angle 0° \text{ amps}$

2.5.7 Millman's Theorem

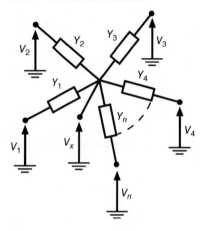

Figure 2.16 Circuit defining Millman's theorem.

Millman's theorem is an extension of Kirchhoff's current law, and in many applications it is so useful that we shall include it here with the other major network theorems. It can be derived from the circuit shown in Figure 2.16.

Millman's theorem provides a method of calculating the unknown potential V_x, assuming all the other potentials and their associated admittances are known. All the potentials in Figure 2.16 are measured with respect to ground, and therefore this terminal is omitted in the phasor nomenclature. Since we know that the sum of all the currents flowing *towards node x* must add to zero then, by applying KCL to this node, we find:

$$(V_1 - V_x)Y_1 + (V_2 - V_x)Y_2 + (V_3 - V_x)Y_3 + \ldots + (V_n - V_x)Y_n = 0 \tag{2.13}$$

On solving Equation (2.13) for V_x we find:

$$V_x = \frac{\sum_{i=1}^{n} V_i Y_i}{\sum_{i=1}^{n} Y_i} \tag{2.14}$$

Equation (2.14) represents Millman's theorem, and although it looks cumbersome it is surprisingly useful in determining an unknown potential, particularly in the presence of three or more voltage sources. If we wish to solve for I_1 and I_2 using Millman's theorem, we must first find the potential V_{AO}, shown in Figure 2.13a.

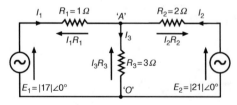

Figure 2.13a Example circuit.

Thus: $V_{AO} = \dfrac{(E_1/R_1 + E_2/R_2 + 0/R_3)}{(1/R_1 + 1/R_2 + 1/R_3)} = 15 \text{ volts}$

Therefore: $I_1 = \dfrac{(E_1 - V_{AO})}{R_1} = 2 \text{ amps}$

and $I_2 = \dfrac{(E_2 - V_{AO})}{R_2} = 3 \text{ amps.}$

2.6 AC Circuit Analysis Examples

In order to demonstrate these analytical techniques and to highlight the use of phasor diagrams as an analytical tool, we now investigate several practical applications, where phasor diagrams provide a visual picture of a circuit's operation.

2.6.1 Example 1: Series and Parallel Circuits.

In this example, we analyse simple series and parallel connected AC circuits, including their phasor representations. In the series case, Figure 2.17a, we find that an impedance representation is convenient since impedances connected in series can be added, whereas in the parallel case (Figure 2.17b) an admittance approach is preferred, since connected parallel admittances can be added.

Beginning with the series connected circuit, the collective impedance can be evaluated as:

$$Z = R + j(\omega L - 1/\omega C) = 15 + j(10.17 - 26.53) = 22.2\angle - 47.5° \, \text{ohms}$$

Since the current is common to each component, we assign it as the reference phasor (Figure 2.18). Accordingly, the voltage will be allocated the same phase angle as the circuit impedance. The current I is therefore given by:

$$I = \frac{110\angle - 47.5°}{19.2\angle - 47.5°} = 4.96\angle 0° \, \text{amps}$$

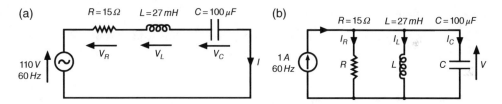

Figure 2.17 (a) Series circuit (b) Parallel circuit.

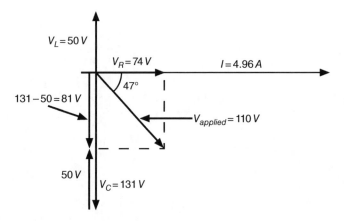

Figure 2.18 Series LRC phasor diagram.

The voltage drop across each component can now be plotted on the phasor diagram. The resistive drop (IR), lies in phase with the current, the inductive drop (IX_L), leads the current by 90° and the capacitive drop (IX_C), lags the current by 90°. These voltage drops can now be added vectorially according to KVL, to yield the phasor corresponding to the applied voltage.

Note that since the capacitive drop is larger than the inductive one, the overall impedance is capacitive, and therefore the current leads the applied voltage, in this case by 47°.

In order to analyse the parallel connected circuit of Figure 2.17b, we compute the total admittance according to:

$$Y = 1/R + j\omega C + 1/j\omega L = 66.7 + j37.7 - j98.3 = 90.1\angle -42° \text{ mS}$$

Since the *voltage* across the combination is common to all elements, we assign it as the reference phasor. Therefore the current must then take on the same phase as the circuit admittance. The voltage is thus given by:

$$V = \frac{I}{Y} = \frac{1\angle -42^0}{90.1\angle -42^0} = 11.1\angle 0° \text{ volts}$$

Knowing the circuit voltage enables the current in each element to be found, which can be plotted on a phasor diagram and summed to yield the applied current, as shown in Figure 2.19.

Since the current flowing in the inductance exceeds that in the capacitance, the overall impedance is inductive, and the current lags the circuit voltage by 42°.

2.6.2 Example 2: Three-Phase Loads

Where three-phase balanced or near balanced loads are involved, it is unnecessary to represent all phases on a phasor diagram. Instead one phase is usually drawn, since it is representative of the other two. The circuit shown in Figure 2.20a represents one phase of a near balanced industrial load. In practice such a load might comprise motor driven machinery, water heating, air conditioning or refrigeration equipment, together with an assortment of single phase plug loads. Electrically all these can usually be represented by an inductive-resistive combination, since industrial loads are generally partially inductive.

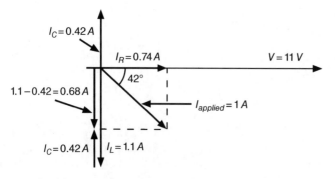

Figure 2.19 Parallel LRC circuit phasor diagram.

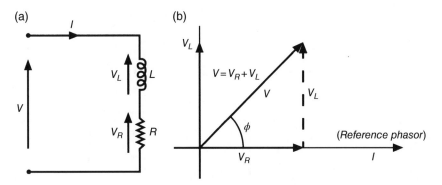

Figure 2.20 (a) Inductive load equivalent circuit (b) Phasor representation.

Since it is common to both components, the *current* will be chosen as the reference in the phasor diagram shown in Figure 2.20b.

The voltage developed across the resistive element lies in phase with the current, while that associated with the inductive element leads the current by 90°. Kirchhoff's voltage law requires that the sum of the voltage drops around a circuit equals the applied voltage, so we may write:

$$V = V_R + V_L$$

In the phasor diagram of Figure 2.20b these voltages are added vectorially. The resultant voltage *V*, can be seen to lead the current by an angle ϕ, or alternatively the current lags the voltage by the same angle. As we shall see in the next chapter, the cosine of this angle is called the *power factor* of the load, and for reasons that will be explained, supply authorities prefer customers to ensure that their load operates at a high rather than a low power factor – that the load current lags the voltage by only a *small* angle.

2.6.3 Example 3: Power Factor Correction

The circuit of Figure 2.21 also shows an inductive load to which has been added a parallel connected capacitor, *C*, provided for the purpose of power factor correction. By adding a device that consumes a leading current, some of the lagging effects of an inductive industrial load can be offset, and an increase in the power factor of the combination results.

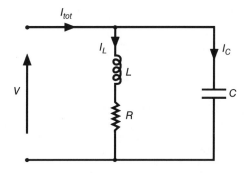

Figure 2.21 Inductive load with power factor correction.

In this circuit the total current splits into two components, one flowing in the load I_L, and the other in the power factor correction capacitance, I_C. Since the applied voltage is common to both branches of this circuit, it will now be used as the reference phasor.

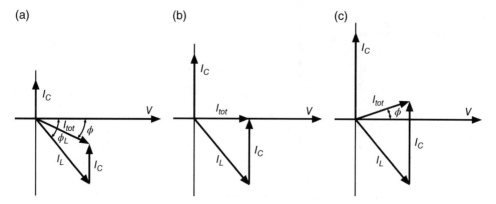

Figure 2.22 (a) Under-compensated (b) Ideal compensation (c) Over-compensated.

By analogy with Example 2, the inductive current lags the applied voltage by an angle ϕ_L. In contrast, the capacitive current leads the applied voltage by 90°, as shown in Figure 2.22a. The total current, can be found by vector addition of these two phasors. The resultant current I_{tot} can now be seen to lag the voltage by a smaller angle ϕ. Because the new phase angle is less than the old, the power factor has been increased and thus improved, but the circuit remains under-compensated. Figure 2.22b depicts the same load with sufficient capacitance applied to make the total current appear resistive, (i.e. $\phi = 0$), so with $\phi = 0$ the circuit is now ideally compensated. Finally, Figure 2.22c shows the case where an excessively large capacitive compensation has been applied, and the total current now leads the applied voltage. The circuit is therefore over-compensated.

2.6.4   Example 4: Capacitive Voltage Rise

Whenever a capacitance is connected to an AC bus, there will usually be an accompanying rise in the bus voltage. This is generally only a few per cent, and is due to the fact that the supply impedance is at least partially inductive. We can demonstrate this effect by considering the single phase Thévenin representation of a three-phase AC bus, as shown in Figure 2.23a. The bus impedance is represented by an inductive component X, and a resistance R, both of which are quite small, in addition to which R is generally considerably smaller than X.

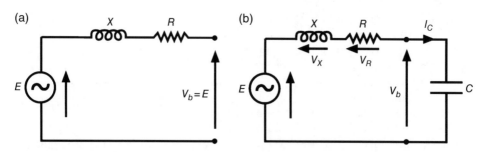

Figure 2.23 (a) AC bus Thévenin equivalent circuit (b) Capacitive loading of an AC bus.

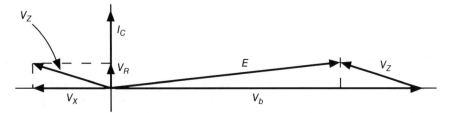

Figure 2.24 Capacitive voltage rise. (Note: the voltage drops V_R and V_X have been considerably exaggerated in scale for reasons of clarity).

When the system is unloaded, the bus voltage V_b is equal to the source voltage E. However, when loaded, V_b and E are no longer equal, due to the potential dropped across the source impedance, $Z = R + jX$. The connection of a capacitor, as shown in Figure 2.23b, results in a leading current, causing voltage drops across both R and X, depicted in the phasor diagram in Figure 2.24. Here we have assumed that $X > R$, which is frequently the case.

The voltage dropped across the source resistance R, lies in phase with the capacitive current, while that across the inductive impedance leads this current by 90°. Kirchhoff's voltage law requires that $E = V_b + V_X + V_R$, and an inspection of Figure 2.24 shows that as a consequence of this summation V_b has increased in magnitude when compared to E. (Remember that the Thévenin equivalent voltage E, is fixed.) In effect, the capacitive current I_C creates a voltage drop across X that adds to V_b. It can also be seen that because V_X and V_R are in *quadrature* (i.e. they are 90° apart), V_X has a much larger effect on the magnitude of V_b than does V_R, so we may write:

$$|V_b| \approx |E| + |V_X| = |E| + |I_C| X$$

If we represent the increase in V_b by δV, we can write:

$$\delta V = |V_b| - |E| \approx |I_C| X = \frac{(|E| + \delta V) X}{X_C}$$

where X_C is the reactance of the connected capacitance. We can express the *fractional rise in voltage* (also known as the *per-unit voltage rise*) as:

$$\frac{\delta V}{|E|} \approx \frac{X}{(X_C - X)} \tag{2.15}$$

By way of example, suppose that the source impedance of a 6.6 kV three-phase bus is $0.07 + j0.6$ ohms per phase, and that a capacitor of 30 ohms reactance is connected between each phase and the neutral terminal. The resulting fractional increase in the bus voltage will be about 2% or 130 V.

For this reason, capacitors are often connected to MV and HV networks to provide voltage support during times of heavy demand. Equation (2.15) can also be used to show that the bus voltage will fall when loaded with an inductive impedance, X_L. This can be achieved by replacing X_C with $-X_L$ to yield:

$$\frac{\delta V}{|E|} \approx -\frac{X}{(X_L + X)}$$

So a 30 ohm *inductive* impedance connected to the same bus, will result in a *fall* of about 2% in the bus voltage.

2.6.5 Example 5: Phase Sequence Indicator Analysis

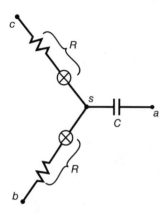

Figure 2.25 Simplified phase sequence indicator.

We now consider the design of a phase sequence indicator, which is a simple device capable of indicating the phase sequence of a three-phase supply. There are many versions of this circuit in existence, all of which operate on the same principle. The basic circuit is shown in Figure 2.25, and consists of two resistive lamps and one capacitor, connected in a star (or wye) configuration.

The phase sequence indicator is connected between the line voltages of the supply in question, so no neutral connection is required. The circuit generates its own star-point voltage Vs, one that is quite different from that of the unused neutral terminal, which represents zero volts. We will initially analyse the circuit using Millman's theorem, and once we have a clear understanding of its operation, we will re-analyse it using phasor analysis.

For the circuit to operate correctly we require that the magnitude of the capacitive admittance connected to the 'a' phase, be equal to the effective lamp resistances on the other two:

$$\left|Y_{cap}\right| = \frac{1}{R} \tag{2.16}$$

Recall that Millman's theorem, as applied to this circuit, states that:

$$V_s = \frac{V_a Y_a + V_b Y_b + V_c Y_c}{Y_a + Y_b + Y_c} \tag{2.17}$$

where Y_i is the admittance associated with phase i and V_a, V_b and V_c are the phase voltages. Let us assume that the phase voltages each have a magnitude of 1 volt, and therefore they can be expressed as:

$$V_a = 1\angle 0, V_b = 1\angle -120° \text{ and } V_c = 1\angle 120°$$

Similarly, let us assign a magnitude of 1 siemen to each admittance, therefore we can write:

$$Y_a = 1\angle 90 = j1, Y_b = 1\angle 0 \text{ and } Y_c = 1\angle 0$$

Substituting these values into Equation (2.14a) we find that $V_s = V_{sn} = 0.632\angle 108.4°\,V$, as depicted in the phasor diagram of Figure 2.26. Here we see that there is a much larger potential dropped across the 'b' phase lamp than across the 'c' phase one. Therefore we expect that the 'b' phase lamp will burn more brightly than that on the 'c' phase.

So how does this circuit indicate the phase sequence of the connected supply? From the foregoing, the brightest lamp is supplied from the phase following that connected to

Figure 2.26 Phase sequence indicator phasor diagram.

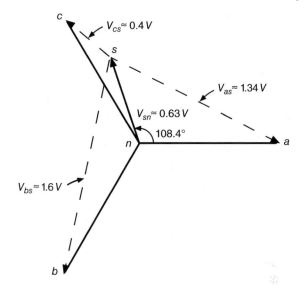

the capacitor, which in this case is the 'a' phase. Therefore in the example above, with an *abc* phase sequence, the 'b' phase lamp will burn the brightest. However, if we reverse the phase sequence, by swapping the incoming 'b' and 'c' phases, then the 'c' phase lamp will be the brightest, since it now follows the 'a' phase, thus indicating an *acb* phase sequence.

This circuit finds application in the connection of three-phase motors, where the correct phase sequence must be provided in order to ensure the desired direction of rotation. It is also used in metering applications to ensure that the energy meter sees the correct voltage phase sequence, although this requirement is more important with induction-disc meters than with electronic instruments.

Now that we have an understanding of how the phase sequence indicator works, let us analyse its operation using a phasor diagram. Essentially all we have to do is to locate the star-point *s* to complete the diagram for the sequence indicator.

We do need a little information about the circuit, so we begin by applying KCL to the currents flowing towards the star point, *s*. Thus we find:

$$(V_a - V_s)/Z_a + (V_b - V_s)/Z_b + (V_c - V_s)/Z_c = 0$$

Therefore $(V_a/Z_a) + (V_b/Z_b) + (V_c/Z_c) = (V_s/Z_a) + (V_s/Z_b) + (V_s/Z_c)$
We can rewrite this equation in the form of two equal currents, I' and I'', where:

$$I' = (V_a/Z_a) + (V_b/Z_b) + (V_c/Z_c)$$

And $I'' = (V_s/Z_a) + (V_s/Z_b) + (V_s/Z_c)$

The current I' consists of a component from each phase that can be drawn on a phasor diagram as shown in Figure 2.27. Here each of the three-phase currents have been added vectorially to generate I'. Since all the admittance values have a magnitude of unity, then the magnitude of each current component will also be unity. Finally, since we have chosen unity voltages as well, both the current and voltage scales on the phasor diagram will be the same.

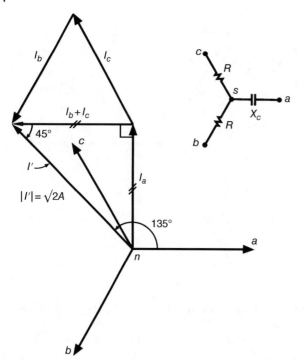

Figure 2.27 I' phasor diagram.

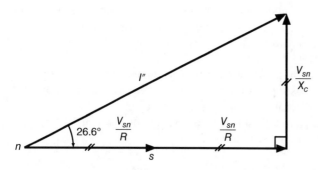

Figure 2.28 I'' phasor diagram.

As shown in Figure 2.27, the current I' leads the 'a' phase voltage by 135° and has a magnitude of $\sqrt{2}$ amps.

We can similarly construct a phasor diagram for I'', but since at this time we do not yet know the scale of this diagram or how to orient it, we will draw V_{sn} in the reference direction for convenience. I'' will also have three components and, as shown in Figure 2.28, it will lead V_{sn}, by 26.6°.

Since we know the currents I' and I'' are equal, we must rescale Figure 2.28 so that I'' also has a magnitude of $\sqrt{2}$ amps and superimpose the two phasor diagrams, so that I'' lies on top of I'. However, this must be done in such a way that the current I'' leads V_{sn}, since this circuit is capacitive. The location of the star point voltage V_{sn} on the

Figure 2.29 Overall phasor diagram.

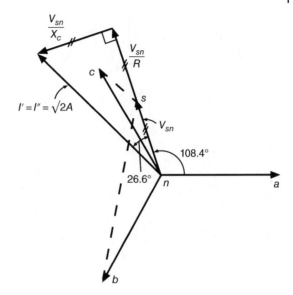

phasor diagram will then be revealed. This has been done in the diagram of Figure 2.29, where we again find that $V_{sn} = 0.632\angle108.4°$ volts.

Our phasor analysis of the phase sequence indicator is probably no simpler than using an analytical approach, but it does illustrate what can be achieved using phasor diagrams. In more complex circuits the phasor approach can save considerable calculation while providing the benefit of a visual appreciation of the circuit's operation. The next example illustrates such a case.

2.6.6 Example 6: Induction Furnace Operation

This example concerns a low frequency induction furnace of the type used to melt and cast aluminium or zinc. It operates on the principle that molten metal within the furnace forms a shorted secondary winding on the furnace transformer (known colloquially as an *inductor*). Our interest in this furnace relates to the way in which a single-phase furnace is supplied from a three-phase transformer, in such a way that all phases are loaded equally and at unity power factor. This is achieved through the clever use of a line-connected balancing reactor and capacitor, as shown in the furnace schematic of Figure 2.31. An arrangement of this kind is necessary because an induction furnace is too large a load to be supplied from one phase only.

Figure 2.30 shows a cutaway view of a small induction furnace. In particular, it depicts the location of the furnace transformer, the core of which is mounted horizontally beneath the crucible.

The transformer core and its associated primary winding are enclosed within a refractory material capable of withstanding the temperature of the molten metal. Cast into this are channels through which the metal forms two *shorted turns* around the core. These channels form the *throats* of the furnace in which the current flowing rapidly heats and expands the metal. A strong thermosyphon action draws metal into the centre throat and expels it from those on either side, thereby ensuring an even temperature distribution throughout the melt, as depicted in Figure 2.30.

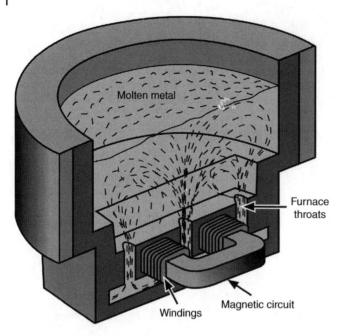

Figure 2.30 Sectional view of an induction furnace showing the arrangement of the furnace transformer. (Ref: 'Industrial Electric Furnaces and Appliances' Paschkis & Persson 1960).

The power delivered to the melt depends on the dimensions of the throats and also of the resistance of the molten metal within them, as well as the leakage inductance that exists between the primary windings and the metallic secondary winding. The cooling of the transformer's primary winding is critical, and in this case this is achieved by a fan forcing cool air around the winding. Alternatively de-ionised water pumped through the winding may be employed instead.

Figure 2.31 is a simplified schematic diagram of the furnace. Power is supplied from an 11 kV:460 V three-phase transformer, where the high voltage winding is delta connected, and the low voltage is star (or wye) connected. The LV windings supply the furnace's own three-phase star connected autotransformer, which provides a tapping range from 90 to 600 volts. The power delivered to the furnace can thus be adjusted using the on-line tap-changer, enabling precise temperature control of the melt.

The capacitors shown are of two types, each specified according to the quantity of reactive power (V_C^2/X_C, expressed in volt-amps-reactive, VAr), that it consumes from a 600 volt AC source. Type 'A' supplies 200 kVAr @ 600 volts, and comprises 2×100 kVAr capacitors, while Type 'B', (used only for trimming purposes), supplies 87.5KVAr @ 600 volts, and comprises 1×12.5KVAr, 1×25 kVAr and 1×50 kVAr capacitor.

The furnace transformer, together with its seven type 'A' power factor correction (PFC) capacitors is connected between phases 'a' and 'c'. Power factor correction is required due to the leakage impedance of the furnace transformer, which adds a substantial inductive element in series with the melt resistance, when seen reflected into the primary winding. The power factor correction capacitors are dimensioned so as to restore the furnace transformer's phase angle to zero, and therefore its power factor to unity. (The type B

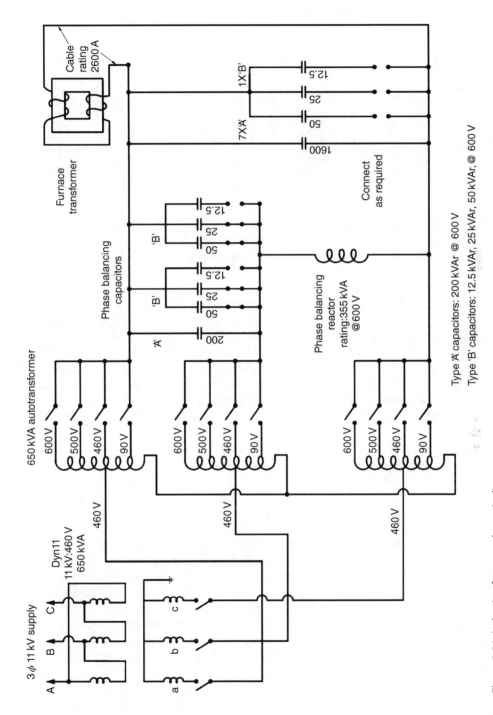

Figure 2.31 Induction furnace schematic diagram.

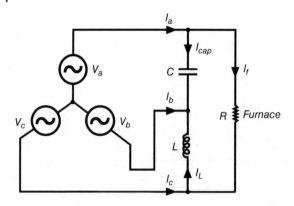

Figure 2.32 Simplified furnace schematic.

capacitor is provided for trimming the value of the type A capacitors.) Thus the transformer, its PFC capacitance and the melt resistance, can all be replaced with a single equivalent resistive element connected between phases 'a' and 'c'.

The air-cored phase balancing reactor is cooled by de-ionised *water*, pumped through hollow windings. It is rated at 355 kVAr @ 600 volts and is connected between the 'b' and 'c' phases. Finally the phase balancing capacitors are connected between phases 'a' and 'b' and consist of one type A capacitor and two type B, which are adjusted to provide a capacitive reactance exactly equal to that of the balancing reactor.

With the exception of the 355 kVAr rating of the balancing reactor, the technical data presented above is not essential to understanding the phasor analysis of this furnace. In the simplified schematic shown in Figure 2.32, the tap-changers have been omitted and the furnace transformer and its associated PFC capacitors have been replaced with an equivalent resistor representing the melt, while the balancing reactor and capacitors are unchanged.

We will show that, as a result of these balancing components and the PFC applied to the furnace transformer, the incoming phase currents are balanced, and that each lies exactly in phase with its associated phase voltage. Therefore the power demanded by the furnace is shared equally across all three phases of the supply transformer. We will also use our phasor diagram to determine the maximum power rating of the furnace.

To do this we must plot the phase currents (I_a, I_b and I_c) on a phasor diagram and compare them with the associated phase voltages, making any adjustments necessary to the magnitude of the furnace current, to ensure that a unity power factor is achieved. We begin by plotting the furnace, the reactor and the capacitor currents alongside their associated line voltages, as shown in Figure 2.33. The furnace current therefore lies in phase with V_{ac}, the reactor current lags V_{cb} by 90° and the capacitor current leads V_{ab} by 90°.

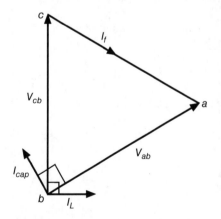

Figure 2.33 Furnace, inductor and capacitor currents.

Next, from Figure 2.33 we write an equation for each of the three phase currents:

$$I_a = I_f + I_{cap}$$
$$I_b = -\left(I_{cap} + I_L\right)$$
$$I_c = I_L - I_f$$

These phase currents are plotted on the phasor diagram in Figure 2.34, where we see that in order to achieve a unity power factor on each phase, the resistive furnace current must be $\sqrt{3}$ times as large as either the reactor or the capacitor current. Therefore the power rating of the furnace will be $\sqrt{3}$ times the VAr rating of the balancing reactor, thus: $P = \sqrt{3} \times 355\text{kVAr} = 615\text{kW}$.

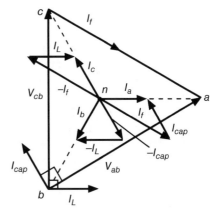

Figure 2.34 Phase current summations.

The resultant current flowing in each phase is equivalent to supplying a balanced resistive load, in which the power delivered by each phase is one third of that required by the furnace. From our phasor diagram, we see that the current flowing in each phase has the same magnitude as that in either the reactor or the capacitor, and on the maximum tap this is equal to $355\,\text{kVar}/600\,\text{V} = 592\,\text{A}$. Therefore the total power delivered by the supply transformer will be given by: $P = \sqrt{3} \times 600\,\text{V} \times 592\,\text{A} = 615\text{kW}$ as derived above.

Thus, through a clever connection of balancing components, the 'single phase' furnace load has been distributed equally across all three phases of the supply transformer. Without such a connection it would not be possible to operate this furnace, since 615 kW is too large a load to apply to one phase alone and would severely unbalance a local LV network.

As well as providing an equation for the rating of the furnace, the phasor diagram can also provide us with a good idea of just how this circuit works; this can be shown as follows. Since the furnace current I_f is in phase with the line voltage V_{ac} it lags the 'a' phase voltage by 30° while its inverse $-I_f$ leads the 'c' phase voltage by the same amount. In order to bring these line currents into phase with their respective voltages, the 'a' phase requires an additional capacitive current component, while the 'c' phase requires an inductive one.

The capacitive current is provided to the 'a' phase by the balancing capacitor, connected across V_{ab}. If the magnitude of this current is equal to $1/\sqrt{3}$ times that of the furnace current, then the resultant 'a' phase line current will lie in phase with the 'a' phase voltage, as required. Similarly, the 'c' phase furnace current $-I_f$ leads the 'c' phase voltage by 30° (Figure 2.34) and therefore an inductive current component is required to bring the 'c' phase current back in phase with the 'c' phase voltage. The reactor, connected across V_{cb}, will inject such a component, and if the magnitude of this current is equal to $1/\sqrt{3}$ times that of the furnace current, then the resultant 'c' phase current will lie in phase with the 'c' phase voltage.

Finally, the 'b' phase line current is equal to minus the sum of the balancing capacitor and reactor currents, $-(I_L + I_C)$, and this current conveniently falls in line with the 'b' phase voltage, as shown in Figure 2.34. Therefore through the use of a phasor diagram, we have not only been able to determine the furnace rating, but have also obtained an insight into the operation of the phase balancing circuitry.

2.7   Resonance in AC Circuits

Resonance arises as a result of the fact that the complex impedance of a capacitor and that of an inductor are of opposite sign. When connected in *series* and excited at the *resonant frequency*, the impedance of an inductor cancels that of a capacitor, leaving a connection which presents a very low impedance to the rest of the circuit and therefore the potential developed across the combination will be low. This is not to say, however, that large voltages cannot be generated *within* the series combination. Indeed, the current flowing through a series resonant circuit can generate very high voltages across both the inductor and the capacitor. However, because these are 180° out of phase with each other, the resultant voltage is usually very small. The circuit is said to be *resonant* at this frequency, since the applied voltage and the resulting current are in phase with each other. Should a series resonance of this kind occur within a power system, it is possible that the voltage across individual components may reach a level sufficient to cause damage. The likelihood of this occurring depends on the level of *resistive damping* present. This is quantified in a parameter known as the *quality factor* Q of the circuit, which is proportional to the ratio of the *energy stored* within the circuit to that *dissipated*, during a cycle.

Similarly, when an inductor and a capacitor are connected in *parallel* and excited at their resonant frequency, the *admittance* of the inductor cancels that of the capacitor, leaving a total admittance close to zero, thus the parallel combination effectively presents a very *high impedance*. However, although the current flowing *into* the parallel combination in this situation is small, there may be a *large* current circulating *between* the inductor and capacitor. Again, the likelihood of this depends upon the quality factor of the circuit concerned.

Natural resonances like these occasionally arise in power systems, usually at frequencies above the system frequency. When these occur at or near harmonic frequencies that exist in the network, the generation of destructive voltages or currents may result. This can cause the failure of one or more items of primary plant, and therefore underdamped network resonances are to be avoided, especially near active harmonics.

For example, when voltage support capacitors are introduced into MV and HV networks, care must be taken to avoid an adverse parallel resonance condition from arising between these components and the system impedance (which is usually inductive). Were such resonances to occur near an active network harmonic, large circulating currents could arise, sufficient to cause damage to equipment within the resonant current path. In order to avoid such resonances, *de-tuning reactors* are usually installed in series with these capacitors to force the the parallel resonant frequency away from odd harmonic frequencies.

Another example of parallel resonance arises in the power factor correction (PFC) circuits briefly discussed above. Here the PFC capacitance is resonant or nearly so, with the inductive component of the load impedance that it is correcting. This near resonance condition generally has no adverse effects, although depending on the size of the load, large circulating currents may exist between the two. For this reason the PFC equipment should be placed as near to the offending load as possible.

Series Resonance

We consider the analysis of series resonance and parallel resonance separately, although the two connections generate very similar algebraic results. The partial cancellation of impedances in a series circuit, when operating at a frequency above resonance, is shown in the phasor diagram of Figure 2.35. Since the current flowing is common to all three components, it has been chosen as the reference phasor.

In this case, we have included the small resistance R, in series with the winding, to represent the core and winding losses of the inductor. The capacitor on the other hand may be considered virtually lossless, since any tiny resistive or dielectric losses will generally be negligible compared to loss within the inductor. Since Kirchhoff's voltage law requires that $V_R + V_L + V_C = V$, in Figure 2.35 we see that the supply voltage V consists of a resistive component in quadrature with a residual component equal to $V_L + V_C$. For frequencies above resonance the circuit appears inductive, since the current lags the applied voltage, while for frequencies below resonance it is capacitive, since the current leads the applied voltage.

Figure 2.35c shows the situation when the circuit operates *exactly* at its resonant frequency; here the voltage drops across L and C completely cancel, leaving only the resistive component, across which is dropped the applied voltage V. Resonance therefore occurs when $X_L = X_C$, i.e. when $\omega = \dfrac{1}{\sqrt{LC}}$ radians/sec.

Under resonant conditions the circuit impedance becomes equal to R, the inductor's winding resistance, and so there is usually little to limit the flow of current in the circuit. This is not to say that there is no voltage dropped across both L and C; it's just that the voltage drops across these elements cancel in the summation. As shown in Figure 2.35c, there can be quite large voltage drops across both L and C, – frequently well above the applied voltage V – so these components must be rated appropriately. These resonant voltage drops can be shown to be equal to the quality factor Q multiplied by the applied voltage V.

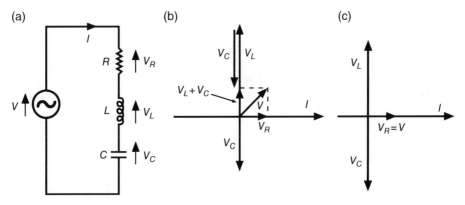

Figure 2.35 (a) Series tuned circuit (b) Phase diagram above resonance (c) Phase diagram at resonance.

2.7.1 Quality Factor 'Q'

The quality factor is a parameter that describes how well a tuned circuit performs close to resonance. High Q circuits exhibit rapid changes in response around the resonant frequency, while in low Q circuits the response changes much more gradually. The Q of a circuit is defined as:

$$Q = 2\pi \frac{\textit{Peak energy stored during each cycle}}{\textit{Energy dissipated during each cycle}} \tag{2.18}$$

Therefore a high Q circuit will exhibit very low loss per cycle and, as we shall see, it will have a sharp peak in its amplitude response at resonance. The peak energy stored in the circuit can be expressed in terms of the *peak current* flowing $\hat{\imath}$, as $1/2L\hat{\imath}^2$ and the energy lost per cycle is found by integrating the power loss $i(t)^2 R$, over a complete cycle. Thus we may write:

$$Q = 2\pi \frac{1/2L\hat{\imath}^2}{\int\limits_0^T \hat{\imath}^2 R\cos^2(\omega t)\,dt}$$

$$Q = 2\pi \frac{1/2L\hat{\imath}^2}{\dfrac{1}{\omega}\int\limits_0^{2\pi} \hat{\imath}^2 R\cos^2(\omega t)\,d\omega t}$$

$$Q = 2\pi \frac{1/2L\hat{\imath}^2}{\dfrac{1}{2\omega}\int\limits_0^{2\pi} \hat{\imath}^2 R(\cos 2\omega t + 1)\,d\omega t}$$

$$Q = 2\pi \frac{1/2L\hat{\imath}^2}{\dfrac{\hat{\imath}^2 R}{2\omega}\,2\pi} = 2\pi \frac{1/2L\hat{\imath}^2}{\hat{\imath}^2 RT/2} = \frac{\omega L}{R} = \frac{X_L}{R} \tag{2.19}$$

Therefore the Q of a series resonant circuit is simply the reactance of the inductor ωL, divided by its associated series resistance R, and thus a small resistive loss will give rise to a large Q value. It is interesting to note that the Q also rises with the applied frequency ω. This is because at higher frequencies a shorter cycle time is available for energy to be consumed within the resistive element. The Q of a tuned circuit is often expressed in terms of the resonant frequency of a circuit ω_o, and since we have seen that for a series circuit $\omega_o = \dfrac{1}{\sqrt{LC}}$ (or $f_o = \dfrac{1}{2\pi\sqrt{LC}}$) at resonance we may write: $Q_0 = \dfrac{\omega_o L}{R} = \dfrac{X_0}{R} = \dfrac{1}{R}\sqrt{\dfrac{L}{C}}$.

2.7.2 Parallel Resonance

A parallel resonant circuit can be constructed by connecting the inductor and capacitor in parallel, as shown in Figure 2.36a. Here the small resistance R still appears in series with the inductance. In this form, the parallel circuit lacks the symmetry of the series circuit, and Figure 2.36b shows a simplification of this circuit, through the introduction of a new resistive component R', in parallel with L, in place of the resistance R in series

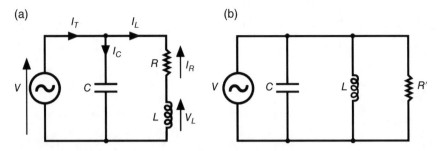

Figure 2.36 (a) Parallel tuned circuit (b) Approximate parallel circuit

with it. In this form, the parallel circuit is easier to analyse, although it should be stressed that this change represents an approximation to the circuit of Figure 2.36a, albeit a quite good one, especially for larger Q values.

The value of the new component R' must be chosen so that the same power loss occurs in each circuit, thus:

$$\frac{V^2}{R'} = I_L^2 R$$

This means that:

$$\frac{V^2}{R'} = \frac{V^2}{\left(\omega L\right)^2 + R^2} R$$

And for $Q > 3$ we find:

$$R' = R\left\{\left(\omega L/R\right)^2 + 1\right\} = R\left(Q^2 + 1\right) \approx RQ^2 \tag{2.20}$$

Thus for larger Q values we can replace the series connected resistance R with an equivalent parallel resistance R', where $R' \approx Q^2 R$.

2.7.3 Parallel Tuned Circuit Quality Factor

The Q value for a parallel circuit may also be expressed in terms of C and R' from the definition in Equation (2.18):

$$Q = 2\pi \frac{Peak\ energy\ stored\ during\ each\ cycle}{Energy\ dissipated\ during\ each\ cycle}$$

$$Q = 2\pi \frac{1/2C\hat{v}^2}{\displaystyle\int_0^T \frac{\hat{v}^2}{R'}\cos^2\left(\omega t + \phi\right)dt}$$

Where ϕ is the angle between the circuit voltage and the current in the inductor:

$$Q = 2\pi \frac{1/2C\hat{v}^2}{\displaystyle\frac{1}{\omega}\int_0^{2\pi} \frac{\hat{v}^2}{R'}\cos^2\left(\omega t + \phi\right)d\omega t}$$

$$Q = 2\pi \frac{1/2C\hat{v}^2}{\frac{1}{2\omega} \int\limits_0^{2\pi} \frac{\hat{v}^2}{R'} \left[\cos(2\omega t + 2\phi) + 1\right] d\omega t}$$

$$Q = 2\pi \frac{1/2C\hat{v}^2}{\frac{1}{2\omega R} \hat{v}^2 2\pi} = \omega CR'$$

$$Q = \omega CR' \tag{2.21}$$

2.7.4 Series Resonant Circuit Normalised Frequency Response

It is often useful to express the magnitude of the current flowing in a series tuned circuit I, at a frequency ω, by normalising it with respect to the current flowing at resonance I_o, assuming a constant excitation potential. Therefore:

$$\left|\frac{I}{I_o}\right| = \frac{V}{Z} \frac{Z_o}{V} = \frac{Z_o}{Z}$$

$$\left|\frac{I}{I_o}\right| = \frac{Z_o}{Z} = \frac{R}{\sqrt{R^2 + \left\{\omega L - \frac{1}{\omega C}\right\}^2}}$$

$$\left|\frac{I}{I_o}\right| = \frac{1}{\sqrt{1 + \left\{\frac{\omega L}{R} - \frac{1}{\omega CR}\right\}^2}}$$

$$\left|\frac{I}{I_o}\right| = \frac{1}{\sqrt{1 + \left\{\frac{\omega}{\omega_o} \frac{\omega_o L}{R} - \frac{\omega_o}{\omega} \frac{1}{R}\sqrt{\frac{L}{C}}\right\}^2}}$$

$$\left|\frac{I}{I_o}\right| = \frac{1}{\sqrt{1 + Q_0^2 \left\{\frac{\omega}{\omega_o} - \frac{\omega_o}{\omega}\right\}^2}} \tag{2.22}$$

where $Q_0 = \frac{1}{R}\sqrt{\frac{L}{C}}$.

Equation (2.22) shows that the response at an arbitrary frequency ω, is always less than that at the resonant frequency, ω_o. Figure 2.37 shows the normalised frequency response for a series resonant circuit, for various values of Q_0.

For circuits with low Q_0 values, the response around the resonant frequency is not particularly sharp, but as Q_0 increases so does the sharpness of the response around the resonant frequency. The maximum normalised response is unity, occurring when $\omega = \omega_o$. The series resonant circuit is thus capable of passing signals near its resonant frequency, i.e. it is acting as a band-pass filter.

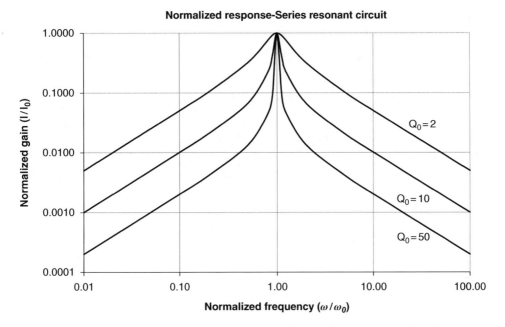

Figure 2.37 Normalised frequency response of a series resonant circuit with constant excitation.

2.7.5 Parallel Resonant Circuit Normalised Response

We can analyse the parallel resonant circuit's normalised response in a similar way to the series circuit. We shall begin with the symmetrical arrangement of the three elements, L, C and R', connected in parallel, which suggests that we approach the task from an admittance perspective, since admittances in parallel can be summed. Again we shall consider the total current flowing into the circuit from the source, and as before we shall calculate the magnitude ratio $\left|\dfrac{I}{I_o}\right|$.

Thus:
$$\left|\frac{I}{I_o}\right| = \left|\frac{Y}{Y_o}\right| = \frac{\sqrt{\left\{\dfrac{1}{R'}\right\}^2 + \left\{\omega C - \dfrac{1}{\omega L}\right\}^2}}{\dfrac{1}{R'}}$$

$$\left|\frac{I}{I_o}\right| = \sqrt{1 + \left\{\omega CR' - \frac{R'}{\omega L}\right\}^2}$$

$$\left|\frac{I}{I_o}\right| = \sqrt{1 + \left\{\frac{\omega}{\omega_0}\omega_o CR' - \frac{Q_0}{\omega L}\sqrt{\frac{L}{C}}\right\}^2} \qquad \left(R' = Q_0\sqrt{\frac{L}{C}}\right)$$

$$\left|\frac{I}{I_o}\right| = \sqrt{1 + \left\{\frac{\omega}{\omega_0}Q_0 - Q_0\frac{\omega_o}{\omega}\right\}^2}$$

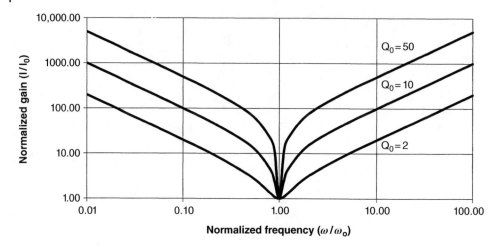

Figure 2.38 Normalised frequency response of a parallel resonant circuit with constant excitation.

$$\left|\frac{I}{I_o}\right| = \sqrt{1 + Q_0^2\left\{\frac{\omega}{\omega_0} - \frac{\omega_0}{\omega}\right\}^2} \qquad (2.23)$$

Equation (2.23) is similar in form to Equation (2.22), but at frequencies away from resonance the value of $\left|\frac{I}{I_o}\right|$ is now considerably greater than one, since the LC combination no longer appears as an open circuit. For example, at low frequencies the inductive admittance is large, generating a high current flow. Similarly, at high frequencies the capacitive admittance is large, allowing high currents to flow. Figure 2.38 shows the normalised response plotted for several Q_0 values and, as expected, as the frequency moves further from resonance, the response of the circuit increases, and as the circuit Q_0 increases, so does the sharpness of the response around resonance.

The parallel tuned circuit is capable of attenuating the signals near its resonant frequency, i.e. it is acting as a *band-stop filter*.

Current Flowing in a Parallel Tuned Circuit at Resonance

Even though the total current flowing in a parallel resonant circuit is quite small, there can be a very large current circulating between the inductor and the capacitor at resonance, for which these components must be adequately rated. The magnitude of this circulating current is Q_0 times the total current flowing, i.e. that in the resistive element R'. This can be easily shown as follows. The current circulating between the capacitor and the inductor is given by $I_c = V\omega_o C$, and the total current flowing in the circuit at resonance I_o, is given by $I_o = \frac{V}{R'}$, where V is the applied voltage. Therefore the ratio of these quantities becomes $\frac{I_c}{I_o} = \omega_o CR' = Q_0$. Thus $I_c = Q_0 I_o$. So the circulating current at resonance is Q_0 times as large as the total current flowing in the parallel combination.

2.7.6 Resonant Circuit Bandwidth and Cut-Off Frequencies

The characteristic shown in Figure 2.37 is typical of a band-pass filter, while that of Figure 2.38 is characteristic of a band-stop filter. It is instructive to evaluate the upper and lower cut-off frequencies for both the series and parallel resonant circuits – those frequencies at which the magnitude of the normalised response falls by $1/\sqrt{2}$. This corresponds to a change in the power dissipated in the circuit of a factor of 2, and as a result the cut-off frequencies are also known as the *half-power frequencies*, (or alternatively as the *3 dB frequencies*).

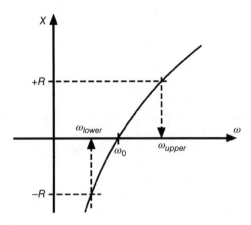

Figure 2.39 Upper and lower cut-off frequency definition.

From Figure 2.35, the current in a series circuit at resonance is V/R, occurring when the reactance term X equals zero. As the frequency changes the reactance will vary as depicted in Figure 2.39. The magnitude of the normalised response will fall by a factor of $1/\sqrt{2}$ when $X = R$, occurring at the upper cut-off frequency, or when $X = -R$ at the lower cut-off frequency. The upper frequency can therefore be found from the condition $\left(\omega L - 1/\omega C\right) = R$, and the lower one from $\left(\omega L - 1/\omega C\right) = -R$. Each of these quadratic equations will yield two roots, only one of which will be positive and will represent the respective cut-off frequency.

The quadratic equation for the upper cut-off frequency may be written in the form:

$$\left(\frac{\omega}{\omega_0}\right)^2 - \left(\frac{\omega}{Q_0\omega_0}\right) - 1 = 0$$

Which leads to the result:

$$\omega_{upper} = \omega_0\left(1 + 1/2Q_0\right) \tag{2.24}$$

Similarly, the lower cut-off frequency is obtained from the equation:

$$\left(\frac{\omega}{\omega_0}\right)^2 + \left(\frac{\omega}{Q_0\omega_0}\right) - 1 = 0$$

which yields:

$$\omega_{lower} = \omega_0\left(1 - 1/2Q_0\right) \tag{2.25}$$

So for large Q values the upper and lower cut-off frequencies lie close to the resonant frequency ω_o, which is consistent with the sharp band-pass response we have observed in such circuits. The bandwidth of the circuit (radians/sec) is defined as the range of frequencies between ω_{lower} and ω_{upper} which, from Equations (2.24) and (2.25), may be expressed in the form:

$$Bandwidth = \frac{\omega_0}{Q} \tag{2.26}$$

Equation (2.26) is often used as an alternate definition for the quality factor Q. The same analysis when applied to a parallel resonant circuit generates identical results.

2.7.7 Parallel Tuned Circuit Resonant Frequency

The analysis thus far has suggested that the effects of the inductor's winding resistance R can be accounted for by placing a large resistance R' in parallel with L and C (Figure 2.36). While this approximation is generally quite good, especially for high Q circuits, it fails to predict a slight downward shift in the resonant frequency of the parallel circuit when compared with that of the series circuit.

The approximate circuit suggests that the resonant frequency will be the same for both series and parallel circuits, but this is not quite correct. This assertion can be demonstrated using a phasor diagram, as shown in Figure 2.40.

This phasor diagram can be constructed by first choosing the inductive current as the reference phasor, i.e. lying along the positive horizontal axis. Once this current has been drawn, both the resistive and inductive voltage drops can be included (V_R and V_L respectively). (Remember: V_L leads this current by 90°, while V_r is in phase with it.) From these phasors, the total voltage V can be found, since $V = V_L + V_R$.

The capacitive current can be included next, since it leads the total voltage by 90°. Finally, the total current phasor can be constructed, since $I_T = I_L + I_C$. This phasor diagram has been constructed when the circuit is operating at resonance, which occurs when the applied voltage and the resulting current are in phase. From our knowledge of AC circuit theory, we can write down equations for some of the important quantities in the circuit.

Specifically: $|I_L| = \dfrac{V}{\sqrt{(\omega L)^2 + R^2}}$ and $|I_C| = V \omega C$ and $\cos(\varphi) = \dfrac{\omega L}{\sqrt{(\omega L)^2 + R^2}}$.

From the geometry of the phasor diagram at resonance we can see that:

$$|I_L|\cos(\phi) = |I_C|$$

therefore: $\dfrac{V}{\sqrt{(\omega_p L)^2 + R^2}} \dfrac{\omega_p L}{\sqrt{(\omega_p L)^2 + R^2}} = V \omega_p C$

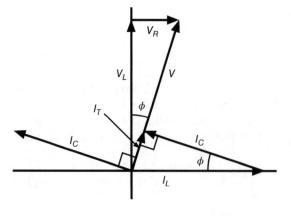

Figure 2.40 Parallel tuned circuit phasor diagram.

where ω_p is the parallel resonant frequency.

Thus:

$$\left(\omega_p L\right)^2 + R^2 = \frac{L}{C}$$

$$\omega_p^2 = \frac{1}{LC}\left[1 - R^2 \frac{C}{L}\right]$$

$$\omega_p = \omega_s \sqrt{1 - \frac{R^2 C}{L}} = \omega_s \sqrt{1 - \frac{1}{Q_0^2}}$$

Here ω_s is the series resonant frequency $= \dfrac{1}{\sqrt{LC}}$ and $Q_0 = \omega_s L / R$.

Therefore the parallel resonant frequency, ω_p is often just a little less that the series resonant frequency, particularly when the circuit Q_0 is low, say $Q_0 < 3$

By way of example, a 3 mH inductor with a winding resistance of 7 ohms is series resonant with a 1 μF capacitance at 2.91 kHz. When these components are connected in parallel then according to the above equation, the resonant frequency becomes 2.87 kHz. For a high Q circuits like this one the difference is not particularly great and the approximation is quite reasonable. However, for low Q circuits, the parallel resonant frequency can be significantly different. If, for example, the same inductor is resonated with 10 μF, the series resonant frequency would become 919 Hz while the parallel resonant frequency would only be 840 Hz – quite different!

2.8 Problems

1 An unbalanced three-phase load is connected to a balanced 230 V three-phase supply, shown in Figure 2.41.

A Use Millman's theorem to find V_{SN}, the star point voltage in Figure 2.41.
(Answer: $V_{SN} = 0$ volts)

B Use your result from part (a) and further calculations to determine the Thévenin equivalent circuit between the star point and the neutral terminal.
(Answer: $V_{Th} = 0$ volts, $Z_{Th} = 230\angle 0$ ohms)

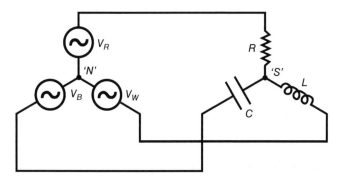

Figure 2.41 Diagram for question 1. ($R = 230\,\Omega$, $L = 1.267$ H, $C = 8\,\mu$F)

C Find the neutral current that would flow if the load star point and the neutral terminal were connected. The phase voltages are each 230 V.
(Answer: 0 amps. Note that this does not mean that the load is necessarily balanced; it's just that in this case the phase currents happen to sum to zero.)

2 Repeat question 1 if the inductor and capacitor are swapped, i.e. the capacitor is connected to the white phase and the inductor to the blue, as shown in Figure 2.42. (This is the equivalent of reversing the phase sequence.)
   A (Answer: $V_{SN} = 460\angle 0$ volts)
   B (Answer: $V_{TH} = 460\angle 0$ volts, $Z_{Th} = 230$ ohms)
   C (Answer: $I_N = 2$ amps)

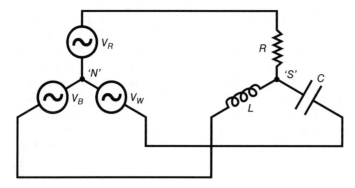

Figure 2.42 Diagram for question 2. ($R = 230\,\Omega$, $L = 1.267\,H$, $C = 8\,\mu F$)

3 Use superposition to find the neutral current that would flow if the star point 'S' and the neutral terminal in Figure 2.42 were connected. Draw the complete phasor diagram, showing the phase voltages and currents and their summation to produce the neutral current.
(Answer: $I_N = 2$ amps).

4 Figure 2.43 shows two possible three-phase load connections, star and delta. If the power consumed by each load is to be the same, determine the relationship between r and R. (The average power in a resistive load R, is given by $P = VI = V^2 R = I^2 R$).

Figure 2.43 Star and delta connections.

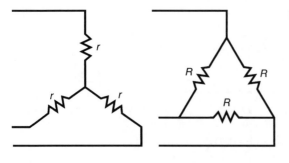

5 A *balanced* three-phase industrial load is represented by the single phase circuit of Figure 2.44.

A Draw the complete phasor diagram showing all the phase voltages and currents.

B Determine the phase angle ϕ between each phase voltage and its associated current.

(Answer: 14° lagging)

C Show on your phasor diagram how each phase current would change if a small capacitance were connected in parallel with the load. What value of capacitive reactance would be required in order to make $\phi = 0$?

(Answer: 40 ohms)

Figure 2.44 Diagram for question 5.
($R = 10$ ohms, $X_L = 40$ ohms)

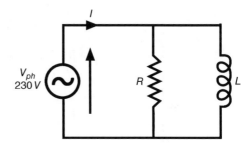

6 Show that a series L-R circuit can be considered equivalent to a parallel L-R circuit, as shown in the Figure 2.45, if the relationships between there quantities are as follows:

$$r = \frac{RX^2}{\left(R^2 + X^2\right)}; x = \frac{R^2 X}{\left(R^2 + X^2\right)} \text{ and } R = \frac{\left(r^2 + x^2\right)}{r}; X = \frac{\left(r^2 + x^2\right)}{x}$$

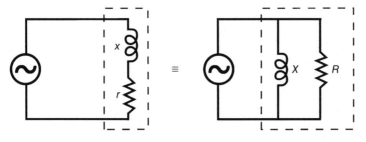

Figure 2.45 Series and parallel circuit equivalence.

7 The schematic diagram shown in Figure 2.46 relates to an induction furnace having *two* furnace transformers (inductors). It is also designed to spread what effectively is a two-phase load evenly over all three phases, in such a way that each sees a resistive load. Each furnace transformer is capable of delivering 550 kW to the melt when supplied with 660 V. In this circuit, no balancing reactor is employed; instead each furnace transformer is provided with a different set of power factor correction capacitors.

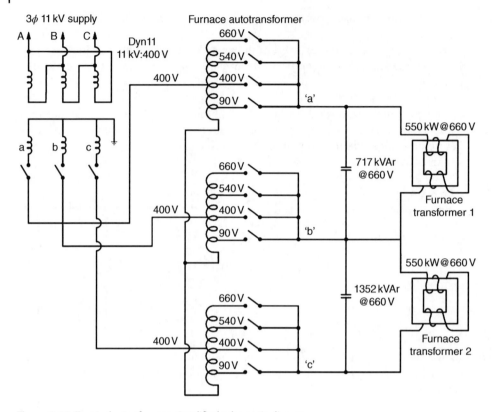

Figure 2.46 Two inductor furnace simplified schematic diagram.

Calculate:
A The effective melt resistance, seen looking into the primary winding of each furnace transformer.
(Answer: 0.792 ohms)
B The reactive impedance that exists in parallel with the melt resistance, if the primary circuit has a power factor of 0.5.
(Answer: 0.457 ohms)
C The total current flowing on the supply side of each PFC capacitor.
(Answer: transformer 1: 962 A, 30° lagging, transformer 2: 962 A, 30° leading.)
D The minimum rating required for the flexible cables supplying each furnace transformer.
(Answer: 1667 A)
Plot the phase currents on a phasor diagram along with the phase voltages, and hence determine the VA rating for the delta-star connected supply transformer. (You may assume that the autotransformer is lossless.) (Answer: 1100 kVA, all phases resistively loaded.)

8 The induction furnace schematic in Figure 2.31 shows that the flexible cables supplying the furnace transformer are rated at 2600 A. If the furnace transformer is

modelled by an inductance L, in parallel with the equivalent furnace resistance R, and complete PFC is provided by seven type 'A' capacitors, calculate the current that will flow to the furnace transformer when it operates at maximum power. Compare your result to the cable rating.
(Answer: 2548 A)

9 A coil with resistance of $4.5\,\Omega$ but unknown inductance is connected in series with a $0.12\,\mu\text{F}$ capacitor to form a series resonant circuit. The measured centre frequency (resonant frequency) is 7.2 kHz. The same coil is now to be used in a parallel resonant circuit with a new capacitor, to be resonant at 5 kHz.
   A What parallel capacitance is required?
      (Answer: $C = 0.249\,\mu\text{F}$)
   B Estimate the Q and the bandwidth of the parallel circuit.
      (Answer: $Q = 28.4$, $BW = 176\,\text{Hz}$)
   C If a bandwidth of 500 Hz is required for the parallel circuit what value of resistance would be needed:
      1. If an extra resistance is connected in parallel with the coil?
          (Answer: $R_{extra} = 1.97\,\text{k}\Omega$)
      2. If the extra resistance is connected in series with the coil?
          (Answer: $R_{extra} = 8.3\,\Omega$)

10 A Calculate the cut-off frequencies of the circuit in Question 9A.
    B What will these frequencies become when the bandwidth is increased as in 9C?
       (Answer a: 7.32 kHz and 7.07 kHz, b: 7.70 kHz and 6.94 kHz)

11 An LC parallel resonant circuit shown in Figure 2.47 is to be used in a metal detector circuit. The inductance value is $300\,\mu\text{H}$ with a winding resistance of $1.0\,\Omega$. The circuit is to be resonant at 10 kHz and is to be excited from a 5 V, 10 kHz supply.

Figure 2.47 Parallel resonant circuit.

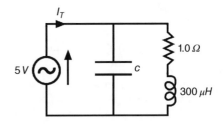

Find:
   A The capacitance value required
      (Answer: $0.845\,\mu\text{F}$)
   B The circuit bandwidth
      (Answer: 531 Hz)
   C The 3 dB frequencies
      (Answer: 10.26 kHz, 9.73 kHz)

D The total current drawn from the supply at resonance.
(Answer: 14 mA)

12   The metal detector circuit described in question 11 is operated at a frequency of
9 kHz. When a non-ferrous metal object enters the magnetic field surrounding the
inductor, energy is absorbed from the field by the eddy current losses within the
metal. This is reflected as an increase in the power supplied to the circuit from
the 5 volt exciting voltage. If a particular metal object, when placed near the inductor,
increases this power by an additional 50 mW calculate:
A The new Q value for the circuit in the presence of the metal object
(Answer: 11)
B The total current now drawn from the 5 volt supply.
(Answer: 24 mA)

13   When a particular ferrous metal object is brought near the same inductor, an addi-
tional 40 mW of power is consumed *and* its inductance value increases by 20%, due
to the reduction in reluctance of the magnetic circuit. (The presence of ferrous
material within the coil's field enables a larger magnetic flux to be established and,
as a result, its inductance value increases.)
Calculate:
A The new Q value for the circuit
(Answer: 12)
B The new resonant frequency
(Answer: 9.13 kHz)
C The total current now drawn from the supply.
(Answer: 60 mA)

14   Find the series resonant frequency of a 500 μH inductor and a 5 μF capacitor. If the
inductor has a winding resistance of 4 Ω, find the parallel resonant frequency of the
combination.
(Answer: Series: 3.18 kHz, Parallel: 2.91 kHz)

2.9   Practical Experiment

The schematic in Figure 2.48 shows an improved phase sequence detector. Here the
incandescent lamps have been replaced with neon lamps. These 'fire' when the applied
voltage exceeds about 90 V and thereafter present a low impedance, hence the need for
the series resistors. By using *two* neon lamps per phase on a 400 V AC supply, the potential across the low voltage pair is insufficient to illuminate them both, thereby removing
any ambiguity as to which lamp is the brighter.

On some occasions the phase sequence needs to be determined from the secondary
side of a *voltage transformer* where the standard line voltage is around 110 V. In this
event, only *one* lamp is required in each phase, and the switch *S1* in Figure 2.48 is provided for the purpose of bypassing one lamp in each pair. In this way, the circuit will
function equally well from either a 400 or a 110 volt AC supply.

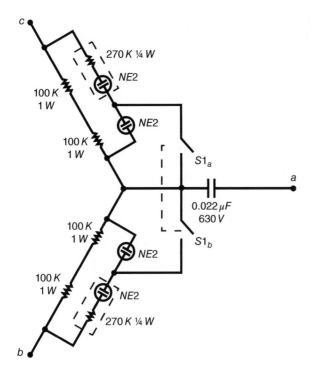

Figure 2.48 Improved phase sequence detector.

Procedure:

1) Construct this circuit and observe its behaviour when the applied phase sequence is reversed and it is supplied from both a 400 volt and a 110 volt source.

2) What potential appears at the star point in each case, relative to the neutral terminal? Is this approximately as you would expect?

3

Active Power, Reactive Power and Power Factor

In this chapter we will consider the concept of electrical power as it applies to AC circuits. We shall see that unlike DC circuits, where the power dissipated can be found from the product of the voltage and current, in the AC case there are two forms of power to be considered, *active power* and *reactive power*, each of which must be generated somewhere within the AC network.

But before we begin, it is necessary to clearly distinguish between the concepts of *power* and *energy*. Energy (E) is added to a system the result of performing *work* on it; the unit of energy is the joule. For example, the energy stored within a capacitor is proportional to the square of its terminal voltage. To increase this energy, additional charge must be accumulated, and thus work must be done to increase the stored charge, against the force of the existing electric field.

Power is defined as the *rate of delivery of energy*, or the rate of doing work, and is therefore measured in joules per second, or watts. We can summarise this in the following simple equation:

$$P(t) = \frac{dE(t)}{dt}$$

Power is *positive* when work is done *on* a system and therefore its energy content increases, (e.g. increasing the charge on our capacitor) and *negative* when work is done *by* the system, which is accompanied by a loss of system energy (discharging our capacitor into a resistor).

Suppose a room is warmed by adding heat at a rate of 1000 watts (1 kW) for one hour, as evidenced by a rise in its temperature. This is equivalent to adding 3,600,000 joules over this period, or in electrical terms, by adding the energy equivalent of one *kilowatt-hour (kWh)*. Electrical energy is traditionally sold in kilowatt hours, whereas the energy stored in LP gas is sold in either in *megajoules (MJ) or gigajoules (GJ)*.

3.1 Single-Phase AC Power

The power in an AC circuit can be evaluated by considering the average of the product of the applied AC voltage and the resulting current. However, because of the sinusoidal nature of both these quantities, the AC analysis is considerably more complex than the

AC Circuits and Power Systems in Practice, First Edition. Graeme Vertigan.
© 2018 John Wiley & Sons Ltd. Published 2018 by John Wiley & Sons Ltd.

DC one. We will initially consider this product in the case of a single-phase circuit, and then we will generalise our analysis to include the three-phase case.

Consider the product of a sinusoidal voltage and current, delivered to a single-phase load and displaced from one another by a *phase angle* ϕ.

Thus if

$$V(t) = \hat{v}\sin(\omega t)$$

we can write

$$I(t) = \hat{i}\sin(\omega t - \phi)$$

where ω is the angular frequency $(2\pi f)$, $\hat{v}$ and $\hat{i}$ are the peak values of the voltage and current respectively, and ϕ is the angle by which the current lags the voltage.

The product of the voltage and current can be written:

$$V(t)I(t) = \hat{v}\hat{i}\sin(\omega t)\sin(\omega t - \phi)$$

$$V(t)I(t) = \hat{v}\hat{i}\left[\sin^2(\omega t)\cos(\phi) - \sin(\omega t)\cos(\omega t)\sin(\phi)\right]^1$$

$$V(t)I(t) = \hat{v}\hat{i}\left(\frac{[(1-\cos(2\omega t)]\cos(\phi)}{2}\right) - \hat{v}\hat{i}\left(\frac{(\sin(2\omega t))\sin(\phi)}{2}\right)$$

$$V(t)I(t) = V_{RMS}I_{RMS}\left[(1-\cos(2\omega t)]\cos(\phi) - V_{RMS}I_{RMS}\sin(\phi)\sin(2\omega t) \qquad (3.1)$$

3.2 Active Power

The first term in Equation (3.1) varies at *twice* the system frequency, as shown in Figure 3.1, since whenever either the voltage or the current pass through zero, the power waveform changes sign.

The average value of the $V(t)I(t)$ waveform can be expressed in terms of the RMS voltage and current as $V_{RMS}I_{RMS}\cos(\phi)$, and is called the *active power*, for which we use the symbol P. It represents the average rate at which energy is flowing from the voltage source to the load and is expressed in joules per second or more usually in watts.

The $V(t)I(t)$ product in Figure 3.2 is negative whenever either the voltage or the current waveforms (but not both) are negative; at all other times this product is positive. Its average value depends on the angle between the voltage and the current, usually called the *phase angle*, ϕ. In this example, the voltage has a magnitude of $230\,V_{RMS}$ and the current $1\,A_{RMS}$. In addition the current lags the voltage by $60°$ – it goes through its positive going zero crossing $60°$ after the voltage does. So the active power flow P, is equal to $230\,Cos(60) = 115\,W$; equal to the average value of the trace in Figure 3.1.

---

1 Recall that: $\sin(A \pm B) = \sin(A)\cos(B) \pm \cos(A)\sin(B)$
And $\sin^2(A) = [1 - \cos(2A)]/2$

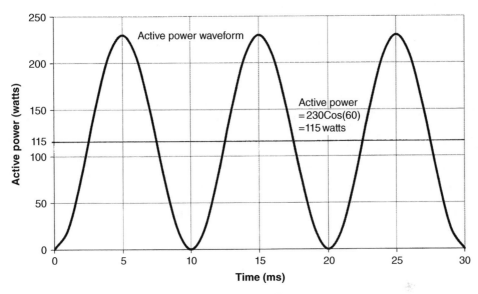

Figure 3.1 Active power waveform: $V_{RMS}I_{RMS}\left[(1-\cos(2\omega t)\right]\cos(\phi)$. ($V_{RMS}=230$ V, $I_{RMS}=1$ A and $\phi=60°$).

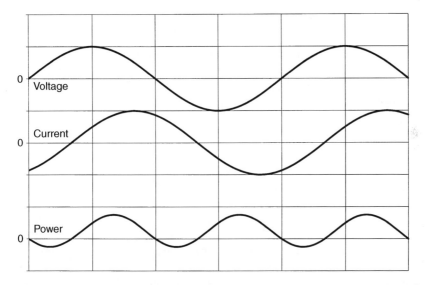

Figure 3.2 Instantaneous $V(t)I(t)$ product (bottom trace). ($V_{RMS}=230$ V, $I_{RMS}=1$ A, $\phi=60°$ lagging).

3.3   Reactive Power

During the negative portion of the power waveform in Figure 3.2, energy is being returned to the supply. This suggests that the current flowing in the circuit is actually larger than would be required to deliver 115 W to the load if the phase angle were able to be made equal to zero. A little reflection will show that this is indeed the case; with zero phase angle a current of only ½ A_{RMS} is required for the same power transfer.

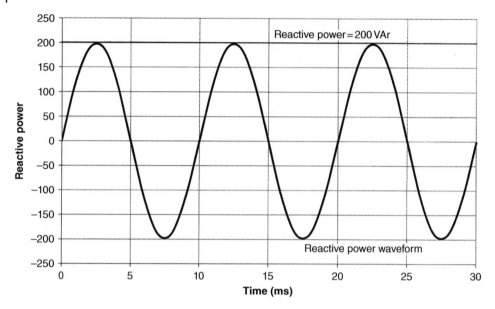

Figure 3.3 Reactive power waveform: $V_{RMS}I_{RMS}\sin(\phi)\sin(2\omega t)$. ($V_{RMS}=230\,V$, $I_{RMS}=1\,A$ and $\phi=60°$).

The second term in Equation (3.1), $V_{RMS}I_{RMS}\sin(\phi)\sin(2\omega t)$, reflects this effect. This also varies sinusoidally at twice the system frequency, as shown in Figure 3.3; however, this waveform has a *zero* average – there is no net power flow. We call the magnitude of this sinusoid the *reactive power*, for which we use the symbol Q. Reactive power has the dimension volt-amps, but to highlight its reactive nature we define the units of Q as volt-amps reactive, or VArs. Reactive power reflects the fact that with either lagging or leading phase angles there is a continuous and lossless exchange of energy between the source and the load, in addition to the active power flowing. As a result a larger current is required than would be needed to support the active power flow alone.

We can therefore write:

$$P = V_{RMS}I_{RMS}\cos\left(\phi\right) \tag{3.2}$$

$$Q = V_{RMS}I_{RMS}\sin\left(\phi\right) \tag{3.3}$$

Since the VAr flow in Figure 3.3 averages zero there is no net energy associated with it; however, this does not mean it can be ignored. Both P and Q must be generated somewhere in the electrical system. The active power must be produced by a generator, which requires an external energy source. On the other hand, the generation of reactive power does not require a supply of energy; it will usually be generated within the electrical network itself, partially by the distributed capacitances within the system, but more so by the action of grid connected capacitors that are switched into the network as the need for reactive power arises. Finally, some of the required reactive power required will be produced by generators, although this will generally be at the expense of the production of active power. As we shall see in Chapter 6, the provision of sufficient reactive power is essential for maintaining the voltage level throughout the AC network.

There are two cases where reactive power occurs without any the need for active power to be generated. The first case relates to the current drawn by a pure inductance. As we saw in Chapter 2, an inductor's current lags the applied voltage by 90°, ($\phi = -90$) and therefore according to Equations (3.2) and (3.3) the active power transferred will be zero, while the reactive power is equal to the product of the bus voltage and the inductor current, $V_{RMS}I_{RMS}$, as shown in Figure 3.4.

The second case, although similar to the first, relates to the current drawn by a capacitor which leads the voltage by 90°, ($\phi = +90$). According to Equations (3.2) and (3.3) the active power will equal zero, while the reactive power will equal $V_{RMS}I_{RMS}$, as shown in Figure 3.5.

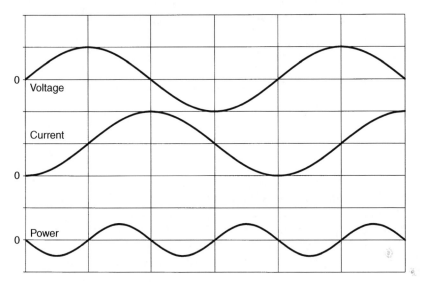

Figure 3.4 Inductive voltage, current and power waveforms.

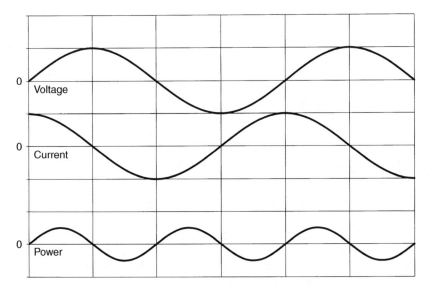

Figure 3.5 Capacitive voltage, current and power waveforms.

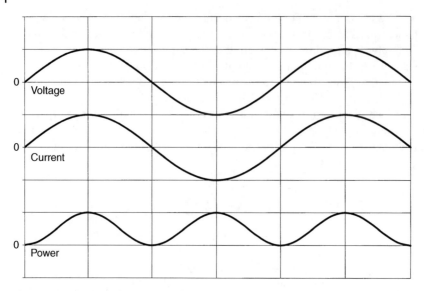

Figure 3.6 Resistive voltage, current and power waveforms.

There is one vital difference between these cases, however, which relates to the relative direction of these power flows. According to Equation (3.3) the reactive power demanded by the inductor has the opposite sign to that required by the capacitor, since the phase angles are of opposite sign. Therefore the reactive power consumed by an inductor can be generated by a capacitor. This can be readily demonstrated by connecting a capacitor and an inductor in parallel across an AC source. If $X_C = X_L$ at the system frequency, then the capacitor will exactly generate the reactive power required by the inductor and, as a result, there will be a lossless exchange of reactive power between the two, while virtually no current is drawn from the source. (This is precisely the operation of a parallel resonant circuit.)

Finally, in a purely resistive circuit, the voltage and current lie in phase and $\phi = 0$. Figure 3.6 shows the associated current and voltage waveforms as well as the power waveform. Here the amplitude of the power waveform is $\hat{v}\hat{i}$ or 460 W, while the average power flowing is half this figure, or $(\hat{v}\hat{i})/2 = V_{RMS}I_{RMS} = 230$ W. The lack of reactive power means that, from the network perspective, this represents the most efficient way of delivering power to a load.

3.4 Apparent Power or the volt-amp Product, S

Most loads on the electrical system draw currents that lag the supply voltage. This is due to the inductive nature of electromagnetic devices such as motors and transformers. Figure 3.7 illustrates that a lagging current can be resolved into two components: one in phase with the voltage, the other in quadrature with it, (i.e. 90° out of phase). The in-phase component, equal to $I\cos(\phi)$, is associated with the active power P, while the quadrature component, equal to $I\sin(\phi)$, is associated with the reactive power Q demanded by the load.

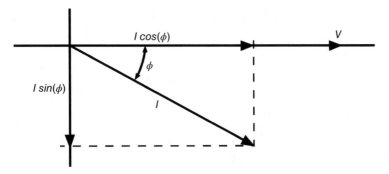

Figure 3.7 In-phase and quadrature current components.

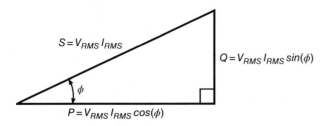

Figure 3.8 Power triangle.

Figure 3.7 suggests that P and Q are also in quadrature, and therefore their sum, called the *complex power*, can be written as:.

$$Complex\ power = V_{RMS} I_{RMS} \cos(\phi) + j V_{RMS} I_{RMS} \sin(\phi)$$
$$= P + jQ \quad\quad\quad (3.4)$$

Thus the apparent power S is simply the VI product, thus $S = V_{RMS} I_{RMS}$. It has the unit *volt-amps* expressed as VA, and it therefore follows that:

$$\left| S^2 \right| = \left| P^2 \right| + \left| Q^2 \right| \quad\quad\quad (3.4a)$$

We summarise this relationship in the form of the *power triangle* shown in Figure 3.8.

3.5   Three-Phase Power

Let us re-evaluate the VI product, this time in the case of a three-phase supply delivering balanced currents, perhaps to a synchronous motor or an induction motor. In this case, there will be three independent phase voltages each delivering a component of power to the load. As we will see, because of the 120° phase shift between these, the total power flowing to the load is now constant.

The overall $V(t)I(t)$ product, includes a component from each phase. We assume that all voltages and currents are expressed as RMS quantities, and so we will henceforth drop this suffix; therefore $V = V_{RMS}$ and $I = I_{RMS}$. The three-phase $V(t)I(t)$ product may be written in the form:

$$
\begin{aligned}
\Sigma V(t)I(t) &= \hat{v}\hat{i}\sin(\omega t)\sin(\omega t - \phi) + \hat{v}\hat{i}\sin(\omega t - 2\pi/3)\sin(\omega t - 2\pi/3 - \phi) \\
&\quad + \hat{v}\hat{i}\sin(\omega t + 2\pi/3)\sin(\omega t + 2\pi/3 - \phi) \\
&= \hat{v}\hat{i}\left\{\sin^2(\omega t)\cos(\phi) - \sin(\omega t)\cos(\omega t)\sin(\phi)\right\} \\
&\quad + \hat{v}\hat{i}\left\{\sin^2(\omega t - 2\pi/3)\cos(\phi) - \sin(\omega t - 2\pi/3)\cos(\omega t - 2\pi/3)\sin(\phi)\right\} \\
&\quad + \hat{v}\hat{i}\left\{\sin^2(\omega t + 2\pi/3)\cos(\phi) + \sin(\omega t + 2\pi/3)\cos(\omega t + 2\pi/3)\sin(\phi)\right\} \\
&= VI\left\{\left[1 - \cos(2\omega t)\right]\cos(\phi) + \sin(\phi)\sin(2\omega t)\right\} \\
&\quad + VI\left\{\left[1 - \cos(2\omega t - 4\pi/3)\right]\cos(\phi) + \sin(\phi)\sin(2\omega t - 4\pi/3)\right\} \\
&\quad + VI\left\{\left[1 - \cos(2\omega t + 4\pi/3)\right]\cos(\phi) + \sin(\phi)\sin(2\omega t + 4\pi/3)\right\}
\end{aligned}
$$

This leads to:

$$
\begin{aligned}
\Sigma V(t)I(t) &= 3VI\cos(\phi) \\
&\quad + VI\sin(\phi)\sin(2\omega t) + VI\sin(\phi)\sin(2\omega t - 4\pi/3) \\
&\quad + VI\sin(\phi)\sin(2\omega t + 4\pi/3)
\end{aligned}
\tag{3.5}
$$

The first term in Equation (3.5) is the active power P, comprising an identical component from each phase. It is important to note that in this balanced three-phase case, P is constant; it does not oscillate about its mean at twice the system frequency, as it does in the single-phase case.

Important examples of balanced loads are three-phase induction or synchronous motors. Since the torque produced by these machines is equal to the output power divided by the shaft speed in radians per second, then if the power delivered to the machine is constant, so will be the torque it produces. This is an important advantage over a single-phase motor, where, as we have seen, both the power and the torque vary sinusoidally about the mean. Therefore we may write:

$$
P_{Three\ Phase} = 3VI\cos(\phi)
\tag{3.6}
$$

The remaining VI product terms in Equation (3.5) relate to the reactive power required by the load. From these we can see that there is a reactive current flowing in each phase at twice the system frequency, just as in the single-phase case. The amplitude of each of these currents is $I\sin(\phi)$, and while collectively they sum to zero, each one constitutes a component of phase current that must be generated and supplied to the load. These reactive currents when multiplied by the phase voltage V, constitute the three-phase reactive power, given by:

$$
Q_{Three\ Phase} = 3VI\sin(\phi)
\tag{3.7}
$$

Equations (3.6) and (3.7) have exactly the same form in the three-phase case as they do in the single-phase, and therefore in any circuit where the phase angle is non-zero, reactive power will exist in addition to any active power flow.

When the load is unbalanced, the contributions to both P and Q will be different in each phase, in which case P and Q can be expressed as:

$$P = |V_a||I_a|\cos(\phi_a) + |V_b||I_b|\cos(\phi_b) + |V_c||I_c|\cos(\phi_c)$$
$$Q = |V_a||I_a|\sin(\phi_a) + |V_b||I_b|\sin(\phi_b) + |V_c||I_c|\sin(\phi_c)$$

Here $|V_i|$ and $|I_i|$ represent the magnitude of the phase i voltage and current, in volts and amps RMS. This notation is used to distinguish between the magnitude of these quantities and their phasor representations V_i and I_i.

The power triangle concept is still useful in the case of an unbalanced load; however, in this case we must define S according to $\sqrt{P^2 + Q^2}$. Similarly, the *effective phase angle* ϕ is also useful in the unbalanced case, and is defined by Equation (3.9), as explained below.

3.6   Power Factor

As we have seen, active power is generated by the in-phase component of load current, $I\cos(\phi)$, and therefore $\cos(\phi)$ is an important quantity. When it approaches unity (i.e. $\phi \approx 0$) the load current is largely associated with the active power flow, since very little reactive power is demanded. In such cases, the network supplying the load is being used most efficiently, and the magnitude of the load current is close to the minimum value required for the active power transferred.

On the other hand, when the phase angle is particularly large $(\phi > 45°)$ much of the supply current is associated with a reactive power flow, and its magnitude is considerably larger than would be required by a purely resistive load consuming the same active power.

We define the circuit's *power factor* as the ratio of the active power (P) to the apparent power, (S).

$$PF \equiv \frac{Active\ power}{V_{RMS}I_{RMS}} = \frac{P}{S} \tag{3.7a}$$

When the circuit voltages and currents are purely sinusoidal (i.e. no harmonics exist), and the currents are balanced, the power factor becomes equal to the cosine of the angle ϕ.

$$PF = \frac{P}{\sqrt{P^2 + Q^2}} = \cos(\phi) \tag{3.7b}$$

Loads that are not well balanced may have a different angle associated with each phase, and the effective power factor, averaged over a period of time, can be defined in terms of the active and reactive energies summed over all three-phases, as follows:

$$PF = \cos\left(\tan^{-1}\left[\frac{VArh}{Wh}\right]\right) \tag{3.8}$$

In Equation (3.8) both the total VAr-hours and the total watt-hours are measured over the same time interval, which may be as short as a metering interval (typically 15 or 30 minutes), or as long as a billing period, which may be between 1 and 3 months. In the former case, Equation (3.8) provides a dynamic measure of a load's power factor. This parameter is frequently defined in the network connection agreements and, for reasons that will soon become apparent, usually must not fall below a prescribed limit.

We can also use Equation (3.8) to define an *effective phase angle* in unbalanced three-phase circuits:

$$Effective\ phase\ angle = \phi_{eff} = \tan^{-1}\left[\frac{VArh}{Wh}\right] = \tan^{-1}\left[\frac{Q}{P}\right] \tag{3.9}$$

The effective phase angle is the angle between S and P in the power triangle of an unbalanced circuit, where P and Q each represent the summation of the watt and VAr contributions from each phase.

Typical power factors associated with industrial loads lie in a range 0.75 to 1.0. Loads that operate at low power factors (<0.85) impose unreasonable burdens on the power system, since the volt-amps they demand is large in comparison to the active power delivered, as can be seen from the power triangle in Figure 3.8.

The reactive current consumed by such loads creates additional losses within the distribution network when compared to those of a purely resistive load, and therefore the network's current capacity must be increased if it is to accommodate heavily reactive loads. By way of illustration, a load with a power factor of 0.5 will generate *four* times the system losses of one operating at unity power factor for the same power transfer! Supply authorities therefore discourage heavily inductive loads by imposing penalties on customers whose loads have poor power factors. This is often achieved either directly, through a contract limit on the load power factor, or via either a *maximum demand charge*, (often levied on the maximum kVA consumed)[2], or a *reactive demand charge*. The latter may be levied on the reactive energy consumed, should this exceed a prescribed limit.

3.7 Power Factor Correction

We saw in Chapter 2 that a capacitor, when connected to an AC bus, can generate some or all of the reactive power required by an inductive load. This is the principle of power factor correction (PFC), and it is applied to inductive loads with inherently poor power factors. Capacitors are connected in parallel with an offending load and supply most of the reactive power demanded, leaving a largely flat power triangle, where Q lies close to zero. Therefore the total load has a power factor considerably nearer unity.

The circuit of Figure 3.9 depicts a three-phase lagging load, in this case one represented by a parallel connected inductance and resistance, to which has been added a power factor correction capacitance. We repeat the analysis presented in Chapter 2, in this case modelling the load with a parallel connected circuit, as shown in the phasor diagram in Figure 3.10.

The total current flowing into the combination comprises three components, one flowing in the resistor (generating the active power delivered to the load), one flowing in the inductance (generating the reactive power consumed by the load) and finally the compensation current flowing in the PFC capacitance. Since the bus voltage V is

---

2 See section 3.10.3.

Figure 3.9 Inductive load with power factor correction.

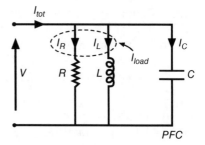

Figure 3.10 Phasor diagram.

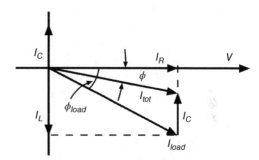

common to all circuit elements it will be used as the reference phasor. The load current lags the applied voltage by an angle ϕ_{load}, while the capacitive current leads the voltage by 90°, as shown in Figure 3.10. The total current can be found by vector addition of these phasors. The resultant current I_{tot} can be seen to lag the voltage by a smaller angle ϕ and thus we observe that the power factor has been increased, i.e. it has been improved. As a consequence, we also note that the magnitude of the customer's load current has been reduced by the inclusion of the power factor correction equipment. This means that the network losses incurred by this customer will be reduced accordingly.

If the capacitor is chosen so that the current it consumes is equal in magnitude to the inductive current I_L, then the total current will lie exactly in phase with the bus voltage V, and the resulting PF will equal unity. Should an over-large capacitor be chosen, then the circuit will be over-compensated and the total current will lead the bus voltage; the load is then said to have a leading power factor. For reasons that will be explained in the next section, supply authorities generally discourage leading power factors.

3.7.1 Capacitive Voltage Rise

Sometimes it is possible that excess power factor correction may be inadvertently applied, usually as a result of the load falling in the presence of fixed capacitive compensation. Under these conditions, the power factor will become leading and a small rise in the local bus voltage will also occur.

It should be noted that any time a capacitor is connected to a bus the voltage will rise slightly. However, if the entire current delivered by the bus is permitted to become leading, then it is possible that the associated rise may cause the bus voltage to exceed its normal limits. Should this occur on an LV bus, where on-load tap-changing facilities are not available, then the network operator has no easy means of correcting the problem.

Supply authorities therefore generally discourage leading loads to avoid the danger that over-voltage poses to all the equipment connected, not only that of the customer concerned but of others as well.

3.7.2 Power Factor Correction Equipment

Large LV customers such as supermarkets, shopping centres and industrial sites frequently install power factor correction (PFC) equipment of the kind pictured in Figure 3.11 to overcome the reactive effects of air conditioning, refrigeration equipment and fluorescent lighting. Usually the capital cost of this equipment and its installation, can be repaid within a short time from savings in the maximum demand charge levied by the retailer or the network owner.

Low voltage PFC installations generally consist of delta connected capacitors, since the reactive energy provided is proportional to the square of the applied voltage and there is no difficulty or excessive cost in building LV capacitors capable of supporting voltages up to $500\,V_{RMS}$. On the other hand, voltage support capacitors used in MV and HV networks are generally wye (star) connected, since these usually have a limited withstand voltage capability, and therefore a phase potential is more convenient. In both cases, however, each capacitor is connected in series with a *de-tuning inductance*, to ensure that there is no adverse harmonic resonance created between the capacitors themselves and the local network's inductive source impedance.

Power factor correction equipment often contains two or more sets of capacitors which can be switched onto the local bus incrementally so that capacitive VArs can be progressively added, so as to offset any increase in the inductive VAr demand. The capacitor switching is controlled by a reactive power control relay, like the one shown in Figure 3.11a. This device assumes a balanced load, and monitors the line current and voltage on one

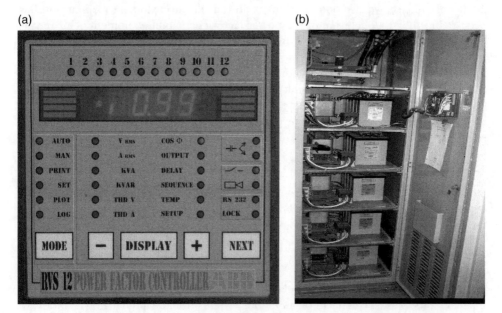

Figure 3.11 (a) PFC control relay (b) Low voltage PFC cabinet.

phase, via a dedicated current transformer, and computes the power factor of the load concerned. As the reactive demand increases, the controller switches in more capacitive compensation, thereby maintaining a nearly constant power factor. Figure 3.11b shows a PFC cabinet containing five identical stages. Sometimes 'n' capacitive stages are provided and arranged in a binary sequence, so that 2^n increments can be generated thereby enabling the inductive demand of the load to be accurately compensated.

Capacitors are switched either with metallic contacts, or where bus disturbances must be avoided, with the use of thyristors, which can be fired at voltage zero crossings, thereby minimising the disturbing effects of capacitive inrush currents.

The effectiveness of PFC can be seen in Figure 3.12, where the load profile of an LV load is shown; the lower traces represent the active power and the kVA demanded. The consumption cycles on a daily basis, and on one Monday afternoon a new power factor correction panel is switched into service for the first time, as evidenced by a jump in the load power factor, (top trace). Simultaneously the VA demand falls to a value just a little above the active power consumed. As a result the power triangle of Figure 3.8 has now become quite flat, since the reactive power Q supplied by the bus, is now approximately zero. Consequently, $\cos(\phi)$ rises to about 0.97 from its previous value of around 0.78.

3.7.3 Sizing of Power Factor Equipment

It is generally unnecessary to fully correct the reactive demand to zero when applying power factor correction. There are two reasons for this. Firstly, it makes the possibility of a leading power factor more likely, and secondly there is a diminishing return between with money spent on PFC equipment and achieving a particularly high power factor.

We can demonstrate this effect using the following analysis. We begin by expressing the PF in terms of the load's active and reactive power:

$$PF = \cos\left(\tan^{-1}\left\{\frac{Q}{P}\right\}\right)$$

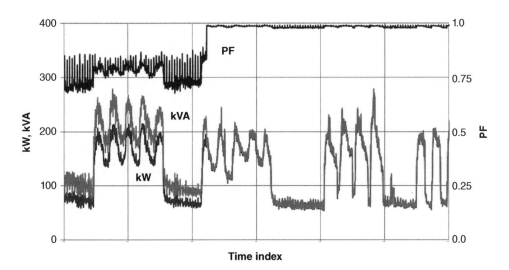

Figure 3.12 LV customer load profile and power factor.

With power factor correction applied, the reactive load remaining Q can be written:

$$Q = (Q_L - Q_C)$$

where Q_L is the maximum value of the load's reactive demand, and Q_C is the maximum value of the capacitive correction to be applied. Thus we can write:

$$PF_{corrected} = \cos\left(\tan^{-1}\left\{\frac{Q_L - Q_C}{P}\right\}\right) = \cos\left(\tan^{-1}\left\{\frac{Q_L(1-k)}{P}\right\}\right)$$

Here we write $Q_C = kQ_L$, where k is the fraction of the load's reactive demand to be corrected. Thus if $k=1$ then the load will be fully corrected and the final PF will equal unity; if $k=0$ then there will be no improvement in the load's PF.

If we plot the corrected PF as a function of k we obtain the set of curves shown in Figure 3.13, where we see that as the initial (uncorrected) PF becomes worse, (i.e. lower) we need to correct a progressively larger percentage of the reactive demand in order to achieve an acceptable final PF. Note that all these curves flatten as k tends towards 100%, suggesting that in order to achieve very high power final factors, a disproportionately high degree of PF correction will be required.

For example, if the load's initial PF is 0.6 and we aim for a final power factor of 0.95, then from Figure 3.13, we will require a capacitive correction equal to about 75% of the load's maximum reactive demand. However, if we were to aim for a final PF of unity, 100% correction will be required. Thus the required PFC capacity must be increased by 25% in order to raise the PF from 0.95 to unity. This is both expensive and unnecessary; a power factor of 0.93 or 0.95 is generally quite sufficient to provide the customer with substantial saving in demand charges, while satisfying the requirements of most supply contracts.

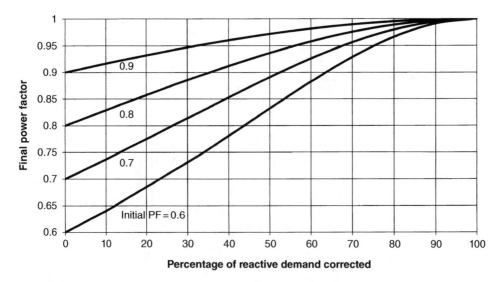

Figure 3.13 Final PF versus the percentage of the original reactive load corrected.

The step size of the capacitive compensation to be provided is also worth mentioning. If the PFC manufacturer offers a choice in this, it is wise to choose a step size a little smaller than the load's minimum reactive demand so that at minimum load the capacitive correction can still be applied. However, this choice is not critical and will not affect the MD charges levied. In LV applications the increment size usually lies in the range 25–50 kVAr, and most controllers will not allow the final PF to become leading.

3.8 Typical Industrial Load Profiles

3.8.1 Small Industrial Loads

An actual load profile of a small food processing plant, supplied from a 22 kV distribution feeder under a kVA based maximum demand (MD) tariff, appears in Figure 3.14. The load profile reveals several interesting things about the business. Firstly, it obviously only operates during normal working hours, as evidenced by the cyclic daily demand. During the evenings and at weekends the site load falls to a low level, probably due to background refrigeration and water heating loads.

Secondly, both the active and reactive power demands occur in approximately the same ratio. This means that the site's power factor lies in a relatively restricted range, in this case from about 0.65 to 0.8, with PF values at the peak of the site's demand close to 0.78. This suggests that the management probably does not understand the concept of a kVA based maximum demand tariff, otherwise power factor correction equipment might have been installed and a considerably lower demand charge would have been achieved.

Under this particular kVA maximum demand tariff, the customer pays the network owner an amount based on the maximum kVA demand measured in any metering interval (usually 15 or 30 minutes), during the previous calendar year. This is to cover the cost of the network and the connection assets provided on the customer's behalf. In this way, the network owner achieves a return on the assets provided for this particular

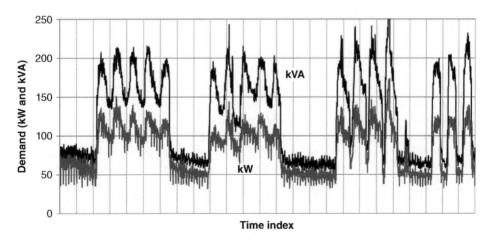

Figure 3.14 Small low voltage industrial customer's load profile.

customer. By basing the MD tariff on the maximum kVA consumed by the load, rather than the maximum kW demand, the network owner is subtly enticing customers to operate their equipment at as high a power factor as economically feasible.

Unfortunately, relatively few small industrial customers understand the concepts of power factor and kVA demand, and even fewer know what steps to take in order to minimise their maximum demand charges, which can amount to more than half of their total electrical expenditure. Further, some electricity retailers are not forthcoming with the information necessary for customers to make informed decisions, as it is seldom in their interest to do so.

For example, in the case of this particular business if we assume that its maximum demand for the year was 250 kVA (as suggested in Figure 3.14) then it would pay an MD charge based on this, despite the fact that the peak power demand on most days is only about 130–140 kW. Effectively the business is hiring assets capable of delivering 250 kVA, when its daily maximum is much less that this. This does not represent good value for the customer.

There are two areas in which the demand charge can usually be reduced without impinging on production. Firstly, where this charge relates to the kVA demand (rather than the kW demand), installing power factor correction equipment will see the site's demand fall to a level very close to its kW demand, thereby potentially achieving a considerable saving in maximum demand charges. In the case of this business, the application of power factor correction will result in a reduction in MD from 250 kVA to around 170 kVA.

Secondly, by installing a dynamic load controller and flattening the load profile, the business can reduce its MD charges even further. A dynamic load controller automatically reduces those sacrificial components of a load (such as water heating, air-conditioning and possibly some refrigeration), as the critical process load increases, and restores them as the critical loads diminish, thereby flattening the load profile and reducing the site's MD, without significantly reducing production. In the example above, on most days the business's maximum power demand is less than 140 kW, but occasionally it rises to around 170 kW. If this business can operate with an MD of 140 kW during most weeks, then it should be possible to manage its load so that this occurs every week. This is the function of a dynamic load controller, and by installing one a further MD charge saving may well be possible.

3.8.2 Large Industrial Loads

Large industries such as aluminium, zinc and nickel refineries, as well as paper and automotive manufacturers, all operate continuously and consume very large quantities of electricity. They form part of a power system's base load and are usually connected directly to the transmission network and are therefore supplied at either HV or EHV potentials. They also generally have very specific connection agreements with the transmission company, including agreed maximum demand limits on the active and reactive power that they can consume. These figures are instrumental in determining the VA capacity of the connection assets that must be provided by the transmission company, and it is therefore important that this capacity should not be exceeded, lest this equipment becomes overloaded. As a deterrent, the connection agreement usually includes steep financial penalties if either of these limits is violated.

Large industries like these adopt an entirely different approach to their electricity management: they have access to significant engineering expertise and well understand the supply contract requirements. Electrical demand is usually tightly controlled, and unless site requirements prevent it, the load imposed on the network usually lies just below the agreed maximum demand. In this way, the money spent on MD charges is best utilised.

The load profile shown in Figure 3.15 is typical of a large industrial plant operating continuously, in this case one with a 100 MVA maximum demand. The plant's MVA demand is generally carefully managed so that for most of the time it lies just below the contract MD. It may occasionally fall for process reasons, but it will be restored as quickly as possible in order to maintain production. The power factor of such loads tends to remain largely constant and close to unity, in this case 0.95.

In the case of customers like these the concept of a load factor is useful and is frequently measured. It is a way of quantifying the 'peakiness' of a customer's load, or alternatively it may be considered a measure of the degree to which a plant is operated at full capacity. It is defined as follows:

$$Load\ Factor = \frac{Total\ kWhrs\ Consumed\ during\ the\ Billing\ Period}{Maximum\ Demand\ (kW) \times Hours\ in\ the\ Billing\ Period}$$

Thus if a load profile is constant throughout the billing period, and equal to the contract MD then the load factor will equal 1.0. This is the ideal situation but it rarely occurs in practice. On the other hand, if the load profile varies considerably below the contract MD then the load factor will be much less than one, and the customer would potentially be paying the network provider for supply capacity that isn't being used effectively. Looked at from another perspective, since the MD is usually chosen to be equal to or just a little above a plant's electrical rating, operating at load factors significantly below unity means that the site is not operating at its full capacity. Most large industrial customers therefore strive to achieve a load factor as close to unity as possible.

In addition to MD limits, connection agreements often specify other parameters which define the quality of the connected load, such as its harmonic content and the permitted unbalance between phase currents.

3.9 Directional Power Flows

The directions in which P and Q flow are controlled by the phase angle between the voltage and current, ϕ. We know that for a balanced three-phase load $P = 3VI\cos(\phi)$ and $Q = 3VI\sin(\phi)$, therefore the sign of these quantities determines their direction. The direction of the active power flow changes sign with $\cos(\phi)$, while that of the reactive power flow changes with $\sin(\phi)$.

As discussed previously, most customer load involves lagging phase currents, and therefore if we take the A phase voltage as a reference phasor, the A phase current will lie in the fourth quadrant, as shown in Figure 3.16. Currents in the fourth quadrant correspond to loads which import both active and reactive power from the bus. If we consider the power flow from the perspective of the load, then we can define both these quantities as positive in this quadrant.

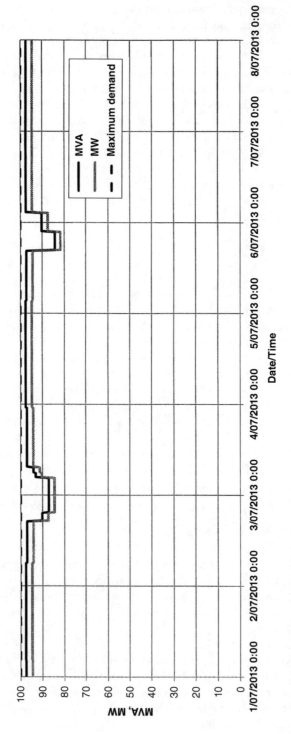

Figure 3.15 Typical large industrial load profile.

Figure 3.16 Power flows in the 4th quadrant are defined as positive.

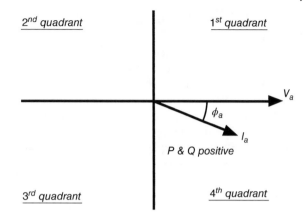

Figure 3.17 Power flow definitions from the perspective of the load.

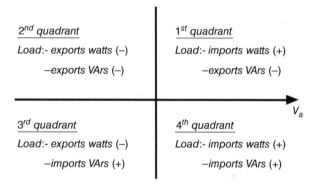

We considered the case where excessive capacitive compensation results in a leading current, which lies in the first quadrant. In this situation, the load is exporting reactive power to the bus, and we define this VAr flow as negative; however, the active power flow remains unchanged. Therefore reactive power is considered positive when ϕ lies between 0 and −180°, i.e. in quadrants 3 and 4 and negative when ϕ lies between 0 and +180°, i.e. in quadrants 1 and 2.

Similarly, we find that the active power flow is positive when ϕ lies between −90 and +90°, i.e. in quadrants 1 and 4, where the load imports power from the bus. It is negative in quadrants 2 and 3 when ϕ is greater than +90 and less than −90°, where the load exports active power to the bus. It should be noted that in order to export power, the load must have some generating capacity. These power flow directions are summarised in Figure 3.17, where it is evident that it is quite possible to have active and reactive power flows occurring simultaneously in opposite directions.

The terms *import* and *export* used above relate the perspective of the load. Sometimes the power flow directions are described from the perspective of the bus, and then the terms import and export take reciprocal meanings. It is therefore important to clearly understand the context in which these terms are used.

3.10   Energy Retailing

In this section we will consider some of the different charges that commercial and industrial electricity users must pay for their electricity. From an engineering point of view we will not consider residential customers, although they do generally have much simpler bundled price structures than do commercial customers.

There are many different charging schemes in use throughout the world, each slightly different, often depending on regional requirements. The following discussion should therefore be seen only as a general approach to electricity pricing, and the charges outlined here may not necessarily be indicative of all countries.

3.10.1   Energy Charges

All commercial and industrial customers will pay for the energy they use. This will be recorded in kWh for each metering interval throughout the billing period. The energy meter will aggregate the energy metered in each interval throughout the billing period, providing the energy retailer with the total number of kWh used by the customer. Energy may be charged at a flat rate, or its price may vary with the time of use. Time of use tariffs seek to modify customer behaviour by moving consumption away from peak periods (where energy is expensive), to off peak periods where it is considerably cheaper.

The energy meter usually also records the reactive energy demand of commercial and industrial customers, in kVArh. This is generally to ensure that the customer is satisfying any reactive demand requirements of their contract. There is usually no requirement to pay for the reactive energy provided, although some jurisdictions do impose financial penalties if the customer's power factor becomes too small.

3.10.2   Transmission and Distribution Loss Factors

Both the transmission and distribution networks incur losses as a result of the transmission of energy. These losses must be paid for, since they consume a small portion of the energy that has been dispatched by generators into the power system. Electricity market operators usually prepare distribution and transmission loss factors (DLFs and TLFs) for each region and for each potential in their electricity network. These loss factors are applied to the energy measured by each customer's energy meter, inflating the kWh count typically by between 5 and 15%. Thus each customer pays not only for the energy seen by their electricity meter, but also for the associated losses incurred in the transmission and distribution networks on their behalf as well.

The magnitude of these loss factors depends upon where a particular customer is located in the network, and the potential at the connection point. For example, a customer residing a considerable distance from generation assets will have a higher transmission loss factor than those situated close by. Similarly, customers connected to the distribution network at say 22 kV will generally have a smaller distribution loss factor than those connected at 230/400 volts, simply because they use less of the network.

3.10.3   Network Charges

Network charges are levied on all customers to cover the network cost of transporting the energy from the electricity generators. Residential customers will generally have these costs bundled together with the energy cost into a single kWh charge. For commercial and industrial customers network charges can amount to more than half the total electricity expenditure. In the case of smaller businesses, these charges are often not well understood, and neither are the steps that can be taken in order to reduce them.

Network charges generally contain three components:

1) A *daily charge*, sometimes called an *access charge* or a *standing charge*. This component usually represents a minor component of the total network charge.
2) An *energy component*; this is a charge per kWh levied for the customer's for use of the network, and as such it is in direct proportion to the customer's load. Sometimes this component is split into separate transmission and distribution components, but both are aggregated as part of the network charge, and collectively relate to the energy provided to the customer. This energy component can also sometimes depend upon the time of use (ToU), rising during times of peak demand.
3) A *maximum demand component*; this relates to the maximum demand (MD), that a customer imposes on the network. The maximum demand is defined as the maximum power (either P or S) consumed during any metering interval in the billing period. The maximum demand charge relates to the size of the connection assets that must be provided by the local network owner, in order to safely deliver energy to the customer.

3.11 Problems

1 *Power Factor Correction*
A small industrial electricity site has a typical daily consumption profile shown in Figure 3.18. The owner is keen to reduce the local network supply charge by reducing the maximum demand. The site's highest daily demand over the billing interval is shown in Figures 3.18 and 3.19.
A What options are available to the owner?
B Estimate the maximum capacity of the power factor correction equipment required to increase the worst case power factor to 0.95.
(Answer: $\approx 100\,\text{kVAr}$)
C If the owner pays $0.5 per kVA per day in maximum demand charges, and the PFC equipment can be installed for $12,000, what will be the payback period on the purchase of this asset?
(Answer: $\approx 1.5\,\text{years.}$)
D Estimate the current weekday load factor. How would this change following the improvement in power factor?
(Answer: $\approx 50\%$ in each case.)
E In what other ways might the MD charge be further reduced? What additional annual savings might be possible?
(Answer: $\approx \$10,000$ per year)

2 A small industrial three-phase 60 Hz load is supplied from a transformer with an inductive impedance of 0.1 ohms. At full load it consumes 400 kVA with a power factor of 0.75 and sees a phase voltage of 375 V.
A Calculate the open circuit winding voltage assuming that the source impedance on the MV side of the transformer is negligible.
(Answer: $398\,\text{V}_{\text{phase}}$ or $690\,\text{V}_{\text{line}}$)

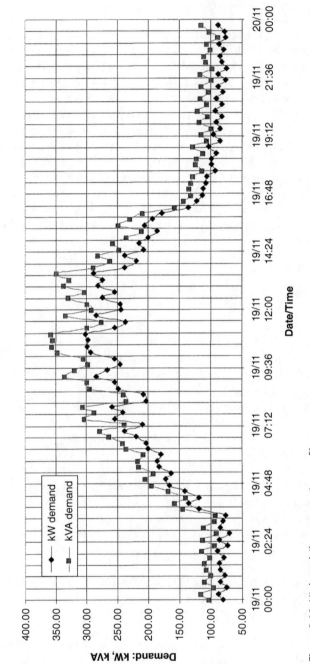

Figure 3.18 Highest daily consumption profile.

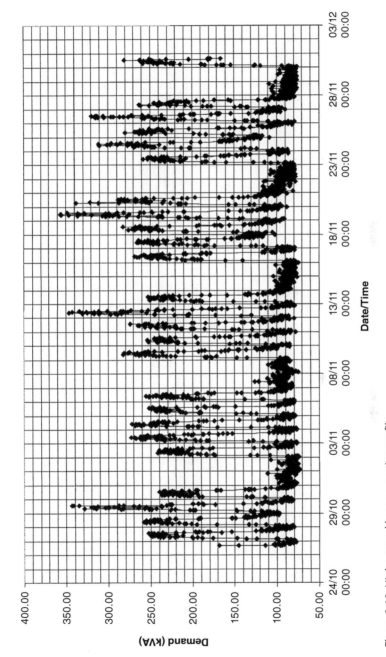

Figure 3.19 Highest weekly consumption profile.

B Find the size of the power factor capacitance required in parallel with each phase to restore the full load power factor to 0.95. Assume that the load impedance is invariant with voltage changes.
(Answer: 1260 μF)

C Calculate the full load voltage seen at the load in the presence of this capacitance.
(Answer: 389 V)

D If at low load the load admittance falls to 20% of its full load value, while maintaining a 0.75 power factor, calculate the phase voltage in the presence of this fixed PFC capacitance.
(Answer: 409 V)

3 An unbalanced three-phase load exhibits the phasor diagram shown in Figure 3.20.
A Calculate the load's effective power factor. Compare your result with that obtained from the average of the three-phase angles.
(Answer: 0.985, 0.802)

B Demonstrate that when the phase currents all have the same magnitude, the effective phase angle for the total load equals the average of the individual phase angles, despite the remaining phase unbalance.

4 Estimate the load factor for the load profiles in Figures 3.14 and 3.15.

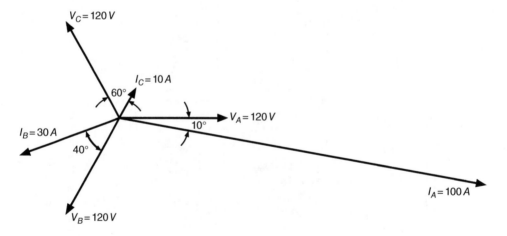

Figure 3.20 Unbalanced three-phase load.

5 Use Figure 3.13 to estimate the size of the PFC unit required in Figure 3.12.
(Answer: Approximately 120 kVAr)

4

Magnetic Circuits, Inductors and Transformers

In this chapter we examine the theory of transformers, since they comprise a major part of the power system and were, to a large extent, the reason that alternating current became the generally accepted medium for the transmission of electrical energy. The ability to easily change from one AC potential to another, makes it possible to transmit large amounts of energy across considerable distances at HV potentials with minimal transmission loss. We begin with an analysis of magnetic circuits, inductors and single-phase transformers and explore the apparent difference in behaviour between *current transformers* and *voltage transformers*. In Chapter 7 we consider the characteristics of three-phase transformers.

In this chapter we will also introduce the concept of the *per-unit system* of expressing electrical quantities. This is a convenient way to perform calculations across trans-former boundaries because per-unit quantities do not change when moving from one potential to another.

We begin our study of transformers with a discussion of magnetic circuits and their similarities to electric circuits.

4.1 Magnetic Circuits

Just as in an electric circuit where the current flows around a closed path, so lines of magnetic flux in a magnetic circuit also follow a closed path. Indeed, there is more than a passing similarity between these two quantities. Electric circuits consist of materials capable of supporting the passage of an electric current, while magnetic circuits contain materials capable of supporting the passage of a magnetic flux. As we shall see, the equations defining the behaviour of electric circuits have magnetic analogies. Consider the magnetic circuit shown in Figure 4.1 where the current i, flowing in the winding has excited a magnetic flux Φ 'flowing' within a rectangular magnetic circuit, which we shall henceforth call a *magnetic core*.

Magnetic cores are generally constructed by stacking thin laminations of transformer steel, each electrically insulated from the next by a minute oxide layer grown upon its surfaces. This technique is used to minimise losses arising from the eddy currents that would otherwise flow within the core. These losses occur because the core is electrically as well as magnetically conductive, and as a result of the time varying magnetic field, currents are induced within the cross-section of the core itself.

AC Circuits and Power Systems in Practice, First Edition. Graeme Vertigan.
© 2018 John Wiley & Sons Ltd. Published 2018 by John Wiley & Sons Ltd.

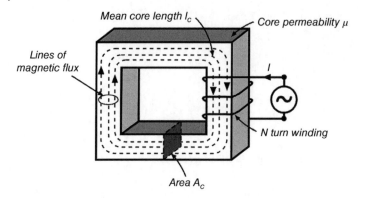

Figure 4.1 Magnetic circuit.

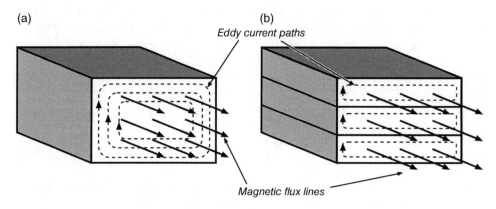

Figure 4.2 (a) Non-laminated core (b) Laminated core.

A laminated core is depicted in Figure 4.2b; here the eddy current paths are shown dotted around the flux vectors. Since substantial eddy currents cannot flow *across* lamination boundaries, they are constrained to flow *within* single laminations, as shown. Each lamination links only a small fraction of the total flux and because the average internal current path length is long, the resulting eddy current losses can be kept acceptably small, provided suitably thin laminations are used. (The lamination thickness shown in Figure 4.2b has been considerably enlarged for reasons of clarity.)

On the other hand Figure 4.2a shows the effect of a *non-laminated* core carrying the same total flux. Here there are many eddy current paths throughout the core material, most of which link a substantial fraction of the flux. As a result, such a core would suffer considerable eddy current losses especially when operating at high flux densities, high frequencies or both.

4.2   Magnetic Circuit Model

In a metallic conductor we relate the current density J to the conductivity σ and the electric field E set up within it, according to:

$$J = I/A = \sigma E$$

A corresponding equation applies in magnetic circuits and relates the *magnetic flux density B* (also called the *magnetic induction* and measured in tesla or webers/m^2), to the *magnetic field strength H*, which is proportional to the *ampere turns* exciting the core. Thus for a *linear* magnetic material we may write:

$$B = \Phi/A_c = \mu H \tag{4.1}$$

where A_c is the cross-sectional area of the core and μ is its *magnetic permeability*. This is a measure of the ease with which the material can be made to support a magnetic flux Φ. Just as the electric field E, driving a current is aligned with the direction of flow in a conductor, so the magnetic field strength H is also aligned with the direction of the *magnetic flux* Φ. Ampere's equation relates the magnetic field strength H, integrated around the magnetic circuit, to the ampere-turns in the winding, according to:

$$\oint H.dl = NI$$

Here N is the number of turns on the winding and I is the current flowing through them. Therefore if we sum the elements of H dropped around a magnetic circuit of uniform cross-section, whose length is l_c, we find that:

$$Hl_c = NI$$

or:

$$H = NI/l_c = MMF/l_c \tag{4.2}$$

where Ni is the magnetomotive force (MMF) that establishes the flux Φ. Therefore the magnetic field strength H, can be expressed as the MMF per unit core length, in a similar way to which the electric field strength, E is expressed in *volts per metre*.

If we combine Equations (4.1) and (4.2) we find that the applied MMF is proportional to the flux Φ, and if the current is a function of time then so will be the resulting flux. Therefore we may write:

$$NI(t) = \frac{l_c}{\mu A_c} \Phi(t) \tag{4.3}$$

The proportionality constant $l_c/\mu A_c$, is called the *reluctance* ($\mathfrak{R}$) of the magnetic circuit. This parameter is analogous to the resistance in an electric circuit. We can therefore rewrite Equation (4.3) in the form:

$$MMF = \Phi\mathfrak{R} \tag{4.4}$$

Equation (4.4) suggests that magnetic circuits can be represented by an electrical analogy, as shown in Figure 4.3 and summarised in Table 4.1.

Consider now the inclusion of an air gap in the magnetic circuit, as depicted in Figure 4.4. Air is much more difficult to magnetise than an iron-based material, and as a result its permeability is much lower. We give the *permeability of free space* (air) the symbol μ_0 ($= 4\pi \times 10^{-7}$), and the permeabilities of magnetic materials are frequently expressed relative to that of air, through the concept of a relative permeability μ_r, where:

$$\mu_r = \frac{\mu}{\mu_0}$$

Table 4.1 Electric–magnetic analogy.

Electrical circuit	Magnetic circuit
Voltage (V)	Magnetomotive force (MMF)
Current, $I\,(=V/R)$	Flux, $\phi\,(=MMF/\mathfrak{R})$
Resistance $R\,(=l/\sigma A)$	Reluctance $\mathfrak{R}\,(=l_c/\mu A_c)$

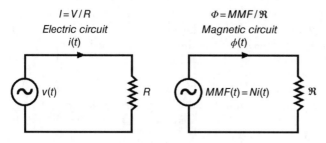

Figure 4.3 Electrical and magnetic circuits.

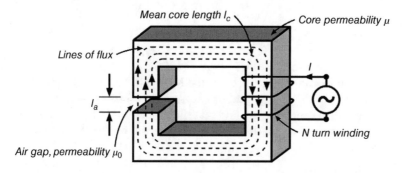

Figure 4.4 A magnetic circuit containing an air gap.

If we analyse this magnetic circuit (much as we would an electric circuit by applying Kirchhoff's voltage law), we can write:

$$MMF = Ni(t) = \Phi(t)\left(\mathfrak{R}_a + \mathfrak{R}_c\right) = \Phi(t)\left(\frac{l_a}{\mu_0 A_a} + \frac{l_c}{\mu_0 \mu_r A_c}\right) \qquad (4.5)$$

where $\mathfrak{R}_a$ is the reluctance of the air gap, $\mathfrak{R}_c$ is the reluctance of the core, l_a is the length of the air gap, l_c is the length of the magnetic core, A_c is the cross-sectional area of the core material, and A_a is the cross-sectional area of the air gap.

The *total reluctance* in the circuit is therefore given by:

$$\mathfrak{R}_t = \mathfrak{R}_a + \mathfrak{R}_c = \frac{l_a}{\mu_0 A_a} + \frac{l_c}{\mu_0 \mu_r A_c}$$

Because the relative permeability of the magnetic material is generally much greater than unity, we find that even for relatively small gap lengths, the reluctance of the air gap can dominate, and therefore most of the applied MMF will often be required to excite the air gap. Extending the electrical analogy, we say that this MMF is *dropped across the air gap*. If this is the case, then $\mathfrak{R}_t \approx \mathfrak{R}_a$ and we can write:

$$NI(t) \approx \Phi(t)\left(\frac{l_a}{\mu_0 A_a}\right)$$

4.3 Gapped Cores and Effective Permeability

By writing Equation (4.5) in terms of the flux density in the core and the applied magnetic field strength, the effective permeability of the core and the air gap can be found.

$$B = \mu_0\left[\frac{\mu_r}{\mu_r l_a / l_c + 1}\right]H$$

Where μ_r is the relative permeability of the core material.

Thus the effective permeability μ_{eff} for the complete magnetic circuit is:

$$\mu_{eff} = \frac{\mu_0 \mu_r}{\mu_r l_a / l_c + 1} \tag{4.5a}$$

and therefore the effective **relative** permeability μ_{reff} becomes:

$$\mu_{reff} = \frac{\mu_r}{\mu_r l_a / l_c + 1}$$

When the length of the air gap is sufficiently large, so that $\mu_r l_a / l_c \gg 1$, the effective relative permeability becomes the ratio between the length of the core and that of the gap, l_c / l_a.

From the foregoing, there is no reason to assume that the area of the gap is any different from that of the core, and therefore we might expect that $A_a = A_c$. However, in practice this is not quite the case. The lines of flux spread out into the adjacent air as they pass through the gap, a phenomenon known as *fringing*, and illustrated in Figure 4.5. As a result, the effective area of the gap is slightly larger that of the core, and as a result the flux density in the gap is slightly smaller than that within the core. Just how much smaller is difficult to accurately predict, since the fringing effects increase with the length of the gap.

As a rough rule of thumb for short gaps, the gap length can be added to *both* the width and depth dimensions of the core, from which an approximate effective cross-sectional area for the gap can be obtained and from this, estimates of the gap flux density and reluctance can be found.

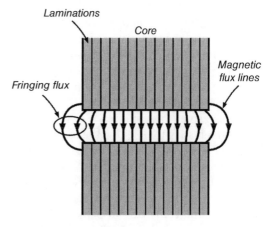

Figure 4.5 Air gap fringing flux.

4.4   Inductance Calculations

We saw in Chapter 2 that the *inductance* of a winding is given by:

$$L = N \frac{d\Phi}{di}$$

From Equation (4.5) the flux waveform $\Phi(t)$ can be expressed as $\Phi(t) = \dfrac{NI(t)}{\mathcal{R}_t}$ and therefore the inductance of the winding is $L = N^2/\mathcal{R}_t$. In the case where a dominant air gap exists in the magnetic circuit, the inductance of the winding can set by adjusting the gap length. This can be expressed in terms of the air gap dimensions A_a and l_a:

$$L \approx N^2 \mu_0 A_a / l_a$$

Fringing effects mean that the gap reluctance is smaller than that predicted from the dimensions of the core, and therefore the inductance value will be larger than expected. This effect becomes more pronounced as the length of the gap increases. As we shall see, the inclusion of one or more air gaps in a magnetic circuit provides the beneficial effect of making the resulting B–H characteristic more linear. Their inclusion also reduces the effective permeability and reduces the peak flux density attained for a given excitation current. Even a short air gap can have a major effect on the reluctance of the entire magnetic circuit, since the relative permeability of the core material can be hundreds or thousands of times larger than that of air.

Non-magnetic (and non-conductive) media are frequently inserted in magnetic circuits to provide air gaps of known thickness, but these materials must be capable of withstanding the compressive forces experienced within the core, without deformation.

One disadvantage of the fringing phenomenon lies in the fact that the flux leaves and re-enters the magnetic core through the faces of the laminations on two of its four sides. Because these lines of flux pass through a relatively large cross-sectional area, additional eddy current losses occur within the material there, especially when the air gap is wide, and relatively large flux densities are required.

So as to minimise this effect, when iron cored inductors (often called *reactors*) are required for power factor correction or harmonic suppression equipment, multiple small air gaps are incorporated into the magnetic circuit, usually within the limbs of the core supporting the windings, where they can be tightly held in compression. This technique requires the fabrication of small packets of magnetic material, generally between 25–50 mm in height, between which the gaps can be inserted, as shown in Figure 4.6b. The use of distributed gaps reduces the effects of fringing, together with the additional eddy current losses it creates, since each one is considerably shorter than an equivalent single gap. The presence of close-fitting windings further suppresses fringing effects, since its magnetising force tends to confine the fringing flux closer to the core.

(It is interesting to note that the reactor core in Figure 4.6a is isolated from ground by resin support insulators. A careful inspection shows that the core is also strapped to the midpoint of the reactor winding. This considerably eases the insulation requirement between the core and windings.)

Reactor windings frequently consist of multiple strands of separately insulated copper or aluminium, connected in parallel. The idea of this is to present a relatively small cross-sectional area in each one (similar to laminating a transformer core), with a view

(a) (b)

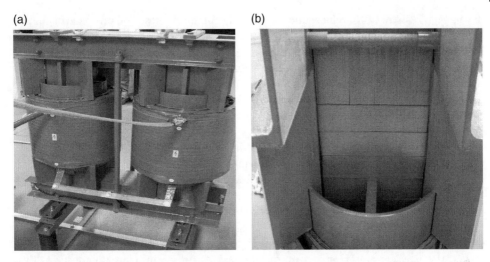

Figure 4.6 (a) 22 kV power factor correction de-tuning reactor (b) Distributed air gaps.
(Note that the reactor core in (a) is insulated from the ground and is connected to the midpoint of the inductor winding. Also, the laminated 'packets' in (b) provide multiple small gaps throughout the core, minimising the effects of fringing).

to minimising any eddy current losses created within the winding itself. This arrangement also assists in reducing the winding's AC resistance by minimising the *skin effect*, whereby high frequency currents tend to flow on the outside of a conductor. By spreading the total current between several small conductors, better use can be made of each one, and a reduction in the overall resistance of the winding can be achieved. This effect is more important in harmonic filtering applications, where a high quality factor (*Q*) is required at the target harmonic.

4.5 Core Materials

Magnetic materials used in the design of reactors and transformers are all iron-based alloys. The most common of these are known as *grain oriented silicon steels* (GOSS), also known as *electrical steels*. It has been discovered that these alloys are easier to magnetise along the direction of rolling rather than perpendicular to it and hence laminations are cut so that the intended flux direction is aligned in the direction of rolling. Cold rolling aligns the grain boundaries and the crystalline structure within them, in the direction of rolling. Cold-rolled silicon steels are also called cold-rolled grain oriented (CRGO), and these are frequently used in the production of power and distribution transformers.

Grain oriented silicon steels are a family of iron alloys containing between 1 and 6% silicon. It is desirable for a transformer steel to be easily magnetised (i.e. to have a high permeability, μ), to exhibit a low *hysteresis loss* (i.e. to have a narrow B–H loop[1]) and to

---

1 The energy required to magnetise the core every cycle, first in one direction and then in the other, is proportional to the area contained within the B–H loop. Hence a narrow loop implies lower hysteresis loss.

have a high resistivity, thereby reducing the eddy current losses. The presence of silicon delivers all these attributes, although higher concentrations make the material hard and brittle. The presence of oxygen, sulphur, nitrogen and particularly carbon in the iron alloy, increases the hysteresis loss, while decreasing the permeability of the material, therefore care must be taken to remove these elements. It is also usual to anneal the material after machining to remove any carbon present.

GOSS steels can be graded according to the hysteresis and eddy current loss per kilogram of the material or by the percentage of silicon in the alloy.

For applications where the flux orientation is not constant, cold rolled *non*-grain oriented (CRNGO or NGOSS) steel is produced. This is used in the production of motors and generators where the flux direction changes due to the rotating nature of these machines. CRNGO materials are cheaper than CRGO materials and are said to be *isotropic*, since they have the same magnetic properties in all directions.

Amorphous steels are more expensive magnetic media that are glass-like, and are produced by pouring the molten alloy onto a chilled spinning plate that cools the metal so fast that the crystalline structure of GOSS does not form. Cores made from this material suffer only about one third the losses of GOSS, but are about twice as expensive to produce. The reduction in magnetic loss makes amorphous steel attractive for the production of large transformers when low loss is called for in the specification.

4.6 Magnetising Characteristics of GOSS

The *B–H* characteristic typical of GOSS is shown in Figure 4.7. This is a plot of the applied magnetic field strength *H*, against the resulting flux density *B*, for a GOSS magnetic circuit having no air gap. The graph shows several B–H loops, each plotted for a different peak flux density. The existence of these loops implies that the flux density traverses a different path on the *BH* plane when the flux density is increasing, from that taken when it is decreasing. This effect is called *hysteresis*, and in order to coerce the core's operating point to move completely round a BH loop, a small amount of energy is absorbed by the core from the circuit exciting it. Therefore in the case of sinusoidal excitation, a component of the exciting current (or the magnetising current) must lie in phase with the applied voltage.

At low flux densities (0.1–0.5 tesla) the magnetising impedance is largely constant and approximately linear, and characterised by an elliptical B–H loop. Thus the magnetising impedance can be modelled by an equivalent magnetising resistance R_m, and a magnetising reactance X_m, parallel connected across the applied potential. As a result, the magnetising current is largely sinusoidal in shape, lagging the applied voltage as shown in Figure 4.8.

As we shall see later in this chapter, current transformers (CTs) are designed to operate at low flux densities like these, where the magnetising impedance is linear and relatively large in magnitude. As a result, current transformers consume very little magnetising current, which helps considerably in reducing the measurement error.

Reactors (or inductors) are also generally designed to operate at relatively low flux densities, well away from the saturation region. This is usually achieved by including one or more air gaps in the magnetic circuit, which provide a near-linear impedance, and as a result an iron-cored reactor can be represented by the equivalent circuit shown

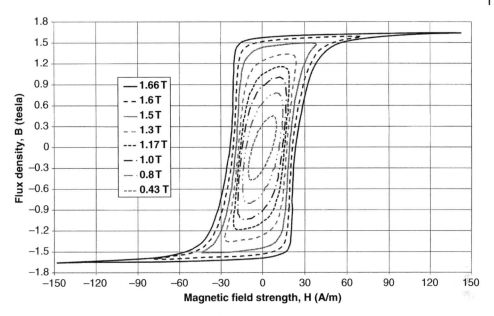

Figure 4.7 B–H characteristics of grain oriented silicon steel as a function of the peak flux density.

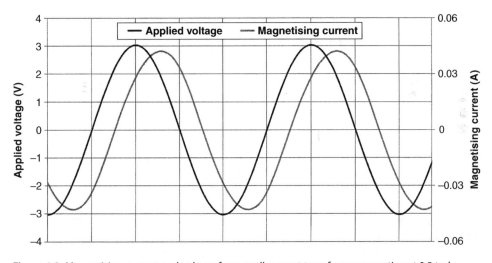

Figure 4.8 Magnetising current and voltage for a small current transformer operating at 0.2 tesla.

in Figure 4.9. Provided the flux density is kept sufficiently small, the magnetising losses can be kept low. Therefore R_m will appear high, and the in-phase component of current drawn can be made almost negligible.

At higher peak flux densities (1.4–1.6 T) the B–H loop tends to become slightly squared and the magnetising current starts to become non-sinusoidal, as shown in Figure 4.10. Power transformers are generally designed to operate at these flux densities (i.e. about 15–20% below saturation). By ensuring that the core is fully utilised magnetically, a larger VA rating can be achieved while keeping the size of the transformer to a

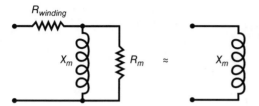

Figure 4.9 Equivalent circuit of an iron-cored reactor.

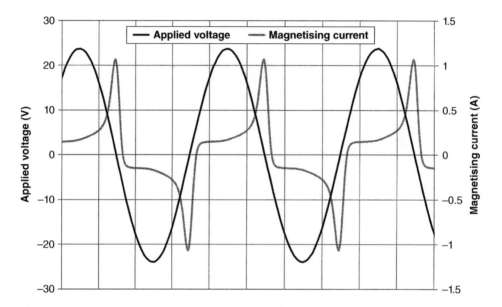

Figure 4.10 Magnetising current and voltage for a small current transformer operating at 1.7 tesla.

minimum. As a result of the higher working flux density the magnetising current is generally non-sinusoidal and therefore contains some harmonic content. However, since this current is generally less than 3–5% of a transformer's rated load current, this effect is not usually significant.

For flux densities around 1.9–2.0 tesla, the B–H loop exhibits a fully square characteristic, where the peak flux density changes little, despite very large increases in H. In this region the core is said to be saturated. The magnetising current becomes many times larger and takes on extremely high values coincident with the peak flux density. As a result, it contains a significant harmonic content.

If we consider a magnetic circuit in which the peak flux density extends well into the saturation region, we can infer from the square B–H loop and the shape of the resulting magnetising current that its permeability μ must take one of two values. The first and relatively large value applies to the vertical sides of the B–H loop, when the core is operating below its saturation flux density. This provides a large magnetising inductance and, as a result, only relatively small magnetising current flows during this time.

The second value occurs when the core is operating in the saturation region, where the B–H loop is flat. Here the permeability is very low, and the incremental magnetising inductance tends towards zero, as evidenced by the very large and peaky magnetising current, similar to that shown in Figure 4.10.

The consequences of operating a magnetic core at or very near to saturation can be severe. Firstly, the hysteresis losses can become very large. Secondly, the magnetising current can take on excessively large peak values, sufficient to cause severe heating within the winding and capable of impressing harmonic distortion onto the local voltage supply. And thirdly, the noise made by the core can become unpleasantly loud, due to the extreme effects of *magnetostriction* which causes the laminations to undergo minute changes in dimension as a consequence of strong magnetisation. This change occurs twice per cycle and therefore transformers tend to 'hum' at twice the system frequency.

4.7 Energy Stored in the Air Gap

The hysteresis loss in a gapped magnetic circuit relates only to the magnetic material. While air gaps may demand the majority of the applied ampere-turns, they do not exhibit the hysteresis characteristics of GOSS, nor do they contribute to the hysteresis loss of the device. This is because the B–H relationship of an air gap can be fully described by the linear equation $B = \mu_0 H$. However, an air gap (or a gap filled with any non-magnetic material) is capable of storing energy in its magnetic field, which is released when the field collapses.

We can illustrate this by calculating the energy delivered to the magnetic circuit as the current in the winding increases from zero, according to the integral:

$$Field\ energy = \int_{0}^{t'} E(t) I(t) dt = \int_{0}^{t'} L \frac{dI(t)}{dt} I(t) dt$$

Changing the variable of integration and letting $I(t') = I'$ yields:

$$Field\ energy = L \int_{0}^{I'} I dI = \frac{1}{2} L I'^2 \tag{4.6}$$

Thus the energy stored in the magnetic field depends only on the *final* value of the current flowing in the winding I' and not on the trajectory taken to achieve it. If we express the energy in terms of the magnetic circuit parameters, we find:

$$Field\ Energy = \frac{1}{2} \frac{N^2}{\mathfrak{R}_t} I'^2 = \frac{1}{2} \frac{N^2 I'^2}{(\mathfrak{R}_a + \mathfrak{R}_c)} \tag{4.7}$$

In a well-designed reactor the reluctance of the air gap is generally much larger than that of the magnetic core; therefore most of the applied MMF is consumed in supporting the flux in the gap, thereby forcing the resulting B–H loop to become considerably more linear. Thus we can write:

$$Field\ energy \approx \frac{1}{2} \frac{N^2}{\mathfrak{R}_a} I'^2 = \frac{1}{2} \left[\frac{(NI')}{l_a} \right]^2 \mu_0 l_a A_a = \frac{1}{2} H_a{}^2 \mu_0 l_a A_a$$

Since the air gap the flux density B_a, is related to the magnetic field strength H_a according to $B_a = \mu_0 H_a$ we may write:

$$Field\ energy \approx \frac{1}{2} B_a H_a l_a A_a = \frac{1}{2} \frac{B_a{}^2}{\mu_0} l_a A_a \tag{4.8}$$

Equation (4.8) suggests that an air gap is a necessary requirement for efficient energy storage in a magnetic field. Alternatively the energy density in the gap can be expressed as:

$$Energy\ density = \frac{Field\ energy}{Gap\ volume} \approx \frac{1}{2}\frac{B_a^2}{\mu_0}\ joules/m^3 \tag{4.9}$$

Equation (4.8) is also useful in calculating the *magnetic force* between the pole pieces adjacent to an air gap. Imagine trying to pull apart the halves of a C core (like that in Figure 4.11), in which a flux density B exists. The incremental work required to increase the length of the air gap by δx is equal to the force between the pole pieces F, multiplied by the total displacement δx. This is also equal to the energy *gained* by the magnetic circuit and stored in the gap. Therefore we can write:

$$F\delta x = \frac{1}{2}\frac{B^2}{\mu_0}A_a\delta x$$

Thus the force required is:

$$F = \frac{1}{2}\frac{B^2}{\mu_0}A_a\ newtons \tag{4.10}$$

This analysis assumes that the increase in gap length δx is sufficiently small so as not to appreciably change the flux density in the core. For larger displacements, the flux density will fall rapidly as the gap length grows, and with it the force between the pole pieces. So if a small gap can initially be created, progressively less force will be required to enlarge it. Equation (4.10) can also be used to calculate the force created by an electromagnet in supporting a ferrous object, in which case the area A_a represents the contact area between the pole pieces and the object in question.

4.8 EMF Equation

In view of the dangers of operating a magnetic circuit too close to saturation, it is useful to derive a simple equation that relates applied voltage, the core dimensions and the number of turns on the winding, to the resulting *peak flux density*. This is called the *EMF equation*, and it can be derived from Faraday's law of induction, assuming sinusoidal excitation.

We know that the voltage $E(t)$ induced within a winding of N turns, is given by:

$$E(t) = N\frac{d\Phi(t)}{dt} \tag{4.11}$$

If we assume that the applied voltage is sinusoidal, then from Equation (4.11) so will be the flux in the core. Accordingly we can express the flux waveform as:

$$\Phi(t) = \hat{\Phi}\sin(\omega t)$$

Therefore the voltage induced within the winding is:

$$E(t) = N\hat{\Phi}\omega\cos(\omega t) = N\hat{B}A2\pi f\cos(\omega t) \tag{4.12}$$

Where $\hat{B}$ is the peak flux density and A is the cross-sectional area of the core. Therefore the RMS voltage can be expressed as:

$$E_{RMS} = \frac{2\pi}{\sqrt{2}} N\hat{B}Af = 4.44 N\hat{B}Af \tag{4.13}$$

Equation (4.13) shows that the peak flux density in the core is proportional to the *volts per turn*. Once the operating frequency, the core area and the number of turns have been fixed, the peak flux density is solely determined by the voltage applied to the winding. Since an excessively high applied voltage will result in magnetic saturation, reactors and transformers are designed to operate at or below a defined maximum voltage rating.

4.9 Magnetic Circuit Topologies

There are many ways to construct magnetic circuits suitable for transformers. With the exception of *toroidal* construction, each must enable the core to be assembled through the coils, while minimising the effects of any residual air gaps that may arise at each joint. The simple *butt joint* technology of the type shown in Figure 4.11 is a relatively crude construction technique, since there is always some residual air gap as the pole pieces can never mate perfectly. This decreases the magnetising inductance and increases the magnetising current, and while this topology is used for the construction of small reactors, it is rarely used in transformer construction.

There are other topologies that do a much better job of avoiding the adverse effects of air gaps. The *toroidal wound* cores shown in Figure 4.12, for example, are frequently used in the construction of current transformers and in the production of small single-phase power transformers. These consist of a single strip of magnetic material close wound to form a core. Lines of flux follow a circular path and hence must eventually pass from one lamination to the next, but since these share a large common area, the reluctance of the gap between them is very small. This leads to a particularly low reactive core loss.

Figure 4.11 'C' core and three-limb 'E' core structures containing butt joints.

Figure 4.12 Small toroidal power transformer (left) and toroidal 200:5 amp current transformer (right).

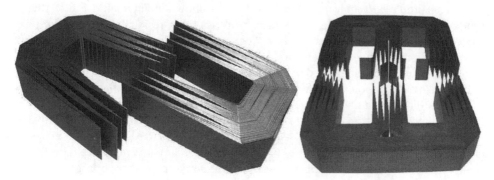

Figure 4.13 Distributed gap unicores. (Photos courtesy AEM Cores, South Australia).

Where single- or three-phase cores with minimal residual air gaps are required, the laminations can be interleaved, as shown in Figure 4.13. Here the length and positions of the folds and cuts of the laminations must be very accurate, in order for the core to fit together as required, a manufacturing process developed by a South Australian company, AEM Cores Pty Ltd. By interleaving the legs of the core in this way, the cross-sectional area of any residual gap can be made much larger than that of the core itself, substantially reducing the gap reluctance. Further, by providing the multiple 'fingers' shown in Figure 4.13, the fraction of the total flux passing across any each residual gap is small. In this way, the effects of residual air gaps can almost be eliminated. These *distributed gap* wound cores (known as *unicores*) are used for the construction of voltage transformers and distribution transformers up to about 2.5 MVA.

Finally, for larger three-phase power transformers designers interleave the horizontal yoke laminations with those of each vertical limb to form a stacked core in order to minimise the effects of residual air gaps. Figure 4.14 shows part of the core of a distribution transformer. Each limb is approximately round in cross-section, in order to efficiently fill the space within the winding and minimise the required turn length.

Figure 4.14 Section of a 500 kVA stacked core three-phase transformer core (left) Enlarged view (photographed through oil) showing the interleaving of vertical and horizontal laminations (right).

Where the vertical limbs meet the upper and lower yokes, the laminations have been interleaved as shown, a technique known as *step lapping*. While this reduces residual air gaps to an acceptable degree, the flux in each joint must turn through 90°, and therefore it cannot remain entirely aligned with the rolling direction of the laminations. As a result, the losses are higher in a step lapped joint than those in a distributed gap joint.

Wound cores are predominantly used in the construction of small single- and three-phase distribution transformers in North and South America, Australia, New Zealand, Ireland, Greece, India, Japan, South Africa and Taiwan, whereas stacked cores are dominant throughout Europe, China and Thailand. Stacked cores, however, are almost universally used for distribution transformers with ratings in excess of 5 MVA.

4.10 Magnetising Losses

The hysteresis energy consumed in each cycle is given by the time integral of the applied voltage multiplied by the magnetising current, thus:

$$Hysteresis\,Energy\,per\,Cycle = \int_0^T E(t) I_{mag}(t)\,dt = \int_0^T NA\frac{dB}{dt} H(t)\frac{l}{N}\,dt$$

so: $$Hysteresis\,Energy\,per\,Cycle = Al\int_{\overset{\smash{.}}{H}}^{\overset{\smash{.}}{H}} H(t)\,dB$$

where A is the effective cross-sectional area of the core and l is its effective length.

Therefore the hysteresis energy per unit volume of core material is equal to the area contained within the B–H loop. Therefore the power consumed as the core is cycled around the B–H loop is equal to this quantity multiplied by the operating frequency. As the flux density increases, the area contained within the B–H loop increases substantially and therefore so does the hysteresis loss. This effect can be seen in Figure 4.7. The equation above does not include the effects of eddy current losses occurring within the core's laminations. Therefore the total magnetising loss (often called the *iron loss*) will be higher than the B–H loop alone suggests.

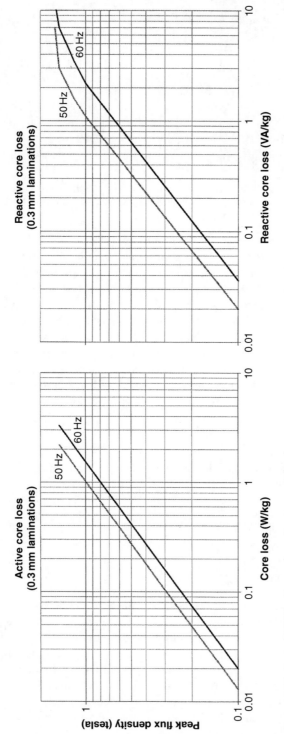

Figure 4.15 Typical GOSS active and reactive magnetising loss.

The makers of transformer steels publish graphs of both the active and reactive loss per kilogram of material for a given lamination thickness and operating frequency, as a function of the peak flux density in the core. Two typical graphs appear in Figure 4.15, and these can be used to estimate the magnetising losses in a transformer when the above parameters are known. The active and reactive losses are roughly proportional to the square of the flux density; Figure 4.15 shows that they increase by about two orders of magnitude for a tenfold increase in flux density.

The production of this data is carefully managed to ensure that it represents the *minimum* loss achievable. This is done by arranging the test core is so that there are no air gaps and the flux is always oriented parallel to the rolling direction.

In practice, real transformer cores are seldom so perfect. They contain tiny air gaps between laminations, especially in the interleaved joints between limbs, where the flux must change direction and cannot remain aligned with the rolling direction. The losses in these locations increase, and so the magnetising loss in a real transformer is higher than the maker's data would suggest, usually by about 20–40%.

4.11 Two-Winding Transformer Operation

The construction of a simple two-winding transformer is illustrated in Figure 4.16. The windings are wound on an un-gapped laminated core, the purpose of which is to provide tight magnetic coupling between the two. Winding 1 is the primary and is provided with energy from a voltage source V_1 from which the magnetising current flows. The secondary winding delivers energy to the load, usually at a different potential.

Figure 4.16 shows the transformer in its unloaded state, where the secondary current is zero. As a result the only current flowing in the primary is the small magnetising current. This is shown on the vector diagram of Figure 4.17. The magnetising current can be seen to lag the primary voltage. It flows within the primary winding and induces a magnetic flux ϕ_m within the core, capable of supporting the applied primary voltage V_1. Since this voltage is sinusoidal, then according to Equation (4.12) so will be the flux, which is also plotted in Figure 4.17, and this lags the applied voltage by 90°.

As discussed earlier, the magnetising current can be resolved into resistive and inductive in-phase and quadrature components, as shown in Figure 4.17. The in-phase component is associated with the energy required to magnetise the transformer, which is dissipated as heat within the core. The quadrature component of the magnetising current generates the magnetising flux ϕ_m and therefore lies in phase with this flux.

When a transformer is excited from a fixed voltage source, the applied potential must be induced within the primary winding. This is achieved through the action of the magnetising current in establishing a flux in the core, according to Equation (4.14). We assume that the core is infinitely permeable, and therefore no flux exists in the air surrounding either

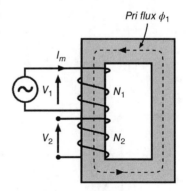

Figure 4.16 An unloaded transformer where only magnetising current flows in the primary winding.

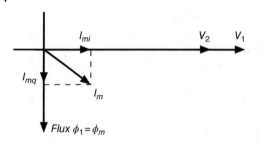

Figure 4.17 Magnetising phasor diagram.

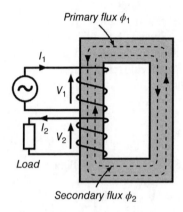

Figure 4.18 Constant flux transformer operation.

winding. Since the secondary winding essentially sees the same flux as does the primary, it has a proportional in-phase voltage induced within it, as shown in Figure 4.16.

$$V_1 = 4.44 \, N_1 \hat{B} A f \tag{4.14}$$

Equation (4.14) shows that the flux density required depends solely upon the applied voltage V_1. This is the basis of the constant flux model of transformer operation, illustrated in Figure 4.18.

The transformer's operation changes slightly when a load is applied to the secondary winding. Since the primary and secondary are tightly coupled, the secondary sees the same flux change as the primary, therefore voltage induced within it lies in phase with that in the primary, and is given by:

$$V_2 = V_1 N_2 / N_1 \tag{4.15}$$

As a consequence of the load impedance, a current I_2 flows from the secondary winding, as depicted in Figure 4.19. The load is assumed to be partly inductive, and therefore the secondary current lags the secondary voltage, in this case by an angle θ. This current also generates a secondary flux within the core, which partially opposes the magnetising flux ϕ_m.

Since the total flux required in the core must equal ϕ_m, in order to support the applied primary voltage, then in response to the secondary current, a primary balance current I_{PB} must flow. This current lies in phase with I_2 and generates a primary balance flux ϕ_{PB} which exactly cancels the secondary flux, leaving the required magnetising flux ϕ_m within the core. The phasor diagram in Figure 4.19 shows this effect. When the primary flux

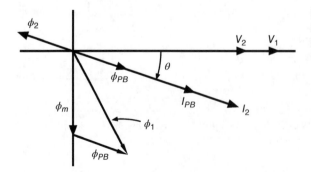

Figure 4.19 Flux balance phasor diagram. (Note: the magnetising current has been omitted for reasons of clarity).

Figure 4.20 Primary phasor diagram (loaded).

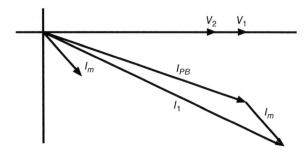

$\phi_1 (= \phi_m + \phi_{PB})$ is added to the secondary flux ϕ_2, the required magnetising flux ϕ_m results. Therefore when a secondary current flows out of the transformer, a primary balance current must flow into it in response.

Because the magnitude of each of these flux components is in proportion to the ampere-turns of the associated winding, we can write:

$$I_{PB}N_1 = I_2 N_2 \tag{4.16}$$

or

$$I_2 = I_{PB}N_1 / N_2 \tag{4.16a}$$

Figure 4.20 shows the phasor diagram for the primary quantities of a loaded transformer. The total primary current I_1 is given by the vector summation of I_{PB} and the magnetising current I_m.

The magnetising current is generally very small when compared to the primary balance current, and has been enlarged for reasons of clarity in this illustration. In most power transformers it is less than 3–5% of the rated primary current.

4.12 Transformer VA Ratings and Efficiency

We saw from Equation 4.13 that the magnetic circuit and the number of turns on the primary winding must be dimensioned with the voltage rating of the device in mind. We might also ask how much power is a given transformer capable of delivering? The answer depends on how much effort we are prepared to make to keep it cool, in order to preserve the integrity of the winding insulation. This may either be enamel insulation in the case of low voltage and current transformers, or kraft paper in the case of power transformers, tightly wrapped around the winding conductors to insulate one from another. These are then immersed in mineral oil to provide a further insulating medium and to enable the heat produced to be transferred to the atmosphere.

All insulation materials have a maximum recommended operating temperature beyond which they will eventually break down. Excessive heat increases the rate of insulation degradation which, given sufficient time, will ultimately lead to a transformer's failure. The efficiency of the cooling system therefore has a major impact on a transformer's rating as well as on its operating lifetime.

The sources of heating in large transformers include hysteresis and eddy current magnetising losses (taking into account any parasitic losses in the tank and clamping structures), the resistive ($I^2 r$) copper loss associated with current flowing within the windings, in addition to losses from eddy currents flowing within the windings themselves. Of these, it is the I^2r copper loss over which the user has most control, through the load connected to the transformer, since the primary voltage will be largely constant and thus the magnetising losses will be as well. The manufacturer therefore applies a maximum current rating to the transformer to ensure that the windings are not permitted to overheat during normal operation.

Once both the voltage and current ratings have been determined, then the VA rating of the device has effectively been determined as well. From Equations (4.15) and (4.16) we can shown that the volt-amps absorbed by the load are equal to those supplied via the primary, thus:

$$V_1 I_1 \approx V_1 I_{PB} = V_2 I_2 \tag{4.17}$$

If this equation is written in terms of the rated voltage and current then it describes the VA rating for a single-phase transformer. In the case of a three-phase transformer, the VA rating can be expressed as:

$$VA \; rating = 3V_{ph} I_{ph} = \sqrt{3} V_{line} I_{ph}$$

It is worth pointing out that a transformer's rating is always expressed in VA and never in watts. This is because both the magnetising and the copper losses are independent of the power factor of the connected load. Therefore the total loss within a transformer is proportional to the apparent power and not to the active power delivered to the load (which of course depends on the load power factor).

On the other hand, the efficiency of a transformer does depend on the power factor of the connected load; it may be defined as:

$$\eta = \frac{P_{out}}{P_{in}} = \frac{V_2 I_2 \cos(\phi_2)}{V_2 I_2 \cos(\phi_2) + Cu \; Loss + Iron \; Loss}$$

We will show in a tutorial question that a transformer's efficiency can be maximised if its copper and iron losses are equal; however, due to the variable nature of most loads, this condition is rarely met. A more useful concept is the all-day efficiency, defined by the ratio of the energy delivered by a transformer to the energy supplied to it, over a 24 hour period:

$$All\text{-}Day \; Efficiency = \frac{Total \; kWh \; Output}{Total \; kWh \; Input} \bigg|_{24 \, Hours}$$

Depending on the profile of the connected load, the magnetising energy may represent a disproportionately large percentage of a transformer's loss, particularly during times of low load. This is because most transformers must remain energised and therefore the magnetising power is consumed all day. For this reason, the magnetising losses are usually kept relatively low.

4.13 Two-Winding Transformer Equivalent Circuit

The transformer equivalent circuit provides a model for the behaviour of a real transformer, including the iron and copper losses as discussed above. It also includes the effects of leakage flux, in which a component of the flux from one winding is set up

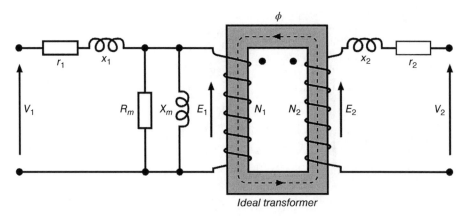

Figure 4.21 Two-winding transformer equivalent circuit.

outside the core, without linking it the other. As we shall see, this gives rise to a series leakage inductance associated with each winding.

The two-winding transformer equivalent circuit appears in Figure 4.21, in which the primary and secondary windings are wound on an ideal core. Further, these windings are assumed to have negligible resistance, and thus the transformer shown in the figure can be assumed to be lossless.

A dot convention is sometimes used to indicate winding polarities. The dot at the upper end of each winding in Figure 4.12 indicates that these points are of the same polarity – they are in phase. Alternatively, this can be shown using arrows, the arrow-head indicating the end of the winding that is instantaneously positive.

The equivalent circuit also contains passive series and shunt elements that collectively represent the losses associated with a real transformer. Of these, R_m and X_m represent the magnetising impedance of the transformer, and when the secondary winding is open-circuited the magnetising current is the only current flowing. Since r_1 and x_1 are small, in this situation the induced voltage E_1, is very close to the applied voltage V_1.

The series elements r_1, x_1 and r_2, x_2 determine transformer's short-circuit impedance. Resistive elements r_1 and r_2 represent the primary and secondary winding resistances respectively, including any skin effects present at the frequency of operation.

The inductive elements x_1 and x_2 represent the leakage inductance associated with each winding. These arise from the fact that in practice not all the flux produced by the primary winding links the secondary. Similarly, not all the flux produced by the secondary current links the primary winding. Although we have thus far assumed that the magnetic flux is confined to the highly permeable core, in practice this is not quite the case. A small proportion of the flux produced by each winding fails to link the other, as shown in Figure 4.22 where not all the flux produced by the primary winding links the secondary.

Because each winding sees a leakage flux component proportional to its own current that is *not* compensated for by the other, a small potential is dropped in each, in proportion to the current flowing. This effect can be modelled by including a small leakage inductance in series with each winding. As a result, the secondary terminal voltage falls slightly when the transformer supplies a resistive or inductive load.

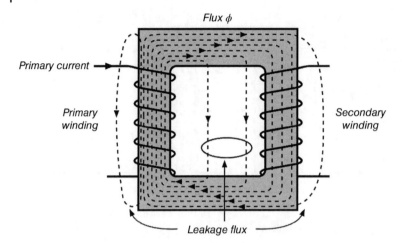

Flux ϕ

Primary current →

Primary winding

Secondary winding

Leakage flux

Figure 4.22 Leakage flux paths permit a portion of the flux created by one winding to avoid linking the other.

Collectively the series resistances and reactances comprise the transformer's short-circuit impedance, which limits the current flowing should the secondary experience a short circuit. This impedance is generally quite small, and short-circuit currents between 10 and 25 times rated the transformer's rated current are likely and therefore unintentional short-circuits should always be avoided. (Under such conditions the electromagnetic force on turns within each winding becomes very large indeed, and in order to prevent mechanical damage, the windings must be clamped tightly in place.)

By dividing Equation (4.15) by Equation (4.16a) a relationship between the impedance connected across the secondary winding and that seen looking into the primary can be obtained:

$$\frac{V_1}{I_{pb}}\frac{1}{N_1^{\,2}} = \frac{V_2}{I_2}\frac{1}{N_2^{\,2}}$$

Since $\dfrac{V_1}{I_{pb}} = Z_1$ and $\dfrac{V_2}{I_2} = Z_2$, where Z_1 and Z_2 are the impedances seen in the primary and secondary windings respectively we can therefore write:

$$Z_1 = Z_2 \frac{N_1^{\,2}}{N_2^{\,2}} = Z_2 n^2 \tag{4.18}$$

where $n = \dfrac{N_1}{N_2}$.

Thus the impedance connected across the secondary winding can be reflected into the primary, scaled according to the square of the turns ratio. Equation (4.18) is also useful to refer the secondary short-circuit impedance elements into the primary winding (or vice versa):

$$Z_{sc\,pri} = \left(r_1 + r_2 n^2\right) + j\left(x_1 + x_2 n^2\right) \tag{4.19}$$

Equation (4.19) shows the short-circuit impedance as referred to the primary winding. Z_{sc} is often simply called the transformer's impedance and appears on its nameplate, expressed as a per-unit[2] quantity, usually in the form of a percentage.

4.14   The Per-Unit System

The per-unit system of expressing electrical quantities was devised in order to simplify manual calculations and while this need has considerably diminished in recent years, it also offers additional benefits that have seen its use continue. Chief among these is the fact that under the per-unit system transformers become transparent. This means that there is no need to scale voltages, currents or impedances when moving from one side of a transformer to the other in the per-unit system, as is the case when using absolute quantities.

The per-unit system also offers the distinct advantage in that while absolute voltages, currents and impedances may vary considerably in magnitude throughout a power system, when expressed in terms of per-unit quantities, this variation is substantially reduced. Therefore with a little experience, typical per-unit values can be used in preliminary calculations in the absence of more specific data.

The basis of the per-unit system lies in expressing electrical quantities as a fraction of a base value, chosen according to rating of the device in question. For example, transformer manufacturers use the rated voltage and the VA of a transformer as its base values, and express its impedance as a per-unit quantity, by dividing it by the transformer's base impedance. A numerical example will help to clarify this idea.

Consider a 100 kVA, 6.35 kV:240 V single-phase transformer. It is customary to assign the rated VA as the base value S_b and the winding voltage as the base voltage. The base current referred to the primary is simply the rated primary current: $I_b = S_b/V_b = 15.75$ A. The base impedance is that which, when connected to the secondary, will generate the rated (base) current. In this case, the base impedance, as referred to the secondary, is given by:

$$Z_b = V_b^2/S_b = 240^2/100{,}000 = 0.576\,\Omega \tag{4.20}$$

And referred to the *primary* winding, the base impedance becomes:

$$Z_b = V_b^2/S_b = 6350^2/100{,}000 = 403\,\Omega.$$

Therefore an impedance of 0.576 Ω connected to the secondary will generate 417 A (rated secondary current). This impedance as seen by the primary winding appears as 403 Ω, where 15.75 A flows.

In the case of a three-phase transformer the base impedance is still given by Equation (4.20), but in this case the line voltage must be chosen for V_b. Consider a three-phase star-star 11 kV:415 V, 300 kVA transformer, with each phase similar to the single-phase example given above. Thus the phase voltages of this transformer are 6.35 kV:240 V and each phase delivers 100 kVA.

---

2   Numerically, the per-unit impedance corresponds to that fraction of a transformer's rated voltage which, when applied to a short-circuited transformer, will generate the rated primary current.

Taking the VA base S_b as 300 kVA, the base current (rated primary current) is given by:

$$I_b = S_b / \sqrt{3} V_{line} = 300 \text{kVA} / \sqrt{3} 11 \text{kV} = 15.75 \text{A (as in the example above)}$$

The base impedance is that which will draw rated current, thus:

$$Z_b = V_b^2 / S_b = 11 \text{kV}^2 / 300 \text{kVA} = 403 \Omega \text{ as seen by the primary.}$$

and $\quad Z_b = V_b^2 / S_b = 415 \text{V}^2 / 300 \text{kVA} = 0.576 \Omega$ as seen by the secondary.

These results are the same as those for the single-phase case. So the same equation for base impedance Z_b applies to the three-phase transformer as for the single-phase, with the proviso that the line voltage is taken as V_b, and the three-phase transformer rating is chosen for S_b.

4.14.1 Transformer Short-Circuit Impedance

Consider the primary short-circuit impedance for each of these transformers. Suppose that $r_{sc} = (r_1 + r_2 n^2) = 6.04 \Omega$ and that $x_{sc} = (x_1 + x_2 n^2) = 36.27 \Omega$, thus the short-circuit impedance can be written as $z_{sc} = 6.04 + j36.27 \Omega$ or $|z_{sc}| = \sqrt{r^2 + x^2} = 36.77 \Omega$. When z_{sc} is expressed as a per-unit quantity, by normalising it with respect to the base impedance, it becomes $36.77/403 = 0.091 \text{pu}$ or 9.1%. It is this figure that is included on a transformer's nameplate. Therefore, in this case, 9.1% of the rated primary voltage, when applied to a short-circuited transformer will generate rated current in both windings. Note that because $X/R \approx 6$ the short-circuit impedance depends most heavily on the transformer's short-circuit reactance, x. This is generally the case for most large transformers.

If the short-circuit impedance is measured from the secondary winding, we find that $z_{sc} = 0.0086 + j0.0156 \Omega$, or $|z_{sc}| = \sqrt{r_{sc}^2 + x_{sc}^2} = 0.052 \Omega$. (These figures can be obtained from those above, using Equation (4.19)). When expressed in per-unit, this impedance becomes $0.052/0.574 = 0.091$ or 9.1%. Thus the per-unit impedance is independent of the base voltage at which it is calculated. This property is one of the principal reasons why per-unit quantities are used. Table 4.2 summarises these results:

The essential equations required when evaluating per-unit quantities for a transformer are summarised in Table 4.3. In future, we will work in terms of per-unit quantities, wherever convenient.

4.15 Transformer Short-Circuit and Open-Circuit Tests

In order to measure the magnetising and short-circuit impedances, open-circuit and short-circuit tests are used respectively. An open-circuit test is used to determine the magnetising elements, R_m and X_m. In this test, the transformer is excited to rated voltage with the secondary winding open circuited, so the applied voltage appears across R_m and X_m (the series elements r_1, x_1 can be ignored in view of the small current flowing). By measuring the applied voltage, the current flowing and the power consumed, R_m and X_m can be evaluated.

The short-circuit test on the other hand is used to evaluate the short-circuit impedance, $z_{sc} = (r_1 + r_2 n^2) + j(x_1 + x_2 n^2)$. In this test the secondary winding is short-circuited and sufficient primary voltage is applied to generate the rated secondary current.

Table 4.2 Single-phase and three-phase transformer per-unit comparisons.

Primary quantities	Single-phase transformer (100 kVA rated)	Three-phase transformer (300 kVA rated)
VA base (S_b)	100 kVA = 1 pu	300 kVA = 1 pu
Base voltage (V_b)	6.35 kV = 1 pu	11 kV = 1 pu
Base current (I_b)	15.75 A = 1 pu	15.75 A = 1 pu
Base impedance (Z_b)	403 Ω	403Ω
Primary short-circuit impedance Z_{sc}	$6.04 + j36.27\,\Omega$, $|Z| = 36.77\,\Omega$	$6.04 + j36.27\Omega$, $|Z| = 36.77\,\Omega$
Primary per-unit impedance	$1.5 + j9.0\%$, $|Z| = 9.1\%$	$1.5 + j9.0\%$, $|Z| = 9.1\%$
Secondary quantities		
VA base (S_b)	100 kVA = 1 pu	300 kVA = 1 pu
Base voltage (V_b)	240 V = 1 pu	415 V = 1 pu
Base current (I_b)	417 A = 1 pu	417 A = 1 pu
Base impedance (Z_b)	0.576 Ω	0.576 Ω
Secondary Z_{sc}	$8.6 + j51.6\,\text{m}\Omega$, $|Z| = 52\,\text{m}\Omega$	$8.6 + j51.6\,\text{m}\Omega$, $|Z| = 52\,\text{m}\Omega$
Secondary per-unit impedance	$1.5 + j9.0\%$, $|Z| = 9.1\%$	$1.5 + j9.0\%$ ($|Z| = 9.1\%$)

Table 4.3 Per-unit parameter evaluation.

Per-unit parameter	Single-phase case	Three-phase case
VA base (S_b)	Transformer VA rating $S_b = V_b I_b$	Transformer VA rating $S_b = \sqrt{3} V_b I_b$
Base voltage (V_b)	Rated **phase** voltage	Rated **line** voltage
Base current (I_b)	Rated current $I_b = S_b / V_b$	Rated current $I_b = S_b / \sqrt{3} V_b$
Base impedance (Z_b)	$Z_b = V_b^2 / S_b$	$Z_b = V_b^2 / S_b$

Generally less than 0.1 pu voltage will be required for this test, and care must be taken not to exceed the current rating of the transformer.

Because the voltages induced in the windings are much smaller than normal, the associated flux density will be small as well. Since the magnetising losses tend to vary roughly as the square of the flux density, the losses encountered during the short-circuit test are very much smaller than those during normal excitation. As a result, the power supplied during the short-circuit test can be reasonably attributed to the resistive component of the short-circuit impedance.

There is one other source of loss, however. Since the winding currents are at or near to their rated values, the *leakage fluxes* will be at their normal levels as well. As a result, there will be associated eddy current losses in those metallic parts of the transformer adjacent to the windings (excluding the core, where the flux density is low). These losses must be supplied in addition to the resistive losses described above, and since it is difficult to

Table 4.4 Short-circuit and open-circuit test results.

	Short-circuit test	Open-circuit test
Applied primary voltage	577 V	6.4 kV
Primary current	15.7 A	207 mA
Power supplied	1490 W	1100 W

determine their magnitude, the short-circuit test can only provide a close approximation of the short-circuit impedance. For most practical purposes this is sufficient.

An example will clarify the calculations required. Our 6.35 kV:240 V, 100 kVA transformer returned the following data from an open and a short-circuit test, as shown in Table 4.4.

Beginning with the short-circuit test results, the power supplied is dissipated in the series resistance $(r_1 + n^2 r_2)$ referred to the primary.

Therefore $\quad 1490 = 15.7^2 (r_1 + n^2 r_2)$

hence $\quad (r_1 + n^2 r_2) = 6.04 \Omega,$

or in per unit terms: $6.04/403 = 0.015 \,\text{pu}.$

The VArs supplied are consumed by the short-circuit reactance $(x_1 + x_2 n^2)$.

And since $\quad Q = \sqrt{S^2 - P^2}$

then $\quad S = 577 \times 15.7 = 9.06 \,\text{kVA}$

and $\quad Q = 8.9 \,\text{kVAr}$

thus $\quad (x_1 + x_2 n^2) = 36.3 \Omega \ \text{ or } \ 0.090 \,\text{pu}.$

The open circuit test results yield the magnetising elements, R_m and X_m, as follows:

$$R_m = V_p^2/P = 6.4 \,\text{kV}^2/1100 \,\text{W} = 37.3 \,\text{k}\Omega \ \text{or } 92.4 \,\text{pu}.$$

Since $\quad Q = \sqrt{S^2 - P^2} = \sqrt{1325^2 - 1100^2} = 738 \,\text{VAr}$

then $\quad X_m = V_p^2/Q = 6400^2/738 = 55.5 \,\text{k}\Omega \ (\text{or } 137.7 \,\text{pu}).$

These tests yield impedance data reflected into the primary. The resistive elements may be apportioned between the windings, since each winding resistance can easily be measured; however, the reactive elements are not so easy to allocate. The series reactance associated with each winding is equal to the square of the turns divided by the associated leakage reluctance. Because these reluctances are difficult to identify, apportionment of the reactive impedance between the primary and secondary is also difficult. Some choose to apportion the reactance equally between the primary and secondary, but on a two-winding transformer, the short-circuit impedance is generally always referred to one winding.

4.16 Transformer Phasor Diagram

The schematic diagram in Figure 4.23 shows the equivalent circuit of a single-phase transformer where all the secondary quantities have been reflected to the primary side. Here it has been assumed that $N_1 = N_2$ so voltages, currents and impedances have the same magnitude on the primary side as on the secondary. This assumption

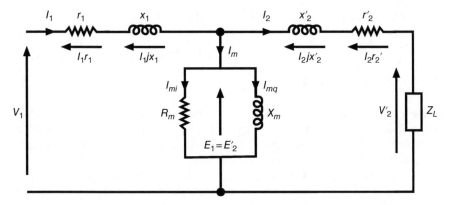

Figure 4.23 Transformer equivalent circuit as seen from the primary side. (Component values are usually expressed in terms of per-unit quantities).

simplifies the phasor diagram without loss of generality.

The phasor diagram in Figure 4.24 relates to the equivalent circuit in Figure 4.23. Here the magnetising flux ϕ_m has been chosen as the reference phasor. The winding voltages E_1 and E_2' lead this quantity by 90° as shown. A secondary current I_2' has been chosen, which lags the secondary terminal voltage by an arbitrary angle of $\theta_2°$. This current generates a voltage drop across the secondary series impedance, which must be subtracted from the E_2' phasor in order to arrive at the secondary terminal voltage V_2'. The resistive voltage drop lies in phase with the secondary current, while the inductive drop leads it by 90°.

The total primary current I_1, equals the primary balance current plus the magnetising current I_m, as shown in the phasor diagram. This current creates the voltage drop across the primary series impedance, which in turn must be added to the winding voltage E_1, in order to determine the applied voltage V_1. Therefore from the point of view of the primary, the total current lags the applied voltage by an angle greater than θ_2, and thus the power factor as seen at the primary is slightly worse than that seen in the secondary. In addition to the secondary and primary currents, the fluxes ϕ_2 and ϕ_{pb} are also shown. The magnetising current in this

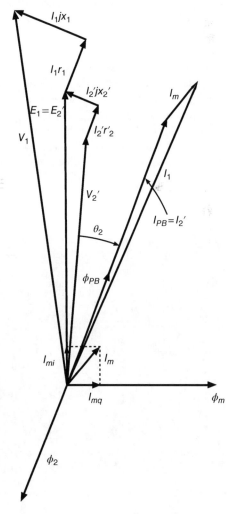

Figure 4.24 Phasor diagram of a loaded transformer.

diagram has been enlarged for reasons of clarity; it is generally less than 5% of the transformer's rated current.

4.17 Current Transformers

Our introduction to transformers would not be complete without a brief discussion of current transformers (CTs), which are used where large currents, or currents at elevated potentials, have to be measured. They provide the ability to generate a scaled version of the target current, while providing the necessary isolation.

Three 2000:5 A metering class CTs are pictured in Figure 4.25, each mounted on an LV busbar (a single-turn primary winding), where the current is being measured for the purpose of energy metering. As the rating suggests, 2000 A flowing in the busbar will result in 5 A in the secondary circuit. The secondary windings of these CTs are tightly wound on a toroidal magnetic core, wound from a continuous strip if GOSS similar to that shown on the right of Figure 4.12.

Our transformer discussion thus far has related to the constant flux model of a voltage transformer's operation, where a fixed voltage source excites the primary winding. However, in the case of a current transformer, a current source provides this excitation, and this appears to make the CT behave in a slightly different fashion.

Diagrammatically this is represented in Figure 4.26, where the customer's load current I_1, is responsible for generating the flux within the core. A current ratio of 2000:5 demands that 2000 ampere-turns in the primary winding generate the same MMF as 5 amps in the secondary; this is achieved with 400 secondary turns. The primary current is determined solely by the customer's load and is entirely independent of the presence of the CT.

Figure 4.25 Busbar mounted 2000:5 metering class current transformers.

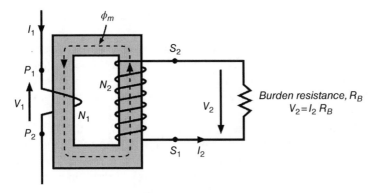

Figure 4.26 Current transformer configuration.

The magnitude of the flux in the core depends upon the load connected to the secondary winding (usually called the *CT's burden*) and to a small extent on the resistance of the secondary winding itself. In Figure 4.26 the burden is resistive and therefore the secondary voltage is given by the product $I_2 R_B$. The peak flux density $\hat{B}$ required to support this voltage is given by the EMF equation:

$$\hat{B} = I_2 R_B / 4.44\, N_1 A_c f \tag{4.21}$$

Thus if the burden impedance is kept small, then so will be the secondary voltage and therefore the flux density required to support it. As a result, the magnetising current required will also be small and sinusoidal, and therefore the great majority of the primary current will form the primary balance current required to offset the secondary flux. The secondary current will therefore be a scaled version of this current, as expressed in Equation (4.22):

$$I_2 = I_{PB} N_1 / N_2 \approx I_1 / N_2 \tag{4.22}$$

This is shown in the phasor diagram of Figure 4.27, where with a resistive burden, the primary balance current lies in phase with the miniscule primary voltage V_1.

Provided that the magnetising current remains acceptably small, the CT will introduce very little error, and the secondary current will be a very accurate representation of that flowing in the primary. For this reason, current transformers can only support small burdens. For example, the CTs in Figure 4.25 have a rated burden of 15 VA, which

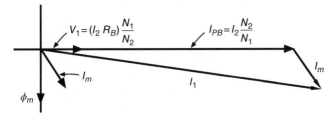

Figure 4.27 Current transformer primary phasor diagram.

means with a secondary current of 5 A, the maximum permitted burden impedance is only $0.6\,\Omega$.

Should the flux density and magnetising current be permitted to rise, usually as the result of a substantial increase in burden, then the secondary current will no longer be a good representation of the primary current, since the magnetising current will be relatively large. Exceeding the rated burden of a CT will therefore result in significant magnitude and phase errors in the secondary current.

It is useful to explore the problem of open circuiting the secondary winding of a CT, which is akin to applying an infinite burden. Since no secondary current can flow in these circumstances, there will be no primary balance current, and therefore the entire primary current will be in the form of magnetising current. This will result in high flux densities and very high secondary voltages will be produced as a result, generally non-sinusoidal.

The EMF equation is useful to demonstrate this. Suppose a 2000:5 CT with a cross-sectional core area of 30 cm² is open circuited with just sufficient primary current to not quite saturate the core (i.e. $\hat{B} \approx 1.8\text{T}$ – this will only require around 200 A.) According to the EMF equation the secondary voltage can be expected to reach:

$$V_2 = 4.44 \times 1.8 \times 0.0030 \times 400 \times 50 = 480 \text{ V}$$

Potentials of this magnitude are dangerous, both to the CT concerned and to maintenance personnel. Had the primary current been greater, considerably more potential would have been generated across the secondary winding, although this would not be of a sinusoidal nature. *It must therefore be stressed that under no circumstances should the secondary winding of a current transformer be open circuited.*

In this respect, a current transformer appears to behave differently from a voltage transformer, whose secondary winding we already know *should never be short-circuited.* In reality, both these transformers operate on the same principles and their behavioural differences simply relate to the way in which they are excited. More will be said about current transformers in Chapter 9.

4.18 Problems

1 A Figure 4.7 shows the excitation characteristic of GOSS, assuming that no air gap exists in the core. Use this information to plot the B–H characteristic for a GOSS core containing a 0.5 mm air gap. Assume a maximum flux density of 1.5 T. The cross-section of this core is 40 mm × 20 mm and its length is 300 mm. What effect has the inclusion of the air gap had on the magnetic characteristics of this core?

(Hint: For each value of flux density B, calculate the ampere-turns required to excite the air gap, and from the graph obtain the ampere-turns required to excite the core steel. The total magnetic field strength H, is then given by the *total* ampere-turns required, divided by the magnetic path length $\approx 0.3\,\text{m}$.)

   B From your graph determine the effective permeability of the composite core. Compare your result with that predicted by Equation (4.5a).
(Answer: $\mu_{eff} = 0.75 \times 10^{-4}$)

C The core in part (a) is fitted with a 200 turn winding. Show that the inductance presented by this can be expressed as:

$$L = \frac{\mu_{eff} N^2 A_a}{l_c}$$

Calculate the inductance of this winding.
(Answer: 83 mH)

D The mass of this core is 1.8 kg. If the magnetising losses are 20% higher than the values shown in Figure 4.15 (due to the additional losses associated with the air gap), estimate the magnetising loss when the peak flux density in the core is 1.0 T. Assume a system frequency of 50 Hz.
(Answer: 2.16 W)

E Calculate the AC voltage that must be applied to the inductor described above, in order to excite a peak flux density of 1.0 T and determine the equivalent magnetising loss resistance that would appear in parallel with the inductive reactance of the winding at this potential.
(Answer: 35.5 V_{RMS}, 583 Ω)

F If the winding has a resistance of 0.8 Ω, determine the overall quality factor (Q) of this inductance at the system frequency.
(Answer: $Q = 13.3$)

G How could the design of this inductor be altered so as to provide a higher Q value, while keeping its impedance constant?

2 In 1976, an Australian inventor patented a sheet metal bending machine, the manufacture of which was subsequently licensed to manufacturers in several countries. It was revolutionary design that used an electromagnetic clamp to hold the sheet to be bent and, as a consequence, it is able to make much more complex and intricate bends than are possible with a conventional pan-brake bending machine. The machine is shown in Figure 4.28, together with a conceptual operating sketch.

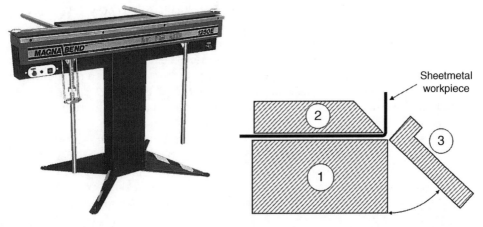

Figure 4.28 Magna-bend magnetic bending machine. (1: Magnet body, 2: Clamp bar, 3: Bending beam) (Reproduced with the kind permission of the inventor Alan Bottomley).

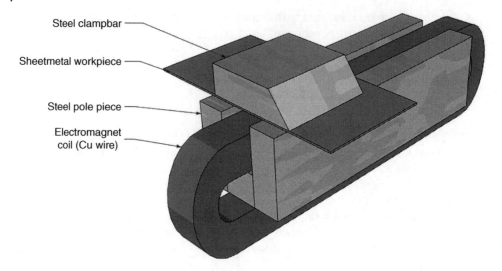

Steel clampbar

Sheetmetal workpiece

Steel pole piece

Electromagnet
coil (Cu wire)

Figure 4.29 Magna-bend 'U' type electromagnet construction. (Reproduced with the kind permission of the inventor Alan Bottomley).

The 'U' type electromagnet (shown in Figure 4.29), holds the sheet in firmly place by applying a large electromagnetic force to the clamp bar. The winding is excited from a DC supply, and applies a magnetic field strength of 3400 ampere-turns/m. The magnetic circuit is made from mild steel rather than GOSS, since the DC flux does not vary significantly in time, and therefore hysteresis and eddy current losses are unimportant. Bends in sheet steel or aluminium up to 1.6 mm thick can be formed; beyond this thickness the force applied to the sheet by the bending beam is sufficient to lift the leading edge of the clamp bar, allowing the sheet to slide backwards.

Our interest in this machine lies in calculating of the clamping pressure it produces, as a function of non-magnetic sheet thicknesses (e.g. when bending aluminium, brass or copper), according to Equation (4.10). The inventor has published measured clamping pressures for various sheet thicknesses, shown in Figure 4.30, with which we will compare our calculations.

The non-linear magnetising characteristics of the mild steel used in its construction are shown in Figure 4.30. In order to determine the flux density when an air gap is created by a sheet of non-magnetic material, it is necessary to plot a magnetic load line on this characteristic. This is a linear relationship between the applied H field and the flux density in the air gap B_{Air}, including the magnetic field strength in the steel H_{Steel}.

An electrical analogy is useful to illustrate how this is done. Figure 4.31 shows the electrical equivalent of the magnetic circuit, including the air gap. The applied MMF (analogous to a source of voltage) is dropped across both the non-linear steel material and the linear air gap. This MMF sets up a magnetic flux around the circuit (analogous to current), which is linearly related to the MMF dropped across the air gap according to:

$$MMF_{Air} = \Phi \Re_{Air} = B_{Air} \frac{l_{Air}}{\mu_0}$$

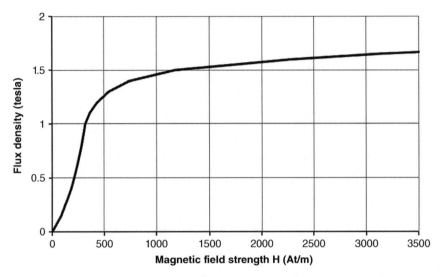

Figure 4.30 Magnetisation characteristics mild steel. (Reproduced with the kind permission of the inventor Alan Bottomley).

Figure 4.31 Electrical analogy of the magnetic circuit.

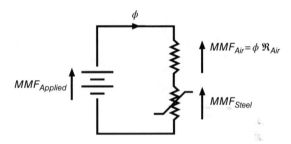

Because of the non-linear nature of the steel, it is not possible to write a simple expression relating the flux in the steel to the MMF dropped across it; however, this information is embodied in the excitation curve. If we plot the magnetic load line on this curve, we can determine both the flux density and the MMF dropped within the steel, since the operating point must lie at the intersection of the load line and the magnetisation characteristic. The equation for the magnetic load line can be obtained from the electrical circuit analogy, using KVL:

$$MMF_{Applied} = MMF_{Steel} + MMF_{Air}$$

$$\frac{(NI)_{Applied}}{l_{Steel}} = \frac{(NI)_{Steel}}{l_{Steel}} + \frac{B_{Air}}{\mu_0}\frac{l_{Air}}{l_{Steel}}$$

$$\text{or} \quad H_{Applied} = H_{Steel} + \frac{B_{Air}}{\mu_0}\frac{l_{Air}}{l_{Steel}} \tag{4.23}$$

A Using the intercept on each axis, plot Equation (4.23) on the magnetisation characteristic, assuming a sheet thickness of 1 mm, and hence determine the resulting flux density in the air gap. (You may neglect any fringing effects.)

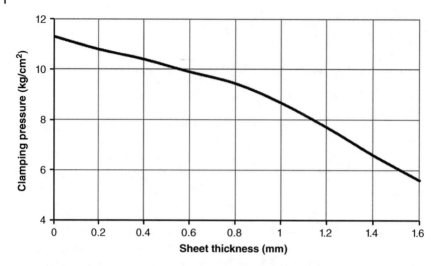

Figure 4.32 Measured clamping pressure as a function of sheet thickness. (Reproduced with the kind permission of the inventor Alan Bottomley).

B Using this result, calculate the force exerted by the clamp bar per m² of core area (i.e. the clamp pressure) in kg/cm². Compare your result with that plotted in Figure 4.32. Calculate the force in tonnes exerted on a sheet of aluminium if the total pole area is $752\,\text{cm}^2$.

(Answer: 6.3 tonnes)

C Repeat items (a) and (b) for sheet thicknesses ranging from 0.2–1.6 mm. Plot your results on Figure 4.32.

3 The efficiency of a power transformer η can be expressed as the ratio between the power supplied to the primary and that removed from the secondary, i.e.

$\eta = \dfrac{P_{out}}{P_{in}} \times 100\%$. We can express the output power as $P_{out} = V_2 I_2 \cos(\phi_2)$ and that sup-

plied to the primary as $P_{in} = P_{out} + I_2{}^2 r + P_{magnetic}$, where V_2 and I_2 are the secondary voltage and current respectively, ϕ_2 is the secondary phase angle, r is the total winding resistance referred to the secondary winding and P_{iron} is the total magnetic core loss (i.e. hysteresis and eddy current loss).

Where a transformer is permanently energised at near constant voltage (as is the case, for example, with a distribution transformer), then the magnetic loss will be roughly constant and independent of the load that the transformer carries. In this event we may write:

$$\eta = \frac{P_{out}}{P_{out} + I_2{}^2 r + P_{iron}} = \frac{V_2 I_2 \cos(\phi_2)}{V_2 I_2 \cos(\phi_2) + I_2{}^2 r + P_{iron}} \tag{4.24}$$

Show that the transformer efficiency described by Equation (4.24) is maximised when the copper loss ($I_2{}^2 r$) equals the magnetic loss (P_{iron}). (Hint: Consider the condition that maximises η, and find the value of I_2 that satisfies this condition.)

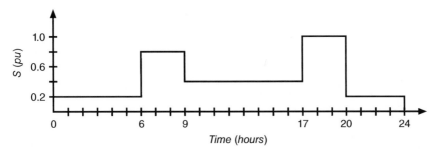

Figure 4.33 Load profile.

4 The result obtained in question 3 is of little use in practice since the load applied to a distribution transformer varies with consumer demand, usually following a daily cycle. Therefore attempting to optimise a transformer's efficiency by running at a constant load is generally not feasible. Instead, another measure of efficiency is used, called all-day efficiency and it is defined as:

$$All\text{-}Day\ Efficiency = \left.\frac{Watt\ hours\ delivered\ by\ the\ transformer}{Watt\ hours\ supplied\ to\ the\ transformer} \times 100\%\right|_{24\,Hours}$$

Manufacturers often design transformers so that the magnetic loss is somewhat less than the full load copper loss, on the basis that the transformer will usually remain energised permanently and therefore the magnetic loss must be continuously supplied, regardless of the applied load. The copper loss will vary with the square of applied load.

Calculate the all-day efficiency for the load profile shown in Figure 4.33, if this load is supplied by a transformer with a winding resistance of 0.015 pu and a magnetic loss of 0.01 pu. Assume the load power factor is 0.8.

(Answer: 95.9%)

5 Use Figure 4.7 to graphically obtain a waveform proportional to the magnetising current, when the core is sinusoidally excited with a peak flux density of 1.66 T. Compare your result with the excitation waveform in Figure 4.10.

6 *Mutual inductance.* In addition to the constant flux model, the operation of a transformer can also be described using the concept of mutual inductance, which relates the voltage induced in one winding due to the current in another. For example, consider two loosely coupled windings, in which only a portion of the flux generated in each links the other, as illustrated in Figure 4.34.

Because windings 1 and 2 are *not* tightly coupled, only a fraction of the flux created by each one links to the other. This fraction, called the *coupling coefficient*, is given the symbol k, where $0 \leq k \leq 1$. If we assume linear magnetic circuits for the primary and secondary of reluctance $\mathfrak{R}_1$ and $\mathfrak{R}_2$ respectively, the total flux seen by winding 1 is given by:

$$\phi_1(t) = \frac{N_1 I_1(t)}{\mathfrak{R}_1} + k \frac{N_2 I_2(t)}{\mathfrak{R}_2}$$

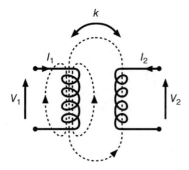

Figure 4.34 Only a portion of the flux generated by winding 1 links winding 2, and vice versa.

and for winding 2: $\phi_2(t) = \dfrac{N_2 I_2(t)}{\mathfrak{R}_2} + k\dfrac{N_1 I_1(t)}{\mathfrak{R}_1}$

A Show that the voltages $V_1(t)$ and $V_2(t)$ may be written as:

$$V_1(t) = L_1\frac{dI_1(t)}{dt} + M_{12}\frac{dI_2(t)}{dt} \text{ and } V_2(t) = M_{21}\frac{dI_1(t)}{dt} + L_2\frac{dI_2(t)}{dt}$$

where L_1 and L_2 are the self inductances associated with windings 1 and 2 respectively.

Find also expressions for the mutual inductances M_{12} and M_{21}.

(Answer: $M_{12} = kN_1 N_2 / \mathfrak{R}_2$ and $M_{21} = kN_1 N_2 / \mathfrak{R}_1$)

Alternatively, these equations may be expressed in matrix form as:

$$\begin{bmatrix} V_1(t) \\ V_2(t) \end{bmatrix} = \begin{bmatrix} L_1 & M_{12} \\ M_{21} & L_2 \end{bmatrix} \begin{bmatrix} \dfrac{dI_1(t)}{dt} \\ \dfrac{dI_2(t)}{dt} \end{bmatrix}$$

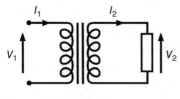

Figure 4.35 I_2 definition.

Let us now consider the operation of a transformer based on the idea of loosely coupled windings on a common magnetic circuit, similar to the arrangement shown in Figure 4.22. As a result, we might expect that each winding will see a substantially similar reluctance $\mathfrak{R}$. We will also redefine the direction of current I_2, as shown in Figure 4.35, so that it is consistent with the constant flux transformer model.

B If we assume that the time functions $V_1(t)$ and $V_2(t)$ are sinusoidal, show that in the steady state we may write the above equations in terms of phasor notation as:

$$\begin{bmatrix} V_1 \\ V_2 \end{bmatrix} = \begin{bmatrix} j\omega L_1 & -j\omega M \\ j\omega M & -j\omega L_2 \end{bmatrix} \begin{bmatrix} I_1 \\ I_2 \end{bmatrix} \qquad (4.25)$$

Figure 4.36 Transformer equivalent circuit.

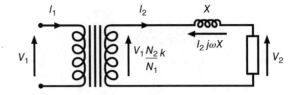

where $M = k\sqrt{L_1 L_2}$.

Note that these equations predict the existence of the inductive component of the magnetising current $V_1/j\omega L_1$. In this simple model we have tacitly assumed a lossless core, hence the resistive current component does not appear.

C By suitably manipulating Equations (4.25), show that:

$$V_2 = V_1 k \frac{N_2}{N_1} - I_2 j\omega L_2 \left(1 - k^2\right) = V_1 k \frac{N_2}{N_1} - I_2 jX \qquad (4.26)$$

Equation (4.26) suggests the transformer equivalent circuit of Figure 4.36, whereby the voltage induced in the secondary winding and the leakage inductance X, both depend upon the degree of coupling between the transformer windings. If the coupling is tight (i.e. $k \approx 1$) then X becomes small and the winding voltage is given by Equation (4.15). On the other hand, if k is small and the coupling loose, then the winding voltage reduces, and the leakage reactance becomes considerably larger. As a result, the transformer's output voltage falls rapidly as the load current increases. Both of these assertions are entirely consistent with the expected behaviour of a transformer, as predicted by the constant flux model.

It is the aim of most transformer designs to provide tight coupling and thus a relatively low leakage reactance commensurate with the desired application. In order to maximise the coefficient of coupling, power transformers are not wound as suggested in Figure 4.22. Instead, both windings are concentrically wound on the *same* limb, as shown in Figure 4.37(a), with sufficient insulation between them to safely support the voltage difference. The inter-winding insulation creates a duct through which leakage flux can pass, and as a result there will always be some degree of leakage reactance; k will never quite equal unity.

It is usual to wind the low voltage winding closest to the core and the HV winding on top (so that any tapping leads may easily be brought out, and the LV winding is closer to the earthed core). In Figure 4.37a the heavier low voltage (LV) winding can be seen closest to the core, while the HV winding, consisting of a number of series connected disc windings, is wound on top. Breaking the HV winding up into segments requires each to only support a defined fraction of the applied voltage, thus avoiding the possibility of the inter-layer insulation breaking down. A further advantage of this arrangement lies in allowing the cooling oil in which the transformer is submerged, to easily penetrate the HV winding, thereby improving its heat dissipation.

(a)

(b)

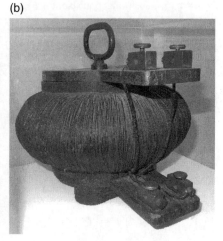

Figure 4.37 (a) Typical distribution transformer winding arrangement (partially disassembled) (b) The world's first transformer, Budapest 1885. (Courtesy Deutsches Museum, Munich).

D Amend your analysis to show that the VA applied to winding 1 is equal to that supplied to the connected load plus that required by the leakage reactance X:

$$V_1 I_1 = V_2 I_2 + I_2{}^2 X$$

Since in most transformers the leakage reactance is relatively low (4–15%) it is usual to assign both the primary and the secondary windings the same VA rating. However, when the impedance of a transformer becomes large (20–35%), as is required in some applications, then the primary winding requires a considerably higher VA rating than can be delivered by the secondary, since a portion of the applied potential is dropped across the inductive impedance of the device.

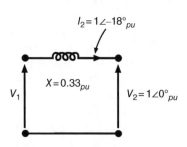

Figure 4.38 Transformer per-unit equivalent circuit.

E Working in per-unit quantities and using the transformer equivalent circuit of Figure 4.38 determine the VA rating required of the primary winding of a transformer, when its fully loaded secondary winding delivers 45 MVA at a power factor of 0.95 to an 11 kV industrial load. Assume that the impedance of the transformer is 33% on a 45 MVA base.

(Answer: Primary VA rating = 51.6 MVA)

F What turns ratio is required of this transformer if its primary potential is 110 kV?

(Answer: Turns ratio = 11.47:1)

G Repeat parts (e) and (f) if the load power factor falls to 0.8.

(Answer: Primary VA rating = 55.2 MVA; Effective turns ratio = 12.27. In practice such a transformer would be fitted with an on-load tap-changer so that the effective ratio can be dynamically altered with changing load characteristics, so as to maintain a constant load voltage.)

4.19   Sources

1 Matsch L, '*Capacitors, magnetic circuits and transformers*', 1964, Englewood Cliffs (NJ): Prentice Hall.
2 Say MG. '*Alternating current machines*' 1976, 2nd ed, Pitman Press, London.
3 Lee R. '*Electronic transformers and circuits*', 1955, 2nd ed, New York: John Wiley.
4 Del Vecchio R, Poulin B, Feghali P, Shah D, Ahuja R. '*Transformer design principles*', 2010 Boca Raton (FL): CRC Press.
5 Ikeda Y, Naito H, Abe M. '*Technological trend in the recent development of shunt reactors*'. Fuji Electrical Review, 1982: Vol 28, pp. 59–67.

5

Symmetrical Components

Generators supplying a distribution network are carefully designed to produce a balanced three-phase supply, which means that each phase voltage has the same amplitude and is separated from its neighbours by exactly 120°. Unfortunately the connected load is seldom exactly balanced, and therefore the magnitude and phase of the load currents may not be quite equal. This causes differing potential drops across the phase impedances, and a proportionally unbalanced supply is seen throughout the network as a result.

The situation becomes considerably worse when faults occur in a network, the most disturbing of which tend to involve short circuits, either between individual phase conductors or between one or more phases and either the neutral terminal or ground potential. Under these conditions the degree of unbalance present in a network becomes extreme, and close to the fault the affected phase voltage(s) may vanish entirely. High fault currents flow as a consequence, and the protection system must quickly detect and clear such faults before permanent damage occurs.

The analysis of balanced three-phase systems is straightforward, and can be done using phasor analysis, but when the degree of unbalance becomes substantial another approach is generally used. This technique, called the *method of symmetrical components*, relies on the fact that the faulted network remains essentially linear. Linear or near linear systems are common in engineering and there are general tools available for their analysis. One of the most powerful of these is the principle of superposition, which can be used to decompose the response of the faulted network into its constituent components, each of which can then be analysed separately. These results can be recombined to determine the voltages and currents that arise in a network as a result of the fault. (Fourier analysis is another example of such a technique.)

Under fault conditions, the degree of unbalance and the current magnitudes can be many times those in the un-faulted network, and it is usual to ignore the currents flowing prior to the fault, and to determine only the currents flowing as a result of the fault, using the symmetrical component approach.

Symmetrical component theory was first described in a famous paper by Charles LeGeyt Fortescue in 1918, entitled "Method of symmetrical co-ordinates applied to the solution of polyphase networks". Fortescue spent his entire working life at the Westinghouse Electric and Manufacturing Company in Pittsburgh, researching numerous aspects of power engineering. He was awarded 180 patents during his career and died at the relatively young age of 60, in 1936. In his paper, he showed that unbalanced

AC Circuits and Power Systems in Practice, First Edition. Graeme Vertigan.
© 2018 John Wiley & Sons Ltd. Published 2018 by John Wiley & Sons Ltd.

polyphase systems could be represented by the sum of three balanced components consisting of a set of *positive sequence* three-phase vectors, a *negative sequence* set and a *zero-sequence* or *co-phasal* set.

Symmetrical component theory has since become the basis for many protection systems, and a sound grasp of these concepts is essential in understanding the operation of modern power systems.

5.1 Symmetrical Component Theory

In its balanced state, a power system contains only positive sequence voltages and currents, and, by convention, these have an *abc* phase sequence. So long as a network is balanced, then positive sequence voltages and currents components exist on their own. However, when phase voltages or currents are not balanced then both negative and zero-sequence components may exist as well. In reality the loads within distribution and transmission networks are almost balanced, and therefore only small negative and zero-sequence voltages co-exist with the main positive sequence voltage. Limits are placed on the magnitudes of these and, should they be exceeded, the offending load must be found and rebalanced. For example, LV distribution companies attempt to spread single-phase residential loads equally over all three-phases, while large industrial or commercial customers will usually have a limit placed on the degree of unbalance permitted in their load currents.

Network unbalance becomes extreme under fault conditions, and the presence of faults is frequently identified by the existence of large negative or zero-sequence currents or voltages. Earth faults for example, are usually detected by the presence of zero-sequence currents, and many protection schemes are based on the application of symmetrical component theory.

5.1.1 Negative Sequence Voltages and Currents

If we accept that *abc* is the positive phase sequence, then the negative sequence will be *acb*. In this analysis we will denote positive sequence voltages and currents with the subscript 1 and negative sequence voltages and currents with the subscript 2. Therefore V_{a1} represents the positive sequence component of the 'a' phase voltage, and V_{a2} is its negative sequence component. Since it is well known that by reversing the phase sequence of voltages supplied to a three-phase machine, its direction of rotation will reverse, it is intuitive to expect that negative sequence voltages might be represented on a phasor diagram simply by reversing the direction of phasor rotation as well. Although this idea might be intuitive, it is actually incorrect.

The reason for this is quite simple. If the coordinates of a rotating phasor are given by Ve^{jwt} then by reversing the direction of rotation we generate a phasor whose coordinates are described by Ve^{-jwt}. When we consider the addition of such a negative sequence voltage set and a positive sequence set, we find that there is no fixed phase relationship between the two.

This can be illustrated by freezing the anticlockwise rotation of the positive sequence phasors and observing that the corresponding negative sequence voltage must now rotate in a clockwise direction at 2ω radians per second relative to them, as shown in

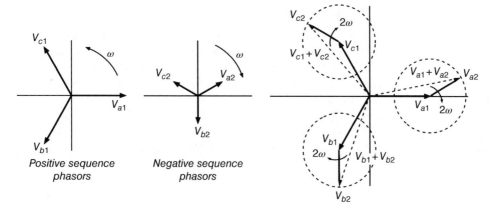

Figure 5.1 Incorrect representation of negative sequence voltages.

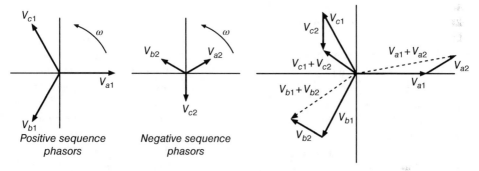

Figure 5.2 Correct representation and addition of positive and negative phase sequence voltages.

Figure 5.1. The vector addition of these quantities results in phasors whose lengths change at a frequency of 2ω radians per second (they are effectively *modulated* at 2ω), and since this situation does not occur in real systems, we must therefore conclude that this method of representation is incorrect.

The correct method of representing a set of negative sequence voltages (or currents) is to swap two phasors (usually the b and c phases), as shown in Figure 5.1b, while maintaining the standard anticlockwise rotation, thereby generating an *acb* phase sequence.

When these voltages are added to a positive sequence set, a fixed phase relationship can be seen to exist between each corresponding pair of voltages.

The phase voltages resulting from the summation of a positive sequence set and a negative sequence set, shown in Figure 5.2, are inherently unbalanced, both in magnitude and phase. However, these voltages do still add to zero, since this property is possessed by both constituents. Therefore we can generate an unbalanced set of voltages by simply adding a negative sequence set to our original positive sequence set. The magnitude and phase of the negative sequence set are quite arbitrary, and these parameters can be chosen to achieve any degree of unbalance, with the proviso that the sum of the resulting unbalanced voltages must add to zero.

5.1.2 Zero-Sequence Voltages and Currents

Figure 5.3 Zero-sequence phasors.

In practice, unbalanced voltages or currents do not always sum to zero; frequently a *residual* voltage (or current) remains. To generate such a degree of unbalance, we need to include a *zero-sequence* component in the summation. Zero-sequence voltages (or currents) are not spaced at intervals of 120°; instead they all share the *same phase angle*, and therefore are said to be *co-phasal* (see Figure 5.3); consequently they do not add to zero. When these components are summed, together with positive and negative components, any degree of unbalance can be achieved (see Figure 5.5). We denote zero-sequence voltages or currents with the subscript 0.

The zero-sequence currents deserve special comment. Because these are co-phasal, they do not add to zero at a star point, as do their positive and negative sequence counterparts. Therefore for zero-sequence currents to exist at all there must be a return path to the supply transformer neutral terminal, either via the neutral conductor, or in the event of an earth fault, via a ground path. Accordingly, zero-sequence currents are usually associated with faults to ground or to neutral.

The currents flowing in an unbalanced network can also be resolved into symmetrical components, with each component current depending upon the impedance seen by its associated sequence voltage. We therefore need to introduce the concept of positive, negative and zero-sequence impedances in addition to negative and zero-sequence voltages and currents. Except in the vicinity of large rotating machines, the positive and negative sequence voltages usually see similar impedances, since these are both balanced networks whose phase voltages sum to zero. However, the zero-sequence impedance is often quite different from the positive and negative sequence impedances. This is due to the fact that zero-sequence voltages do not sum to zero and therefore zero-sequence currents cannot flow towards an ungrounded neutral or star point terminal. Consequently zero-sequence currents generally flow in slightly different paths from those of positive and negative sequence currents. For this reason, and due to the different magnetic coupling seen by zero-sequence currents in transformers and transmission lines, the zero-sequence impedance is usually different from positive and negative impedances.

Fortescue was the first person to realise that, in general, an unbalanced set of voltages consists of a positive, a negative and a zero-sequence set, and he found a way of resolving the magnitude and phase of each of these symmetrical components from an unbalanced set of voltages. To do this he introduced a complex operator α, which is used to phase shift voltages and currents by 120° on a phasor diagram. Alpha has a magnitude of unity and an angle of +120°, thus we can write $\alpha = 1\angle 120° = -0.5 + j0.866$ therefore $\alpha^2 = 1\angle 240°$ and $\alpha^3 = 1$. Finally, $1 + \alpha + \alpha^2 = 0$, as shown in Figure 5.4.

If the complex phase voltages are written V_a, V_b and V_c (where $V_a = |V_a|\angle 0°$, $V_b = |V_b|\angle -120°$ and $V_c = |V_c|\angle 120°$), then, in terms of their symmetrical components V_1, V_2 and V_0, they can be expressed as:

$$V_a = V_1 + V_2 + V_0 = V_{a1} + V_{a2} + V_0$$
$$V_b = \alpha^2 V_1 + \alpha V_2 + V_0 = V_{b1} + V_{b2} + V_0 \qquad (5.1)$$
$$V_c = \alpha V_1 + \alpha^2 V_2 + V_0 = V_{c1} + V_{c2} + V_0$$

Figure 5.4 Showing $\alpha = 1\angle 120°$, $\alpha^2 = 1\angle 240°$, $\alpha^3 = 1$ and therefore $1 + \alpha + \alpha^2 = 0$.

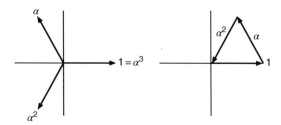

Alternatively, we may express these equations in matrix form as:

$$\begin{bmatrix} V_a \\ V_b \\ V_c \end{bmatrix} = \begin{bmatrix} 1 & 1 & 1 \\ \alpha^2 & \alpha & 1 \\ \alpha & \alpha^2 & 1 \end{bmatrix} \begin{bmatrix} V_1 \\ V_2 \\ V_0 \end{bmatrix} = A \begin{bmatrix} V_1 \\ V_2 \\ V_0 \end{bmatrix}$$

Where V_1, V_2 and V_0 are the positive, negative and zero-sequence voltages respectively associated with the 'a' phase, which is taken as the reference phase.

The positive sequence voltages can be expressed as $V_{a1} = V_1$, $V_{b1} = \alpha^2 V_1$ and $V_{c1} = \alpha V_1$. Similarly the negative sequence set can be written as $V_{a2} = V_2$, $V_{c2} = \alpha V_2$ and $V_{b2} = \alpha^2 V_2$ (see Figure 5.5). Finally, the zero-sequence voltages all share the same magnitude and phase, and each is therefore equal to V_0. *Therefore each phase voltage in an unbalanced network can be considered as having a positive, a negative and a zero-sequence component.*

Unbalanced phase currents can also be represented by symmetrical components through equations analogous to Equation (5.1) thus:

$$I_a = I_1 + I_2 + I_0 = I_{a1} + I_{a2} + I_0$$
$$I_b = \alpha^2 I_1 + \alpha I_2 + I_0 = I_{b1} + I_{b2} + I_0 \quad \text{or} \quad \begin{bmatrix} I_a \\ I_b \\ I_c \end{bmatrix} = \begin{bmatrix} 1 & 1 & 1 \\ \alpha^2 & \alpha & 1 \\ \alpha & \alpha^2 & 1 \end{bmatrix} \begin{bmatrix} I_1 \\ I_2 \\ I_0 \end{bmatrix} = A \begin{bmatrix} I_1 \\ I_2 \\ I_0 \end{bmatrix}$$
$$I_c = \alpha I_1 + \alpha^2 I_2 + I_0 = I_{c1} + I_{c2} + I_0$$

Figure 5.5a demonstrates how an unbalanced set of voltages V_a, V_b and V_c, can be resolved into its symmetrical components, shown in Figure 5.5b.

If we solve Equation (5.1) for the symmetrical components V_1, V_2 and V_0 in terms of V_a, V_b and V_c we find:

$$\begin{aligned} V_1 &= \left(V_a + \alpha V_b + \alpha^2 V_c \right)/3 \\ V_2 &= \left(V_a + \alpha^2 V_b + \alpha V_c \right)/3 \quad \text{or} \\ V_0 &= \left(V_a + V_b + V_c \right)/3 \end{aligned} \qquad \begin{bmatrix} V_1 \\ V_2 \\ V_0 \end{bmatrix} = \frac{1}{3}\begin{bmatrix} 1 & \alpha & \alpha^2 \\ 1 & \alpha^2 & \alpha \\ 1 & 1 & 1 \end{bmatrix} \begin{bmatrix} V_a \\ V_b \\ V_c \end{bmatrix} = \frac{A^{-1}}{3}\begin{bmatrix} V_a \\ V_b \\ V_c \end{bmatrix} \qquad (5.2)$$

and similarly

$$\begin{aligned} I_1 &= \left(I_a + \alpha I_b + \alpha^2 I_c \right)/3 \\ I_2 &= \left(I_a + \alpha^2 I_b + \alpha I_c \right)/3 \quad \text{or} \\ I_0 &= \left(I_a + I_b + I_c \right)/3 \end{aligned} \qquad \begin{bmatrix} I_1 \\ I_2 \\ I_0 \end{bmatrix} = \frac{1}{3}\begin{bmatrix} 1 & \alpha & \alpha^2 \\ 1 & \alpha^2 & \alpha \\ 1 & 1 & 1 \end{bmatrix} \begin{bmatrix} I_a \\ I_b \\ I_c \end{bmatrix} = \frac{A^{-1}}{3}\begin{bmatrix} I_a \\ I_b \\ I_c \end{bmatrix} \qquad (5.3)$$

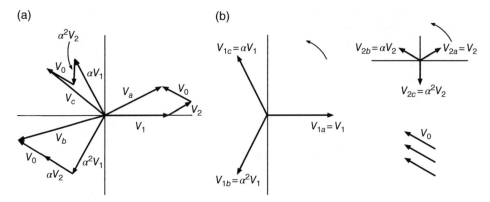

Figure 5.5 (a) Resolution of unbalanced voltages (b) Symmetrical components.

Examination of Equations (5.2) and (5.3) shows that if the original set of voltages or currents is balanced then the negative and zero-sequence components are zero as expected, and only the positive sequence set remains. However, when a fault occurs on the system, asymmetries arise that result in the generation of these components as well (with the notable exception of the inherently balanced three-phase fault, which contains only positive sequence components).

5.2 Sequence Networks and Fault Analysis

If we assume that a short-circuit fault is applied between the 'a' phase and ground, then according to Equation (5.1) we can represent the 'a' phase voltage at the fault as: $V_F = V_a = V_1 + V_2 + V_0$. Similarly, the 'a' phase current can be represented by: $I_F = I_a = I_1 + I_2 + I_0$, as depicted in Figure 5.6.

By a similar process, the 'b' and 'c' phase voltages and currents under fault conditions can also be obtained from a summation of their associated symmetrical components. In the event of a *solid* 'a' phase-to-ground fault for example, the 'a' phase voltage will equal zero, but its individual components, V_1, V_2 and V_0 will not. We might also expect that the current from such a fault will be high, and therefore so will its symmetrical components, I_1, I_2 and I_0. We analyse each component of a fault voltage

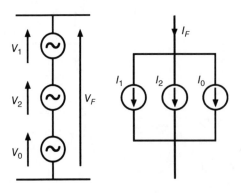

Figure 5.6 Resolution of the 'a' phase fault voltage and current into symmetrical components.

or current according to its sequence network, which relates the sequence voltage to its corresponding sequence current and sequence impedance.

5.2.1 Sequence Networks

The 'a' phase *positive sequence network* can be represented by the Thévenin equivalent circuit of the power system, as seen at the fault (Figure 5.7). This approach seems reasonable when we remember that in a balanced network only positive sequence components exist, in which case network currents are determined from the Thévenin equivalent circuit.

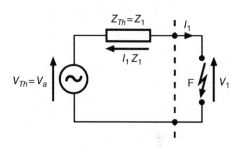

The Thévenin equivalent voltage is therefore equal to the normal positive sequence voltage V_a, seen in the un-faulted network, and the Thévenin equivalent impedance is the corresponding positive sequence impedance (Z_1), seen at the fault.

Figure 5.7 Diagram of the 'a' phase positive sequence network Thévenin equivalent circuit.

From Figure 5.7 the positive sequence component of the 'a' phase voltage at the fault can be expressed as:

$$V_1 = V_a - I_1 Z_1 = V_{a1} \tag{5.4}$$

The 'b' and 'c' phase positive sequence components can also be found from Equation (5.4) as follows:

$$V_{b1} = \alpha^2 V_1 \text{ and } V_{c1} = \alpha V_1 \tag{5.5}$$

Negative Sequence Network
We can similarly construct the negative sequence network as seen at the fault (Figure 5.8), but in this case there is no pre-fault negative sequence driving voltage since the un-faulted network was inherently balanced. Therefore we may write:

$$V_2 = 0 - I_2 Z_2 = -I_2 Z_2 = V_{a2} \tag{5.6}$$

The negative sequence impedance Z_2 is that seen by both negative sequence voltages and currents.

Similarly the 'b' and 'c' phase negative sequence components are found from:

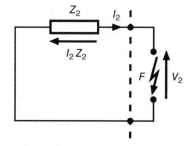

$$V_{b2} = \alpha V_2 \text{ and } V_{c2} = \alpha^2 V_2$$

In most networks the positive and negative sequence impedances Z_1 and Z_2, are approximately equal; they differ only when in proximity to large rotating machines such as generators or induction motors.

Figure 5.8 Diagram of the 'a' phase negative sequence network.

Zero-sequence Network

Since there are also no zero-sequence voltage sources in an un-faulted network, the zero-sequence network is very similar to that of the negative sequence, and can be described by Equation (5.7).

$$V_0 = -I_0 Z_0 \tag{5.7}$$

The zero-sequence impedance Z_0 is that seen by zero-sequence voltages and currents. It is generally not the same as Z_1 or Z_2, depending on the transmission line configurations, transformer winding arrangements and construction, as well as the earthing method used.

5.2.2 Network Sequence Impedances

Table 5.1 shows some common lumped impedance configurations together with their positive, negative and zero-sequence impedances. The first row depicts a series reactor with Z ohms per phase. In this case the positive, negative and zero-sequence impedances all equal Z. While a zero-sequence impedance exists, it is isolated from ground and therefore no zero-sequence current can flow as a result. So this configuration presents an infinite zero-sequence impedance to ground, or Z ohms for through faults, which must flow to ground elsewhere in the network.

Row two of Table 5.1 depicts a *balanced star connected load*, a shunt capacitor bank or a transformer. In this case the positive and negative sequence impedances each equal

Table 5.1 Lumped impedances and their symmetrical component equivalents.

Circuit Configuration	Positive, Negative sequence Circuit	Zero sequence Circuit

Z ohms, and because these network currents add to zero at the star point, these imped-ances can be considered grounded, i.e. positive or negative sequence currents can flow into the star point, yielding both zero current and zero voltage there. The star point is, for this reason, considered to be at ground potential, so the positive and negative sequence impedances there are each equal to Z. The zero-sequence impedance, how-ever, is not connected to ground since co-phasal zero-sequence currents cannot flow into an ungrounded star point, and as a result Z_0 is infinite.

Row three shows a balanced star connected load with the *star point grounded* via earthing impedance Z_n, the presence of which does not change the positive or negative sequence networks, since these currents still add to zero at the star point. However, Z_n does significantly affect the zero-sequence impedance, since zero-sequence currents can now flow to ground. Therefore the total zero-sequence impedance now includes a component equal to $3Z_n$, since a total of $3I_0$ flows in Z_n, i.e. $Z_0 = Z + 3Z_n$.

In the final row a balanced delta connected load is shown. This can be reduced to an equivalent star connected impedance (using the delta to star transformation) in which each star impedance is one-third of the corresponding delta impedance. Hence by anal-ogy with row 2, the positive and negative sequence impedances are $Z/3$, while the zero-sequence impedance is infinite, since no zero-sequence can flow into an open star point.

5.3   Network Fault Connections

The sequence networks described in Section 5.2.1 are used to determine the currents flowing under various fault conditions. How they are interconnected depends upon the type of fault in question. Each fault is uniquely defined by three boundary conditions relating to network voltages and currents at the fault. For simplicity we assume the fault impedance to be zero.

The following analysis of network faults ignores all the normal currents flowing prior to the fault, and considers only those abnormal currents that exist as a result of the fault, since these are generally much greater in magnitude.

5.3.1   Single-Phase-to-Ground Fault

A single-phase-to-ground fault on phase 'a' is shown in Figure 5.9; it can be defined by the following boundary conditions:

$$I_c = 0$$
$$I_b = 0 \tag{5.8}$$
$$V_a = 0$$

We begin the analysis by expressing the symmetrical components of the boundary conditions using Equation (5.3), remembering that $1 + \alpha + \alpha^2 = 0$:

$$I_c = \alpha I_1 + \alpha^2 I_2 + I_0 = 0$$

$$I_b = \alpha^2 I_1 + \alpha I_2 + I_0 = 0$$

and $\quad V_1 = -(V_2 + V_0)$

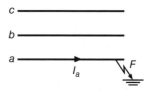

Figure 5.9 Phase-to-ground fault, boundary conditions: $V_a=0$, $I_c=0$ and $I_b=0$.

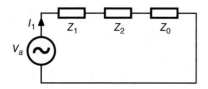

Figure 5.10 Positive sequence network connection for the phase-to-ground fault.

The first two equations can only be true if:

$$I_1 = I_2 = I_0 \tag{5.9}$$

We now write the equations for the sequence networks:

$$V_1 = V_a - I_1 Z_1 = -(V_2 + V_0)$$
$$V_2 = -I_2 Z_2 \tag{5.10}$$
$$V_0 = -I_0 Z_0$$

From Equations (5.9) and (5.10) we find:

$$I_1 = \frac{V_a}{(Z_1 + Z_2 + Z_0)}$$

So the positive phase sequence current is limited by the sequence impedances connected in series. This suggests the sequence network connection depicted in Figure 5.10.

The total fault current is given by the sum of the sequence currents, and since $I_1 = I_2 = I_0$, this can be expressed as: $I_F = \dfrac{3V_a}{(Z_1 + Z_2 + Z_0)}$.

Phasor Diagram of the Phase-to-Ground Fault

It is instructive to construct a phasor diagram for the voltages and currents associated with the phase-to-ground fault. Let us begin with the currents, and assume that all the sequence impedances are equal and wholly inductive, and therefore that $Z_1 + Z_2 + Z_0 = 3Z_1$. Although this is not generally the case, it will simplify our phasor analysis. The phasor diagram for the faulted phase appears in Figure 5.11.

The total fault current lags the 'a' phase voltage by 90°, and each sequence component is equal in magnitude. According to the boundary conditions the currents in phases 'b' and 'c' will be zero. From Figure 5.11 we can see that this is indeed the case, since $I_b = \alpha^2 I_1 + \alpha I_2 + I_0$ and $I_c = \alpha I_1 + \alpha^2 I_2 + I_0$.

The boundary conditions also require that $V_a = 0$. From the positive sequence network we can write:

$$V_1 = V_a - I_1 Z_1 = V_a - \frac{V_a Z_1}{(Z_1 + Z_2 + Z_0)} = \frac{2V_a}{3}$$

Similarly, from the negative and zero-sequence networks we find:

$$V_2 = -I_2 Z_2 = -\frac{V_a Z_2}{(Z_1 + Z_2 + Z_0)} = -\frac{V_a}{3}$$

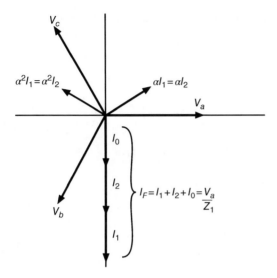

Figure 5.11 Phase-to-ground fault: current phasor diagram.

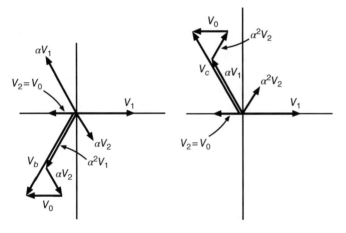

Figure 5.12 Construction of phase 'b' and 'c' voltages from sequence components.

and $\quad V_0 = -I_0 Z_0 = -\dfrac{V_a Z_0}{(Z_1 + Z_2 + Z_0)} = -\dfrac{V_a}{3}$

Therefore Equation (5.1) yields:

$$V_a = V_1 + V_2 + V_0 = \frac{2V_a}{3} - \frac{V_a}{3} - \frac{V_a}{3} = 0$$

The 'b' and 'c' phase voltages are unchanged as a result of the fault. This can also be demonstrated by the phasor diagrams shown in Figure 5.12, since $V_b = \alpha^2 V_1 + \alpha V_2 + V_0$ and $V_c = \alpha V_1 + \alpha^2 V_2 + V_0$.

Thus a symmetrical component analysis has provided an expression for the fault current in terms of the sequence impedances, while satisfying all the boundary conditions.

5.3.2 Phase-to-Phase Fault

A phase-to-phase fault between phases 'b' and 'c' is shown in Figure 5.13 and is characterised by the boundary conditions:

$$I_a = 0$$
$$I_b = -I_c \text{ or } I_b + I_c = 0$$
$$V_b = V_c$$

Expressing the boundary conditions in terms of their symmetrical components we have:

$$I_a = I_1 + I_2 + I_0 = 0$$
$$I_b + I_c = \left(\alpha^2 + \alpha\right)I_1 + \left(\alpha^2 + \alpha\right)I_2 + 2I_0 = 0$$
$$V_b = V_c$$

Therefore:

$$\alpha V_1 + \alpha^2 V_2 + V_0 = \alpha^2 V_1 + \alpha V_2 + V_0$$

These equations can only be true if:

$$I_0 = 0 \text{ and hence } V_0 = 0$$
$$I_1 = -I_2$$
$$\text{and } V_1 = V_2$$

Writing the network equations we find:

$$V_1 = V_a - I_1 Z_1 = V_2 = -I_2 Z_2 = I_1 Z_2$$

So we find:

$$I_1 = \frac{V_a}{\left(Z_1 + Z_2\right)}$$

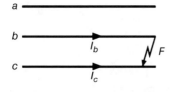

a
b
I_b
F
c
I_c

Figure 5.13 Phase-to-phase fault.

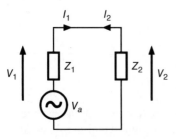

I_1 I_2

Z_1 Z_2

V_1 V_2

V_a

Figure 5.14 Phase-to-phase fault sequence network connection.

This equation is satisfied by a series connection of the positive and negative sequence networks, as shown in Figure 5.14. Since there is no ground or neutral connection, the zero-sequence network is absent from this analysis.

If we consider the phasor diagram for this fault, assuming as before that the positive and negative impedances are equal and wholly inductive, we obtain the diagram in Figure 5.15. Here I_1 lags V_1 by 90° and therefore I_2 leads V_1 by the same amount. The 'b' phase current, driven by the (shorted) line voltage V_{bc}, is given by $I_b = \alpha^2 I_1 + \alpha I_2$ and thus, as expected, has a magnitude equal to $\sqrt{3}I_1 = \dfrac{\sqrt{3}V_a}{Z_1 + Z_2}$. A similar diagram can be drawn for I_c.

Figure 5.15 Phase-to-phase fault, current phasor diagram.

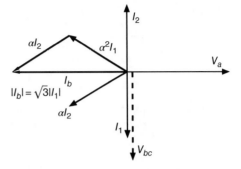

Figure 5.16 Phase voltages resulting from the phase-to-phase fault.

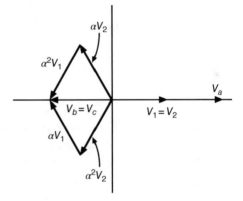

The system voltages resulting from this fault are shown in Figure 5.16. These can be found from the equations:

$$V_a = V_1 + V_2 = (V_a - I_1 Z_1) - I_2 Z_2 = (V_a - V_a / 2) + V_a / 2$$
$$V_b = \alpha^2 V_1 + \alpha V_2 = \alpha^2 V_a / 2 + \alpha V_2 / 2 = -V_a / 2$$
$$V_c = \alpha V_1 + \alpha^2 V_2 = \alpha V_a / 2 + \alpha 2 V_2 / 2 = -V_a / 2$$

From the phasor diagram we see that $V_b = V_c =$ *the average of these voltages seen in the un-faulted network*, while V_a remains unchanged.

5.3.3 Phase-to-Phase-to-Ground Fault

The phase-to-phase-to-ground fault is shown in Figure 5.17 and is characterised by the boundary conditions:

$$I_a = 0$$
$$V_b = 0$$
$$V_c = 0$$

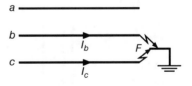

Figure 5.17 Phase-to-phase-to-ground fault.

Writing the boundary conditions in terms of their sequence components we find:

$$I_a = I_1 + I_2 + I_0 = 0$$
$$V_b = \alpha^2 V_1 + \alpha V_2 + V_o = 0$$
$$V_c = \alpha V_1 + \alpha^2 V_2 + V_o = 0$$

These equations require that:

$$V_1 = V_2 = V_o$$
$$I_1 = -(I_2 + I_0)$$

From the network equations we find:

$$I_o = \frac{-I_1 Z_2}{Z_2 + Z_0} \quad \text{and} \quad I_2 = \frac{-I_1 Z_0}{Z_2 + Z_0} \tag{5.11}*$$

and:

$$V_a = I_1 \left[Z_1 + \frac{Z_0 Z_2}{Z_0 + Z_2} \right] \tag{5.12}*$$

(*The derivation of these equations is the subject of a tutorial problem.)

Thus I_1 is found by connecting Z_1 in series with the parallel connection of Z_0 and Z_2, as shown in Figure 5.18, since this connection satisfies the boundary conditions.

$$(V_1 = V_2 = V_o \text{ and } I_1 = -(I_2 + I_0))$$

Finally, the fault current in phases 'b' and 'c' is given by:

$$I_b = \alpha^2 I_1 + \alpha I_2 + I_0$$
$$I_c = \alpha I_1 + \alpha^2 I_2 + I_0$$

As before, if we assume that $Z_1 = Z_2 = Z_0$ then we find that $I_1 = \frac{2V_a}{3Z_1}, I_2 = -\frac{V_a}{3Z_1}$ and $I_0 = -\frac{V_a}{3Z_1}$.

Thus $I_a = 0$, and I_b and I_c can be obtained from the equations above. These currents are shown in the phasor diagrams of Figure 5.19. Note that each current lags its associated driving phase voltage by 90°.

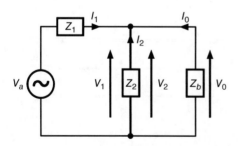

Figure 5.18 Phase-to-phase-to-ground fault sequence network connection.

---

* The derivation of these equations is the subject of a tutorial problem.

(a)

(b)

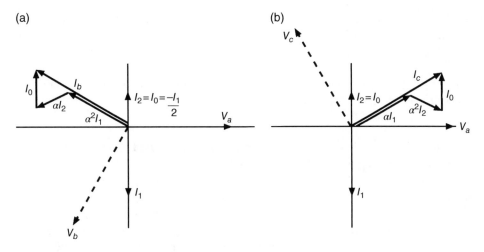

Figure 5.19 (a) Phase 'b' fault currents (b) Phase 'c' fault currents.

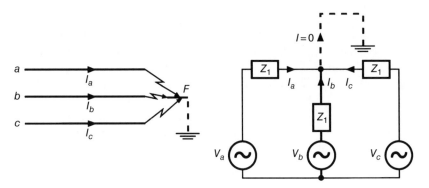

Figure 5.20 Three-phase fault. (Note: $I_a + I_b + I_c = 0$).

5.3.4 Three-Phase Fault

The three-phase fault is perhaps the simplest network fault to be considered. Here all three-phases are connected together; they may also be grounded but this makes no difference to the fault current flowing (Figure 5.20). The three-phase fault is a balanced condition, so we can reasonably expect that only the positive sequence impedances will be present. This is indeed the case, thus we can write:

$$I_a + I_b + I_c = 0$$
$$V_a = 0$$
$$V_b = 0$$
$$V_c = 0$$

It is relatively easy to show that in this balanced fault both I_2 and I_0 must be zero and therefore the fault consists of entirely positive sequence current. The fault current in each phase can therefore be calculated directly from the positive sequence network, since it is limited only by Z_1, thus:

$$I_1 = V_a / Z_1$$

Figure 5.20 shows the phase currents resulting from a purely inductive positive sequence impedance, Z_1. These currents sum to zero at the fault.

5.4 Measurement of Zero-sequence Components (Residual Current and Voltage)

Zero-sequence voltages and currents are also referred to as *residual voltages* or *residual currents*. The residual current in a circuit can be measured as shown in Figure 5.21a, where the secondary currents of a set of current transformers (CTs) are summed. The summation of the symmetrical components from each phase results in a residual current equal to three times the zero-sequence component, since both the positive and negative sequence components independently sum to zero.

Residual currents can also be measured using a *single* current transformer, through which all three-phase conductors pass. Since both the positive and negative sequence fluxes also individually sum to zero, any residual flux relates to a zero-sequence current flow, which induces current in the CT secondary, proportional to $3I_0$. This technique is used in the detection of residual currents in residual current devices (RCDs).

Residual voltages can be measured with an open delta connected voltage transformer (VT), as shown in Figure 5.21b. In this circuit, each primary winding sees a phase voltage and therefore the secondary voltage includes all the associated symmetrical components. Since the positive and negative sequence components individually add to zero, the residual voltage contains only the zero-sequence potential and is therefore equal to $3 V_0$. (This method will also detect the presence of third harmonic voltages, should they be present, as the third harmonic exhibits similar characteristics to zero-sequence voltages.)

The zero-sequence voltage on an LV system can be measured without the need for voltage transformers. This can be done by star connecting three identical impedances,

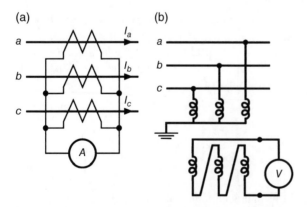

Figure 5.21 (a) Measurement of residual current (b) Measurement of residual voltage.

each connected to a different phase. The zero-sequence component then exists between the star point and the neutral terminal, as can be shown easily by using Millman's theorem.

In the special case where only positive and zero-sequence voltages exist, the zero-sequence component becomes a *neutral displacement potential*, as shown in Figure 5.22. The residual voltage is calculated by adding the phase voltages, and is again equal to $3V_0$.

$$V_{an'} = V_{an} + V_0$$
$$V_{bn'} = V_{bn} + V_0$$
$$V_{cn'} = V_{cn} + V_0$$
$$\text{Thus } V_{residual} = 3V_0$$

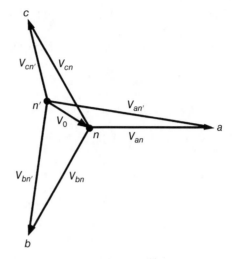

Figure 5.22 Neutral displacement voltage, V_0.

5.5 Phase-to-Ground Fault Currents Reflected from a Star to a Delta Connected Winding

It is useful to consider how a fault on a star connected transformer secondary winding is reflected into a delta connected primary. We will initially consider the phase-to-ground fault in our analysis and assume as before, that the sequence impedances are equal and purely inductive, and that the fault impedance is zero. For simplicity, the transformer is assumed to have a 1:1 turns ratio, as shown in Figure 5.23.

The current in each of the star windings has been expressed in terms of its symmetrical components, with the fault applied to phase 'a'. While the 'b' and 'c' phase currents equal zero, it should be remembered that the individual symmetrical components of these are not zero. As a result of the transformer's operation, the current in each of the

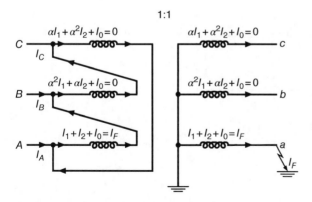

Figure 5.23 Star connected phase-to-ground fault reflected into a delta connected winding.

delta connected windings is equal to that in the corresponding star winding (neglecting any magnetising current).

By applying *Kirchhoff's current law* (KCL) to the node 'A' of the delta winding we find:

$$I_A = I_1 + I_2 + I_0 - \left(\alpha I_1 + \alpha^2 I_2 + I_0\right)$$

And since $I_1 = I_2 = I_0$, we can write:

$$I_A = I_1\left(3 - \left(\alpha^2 + \alpha + 1\right)\right) = 3I_1 = I_F$$

Similarly, by applying KCL at nodes 'B' and 'C' we have:

$$I_B = \alpha^2 I_1 + \alpha I_2 + I_0 - \left(I_1 + I_2 + I_0\right) = -3I_1 = -I_F$$
$$I_C = \alpha I_1 + \alpha^2 I_2 + I_0 - \left(\alpha^2 I_1 + \alpha I_2 + I_0\right) = 0$$

As a check on our analysis, we see that the line currents I_A, I_B and I_C sum to zero, as they should, since there is no return conductor associated with the delta winding.

Therefore a phase-to-ground fault on the star connected side is reflected as a phase-to-phase fault on the delta side. As expected, the line currents sum to zero and contain no zero-sequence component. However, this component must still flow in each of the delta windings, since it flows in the corresponding star winding. Thus I_0 circulates within the delta windings, limited by the total impedance $Z_1 + Z_2 + Z_0$. Therefore while it is impossible for a zero-sequence current to flow towards a delta winding, it is possible for it to be trapped in circulation within one.

5.5.1 Phase-to-Phase Fault Currents Reflected from a Star to a Delta Connected Winding

As a further example, consider the phase-to-phase fault reflected from a set of star connected windings into a delta connected set, as depicted in Figure 5.24. From our previous analysis we know that $I_0 = 0$ and $I_1 = -I_2$. The current flowing in each winding is also shown in Figure 5.24.

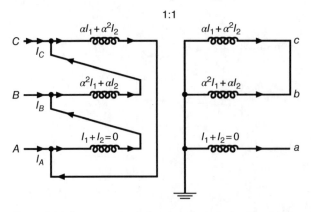

Figure 5.24 Star connected phase-to-phase fault reflected into a delta connected winding.

Applying KCL to node A and remembering that $I_1 = -I_2$, yields:

$$I_A = -\left(\alpha I_1 + \alpha^2 I_2\right) = I_1\left(\alpha^2 - \alpha\right) = -\sqrt{3}I_1$$

and similarly at node B:

$$I_B = -\left(\alpha^2 I_1 + \alpha I_2\right) = I_1\left(\alpha^2 - \alpha\right) = -\sqrt{3}I_1$$

and finally to node C:

$$I_c = \alpha I_1 + \alpha^2 I_2 - \left(\alpha^2 I_1 + \alpha I_2\right) = 2I_1\left(\alpha - \alpha^2\right) = 2\sqrt{3}I_1$$

These equations confirm that $I_A + I_B + I_C = 0$ and they also show that a phase-to-phase fault on the star winding results in fault currents in all three phases of the delta winding. It is also interesting to note that the fault current in one phase is twice that in the others.

5.6 Sequence Components Remote from a Fault

The positive sequence voltage at a fault is generally depressed from its un-faulted value of V_a as a result of the positive sequence current flowing into the fault. Consider the positive sequence voltage on a radial feeder away from the fault, closer to the source. Since the positive sequence impedance decreases towards the source, then we can expect the positive sequence voltage to progressively rise as a result. Thus at locations remote from the fault, the system voltage will almost take on its normal value, as illustrated in Figure 5.25.

As discussed, negative and zero-sequence voltages do not normally exist in an un-faulted network and only arise as a consequence of faults or extremely unbalanced loads. The negative and zero-sequence networks suggest that their magnitudes are largest at the fault and tend towards zero, remote from it, as shown in Figure 5.26. Negative

Figure 5.25 Radial Feeder: positive sequence voltage gradient.

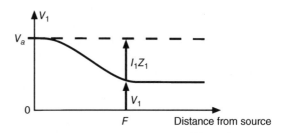

Figure 5.26 Radial Feeder: negative sequence voltage gradient.

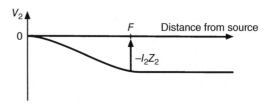

sequence impedances are generally very similar to those of the positive sequence, except in parts of a network close to rotating machines, where $Z_1 \neq Z_2$.

5.6.1 Network Fault Levels

Transmission and distribution companies generally model both the three-phase and single-phase fault levels at major nodes throughout their networks. The three-phase fault levels are expressed either in terms of the fault volt-amps $(= V_{line}^2/|Z_1|)$ or the fault current $(= V_{ph}/|Z_1|)$ at the location in question. In networks remote from generation assets, where $Z_1 \approx Z_2$, these fault levels collectively provide a measure of the positive, negative and zero-sequence impedances.

High three-phase fault levels are characterised by low positive sequence impedances. Such networks are said to be 'stiff' and exhibit similar characteristics to an *infinite bus*, where the voltage varies little with the application of either active or reactive loads. In contrast, low fault levels are representative of a high source impedance, which makes good voltage regulation more difficult to achieve.

Fault levels are determined by the network capacity at each location, so that parts of a network designed to supply large loads have correspondingly high fault levels. In HV networks the positive sequence impedance tends to be small and largely reactive and the X/R ratio is relatively high (typically 10–25). Therefore at major transmission substations the fault level will be very high, possibly as much as 30–40 kA. Fault levels also tend to rise substantially closer to generators, since these offer very low impedance. High fault levels require fast acting protection equipment, and switchgear with sufficient breaking capacity to successfully interrupt the large currents that arise during fault conditions. Therefore the latter must be carefully selected to ensure that prospective fault currents at a particular location can be safely and reliably cleared.

The MVA fault levels in MV distribution networks are lower than those in the adjacent HV system, although the current fault level can be higher. The MV three-phase fault current is largely determined by the positive sequence impedance of the associated HV/MV transformer. The approximate fault current can be calculated by dividing the rated MV transformer current by the transformer's impedance in per-unit, assuming a 1.0 pu bus voltage. For example, a 50 MVA, 132 kV:22 kV transformer has a rated MV current of 1312 amps. If it has an impedance of 0.12 pu, then the three-phase fault current is approximately 1312/0.12 = 10.9 kA. This rough analysis neglects the impedance of the upstream HV network; however, this is usually much lower than that of the transformer itself.

In city or urban areas where MV feeders are short, and the connected load is high, the fault level can be relatively large (10–15 kA), and is frequently matched to the interruption capacity of the downstream switchgear. In addition, the X/R ratio of an MV bus is usually considerably higher than that of the supplying HV network, due to the inductive impedance of the associated HV/MV transformer. MV X/R ratios of 15–30 are not uncommon. High X/R ratios make the bus voltage particularly sensitive to the application of reactive loads; inductive loads will cause the bus voltage to fall, while capacitive loads will cause it to rise. Further, any significant reactive unbalance between phases will result in the generation of negative sequence voltages, which should be avoided since their presence can cause damage to large rotating machines through increased I^2R heating and the production of a retarding torque. Fortunately, at most MV potentials, the load is well balanced and the power factor is usually reasonably close to unity.

On rural MV feeders the fault level falls progressively along the route, due to the accumulated impedance of the feeder itself. Therefore at large distances from the substation, the MV supplies can exhibit quite a low fault levels, sometimes as small as a few hundred amps. This is usually accompanied by a progressive fall in the X/R ratio, as the source impedance becomes more representative of the transmission line and less so of the transformer. As a result the bus voltage becomes sensitive to changes in both active and reactive loads. Voltage regulation on long feeders is generally managed using series-connected regulators, installed at regular intervals along the line.

LV fault levels are relatively low in residential distribution systems (2–6 kA), where many customers are supplied from a small pole- or pad-mounted transformer, via aerial or underground conductors. The size of the wiring within a residential customer's premises is generally sufficient to further reduce the fault level. This is well suited to the fault capacity of the small circuit breakers found on residential switchboards, which is typically 4–10 kA.

On the other hand, large industrial LV networks with dedicated MV/LV transformers can exhibit very large LV fault levels indeed, 40–60 kA being quite possible, depending on the size and impedance of the transformer(s) on the site. In these applications it is common to use split bus arrangements whereby the total load is split into smaller components, each fed by a dedicated transformer, thereby reducing the fault level. In such circumstances it is critical not to parallel transformers and thereby exceed the interruption capacity of the associated switchgear.

5.7 Problems

1 Derive Equations (5.11) and (5.12) for the phase-to-phase-to-ground fault, from the boundary conditions and the network equations.

2 Confirm from the relevant boundary conditions and network equations that for the three-phase fault, no negative or zero-sequence components exist.

3 Reflect a phase-to-phase-to-ground fault on a star connected secondary winding into the delta connected primary on the same transformer, and obtain expressions for the three line currents supplying the delta winding.

4 Table 5.2 contains the maximum published fault levels and sequence impedance data of an Australian transmission substation, having 220, 110 and 22 kV buses. These correspond to maximum bus voltages of 1.1 pu. The simplified bus arrangement of this substation is shown in Figure 5.27. The 22 kV bus is supplied from two 110:22 kV, 50 MVA transformers, feeding a common 22 kV bus. Two transformers have been installed to provide a level of redundancy for the 22 kV bus and the MV feeders it supplies. These transformers have been dimensioned so that the combined 22 kV three-phase fault level is less than 13 kA (which is the rated interruption capacity of the downstream MV switchgear). Therefore the bus can be operated with the coupler closed.

   A Verify the 220 kV, 110 kV and 22 kV three-phase and single-phase fault levels from the tabulated impedance data.

Table 5.2 Maximum published fault level and sequence impedance data.

Bus voltage (kV)	Three-phase fault (kA)	Single-phase fault (kA)	Per-unit positive sequence impedance (100 MVA base)			Per-unit negative sequence impedance (100 MVA base)			Per-unit zero-sequence impedance (100 MVA base)		
			R_1	X_1	X/R	R_2	X_2	X/R	R_0	X_0	X/R
22	10.1	5.6	0.006	0.286	47.667	0.006	0.287	47.833	0.001	0.961	>50
110	14.7	16.5	0.002	0.039	19.500	0.002	0.039	19.500	0.001	0.025	25.000
220	12.9	15.4	0.002	0.022	11.000	0.002	0.023	11.500	0.001	0.010	10.000

(All bus potentials = 1.1 pu)

B Assuming that the 22 kV bus, the 110 kV bus and the transformer cabling represent negligible impedance, estimate the positive (and negative) sequence impedance of each of the 110:22 kV transformers from the data provided. Compare your result with the actual transformer impedance data provided in Table 5.3.

C Repeat (b) for the case of the transformer's zero-sequence impedance.

D Calculate the 22 kV three-phase and single-phase fault levels when one transformer is out of service.
(Answer: Three-phase fault level = 4.7 kA; single-phase fault level = 3 kA.)

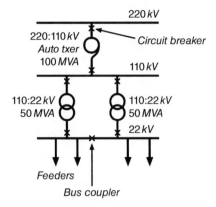

Figure 5.27 Simplified bus arrangement.

5 A 110 kV substation supplies energy to a nearby industrial site at 11 kV. The published fault levels and sequence impedance data at the substation are provided in Table 5.4.

A Verify the published fault levels from the impedance data provided. Notice that the 11 kV single-phase-to-ground fault level is particularly low when compared with the three-phase fault level. This is achieved through the installation of a *neutral earthing resistor* (NER), between the 11 kV neutral terminal and ground, as evidenced by the high R_0 value.

B Using the transformer data in question 4 as a guide, it is likely that the resistive component of the 110/11 kV transformer's zero-sequence impedance will be very low. Assuming this to be the case, estimate the value of the neutral earthing resistor installed in ohms. Estimate the power loss in this resistor during a single-phase fault.
(Answer: 1.95 Ω, 18.1 MW!)

C Estimate the supply transformer impedance if the industrial site is supplied by *three* high impedance 45 MVA, 110:11 kV transformers, connected in parallel. Why might three transformers have been used instead of one?
(Answer: $Z_{TX} = 0.005 + j0.310$ pu on a 45 MVA base.)

6 *Fault location on MV feeders*
The rapid location and repair of faults on feeders is an important task if the resulting outage duration is to be minimised. Fault location can be assisted by using fault current information obtained from protection relays in the source substation, or

Table 5.3 Data for 110 kV:22 kV transformer impedance.

110:22 kV transformer impedance data (50 MVA base)	
Positive and negative impedance	0.004 + j0.257 pu (Tap 5)
Zero-sequence impedance	j0.93 pu

Table 5.4 Industrial substation: maximum fault levels ($V = 1.1$ pu).

Bus voltage (kV)	Three-phase fault (kA)	Single-phase fault (kA)	Per-unit positive sequence impedance (100 MVA base)			Per-unit negative sequence impedance (100 MVA base)			Per-unit zero-sequence impedance (100 MVA base)		
			R_1	X_1	X/R	R_2	X_2	X/R	R_0	X_0	X/R
11	21.3	3.36	0.009	0.270	30.000	0.009	0.272	30.222	4.848	1.162	0.240
110	14.3	12.7	0.005	0.040	8.000	0.005	0.042	8.400	0.009	0.055	6.111

from reclosers positioned along the line. Figure 5.28 shows a typical scenario where a fault occurs on a transmission line some considerable distance from the parent substation. Recloser 1 sees the fault and recloses the line several times before locking itself out if the fault does not clear. After each reclose attempt the fault current seen is reported, so that by the time lock-out occurs operations staff have a good idea of the magnitude and type of the fault.

Since the fault level falls progressively away from the substation, due to the increasing line impedance, the location of the fault can be estimated from the magnitude of the reported fault current. The fault level also depends on the type of fault as well as on the fault impedance. The symmetrical component theory presented above assumed zero fault impedance, but this is not always the case. Feeder faults are usually phase-to-ground, phase-to-phase or occasionally three-phase faults, all of which may involve some form of fault impedance.

For example, when an overhead conductor breaks and falls to the ground, there will usually be a non-zero ground impedance between the fault location and the star point of the remote MV transformer. Such an impedance can reduce the fault current to well below its maximum value, and therefore the magnitude of the fault current alone provides insufficient information from which to locate the fault.

However, since the fault impedance tends to be predominantly resistive, the reactive component of the fault impedance is largely determined by the length of transmission line feeding the fault. Therefore if both the magnitude and phase of the fault current and voltage are known, the reactive impedance can be calculated, and from this the distance to the fault from the measuring device can be found.

A Re-analyse the phase-to-ground fault in terms of its symmetrical components, assuming that a ground impedance R_g exists in the ground path, and hence obtain an expression for the fault current I_F. (Hint: begin by redefining the boundary conditions associated with this fault in terms of R_g and evaluate the fault current accordingly.)

(Answer: $I_F = \dfrac{3V_a}{(Z_1 + Z_2 + Z_0 + 3R_g)}$. This result is intuitive, since when

$|Z_1 + Z_2 + Z_0| \ll R_g$ then $I_F \approx \dfrac{V_a}{R_g}$, as expected.)

B Assuming that the 11 kV line shown in Figure 5.28 has the characteristics shown in Table 5.5, and that a phase-to-ground fault current measured at the parent substation is $203\angle -56.5°$ amps, estimate the distance of the fault from the substation. How many possible fault locations are there? Assume that the voltage at the substation at the time of the fault was 1.05 pu and that the fault current is solely determined by the line impedance and the ground resistance, R_g.

(Answer: 31.5 km, three possible fault locations.)

C What current would you expect to be detected by the recloser? (i.e. relative to the voltage seen at the recloser.)

(Answer: $203\angle -43.9°$ amps)

D Estimate the ground resistance R_g included in the fault path in (a).

(Answer: $10\,\Omega$)

E Repeat (a) for a phase-to-phase fault in the presence of a small fault resistance R_f.

(Answer: $I_F = \dfrac{\sqrt{3}V_a}{(Z_1 + Z_2 + R_f)}$)

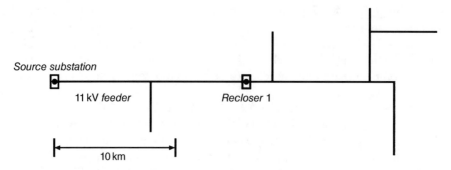

Figure 5.28 Feeder arrangement.

Table 5.5 Impedance and fault level data.

Measurement location	R_1 (Ω)	X_1 (Ω)	R_0 (Ω)	X_0 (Ω)	Distance from source substation (km)	3-phase fault level (A)	2-phase fault level (A)	1-phase fault level (A)
Recloser 1	3.24	6.15	5.95	29.48	16	914	791	437
Source substation	0.04	0.90	0.00	5.00	0	7049	6104	2801

F What fault current would you expect to see at the substation if a phase-to-phase fault were to occur at the same location as in part A? Assume $R_f = 10\,\Omega$.
(*Answer:* $347.5\angle -66.4°$ amps)

7 *Practical experiment: measurement of the positive, negative and zero-sequence voltage components.*
Modern digital protection relays compute symmetrical voltage and current components numerically, but it is useful to be able to observe symmetrical voltage components in real time, in order to obtain an appreciation for their magnitudes, given the degree of voltage unbalance present.

A simple negative sequence filter can be constructed from linear electronic components, capable of extracting the negative sequence voltage from an unbalanced network. The design of such a filter is based on the Equation (5.2) i.e. $V_2 = (V_a + \alpha^2 V_b + \alpha V_c)/3$. We initially consider the quantity $-V_2$, and since $-\alpha = 1\angle -60°$ and $-\alpha^2 = 1\angle 60°$, we can write:

$$-V_2 = 1/3\left[-V_a + 1\angle 60°V_b + 1\angle -60°V_c\right]$$

The equation above describes a negative sequence filter, the output of which contains the summation of the phase voltages, with each phase shifted by a different amount. The 'A' phase is shifted by 180°, while the 'B' and 'C' phases are shifted by +60 and -60° respectively. The block diagram shown in Figure 5.29 will satisfy the equation above. Here each phase voltage is supplied through a 100:1 step-down transformer, which enables the 'A' phase to be shifted by 180° through a winding reversal.

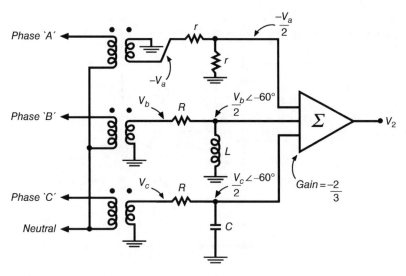

Figure 5.29 Negative sequence filter block diagram.

(a)   (b)

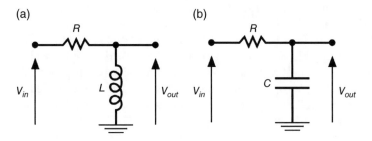

Figure 5.30 (a) 'B' phase inductive high pass filter (b) 'C' phase capacitive low pass filter.

The 'B' phase voltage is phase-shifted using a first-order RL high pass filter (Figure 5.30a), while the 'C' phase voltage is phase-shifted using a first-order RC low pass filter (Figure 5.30b). The transfer function of the 'B' phase high pass filter is given by:

$$\frac{V_o(j\omega)}{V_i(j\omega)} = \frac{j\omega}{(j\omega + \omega_o)}$$

from which the resulting phase shift ϕ is:

$$\phi = 90 - \tan^{-1}(\omega/\omega_o)$$

Where $\omega_0 = \dfrac{R}{L}$ radians per second.

A leading phase shift of 60° requires that $\omega/\omega_o = 1/\sqrt{3}$, where $\omega = 314$ r/s for 50 Hz systems or 377 r/s for 60 Hz systems. Thus $\omega_o = 543.9$ r/s (50 Hz), or $\omega_o = 652.7$ r/s (60 Hz). The magnitude response of the transfer function at ω is ½, which makes it

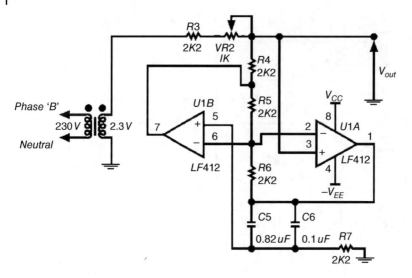

Figure 5.31 Gyrator-based RL filter (Phase B).

very easy to calibrate the filter by adjusting the resistance R, until the output voltage is equal to half that at the input.

If we choose $L = 4.45\,H$ then for a 50 Hz system R will be about 2.4 kΩ, and for 60 Hz $R \approx 2.9$ kΩ. Rather than using heavy and expensive wound inductors, in applications like this it is far easier to use an *Antoniou gyrator* to synthesise this component. A gyrator in this context is a linear operational amplifier circuit that uses a capacitor and several resistors to synthesise an inductor. The only restriction imposed by this topology is that one inductor terminal must be grounded. The schematic of Figure 5.31 shows the gyrator-based RL filter. The inductor appears on the right-hand side of the tuning resistance $VR2$, and the circuit ensures that the current flowing through resistor R_4 lags the output voltage by 90°. The inductor value is given by:

$$L = \frac{R_4 R_6 R_7 \left(C_5 + C_6\right)}{R_5} \approx 4.45H$$

Therefore by adjusting $VR2$, it is possible to tune the filter to the required frequency.

The 'C' phase is similar in concept to the 'B', except that a low pass filter is used to provide the necessary 60° lagging phase shift. In this case, a capacitor is used as shown in Figure 5.30b. The transfer function for this filter is:

$$\frac{V_o\left(j\omega\right)}{V_I\left(j\omega\right)} = \frac{\omega_o}{\left(j\omega + \omega_o\right)}$$

and the phase shift is given by:

$$\phi = -\tan^{-1}\left(\omega / \omega_o\right)$$

Thus for a lagging shift of 60° we find that $\omega / \omega_o = \sqrt{3}$, therefore for 50 Hz systems $\omega_o = 181.3\,r/s$ and for 60 Hz systems $\omega_o = 217.6\,r/s$. Again the magnitude response of

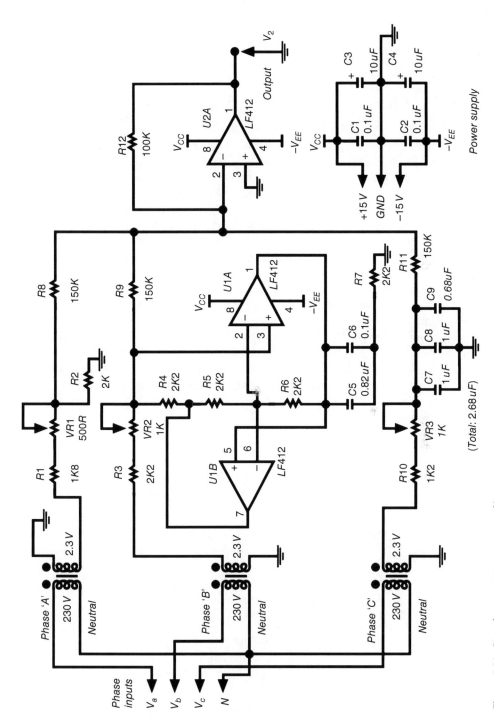

Figure 5.32 Complete negative sequence filter.

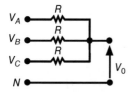

Figure 5.33 Zero-sequence filter.

the filter is ½ at the system frequency, which makes tuning easy. If the capacitor C is chosen to be approximately 2.7 µF, then for a 50 Hz system the resistance value required is 2 kΩ and for 60 Hz it is 1.7 kΩ.

The detailed schematic diagram of the negative sequence filter is shown in Figure 5.32. The summation stage consists of an inverting amplifier with a gain of 2/3, yielding the expected negative sequence voltage. By swapping the 'B' and 'C' phases, this filter can also be used to isolate the positive sequence component instead. The zero-sequence component can be obtained by the simple averaging circuit of Figure 5.33.

Application: These filters can be used to observe the sequence components of an unbalanced voltage network, generated using the connection of *variable transformers* (also known as *variacs*) as shown in Figure 5.34. This arrangement enables the user to unbalance the local three-phase supply in a controlled manner, including the introduction of a neutral shift potential. The outputs of all these filters should be observed on an oscilloscope as the supply is progressively unbalanced.

There are several ways to check that the negative sequence filter is working correctly. Firstly, with a balanced set of voltages the filter should not respond to the introduction of a neutral shift potential, since this represents a zero-sequence component only. Secondly, by beginning with a balanced supply, changing any one phase by δV, should result in a negative sequence voltage equal to $\delta V/3$. The phase of this voltage can be measured by comparing it with the 'A' phase voltage, which is the reference phase and therefore assumed to represent zero degrees. The complete loss of the 'B' phase, for example, will thus result in a negative sequence voltage of $-V_b/3$.

Experimental Procedure

1) Starting with phase voltages of about 120 volts at the variacs, adjust your circuit so that it is balanced, and remove any neutral shift potential that may be present. This can be done by observing the output of the phase sequence filters while adjusting the

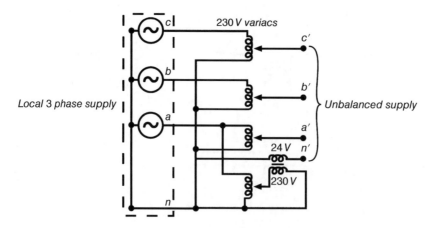

Figure 5.34 Generation of an unbalanced three-phase supply.

phase voltages for minimum filter outputs. Finally, check that your voltages V_a', V_b' and V_c' have the same magnitude and are shifted from one another by 120°.

Introduce an additional 60 volt component into the 'A' phase (i.e. $V_a = 180$ volts) and observe the output of the negative sequence filter. What negative sequence voltage (magnitude and phase) do you expect and why? Was this observed?

2) What do you expect that this unbalance will do to the positive sequence component of the system? Measure this to confirm your calculations.

3) What value of zero-sequence component do you expect this change will have introduced? Explain. Confirm your calculations by measurement.

4) Restore the balance to the system and introduce a known neutral shift potential. Which symmetrical components will this change affect, and why? Confirm your calculations by measurement.

5) Show that each symmetrical component can be considered as the sum of portions arising from several separate system unbalances, i.e. show that the principle of superposition applies. (Vectorial addition will be required when adding symmetrical components, so the angle as well as the magnitude of each must be measured. This can be done using an oscilloscope; take the 'a' phase as the reference phasor, i.e. zero degrees.)

6) Apply an arbitrary unbalance to the system and using an oscilloscope, measure the resulting symmetrical components, both magnitude and phase. Use your measurements and Equation (5.1) to calculate the each phase voltage from these components. Compare your calculated results with actual measurements.

5.8 Sources

1 GEC Alsthom Measurements Limited, *'Protective relays application guide'*, 1990, GEC Alsthom, Stafford.

2 Say, MG, *'Alternating current machines'* 1976, Pitman Publishing Ltd, London.

3 Kennedy, EJ, *'Operational amplifier circuits: Theory and applications'*, 1998, Holt Rinehart & Winston, New York.

6

Power Flows in AC Networks

In this chapter we explore the flow of active and reactive power in AC networks, in particular between network nodes interconnected with transmission lines. These may operate at either HV, MV or LV potentials, and we will see that a line's X/R *ratio* influences to a large degree the use to which it can be put.

Lines that are largely inductive in nature (i.e. having high X/R ratios) are suitable for the transmission of active power, often over great distances. However, if these are used to transmit substantial quantities of reactive power, the voltage drop created quickly becomes excessive and severely reduces the useful distance over which transmission is possible.

High voltage transmission falls into this category, and these lines are generally used for the transmission of active power between generation assets and load centres. Since reactive power can be more easily generated near to a load centre, it seldom appears in large quantities on the transmission network. This choice is also an economic one. High voltage transmission assets are expensive and are generally dimensioned with a priority given to active power transfer capability, without significant coincident reactive power flows.

We shall see that high X/R lines also have the advantage that active and reactive power flows can be controlled independently of one another; they are said to be *decoupled*. In contrast, long MV and LV lines have considerably lower X/R ratios and, as a result, the voltage drop on such circuits is no longer simply a function of the reactive power, but is also strongly influenced by the active power flowing. In this situation these quantities are no longer separately controllable, and are therefore said to be *coupled*.

Since it is not usually feasible to generate all the necessary reactive power close to each load, MV and LV feeders often have to carry much of the required reactive power as well. This fact, coupled with lower X/R ratios, means that the voltage drop along these lines can quickly become considerable. In order to extend MV transmission distances, *step regulators* are inserted in MV feeders at regular intervals to restore the line voltage. They are often based on a single-phase autotransformer, designed to regulate the incoming voltage over a range of ±10%. Regulators are either installed on two phases in an open delta configuration, or on all three, in a closed delta. (More will be said about step regulators in Chapter 7.) Voltage support capacitors are also sometimes installed on MV feeders to provide a local injection of reactive power.

LV transmission networks (particularly LV distribution cables) have very low X/R ratios, frequently less than unity. Here the active power flow determines the voltage drop that quickly accumulates along the line, since this is the principal transmission component. LV distribution networks therefore tend to be very short (generally less than about 400 m), supplying relatively few customers and requiring multiple distribution transformers.

In this chapter we will derive several new equations relating to the flow of active and reactive power in AC networks. We know that we can express both P and Q in terms of the power factor of a load, and here we will see that they can also be expressed in terms of a new angle, known as the *power angle*, sometimes referred to as the *torque angle*.

These equations will show that active power flows between nodes interconnected via high X/R lines, only if there is an appropriate phase shift between their voltages. They will also show that reactive power only flows in such networks when a voltage difference exists between nodes. Our analysis will provide us with a quantitative estimate for the magnitude of the phase shifts and voltage differences likely to be found in power networks.

Finally we analyse power in low X/R networks where active and reactive flows become tightly coupled, to the extent that it is difficult to control one without adversely affecting the other.

6.1 Power Flow Directions

Before we begin our analysis, it is necessary to clearly establish the sign convention which we will adopt for both active and reactive power flow in our discussion. We are initially concerned with the power flowing between a bus and a load, within a power system. We will assume that the bus is infinite, that is it has zero source impedance and therefore its potential remains fixed, in both magnitude and phase, regardless of the load placed upon it.

Our 'load' may consume power (as in the case of an industrial customer) or it may generate power and deliver it to the bus, for use by others. In this chapter we will consider the power flow from the perspective of the load, just as we did in Chapter 3. (This definition however, is not universal and some use its converse, defining the power flows from the perspective of the bus instead.)

Since most loads connected to the power system consume power (both active and reactive), we define the active and reactive *power flowing towards the load as positive*, i.e. when the current lies in quadrant four. We shall therefore continue to use the directional definitions given in Figure 3.17.

6.2 Synchronous Condenser

We begin our analysis with a discussion of the characteristics of a synchronous machine, one capable of operating as either a synchronous motor or as a synchronous generator. Figure 6.1 shows a simplified single-phase equivalent circuit of this machine.

The internal machine voltage, represented by E, is controlled by the DC excitation applied to the field winding mounted on the machine's rotor. This produces a DC magnetic field that is pulled around synchronously through its interaction with the rotating field produced by the stator windings, to which the phase voltages are connected.

Each stator winding has a synchronous reactance X associated with it, as well as a small winding resistance R. However, since the X/R ratio is usually large (>40 for large

Figure 6.1 Synchronous machine connected to a bus, V_B.

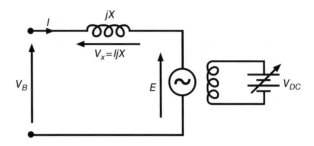

machines), the resistive component can be ignored for the purposes of our present discussion. The machine terminals lie downstream of the synchronous reactance, and are connected to the bus voltage V_B shown in Figure 6.1.

In order for the machine to act as a synchronous motor a load must be connected, one capable of absorbing mechanical power from the machine. Alternately, when operating as a

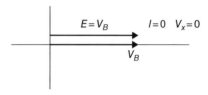

Figure 6.2 Phasor diagram of an idling synchronous machine.

generator a prime mover must drive the machine, to enable it to deliver power to the bus. Since we are only interested in the electrical behaviour of our machine, we will assume that these mechanical components come and go seamlessly, as required.

The current in Figure 6.1 is shown flowing from the bus to the machine, and therefore Kirchhoff's voltage law requires that:

$$V_B = E + i.jX \tag{6.1}$$

The phasor diagram in Figure 6.2 represents an idling synchronous machine, i.e. one that is neither consuming nor generating energy (friction and windage losses neglected). In this case the field current has been adjusted so that the induced voltage E *exactly* matches V_B. Since there is no potential dropped across X, ($V_x = 0$) no current flows as a result, and therefore there is no phase shift between E and V_B.

Under-Excited Synchronous Machine

The previous example is somewhat trivial since no current is flowing to or from the bus; however, the situation changes if the field excitation is altered. Suppose that the field current is reduced slightly. As a result the stator voltage E will also reduce, so that it no longer equals the bus voltage V_B, as shown in Figure 6.3.

This situation requires that the reactive voltage V_X, must make up the short fall between E and V_B. Since the current must *lag* the drop across V_X by 90° it must also lag behind E and V_B as well. Therefore an under-excited machine will behave in a similar manner to an inductance, consuming a lagging current from the bus and therefore absorbing VArs.

Over-Excited Synchronous Machine

If the field excitation is *increased* in our machine then the induced stator voltage E, will also increase, and may exceed V_B in magnitude, in which case we say the machine is over-excited. This situation is depicted in the phasor diagram of Figure 6.4. In this case, the difference between V_B and E appears as V_X, which now lies 180° out of phase with V_B.

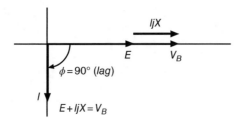

Figure 6.3 Under-excited synchronous machine.

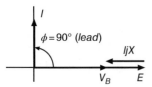

Figure 6.4 Over-excited synchronous machine (synchronous condenser operation).

As a result the machine current must lead both V_B and E by 90°.

In this mode of operation the machine exports VArs to the bus, since it draws a leading current, in a similar way to a capacitor. The machine is therefore said to be operating as a synchronous condenser, and in this mode can be used to support the bus voltage in times of high reactive demand by delivering VARs to the bus.

Synchronous machines can therefore, depending upon their level of excitation, either import VArs from the bus or export them to the bus, independent of any active power flow that may occur. Note that in both cases the VAr flow occurs only when there is a difference between the machine potential and that of the bus.

As a practical example, Figure 6.5a shows an ASEA 110 MVA synchronous condenser, installed in Victoria, Australia. It was made in Sweden in 1970 and features hydrogen-cooled windings. The machine operates at 750 rpm and requires no prime mover, nor does it drive any mechanical load. Its sole function is to either import or export reactive power in response to the demand on the local bus, thereby assisting in controlling the network voltage. These machines are relatively expensive to maintain and operate, and in recent years many have been replaced with static voltage support capacitors of the type shown in Figure 6.5b.

High voltage capacitors are inexpensive and virtually maintenance free, and can perform many of the functions of a synchronous condenser. However, they do suffer from

(a)

(b)

Figure 6.5 (a) ASEA synchronous condenser (b) 33 kV static voltage support capacitor.

three disadvantages. Firstly, they cannot absorb reactive power from a bus and therefore cannot assist in reducing its voltage during times of low reactive demand. Secondly, since the reactive support they provide is proportional to the square of the bus voltage, their effectiveness is reduced when this potential is at its lowest, which is precisely when it is most needed. And thirdly, they can only be switched in stages onto a bus. Despite this, in recent years, voltage support capacitors have been installed throughout many transmission and sub-transmission networks, to provide reactive voltage support.

When it is necessary to be able to import and export reactive power from a network in order to maintain voltage, static VAr compensators (SVCs) are often installed. In addition to capacitive VAr support, these devices use thyristor switched inductors to absorb reactive power from the bus at times of low reactive demand, when the voltage would otherwise become excessive.

6.3 Synchronous Motor

If a mechanical load is applied to our machine then it will behave as a motor, and consume active power from the bus, delivering it in the form of mechanical energy. Since power is consumed, the machine current must now have a component in phase with the bus voltage V_B, i.e. the phase angle (ϕ) must be less than 90°.

This situation is depicted in Figure 6.6 where the machine is both under-excited and motoring. The in-phase component of current is $I\cos(\phi)$ and thus the power consumed by the motor is $V_B I \cos(\phi)$ watts per phase. As before, Kirchhoff's voltage law requires that $V_B = E + IjX$.

Note that the voltage phasors E and V_B are also no longer in phase; instead they differ by an angle δ, known as the *power angle* (or alternatively as the *torque angle*). As the load placed upon the machine increases so does the power angle. As expected, the phase angle ϕ, is now less than 90°, since there is active power flowing from the bus to the machine.

6.3.1 Active Power Equation

A careful inspection of Figure 6.6 will reveal that the component of the machine voltage E, in quadrature with the bus voltage (i.e. 90° away from it) is equal to $|E|\sin(\delta)$. This is also equal to the quadrature component of the potential dropped across the synchronous reactance: $|I|X\cos(\phi)$. Therefore we can write:

$$|I|\cos(\phi)=|E|\sin(\delta)/X$$

Figure 6.6 Synchronous motor phasor diagram.

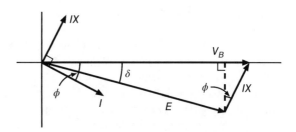

and thus the power P consumed from the bus is given by:

$$P = |V_B||I|\cos(\phi) = \frac{|V_B||E|\sin(\delta)}{X} \tag{6.2}$$

Equation (6.2) shows that an active power flow can only occur if there is a phase shift (δ), between the bus and machine voltages, and it flows towards the lagging potential.

6.3.2 Reactive Power Equation

We can derive a similar equation for reactive power from Figure 6.6 if V_B is be expressed in terms of the in-phase components of the machine voltage E and the potential dropped across the synchronous reactance, V_X. Thus:

$$|V_B| = |E|\cos(\delta) + |I|X\sin(\phi)$$

or $\quad |I|\sin(\phi) = \dfrac{|V_B| - |E|\cos(\delta)}{X}$

therefore:

$$Q = |V_B||I|\sin(\phi) = \frac{|V_B|^2 - |V_B||E|\cos(\delta)}{X} \tag{6.3}$$

Equation (6.3) confirms the fact noted earlier, that reactive power cannot flow between two nodes in a network unless there is an in-phase voltage difference between them. In this case Equation (6.3) expresses the reactive flow from the bus to the machine, and since $V_B > E\cos(\delta)$ then this flow is indeed positive and therefore VArs flow towards the lower potential.

6.3.3 Synchronous Machine Operation in All Four Quadrants

We have already seen that a synchronous machine can either import VArs from the bus or export them to it, depending upon its level of field excitation. In a similar vein, it can also import active power from the bus when motoring, or export it to the bus when generating.

Our synchronous machine can operate with its current in any of the four quadrants, as illustrated by the phasor diagrams in Figure 6.7. The flow of reactive energy, to or from the bus, is quite independent of that of active energy, since these quantities are determined by different parameters.

6.4 Generalised Power Flow Analysis

Equations (6.2) and (6.3) apply not only to synchronous machines, but to any situation where an inductive impedance exists between two nodes in a power network. This includes short transmission lines such as LV feeders, together with MV and HV lines whose lengths are less than about 80–100 km, where the effect of the line's *shunt capacitance*, or *charging capacitance*, is less pronounced. For longer lines where the shunt capacitance becomes significant, the line is represented by a 'π equivalent' circuit.

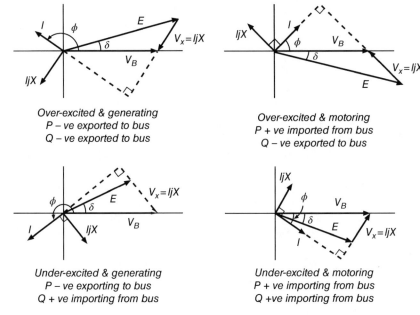

Over-excited & generating
P – ve exported to bus
Q – ve exported to bus

Over-excited & motoring
P + ve imported from bus
Q – ve exported to bus

Under-excited & generating
P – ve exporting to bus
Q + ve importing from bus

Under-excited & motoring
P + ve importing from bus
Q +ve importing from bus

Figure 6.7 Four-quadrant operation of a synchronous machine.

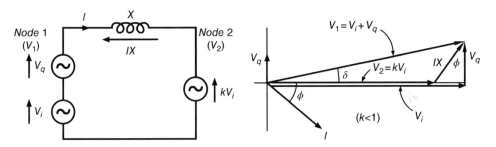

Figure 6.8 Dual node equivalent network and phasor diagram.

This consists of an appropriate series impedance, shunted at each end by a lumped capacitance, representing half the line's total charging capacitance. While the effect of these capacitances does modify the reactive power flow slightly, the following analysis will still provide a useful picture of the way in which the power flow can be controlled.

This general situation is illustrated in Figure 6.8, where power is exchanged between nodes 1 and 2. A phase shift has been introduced between these through the introduction of a quadrature voltage V_q in the potential at node 1. In addition, a difference between the in-phase voltage at nodes 1 and 2 is provided by the parameter k, applied to the voltage at node 2 ($k > 0$).

The reactance X in Figure 6.8 might represent that of a transformer, the reactance of a transmission line or perhaps a synchronous machine as described above. Since this analysis assumes a high X/R ratio, any small resistive impedance will be omitted from the following analysis.

From Figure 6.8 the following equation can be written for the current I flowing:

$$I = \frac{(V_i + jV_q - kV_i)}{jX}$$

Thus:

$$I = \frac{V_q}{X} + j\frac{V_i(k-1)}{X} = i_i + ji_q \tag{6.4}$$

Where i_i is the component of current in-phase with the voltage V_i and i_q is the component in quadrature with V_i.

Note that, in Equation (6.4), the component of the current in phase with V_2 depends on the quadrature voltage, V_q, while the quadrature current component depends on V_i and k. If $k = 1$ then the quadrature component becomes zero. If $k > 1$ then the quadrature current leads the voltage at node 2 by 90°, and if $k < 1$ then the quadrature current lags V_2, by the same angle.

The in-phase current component i_i is responsible for the active power flowing between the nodes, while the quadrature current i_q is responsible for the reactive power flow.

Active Power

From Equation (6.4) we can write equations for the active power flowing into node 2, in terms of the parameters in Figure 6.8:

$$P_2 = |V_2||I|\cos(\phi) = |V_2||i_i| = \frac{kV_iV_q}{X} \tag{6.5}$$

If we apply Equation (6.2) to the circuit of Figure 6.8 we again find that:

$$P_2 = \frac{|V_1||V_2|\sin(\delta)}{X} = \frac{\left(\sqrt{V_i^2 + V_q^2}\right)kV_i}{X}\frac{V_q}{\left(\sqrt{V_i^2 + V_q^2}\right)} = \frac{kV_iV_q}{X}$$

where V_1 is the voltage at node 1 and V_2 is the voltage at node 2.

This equation shows that the active power flow between nodes 1 and 2 depends on the existence of the quadrature voltage V_q. Without it there will be no phase shift between these nodes and therefore no active power flow.

Reactive Power

Equation (6.4) can also be used to obtain the reactive power flowing into node 2, as expressed in Equation (6.6). (The negative sign is included here since we have defined the reactive power flowing into node 2 as positive when the current lags the voltage, and since $\sin(\phi)$ is negative for negative ϕ, the minus sign is required.)

$$Q_2 = -V_2 i \sin(\phi) = -V_2 i_q = \frac{k(1-k)V_i^2}{X} \tag{6.6}$$

From Equation (6.4) we see that if $k < 1$ then a lagging current flows into node 2 which means positive VArs also flow into node 2. On the other hand, when $k > 1$ a leading current results, and the VAr flow is negative, i.e. from node 2 to node 1. These definitions are supported by Equation (6.6).

The reactive power flowing into node 2 can also be derived from Equation (6.3), but we must first adapt it to provide an expression for the reactive power flowing into node 2. Equation (6.3) is an expression for the reactive power flowing from the bus to the synchronous machine. We first rewrite this in terms of V_1 and V_2, expressing the reactive power flowing out of node 1 towards node 2:

$$Q_1 = \frac{|V_1|^2 - |V_1||V_2|\cos(\delta)}{X}$$

Next, we need to adapt this equation to represent the power flowing out of node 2, which we can do by reversing V_1 and V_2. However, since we require the reactive power flowing into node 2, we need to multiply our new equation by -1; hence we obtain:

$$Q_2 = \frac{\left(|V_1||V_2|\cos(\delta) - |V_2|^2\right)}{X} \tag{6.7}$$

From this we can also derive:

$$Q_2 = \frac{\left(|V_1||V_2|\cos(\delta) - |V_2|^2\right)}{X} = \frac{1}{X}\left\{\frac{\left(\sqrt{V_i^2 + V_q^2}\right)kV_i.V_i}{\left(\sqrt{V_i^2 + V_q^2}\right)} - k^2V_i^2\right\} = \frac{k(1-k)V_i^2}{X}$$

This confirms that reactive power will only flow between nodes provided there is a difference in the in-phase voltage between them, and that it flows towards the lower potential.

6.4.1 Typical Values for δ and k

What values might we typically expect for δ and k in a power system? Equation (6.4) can help answer this question. Here we see that the in-phase current component is given by V_q/X:

$$i = \frac{V_q}{X} + j\frac{V_i(k-1)}{X} = i_i + ji_q \tag{6.4}*$$

Since the reactance X may be as small as 0.05 pu in the case of distribution transformers and as high as 0.4 pu in the case of some transmission lines, then in order for the in-phase current component not to exceed 1 pu (nominal current), we find that V_q must be of this order as well. And since the in-phase component of the voltage at node 1 (V_i) is also likely to be around 1 pu, we can obtain an estimate for δ from Figure 6.8, where $\delta = \tan^{-1}(V_q/V_i)$. Over this range of impedances we find that δ roughly lies in a range 3–22°.

The quadrature component of the current expressed in Equation (6.4) i_q, is useful in estimating the typical range of values that k might take. Consider the term V_i/X. If V_i is around 1 pu, and assuming that $X \approx 0.05$ pu, then the value of V_i/X will be about 20 pu. This is a very large current, one that is unlikely to exist in a healthy power system. Equation (6.4) suggests that if i_q is to be around 1 pu, then k must lie in the range $0.95 < k < 1.05$. This provides us with an estimate for the typical range of operational bus

---

* Repeated here for convenience.

voltages that we might expect in a healthy system. According to Equation (6.4) larger impedance values will give rise to a wider range of bus voltages; however, except under fault conditions, these rarely fall outside the range 1.0 ± 0.1 pu, and supply authorities take considerable care to ensure this.

Since we now know that under normal operations system voltages generally lie close to 1 pu, and the power angle δ is relatively small, we can use this information to modify Equations (6.2) and (6.3):

$$P = \frac{|V_1||V_2|\sin(\delta)}{X} \qquad (6.2)*$$

From Equation (6.2) we can express δ in terms of per unit quantities in the form:

$$\sin(\delta) = px$$

and if δ is less than about 0.44 radians (25°), where $\sin(x) \approx x$ we can write:

$$\delta \approx px \, (\text{radians}) \qquad (6.2a)$$

where p and x are the per unit representations of P and X.

(Note that because p and x are dimensionless, this equation evaluates the power angle in radians.)

These two equations suggest that the active power is proportional to the power angle δ and therefore it can be controlled by adjusting this angle.

We can also rewrite Equation (6.3) in terms of the difference between the in-phase potentials.

$$Q = \frac{|V_1|^2 - |V_1||V_2|\cos(\delta)}{X} = \frac{|V_1|\left[|V_1|-|V_2|\cos(\delta)\right]}{X} \qquad (6.3)*$$

If we define $\Delta V = |V_1|-|V_2|\cos(\delta)$ then we can write:

$$\Delta v \approx qx \qquad (6.3a)$$

Where q and Δv are the per unit representations of Q and ΔV.

This equation suggests that the reactive power flow is proportional to the voltage difference between nodes, and therefore it can be controlled by adjusting this difference. Conversely, the voltage dropped throughout the HV network is largely determined by the reactive power flowing, and by keeping this to a minimum, the network voltage can be preserved. Generating reactive power close to the loads that require it greatly assists in this regard.

Reactive flows can be controlled by adjusting the voltage difference between nodes, which may sometimes be accomplished simply by tapping the appropriate transformer. According to Equation (6.2a) such an adjustment will *not* influence the active power flowing, since this is dependent mainly on the power angle δ. As a result we say the P and Q are largely decoupled.

Active and reactive power flows can therefore be controlled independently, by adjusting the phase and the voltage differences across the line. A phase shift can be obtained by introducing a quadrature component into each phase voltage at one end of the line, similar to that shown at node 1 in Figure 6.8. A voltage difference is slightly

---

* Repeated here for convenience.

easier to achieve using a transformer equipped with a voltage tap-changer. *Phase shift transformers* (or *phase angle regulators*), can independently control both these parameters, and hence are used whenever P and Q must be independently regulated.

In transmission and sub-transmission substations where there is a demand for it, capacitor banks are frequently used to generate most of the reactive power required by a load. These are switched on during periods of peak demand, in order to support the network voltage. Failure to inject sufficient reactive power will mean that generators must provide it via the network, and therefore system voltage will fall accordingly. Should a gross imbalance in reactive power arise between local generation and that demanded by the load (particularly during times of high active demand), then network generation will attempt to supply the difference, creating the possibility of a voltage collapse. At these times, the provision of reactive power, whether produced locally or by remote generators, becomes critical in preserving network voltage stability.

6.5 Low *X/R* Networks

The discussion thus far has assumed that network impedances are mainly inductive, which is certainly true in the case of generation and HV transmission assets. However, in MV and LV networks the X/R ratio can be small, in some cases less than unity. In these situations the resistive impedance cannot be ignored when evaluating the voltage drop between interconnected nodes. Figure 6.9 shows the phasor diagram of such a situation, where the internode line drop has been considerably enlarged for clarity.

We assume, as before, that the voltage at node 2 has a magnitude of 1 pu and supplies a lagging current I. The potential at node 1 is then given by:

$$V_1 = V_2 + IR + IjX$$

Expressing the current in terms of its in-phase and quadrature components, as shown in Figure 6.9, we find:

$$V_1 = V_2 + IR\cos(\phi) + IX\sin(\phi) + j\left(IX\cos(\phi) - IR\sin(\phi)\right)$$

The quadrature component of V_1 is usually small with respect to its in-phase component, thus we can write:

$$V_1 \approx V_2 + IR\cos(\phi) + IX\sin(\phi)$$

$$V_1 \approx V_2 + \frac{PR}{V_2} + \frac{QX}{V_2}$$

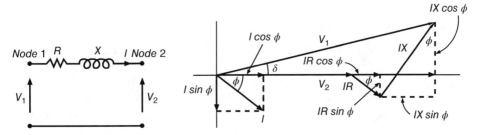

Figure 6.9 Line model and phasor diagram.

where P and Q are the active and reactive power measured at node 2.

Therefore the in-phase voltage drop between these nodes ΔV, can be expressed as:

$$\Delta V \approx \left(V_1 - V_2 \right) \approx \frac{PR}{V_2} + \frac{QX}{V_2} \tag{6.8}$$

Since the magnitude of V_2 is approximately 1 pu, we can express Equation (6.8) in terms of per-unit quantities:

$$\Delta v \approx pr + qx \tag{6.8a}$$

where lowercase symbols p, q, r, x and Δv represent the per-unit equivalents of P, Q, R, X and ΔV respectively.

We can also use Figure (6.9) to derive a similar equation for the power angle δ:

$$\delta \approx px - qr \, (\text{radians}) \tag{6.9}$$

If we solve Equations (6.8a) and (6.9) simultaneously for p and q, we find that:

$$p \approx \frac{\delta x + \Delta vr}{\left(x^2 + r^2 \right)} \tag{6.10}$$

$$q \approx \frac{\Delta vx - \delta r}{\left(x^2 + r^2 \right)} \tag{6.11}$$

These equations confirm that where r is not negligible with respect to x, then p and q are both dependent on Δv and δ; therefore they cannot be controlled independently of one another. In this situation they are said to be *coupled*. We can further highlight this by taking partial derivatives of Equations (6.10) and (6.11) to yield:

$$\frac{\partial p}{\partial \delta} \approx \frac{x}{\left(x^2 + r^2 \right)} \quad \text{and} \quad \frac{\partial p}{\partial \Delta v} \approx \frac{r}{\left(x^2 + r^2 \right)}$$

$$\frac{\partial q}{\partial \Delta v} \approx \frac{x}{\left(x^2 + r^2 \right)} \quad \text{and} \quad \frac{\partial q}{\partial \delta} \approx \frac{-r}{\left(x^2 + r^2 \right)}$$

As long as $x \gg r$ the active power flow is insensitive to voltage changes on the line, and depends only on the phase shift δ. However, when $x \approx r$, then p depends on both these parameters. Partial derivatives of Equation (6.11) show that q is similarly dependent on Δv and δ. Therefore once Δv and δ values are fixed, then so is the relationship between p and q.

6.5.1 Indicative Line Voltage Drops

Figure 6.10 shows the line drop as a function of the receiving end power factor for a 1 pu load, with X/R ratios of 8.0, 1.6 and 0.33. In each case a reactance of 0.02 pu was chosen.

The lowest curve ($X/R = 8.0$) is characteristic of a high voltage line, where the voltage drop remains small as long as the reactive load component is kept low ($PF > 0.95$). The middle curve is typical of LV or MV aerial feeders where $X/R \approx 1.6$. The line drop is

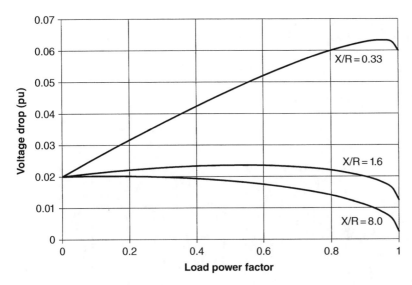

Figure 6.10 Line voltage drop as a function of load PF, for $X/R = 8.0$, 1.6 and 0.33 (1 pu load).

lowest at unity power factor and increases slightly as the load becomes more reactive, in this case peaking at a power factor around 0.6.

Finally, the upper curve is indicative of an LV distribution cable, where the X/R ratio is much less than unity. Consequently the line drop is largest when the load is resistive, falling almost linearly as does the power factor. This is unfortunate, since active power represents a major component of most LV loads and the voltage drop it creates is the reason that LV feeders tend to be short.

6.5.2 Phase Shifting Transformers (Phase Angle Regulators)

Occasionally it is necessary to control the power flowing in a transmission line without unduly disturbing the voltages at either end. In such cases a *phase shifting transformer* (also called a *phase angle regulator)* can be inserted into the line in question. Phase shifting transformers are frequently used to regulate the power flows in transmission lines running between states or across borders, or between distinct portions of a power system, where several parallel transmission paths may exist. In this situation, the power flowing in a given line will be inversely proportional to its impedance, and therefore where natural power flows are disproportionally shared between lines, phase shifting transformers can be used to achieve an acceptable balance.

By adding a quadrature voltage to each phase, a phase shifting transformer can regulate the phase shift between the ends of the line, and thus the active power flow can be controlled. They are also capable of adjusting the voltage at one end of the line relative to that at the other, so that in decoupled circuits active and reactive power flows can be independently controlled. Phase shift transformers can therefore be thought of as having a complex turns ratio, one that influences both the phase and the magnitude of the output voltage.

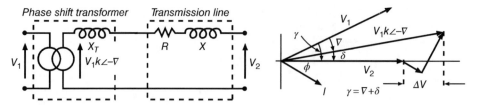

Figure 6.11 Transmission line, phase shifting transformer and associated phasor diagram.

Figure 6.12 Composite phase shifting transformer, 132 kV, 200 MVA (Armidale, Australia).

Figure 6.11 shows a transmission line into which a phase shifting transformer has been incorporated. If the bus voltages V_1 and V_2 are assumed to be fixed, in both magnitude and phase, then the transformer will directly affect the voltage and phase across the transmission line.

Figure 6.12 shows a 132 kV, 200 MVA phase shifting transformer installed in the Armidale substation in New South Wales (Australia). This transformer was installed in 2009 on a transmission line running between Armidale and Kempsey. Armidale is a substation that receives a particularly strong in-feed when energy is imported into NSW from Queensland. During such times this line frequently became overloaded and thereby limited the maximum in-feed permitted. A phase shifting transformer was installed to remove this limitation.

By regulating the phase difference across this line, the active power flowing through it can be controlled. Since this transformer can also control voltage drop in the line, it can also be used to control any reactive power flow without appreciably altering either the Armidale or the Kempsey bus voltages. As can be seen from the photographs, phase shifting transformers are large and expensive items, and therefore they are usually only installed on critical circuits.

This particular transformer comprises two tanks: one contains the series (quadrature) transformer, together with the phase shifting tap-changer, while the other contains the regulating transformer and the voltage tap-changer. The operation of this transformer will be analysed in detail in Chapter 7.

Having established the general relationships between P, Q, V and δ we will now explore two key power system phenomena, both of which are linked to the theory presented thus far. The first of these is the *stability limit* applicable to steady-state power transmission, and the second is the mechanism of *voltage collapse*.

6.6   Steady State Transmission Stability Limit

Equation (6.2) is useful in defining the transmission limit applicable to the power that can be carried by a transmission line. The maximum power that can be accommodated occurs when δ equals 90°.

$$P = \frac{|V_1||V_2|\sin(\delta)}{X} \tag{6.2}$$

Should the power angle attempt to exceed this value, then machines at either end of the line will lose synchronism, and thus 90° represents an absolute stability limit.

In practice, however, transmission lines are never loaded to this extent, and as a result the phase shift across a line and the system impedance at each end, rarely exceed 35–45°. There are several reasons for this. Firstly, allowance must be made for transient perturbations in δ, therefore sufficient headroom must be provided to accommodate momentary increases in δ without loss of synchronism. Secondly, it is common to operate parallel combinations of transmission lines where large quantities of power must be delivered. The unexpected loss of one (or more) of these will increase the loading on those remaining in service, and therefore the associated power angle will also increase.

The loading of a transmission line is also constrained by the thermal limit for the line, which is usually determined by the sag of the line and the minimum permissible ground clearance for the potential involved, together with the need to avoid annealing the conductors. Since the series reactance X increases with the length of the line, Equation (6.2) suggests that lines with a low series reactance can support higher power flows. Long lines therefore reach their stability limits earlier than short ones and, as a consequence, their loading must be proportionally reduced. The thermal limit is therefore generally more applicable to short transmission circuits where power flows are likely to be higher and power angles smaller.

Series capacitive compensation is sometimes inserted at intervals on long lines. This acts to reduce the effective inductive reactance, and therefore the load that can be carried is increased as a result. Equation (6.2) also suggests that the transmission limit is proportional to the square of the network voltage. Therefore long circuits tend to operate at higher potentials in order to achieve the required throughput while maintaining acceptably low power angles.

In order to increase the capacity of parallel connected circuits, long lines are often segmented through the inclusion of transmission substations, where all incoming and outgoing lines terminate on a common bus. The low impedance bus thus created (often characterised as a *high fault level bus*), effectively breaks the line into shorter lengths, each with a lower series impedance and able to provide a higher transmission capacity. Where parallel circuits operate at different potentials, autotransformers are frequently used as a low impedance tie, further increasing the fault level of the substation.

Finally, the voltage drop along a line also presents a limitation on the power that can be transmitted. This generally applies to medium length circuits where the line loading is not severely limited by stability constraints, but instead where the line impedance is of a sufficient magnitude to create a significant potential drop according to Equation (6.8a) ($\Delta v \approx pr + qx$). If, as discussed above, the load on the line is largely resistive the voltage drop can be kept acceptably low (<5%), but when the circuit is forced to supply substantial quantities of reactive power as well, the line drop can become severe and may ultimately lead to a voltage collapse.

6.7 Voltage Collapse in Power Systems

Voltage collapse has been the focus of considerable research in recent years, since when it occurs major outages arise throughout all or substantial parts of a network, requiring considerable effort to restore the system to a stable operating state. The following discussion is presented in order to demonstrate the practical importance of Equations (6.3a) and (6.8a), and to provide the reader with a knowledge of the tasks routinely undertaken by network operators in forecasting the likelihood of a voltage collapse, and the consequent steps taken to prevent one.

Voltage collapse in high voltage networks is usually preceded by a system disturbance. Examples of such events include the loss of an important element within the transmission network (a transmission circuit or a reactive compensation device), the unexpected disconnection of one or more generators, or a rapid and unexpected increase in load. Post disturbance voltages may decay rapidly (seconds) or they may slowly decline over many minutes (sometimes even longer). The post disturbance state of the system and the corrective actions subsequently taken to restore the stability margin will determine whether or not the network suffers from voltage collapse.

In cases where this is likely, the post disturbance network is usually subjected to high levels of active power demand coupled with an inability to supply sufficient reactive power to maintain normal voltage levels. The system response is influenced by the quantity, proximity and characteristics of active and reactive power sources, the characteristics of the connected load and the characteristics of those parts of the network still in service post disturbance. Some of these variables are discussed below.

6.7.1 Reactive Power Sources

Network connected synchronous generators and synchronous condensers typically operate in voltage regulation mode, which is achieved through the operation of an automatic voltage regulator (AVR), the primary function of which is to maintain a fixed voltage at the machine terminals. This can be accomplished by increasing or decreasing the machine voltage E, by varying the DC excitation applied to the rotor winding. In so doing sufficient reactive power is exchanged across the synchronous reactance X in Figure 6.1, such that the bus voltage V_B is held constant. The reactive power generation capacity of generators and synchronous condensers is not adversely affected by low network voltages, and they will therefore increase their reactive power output in response to a falling bus voltage.

We saw in our discussion of synchronous machines that they must be over-excited in order to export reactive power. Most machines have the ability to support short-term overloads in an effort to provide additional reactive power when network voltages are abnormally low. However, there is a limit to the field current that can be applied, and when this is reached the reactive power delivered by the machine will also become limited. In order to protect the machine from overheating, *over-excitation limiters* will eventually return the field current to its rated value and thus any additional short-term reactive support will be lost. This may be detrimental to system stability, depending on the speed of response of other mitigation control actions. The response of excitation limiters has been a contributing factor in several actual events where voltage instability has resulted in major power supply disruptions.

Voltage support capacitors are perhaps the next most common reactive power sources, due to their simplicity and relatively low cost. As mentioned earlier, one negative consequence of fixed capacitors is that the reactive power they provide falls with the square of the network voltage, thus they become less effective when most needed. An excessive reliance on shunt capacitors throughout a network may therefore be detrimental to voltage stability, unless applied in unison with other network control measures, such as under-voltage load shedding.

Static VAr compensators (SVCs) have the same limitations as shunt capacitors, once the maximum level of VAr generation has been reached. Another reactive compensation device is the static voltage source converter (STATCONs) (also known as static compensators, STATCOMs). These devices inject or consume reactive power via a converter driven from a DC source. Given their reliance on semiconductors, the maximum reactive capability of such devices is usually limited by the current rating of the components used. When this limit is reached (possibly after a short period of overload), the reactive current that can be injected remains constant, and the reactive power therefore decreases in proportion to the declining bus voltage.

6.7.2 Load Response Characteristics

The response of different load types to voltage variations can have a significant bearing on voltage stability. Soft loads reduce their active demand as the voltage falls. The power consumed by resistive loads, for example, varies as the square of the voltage and therefore, as it falls so the power they consume falls rapidly as well. This action provides relief to the power system at a time when it is most stressed, and it may even be sufficient to establish a new stable steady state operating point.

Hard loads, on the other hand, effectively consume constant power or near to it, regardless of the voltage supplied. Therefore, as the voltage declines, the current they consume rises in response, thus increasing losses within the network and therefore increasing the likelihood of voltage collapse. Phase controlled rectifiers used in electrowinning of zinc, aluminium or nickel are examples of hard loads. They deliver a constant DC current to the electrowinning cells, and since the process voltage changes little they appear as constant power loads. When the supply voltage falls, these devices respond rapidly by firing their thyristors earlier in each cycle (i.e. the firing angle becomes smaller) thereby increasing the current drawn from the network – at a time when it is most vulnerable.

The reactive characteristics of different load types may often vary quite differently with decaying voltage. For example, depending on the characteristics of the driven load, large induction motors can be particularly sensitive to the supply voltage. As the voltage decreases, the machine slip may increase, resulting in a marked increase in the reactive power consumed. Should the voltage fall sufficiently far for the machine to stall, then the stator windings must support the system voltage in the presence of a locked rotor, until the machine protection operates, during which time the reactive consumption will be very large indeed (several multiples of rated current).

The phase controlled rectifiers cited above behave as soft loads with respect to their reactive consumption. The power factor of these rectifiers is approximately given by the cosine of the firing delay angle. Since this angle reduces when the voltage falls we can conclude that the reactive power consumption will also reduce. Therefore phase controlled rectifiers will ease the reactive burden on the network during times of voltage decline, up to the point at which they operate with zero firing delay, beyond which both their active and reactive demands decline with falling bus voltage.

Numerous industry groups (e.g. IEEE, CIGRE, EPRI) have developed and documented load models that describe the response of different load types to fluctuations in both network voltage and frequency. It is recommended that the reader become familiar with the basic load types and the impact that each may have on power system stability and control.

The action of on-load tap-changers within the transmission and distribution networks can also have a destabilising effect on voltage stability. As the system voltage falls, tap-changers attempt to restore the potential seen by customers connected at lower voltages, thereby partially counteracting the relief offered by soft loads. In doing so, the impedance reflected into the network decreases according to the square of the transformer's turns ratio. As a result higher currents flow, further increasing the voltage dropped throughout the upstream network.

Load shedding is frequently used in networks under threat of voltage collapse and this has been successful in restoring stability in many cases. Provided that the system has not reached or exceeded its maximum power point, reducing both the active and reactive demands will result in an increase in the system voltage. This response can be demonstrated through the use of P–V and Q–V curves. These are standard analysis techniques used throughout the industry to predict the likelihood of voltage collapse and to assist in the selection of mitigating control actions in order to prevent widespread system interruptions caused through loss of voltage control.

6.7.3 Analysis of Voltage Stability (*P–V* and *Q–V* Curves)

The voltage collapse phenomena can be better understood through the use of graphical plots depicting the active power vs voltage (the P–V curve) and reactive power vs voltage (the Q–V curve) at given points in a network. Figure 6.13 shows a set of P–V curves for a particular node in a network. These are generally produced by undertaking a series of load flow studies that utilise computer models of the network, including generation and transmission assets as well as loads representative of the node concerned.

The P–V curve is a plot of the network voltage as a function of the power supplied to the node in question. Several curves are shown, each corresponding to a different load power factor. The maximum power point (or 'nose') on each curve occurs when the voltage dropped along the line equals that seen by the load, or when the load impedance equals that of the network. Beyond the nose of each curve, a decreasing active load corresponds to a continuously declining voltage at the node. By definition, this represents an unstable operating point, which implies a region of voltage collapse.

As the power factor becomes lower, the slope of each P–V curve becomes steeper, due to the increased reactive drop throughout the network. The voltage continues to fall with increasing power, up to the maximum power point (P_{max}), beyond which both the load power and voltage progressively fall with the load impedance, until eventually this

Figure 6.13 *P–V* curve.

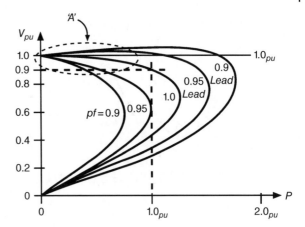

is so small that the voltage tends to zero. The portion of this curve with a positive slope is therefore voltage unstable.

Increasing the power factor of the load has the effect of increasing both the maximum power transferred and the voltage at which this occurs. This is not unexpected, since a local injection of reactive power (provided by a leading load) will help support the voltage as the active power is increased. For typical lagging loads, which consume both active and reactive power, the point of voltage instability occurs earlier, since the network must provide *both* the active and reactive components.

Under normal operating conditions, the network voltage resides around 1 pu and the power supplied is considerably less than P_{max} (region '*A*' in Figure 6.13). In this region the voltage is largely insensitive to the active power flow, and reactive power is responsible for the small voltage drop; the system is therefore decoupled. However, as the load increases towards P_{max}, the voltage becomes sensitive to *both* the active and the reactive power flows, and decoupling is lost.

So long as the operating point lies on that part of the *P–V* curve with a negative slope, the system will remain stable. However, this does not mean that the voltage will necessarily adopt an acceptable value. The post disturbance active and reactive loading may be high, approaching the maximum power point, and as a result the voltage may fall substantially prior to recovery. Figure 6.13 suggests that the injection of reactive power near to the load will assist by increasing the voltage, since this will effectively increase the load's power factor. Therefore *the provision of sufficient reactive power at critical locations in a network is particularly important in restoring voltage stability.*

The *Q–V* curve in Figure 6.14 can assist in demonstrating the benefit of local reactive power injection as well as validating the results from the *P–V* analysis. It is also typically plotted for several different values of active power, and represents the reactive power that must be injected at the node to achieve a given level of voltage. The *Q–V* curve can be understood by considering the zero reactive power line, shown dotted in Figure 6.14, which intersects the load curve P_1 at a node voltage of 0.9 pu. This will be the bus potential when supplying this load without any additional reactive power injection or consumption from the node. In order to increase the steady state operating voltage to say 1 pu, it is clear from Figure 6.14 that an injection of about 0.2 pu of reactive power will be required (i.e. the load power factor must improve).

Figure 6.14 Q–V curve.

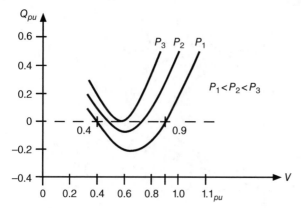

Similarly, if the voltage is to be reduced to say 0.8 pu, then reactive power can be removed from the bus. The practical interpretation of this is that the node can support more reactive load. The Q–V curve corresponding to load curve P_3 in Figure 6.14, represents a *marginally stable* system, since there is no reactive margin available at this operating point.

So long as the slope of the Q–V curve is positive, then the system will be voltage stable. In other words the voltage at the node must increase in response to the injection of reactive power. This ceases to be the case at the base of the curve, where the slope becomes negative, i.e. increasing reactive power injection into the bus results in progressively lower voltages. Therefore voltage stability ceases at the bottom of the Q–V curve.

The additional reactive power that can be supplied by the bus is the difference between the zero power line and the base of the Q–V curve. This quantity is referred to as the *reactive power margin* (or simply the *reactive margin*) and is shown in Figure 6.14. It provides a very useful measure of how close the node is from suffering voltage collapse and is used as a planning tool in many instances. In Australia, for example, the voltage stability criteria defined under the National Electricity Rules requires that each connection point in the network must operate with a reactive margin equivalent to not less than 1% of the maximum three-phase fault level at that point. So, for example, if the three-phase fault level at a connection point is 3000 MVA, then the minimum reactive margin that must be maintained is 30 MVA. This must continue to be the case following the most severe credible contingency event likely at that location.

The preparation of P–V and Q–V curves is time-consuming, since each must represent the network in its normal operating state as well as for various post disturbance conditions. As there are usually many contingencies that require consideration when assessing voltage collapse risks, the number of studies can be significant. Furthermore, such curves may be required for every major node in a network, especially where operational criteria are defined by rules or standards.

Armed with this information, the likelihood of voltage collapse and the need for load shedding can be predicted from the demand for active and in particular reactive power, throughout the network at any given time. By monitoring the state of shunt capacitors, SVCs, excitation limiters on generators and synchronous condensers as well as the temporal variation in the system voltage, the degree of reactive reserve available can be estimated and appropriate steps can be taken to prevent a voltage collapse.

6.8 Problems

1 **A** Use the phasor diagram in Figure 6.9 to show that for the power angle, δ can be expressed as: $\delta \approx (px - qr)$ radians.

 B Show that the line voltage drop $\Delta v \approx pr + qx$ is maximised when the load power factor is equal to $1/\sqrt{1+(X/R)^2}$. Compare this result with Figure 6.10.

2 Redraw the phasor diagram of Figure 6.8 to include a network series resistance R in addition to X. Repeat the analysis used to derive Equations (6.2) and (6.3) including the effects of R, and show that:

$$\frac{|V_1||V_2|\sin(\delta)}{X} = P - \frac{QR}{X} \quad \text{and} \quad \frac{\left(|V_1||V_2|\cos(\delta) - V_2{}^2\right)}{X} = Q + \frac{PR}{X}$$

3 Show that equations derived in question 2 yield the following expressions for P and Q:

$$P = \frac{X|V_1||V_2|_2 \sin(\delta) + R|V_1||V_2|\cos(\delta) - |V_2|^2 R}{\left(X^2 + R^2\right)} \quad \text{and}$$

$$Q = \frac{X|V_1||V_2|\cos(\delta) - R|V_1||V_2|\sin(\delta) - |V_2|^2 X}{\left(X^2 + R^2\right)}$$

4 **A** For the case of a lossless line show that the difference between the reactive power supplied to the line and that received by the load is equal to that consumed by the line itself, I^2X. (Hint: The *cosine rule* will be useful here.)

 B Consider the special case where the load is resistive, and show that all the reactive power supplied is consumed by the line reactance.

 C Figure 6.15 shows the phasor diagram of a lossless transmission line in which the magnitudes of the sending end and receiving end voltages are equal. Show that in this case the load receives no reactive power, and that half the reactive power required by the line is supplied from each end.

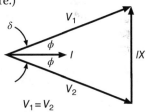

Figure 6.15 Lossless line phasor diagram.

5 A 240 mm² aluminium XLPE (*X-linked poly-ethyl-ene*) 400 V distribution cable has a resistance of 0.162 ohms/km and a reactance of 0.062 ohms/km. It is to be used to distribute 50 kW to a customer whose load has a power factor of 0.72. The installation proposed imposes a current rating of 288 A per phase on the cable, based on the temperature of the insulation. Determine the maximum line length that can be used if the total voltage drop is not to exceed 5%. (Answer: 722 m)

6 Repeat question 5 if the customer corrects their load to a power factor of 0.96. If the cable's voltage drop had not been a constraint, what would be the maximum load this cable could support? (Answer: 888 m, 200 kVA)

7 A 33 kV, 3-core, copper XLPE cable has a resistance, $R = 0.264$ ohm/km and reactance, $X = 0.12$ ohms/km, and is rated at 295 A per phase. What is the minimum cable length that can carry rated current while achieving a voltage drop equal to 0.1 pu, and under what conditions will this occur?
(Answer: 24.2 km, 0.1 pu drop will occur with a resistive load.)

8 Will the transmission distance calculated in question 7 impose any limitations on the use of this cable? Explain.

9 A 22 kV, 60 Hz aerial feeder has a resistance $R = 0.15$ ohms/km and reactance $X = 0.2$ ohms/km. If this line is to be used to carry 15 MW to a remote industrial load with a minimum power factor of 0.9, what maximum separation should be allowed between step regulators, assuming that each receives a potential of 0.95 pu and provides one of 1.05 pu?
(Answer: 13.1 km)

What would be the effect of introducing a series connected capacitance into each phase, midway between each regulator? What value capacitance would be required to extend the regulator spacing to 21.5 km?
(Answer: 617 µF)

10 A 132 kV transmission line runs for 200 km between substations 1 and 2. In substation 1 it is fitted with a 200 MVA phase shift transformer, having an inductive impedance of 13%, as shown in Figure 6.16. The line impedance is $z = 0.05 + j0.3$ ohms/km and in the absence of the transformer, 100 MW and 20 MVAr flow from substation 1 to substation 2.

A Let $V_2 = 1\angle 0$ pu and determine the phase shift and the voltage difference between the busses, in the absence of the transformer.
(Answer: $\delta = 0.333$ radians, $\Delta v = 0.126$ pu)

B Find the phase shift ∇ required from the phase shift transformer to reduce the active power flow to 50 MW, and also the gain k, required to simultaneously drive the reactive flow to zero. Assume that the bus voltages V_1 and V_2 remain as they were in part (a).
(Answer: $\nabla = 0.128$ radians, $k = 0.914$)

C Use Equations (6.10) and (6.11) as a check on the actual P and Q flows in part (b), including the effects of the line's resistance R.
(Answer: $p = -0.249$ pu; $q = -42 \times 10^{-6}$ pu)

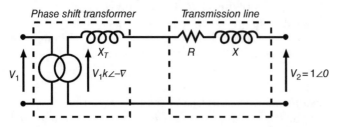

Figure 6.16 Schematic.

D Find the ∇ and k values required to ensure that 70 MW is imported by node 2 and 30 MVAr is imported by node 1.
(Answer: $\nabla = 0.251$ radians, $k = 0.817$)

11 Complete Table 6.1. Use the line's rating as the base VA in each case.

Table 6.1 Line data.

Line type	R (Ω/km)	X (Ω/km)	Rated current and line voltage	X/R ratio	Line rating (MVA)	Line loss per km Line MVA rating (%)	Voltage drop per km (1 pu load @ 0.9pf) (pu)
LV	0.160	0.060	150 A, 400 V	0.37	0.10	?	?
MV	0.150	0.200	400 A, 22 kV	1.33	15	?	?
HV	0.06	0.200	600 A, 220 kV	3.33	230	?	?

Answer:

Line type	R (Ω/km)	X (Ω/km)	Rated current and line voltage	X/R ratio	Line rating (MVA)	Line loss per km Line MVA rating (%)	Voltage drop per km (1 pu load @ 0.9pf) (pu)
LV	0.160	0.060	150 A, 400 V	0.37	0.10	10%	0.106
MV	0.150	0.200	400 A, 22 kV	1.33	15	0.47%	0.007
HV	0.06	0.200	600 A, 220 kV	3.33	230	0.03%	0.0007

6.9 Sources

1 Taylor CW. '*Power system voltage stability*', 1st ed, 1994, McGraw Hill, New York.
2 Kundor P. *Power system stability and control*. 1st ed, 1994, McGraw Hill, New York.
3 IEEE Power System Relaying Committee, Group K12. '*Voltage collapse mitigation report to IEEE power systems committee*' 1996, IEEE Power Engineering Society, New York.
4 Glover JD, Samara M '*Power system analysis and design*' 2nd ed, 1994, PWS Publishing, Boston.
5 Bhaladhare SB, Telang AS, Bedekar P. '*P–V, Q–V curves – a novel approach for voltage stability*'. IJCA, Number 5, Dec 2013, pp. 31–5.
6 Kaur P, Jaiswal M, Jaiswal P. '*Review and analysis of voltage collapse in power systems*', International Journal of Scientific and Research Publications, Jan 2012, Vol 2.

Part II

7

Three-Phase Transformers

Three-phase transformers come in many sizes and are found throughout the power system. Many are designed and built for specific applications, like the transmission transformer in Figure 7.1a, while others are 'off the shelf' designs, used in many standard distribution applications. The small *resin cast* transformer shown in Figure 7.1b is an example of such a transformer; it finds application in interior substations where the risk of fire and spillage from oil-insulated transformers is to be avoided.

Three-phase transformers have essentially the same per-phase equivalent circuit as do single-phase transformers, and they present the same impedance to positive and negative sequence currents, since altering the phase sequence of the supply makes no difference to a transformer's operation. However, the magnetic circuit architecture and the winding arrangements employed can make a large difference to the transformer's zero-sequence impedance Z_0.

In Chapter 4 we saw that the reactive impedance of a transformer was influenced to a large degree by the presence of an insulation filled duct between the windings, through which the leakage flux from one winding can pass, without linking the other. Because this flux is not seen by both windings, it manifests itself as a small inductance in series with the winding that created it. This concept applies equally well to both positive and negative sequence impedances for the reason cited above. The zero-sequence impedance, however, depends upon the transformer's ability to support zero-sequence flux, and this depends upon the architecture of the magnetic circuit and the winding arrangement adopted.

We will consider two core architectures, each widely used in the manufacture of transformers and will see that, depending upon the winding arrangements used, these can result in substantially differing zero-sequence impedances.

7.1 Positive and Negative Sequence Impedance

We derive an approximate expression for the leakage inductance and winding resistance of a two-winding transformer with concentric windings of equal length. Collectively these components comprise the positive or negative sequence impedances. A transformer's leakage inductance consists of two components, one associated with the primary winding and one from the secondary (x_1 and x_2, as shown in Figure 4.21). The leakage flux paths are set up in both the air duct between the two windings and within

AC Circuits and Power Systems in Practice, First Edition. Graeme Vertigan.
© 2018 John Wiley & Sons Ltd. Published 2018 by John Wiley & Sons Ltd.

(a)

(b)

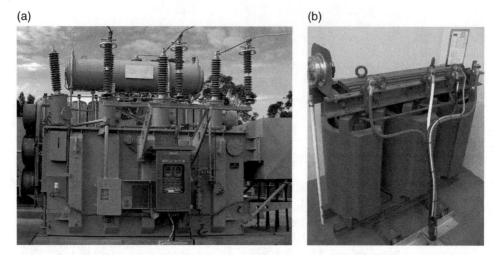

Figure 7.1 (a) 110 kV:33 kV, 60 MVA transformer (b) 22 kV:400 V, 100 kVA transformer.

each winding itself. Approximately equal MMFs suggest that half the duct flux arises from the primary winding and half from the secondary, as shown in Figure 7.2. (The leakage flux from each winding linking itself is not shown in Figure 7.2.)

Consider the duct flux and assume that half is generated by the primary MMF and half by the secondary MMF, as shown in Figure 7.2. The primary leakage inductance is given by:

$$L_p = N_p \frac{d\phi_p}{dI_p} = N_p \left(\frac{d\phi_{duct}}{dI_p} + \frac{d\phi_{winding}}{dI_p} \right)$$

where ϕ_p is the primary leakage flux, ϕ_{duct} is the primary leakage flux set up in the inter-winding duct and $\phi_{winding}$ is the primary leakage flux linking the primary winding itself. The last is a function of the distance from the core leg x (see Figure 7.2), since the number of turns, and therefore the exciting MMF, also vary with x.

We will consider these two flux components separately. ϕ_{duct} is set up in air within the inter-winding duct, and returns via the iron core. Because the duct reluctance is much higher than that of the iron core, we can neglect the MMF dropped across the iron, and assume that all the available MMF is applied to the leakage path length l_w. Therefore we can write:

$$\phi_{duct} = \frac{N_p I_p}{2\mathfrak{R}_{duct}} = \frac{N_p I_p \mu_o \pi d_m S}{2l_w}$$

(Note that only half the duct flux is generated by the primary, and the duct's cross-sectional area is approximately $\pi d_m S$.) The inductance element due to the duct flux is therefore:

$$L_{duct} = N_p \frac{d\phi_{duct}}{dI_p} = \frac{\left(N_p\right)^2 \mu_o \pi d_m S}{2l_w} \qquad (7.1)$$

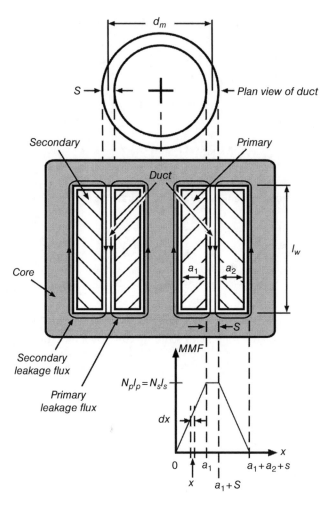

Figure 7.2 Core arrangement and leakage flux paths.

The second leakage flux component $\phi_{winding}$ links the primary winding itself. The distribution of this depends on the location x within the primary winding, since the exciting MMF increases with x, as shown in Figure 7.2.

Let the component of flux in an element of width dx in Figure 7.2 be $d\phi$. The MMF at this point is $N_p I_p x / a_1$, since MMF is assumed to *increase linearly* from zero at $x = 0$ to $N_p I_p$ when $x = a_1$. (In practice this is more likely to be a stepped function since the turns are generally wound in layers.) Thus we can write:

$$d\phi = \left(\frac{N_p x}{a_1} \right) \frac{I_p \mu_o \pi d_m}{l_w} dx$$

The corresponding increment of leakage inductance due to this flux is:

$$dL = \left(\frac{N_p x}{a_1}\right)\frac{d\phi}{dI_p} = \left(\frac{N_p x}{a_1}\right)^2 \frac{\mu_o \pi d_m}{l_w} dx$$

The primary leakage inductance due to all these flux elements can be found by integrating all elements across the winding, thus:

$$L_{winding} = \int_0^{a_1} \left(\frac{N_p x}{a_1}\right)^2 \frac{\mu_o \pi d_m}{l_w} dx = N_p^{\,2} \frac{\mu_o \pi d_m}{l_w}\frac{a_1}{3} \tag{7.2}$$

The total primary leakage inductance is given by the sum of Equations (7.1) and (7.2):

$$L_{Ptotal} = N_p^{\,2} \frac{\mu_o \pi d_m}{l_w}\left[\frac{S}{2}+\frac{a_1}{3}\right]$$

A similar analysis applied to the secondary winding leads to the result that:

$$L_{Stotal} = N_S^{\,2} \frac{\mu_o \pi d_m}{l_w}\left[\frac{S}{2}+\frac{a_2}{3}\right]$$

The total leakage reactance referred to the secondary winding can now be written:

$$x_2' = 2\pi f\left(L_{Stotal} + L_{Ptotal}\left(N_S/N_P\right)^2\right)$$

or:

$$x_2' = 2\pi f\frac{N_S^{\,2}\mu_o \pi d_m}{l_w}\left[S+\frac{a_1}{3}+\frac{a_2}{3}\right] \tag{7.3}$$

This result depends upon the physical layout and length of the windings, the duct width and the number of turns. Although it includes some approximations, it generally provides surprisingly good estimates of a transformer's leakage reactance.

Winding Resistance
The series resistive elements simply reflect the resistive component of the primary and secondary windings. Ignoring the *skin effect* at the system frequency (which sees high frequency currents flowing in the periphery of a conductor), the DC winding resistance can be evaluated from the equation:

$$r = \rho l/A$$

where ρ is the resistivity of the winding material (ohm-m), l is its length (m) and A is the cross-sectional area of the conductor (m^2). In terms of the transformer parameters, we assume that half of the window area is allocated to the primary winding and half to the secondary. We can therefore express the winding resistance as:

$$r_p = \frac{2\rho N_p^{\,2} l_{mtp}}{A_w S_F}$$

where r_p is the resistance of the primary winding, N_p is the number of primary turns, l_{mtp} is the mean turn length of the primary winding, A_w is the window area $\approx l_w(S + a_1 + a_2)$ and S_F is the space factor = the fraction of the window area occupied by copper.

It is interesting to note that where the volume available for the primary winding is constrained, the primary resistance is proportional to the square of the primary turns N_p. Similarly, the secondary resistance can be written:

$$r_s = \frac{2\rho N_s^2 l_{mts}}{A_w S_F}$$

where N_s is the number of secondary turns and l_{mts} is the mean turn length of the secondary winding (m).

Finally, the resistive component of the short-circuit impedance, referred to the secondary is given by:

$$r_2' = \left(r_2 + r_1\left(N_s/N_p\right)^2\right) = \frac{2\rho N_s^2\left(l_{mts} + l_{mtp}\right)}{A_w S_F} \approx \frac{4\rho N_s^2 \pi d_m}{l_w\left(S + a_1 + a_2\right)S_F} \tag{7.4}$$

Equations (7.3) and (7.4) suggest that the X/R ratio is independent of the number of turns on either winding.

7.1.1 Magnetic Circuit Architecture

A three-phase transformer can be constructed from three single-phase devices, each with a separate magnetic circuit, and early in the twentieth century this method was often used, each transformer being enclosed in its own tank. A more common approach, however, is to fabricate a single magnetic circuit capable of supporting the flux generated by each phase, thereby permitting the transformer to be housed in a single tank.

The construction of a three-phase, *single magnetic circuit* transformer can be achieved by merging three single-phase cores as suggested in Figure 7.3. When excited by a set of positive (or negative) sequence voltages, the common leg of the resulting transformer core effectively carries no flux, as the flux components from each of the three phases add to zero at every instant in time.

Therefore when constructing a three-phase transformer, it is not always necessary to provide a common limb. In the very early part of the twentieth century some transformers were built in exactly this manner, as shown in Figure 7.4a, but it was soon realised that the magnetic circuit could also be built in one plane, with only a slight loss of magnetic symmetry, as shown in Figure 7.4b.

The magnetic circuit arrangement shown in Figure 7.4 is known as a *core form* construction, in which one limb (or leg) is provided for each phase, i.e. a three-limb core. These magnetic circuits are

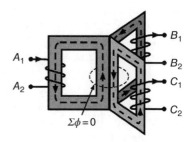

Figure 7.3 Flux cancellation in the centre limb.

Figure 7.4 (a) Very early core form distribution transformer (5 kV:120 V, 150 kVA – the third phase lies behind the other two) (b) A partially dissected core form transformer.

quite capable of supporting positive (or negative) sequence flux, since the three flux components sum to zero in the upper and lower parts of the magnetic circuit (known as *yokes*).

On the other hand, if a zero-sequence current were to flow through each phase winding, the core type magnetic circuit would present high magnetising reluctance, since no metallic return path exists for zero-sequence flux. We might therefore expect that the zero-sequence flux would be very small in such cases, as the return path must be set up in the air surrounding the core. As a result, the back EMF induced in each winding, per amp of zero-sequence current, will be small, as will the zero-sequence impedance (Z_0) of the transformer.

A second magnetic circuit arrangement also used in the manufacture of three-phase transformers, is depicted in Figure 7.5. This arrangement, known as a *shell form* magnetic circuit, has one limb for each phase plus two others which carry no windings, i.e. it is *a five-limb core*.

In the case of the shell form construction, the magnetic circuit for each phase is effectively separate, since a complete flux path exists independent of the other two phases. The five-limb shell architecture can thus be compared to the case where three separate single-phase transformers are used to collectively fabricate a three-phase device, each magnetic circuit being entirely separate from the others. The same is not true, however,

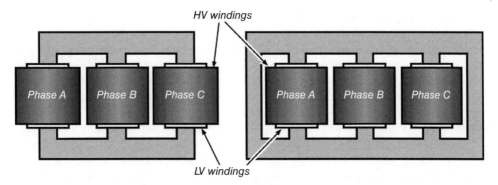

Figure 7.5 Three-phase core and shell form magnetic circuits. (Three-limb core form on the left and five-limb shell form on the right).

of the more common core form construction shown on the left of Figure 7.5, where the three magnetic circuits clearly have a degree of flux linkage.

Consider the operation of a shell form transformer with a zero-sequence current flowing in its windings. Because each phase effectively has its own magnetic circuit, a relatively large zero-sequence flux can exist in each limb (as would be the case in a single-phase transformer), thereby generating a substantial back EMF in each winding. As a result, the zero-sequence impedance presented by a shell form transformer can be large.

7.2 Transformer Zero-Sequence Impedance

The positive and negative sequence impedances of three-phase transformers are identical since, as mentioned earlier, reversing the phase sequence makes no difference to the operation of a transformer. The zero-sequence impedance on the other hand, depends upon both the choice of core architecture and the winding arrangement used.

Shell form cores have the ability to support a zero-sequence flux, provided the winding arrangement employed can pass the necessary zero-sequence current to generate one. In cases where a zero-sequence flux can be generated, a large zero-sequence impedance exists, equal to the *magnetising impedance* of the core concerned. However, if the current becomes large enough to saturate the core, Z_0 will fall to a quite a low value.

In contrast core form transformers cannot readily support zero-sequence flux and therefore generally exhibit low zero-sequence impedances where the winding arrangement permits it. However, the presence of nearby metallic components such as a tank wall or clamping structures can provide a partial return flux path, enabling a small zero-sequence flux to become established within the core, leading to an increase in Z_0.

The ability of a winding to carry *any* zero-sequence current depends on the connection arrangement used. Delta connected windings cannot deliver zero-sequence currents since they provide no return path and hence Z_0 will be infinite.

(A zero-sequence current can however circulate within a delta winding, provided a zero-sequence flux is established within the core limbs to excite one.) The zero-sequence performance of star connected windings depends on whether the neutral is grounded. A grounded neutral terminal provides the necessary return path for zero-sequence currents, enabling a zero-sequence flux to be established within the core. A floating neutral on the other hand, precludes the passage of zero-sequence currents in a similar fashion to a delta connection, yielding an infinite zero-sequence impedance.

Transformers with at least one grounded star winding will provide a local zero-sequence impedance to ground, the magnitude of which depends on the core architecture. Whether a transformer can pass zero-sequence currents from one winding to the other depends on *both* winding arrangements. A star-star wound transformer for example with both neutrals grounded can pass zero-sequence currents from one winding to the other. Such a transformer presents a low impedance to through currents, since flux cancellation between windings removes most of the zero-sequence flux. In this case Z_0 will generally equal Z_1 regardless of the core architecture.

Consider a Yy connected shell form transformer with an earthed HV star point, and a floating LV star point. A high Z_0 value will be presented to HV zero-sequence currents and an infinite one on the LV side. The same transformer when fitted with a closed tertiary delta winding, will exhibit a much lower HV Z_0 value. This is due to the flux cancellation between the zero-sequence current flowing in the HV windings and that circulating within the delta. The zero-sequence impedance to ground will therefore be equal to the sum of the leakage impedances of both these windings.

A set of zigzag connected windings with a grounded star point, as shown in Figure 7.6, is often employed in the construction of earthing transformers. This is because the series connection of phase windings causes a zero-sequence flux cancellation in each limb, yielding a particularly low Z_0 value, equal to the leakage reactance between the two. On the other hand such a cancellation does not occur for positive or negative sequence flux, and therefore a much higher magnetising impedance is presented to these currents. Thus a grounded zigzag winding will easily support positive and negative sequence potentials, while presenting very little impedance to zero-sequence

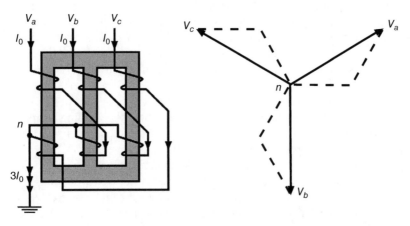

Figure 7.6 Zigzag connected earthing transformer and phasor representation.

currents, which only arise during ground faults. Therefore it is usual to insert a current transformer in the ground connection of an earthing transformer to detect zero-sequence fault currents.

7.3 Transformer Vector Groups

Single-phase transformers generate secondary voltages that are almost exactly in phase with the potential applied to the primary, and while the same is true of three-phase devices, an inherent phase shift arises when line voltages on one winding are used to create phase voltages on the other. For example, a 30° phase shift will exist between the primary and secondary phase potentials in a delta-star (delta-wye) connected transformer shown in Figure 7.7. The delta connected HV windings are excited by line voltages, but phase voltages are generated in the star connected LV windings.

Because of this it is necessary to specify precisely to which *vector grouping* a given transformer belongs. Vector groups are a simple way of representing both the winding arrangements and the associated phase shift inherent in a transformer. An alphabetic code is used to define the winding arrangement, and a numeric suffix denotes the resulting phase shift.

There are principally three winding arrangements that can be used on either the primary or the secondary winding of a three-phase transformer: star (wye), delta and zig-zag. Accordingly it is necessary to clearly identify the appropriate arrangement used in the vector group code. These connections are identified by Y or y, D or d and Z or z, respectively, depending on whether they are HV or LV windings. The connection used on the high voltage winding is specified by a capital letter which appears first in the vector grouping, while the low voltage winding is identified by a lowercase letter which appears second. In cases where there are more than two windings, this convention is repeated. If the neutral terminal of either winding is brought out of the tank, it is noted accordingly in the vector group, N for HV and n for LV.

The phase shift across the transformer is represented by a numerical suffix to the code. The phase shift of the secondary winding is measured with reference to the corresponding primary voltage. A clock face convention has been adopted where a primary

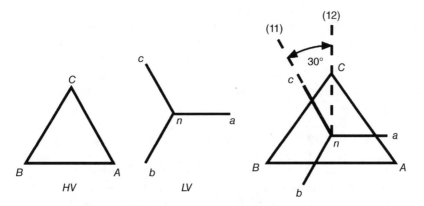

Figure 7.7 Vector grouping for the Dyn11 connection.

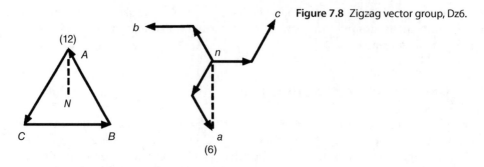

Figure 7.8 Zigzag vector group, Dz6.

phase voltage is assigned the 12 o'clock position. Secondary phase shifts are represented in multiples of 30°, and the phase shift through a transformer is the angle by which the secondary phase voltage lags the corresponding primary one. This is expressed according to the positions of the hours on the face of the clock.

For example the Dyn11 connection represents an HV winding that is delta connected and an LV winding that is star connected, as shown in Figure 7.7. The phase shift arises because although each pair of associated windings is wound on the same leg, the voltage across the LV winding represents a phase voltage while that across the primary is a line voltage. Figure 7.7 shows that if the *C-N* primary voltage is chosen as the reference at the 12 o'clock position, then the *c-n* secondary voltage lies in the 11 o'clock position. In other words, it lags behind the reference phasor by $11 \times 30°$ or 330°.

Figure 7.7 deserves further comment; clearly it is based on a phasor diagram, but it also shows the terminals of each winding, *A*, *B* and *C* for the HV windings, and *a*, *b*, *c* and *n* for the LV windings. Notice that voltages V_{AB} and V_{an} lie in phase with each other; this is because the associated windings are wound on the same core leg. Therefore windings *CA* and *cn* are also on the same core leg, as are windings *BC* and *bn*.

The vector group depends upon which winding is the HV and which is the LV. For example, the Dy11 connection shown in Figure 7.8 will become Yd1 when the HV and LV windings are interchanged. The phase identifiers between such pairs of groups add to either 12 or zero. Vector connections are known as *true* when both windings on the same leg are associated with the same phase. *Untrue* groups arise when windings are transposed from one phase to another.

The zigzag connection uses series connected windings from two core legs, as depicted in Figure 7.8, to support each phase voltage. This introduces an inherent phase shift between primary and secondary, which can be adjusted according to the connection used and the number of turns on each winding. The zigzag connection is frequently used to introduce small phase shifts between transformers.

Figure 7.9 shows some common vector groupings for two winding transformers.

7.4 Transformer Voltage Regulation

Because of its short-circuit impedance a transformer's secondary terminal voltage does not remain constant as load is applied. Moreover, the output voltage depends upon the power factor of the load as well as its magnitude in VA. Transformer voltage regulation is defined by Equation (7.5).

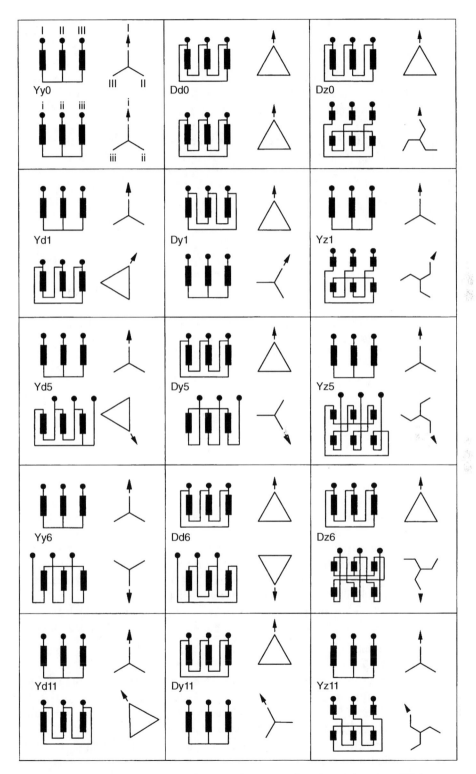

Figure 7.9 Some common vector groupings. (IEC 60076-1 ed.3.0 Copyright © 2011 IEC Geneva, Switzerland. www.iec.ch).

$$\text{Voltage regulation} = \frac{|E_2| - |V_2|}{|V_2|} \times 100\% \tag{7.5}$$

where $|E_2|$ is the magnitude of the open circuit output voltage and $|V_2|$ is this voltage under loaded conditions (see Figure 8.6). The regulation is equal to the per-unit drop in output voltage with respect to the loaded output voltage, and is expressed in per cent. It is a scalar quantity.

The output voltage differs from the secondary winding voltage by the drop across the short-circuit impedance. If we consider the primary series impedance elements r_1 and x_1 reflected to the secondary side, we will see in Chapter 8 that:

$$\frac{|V_2|}{|E_2|} \approx \frac{R}{(R+r)} \frac{X}{(X+x)} \tag{7.6}$$

where R is the resistive component of the load impedance and X is its reactive component. In this analysis the load consists of elements R and X connected in parallel, and r and x are the real and reactive components of the short-circuit impedance, i.e. $r = (r_1' + r_2)$ and $x = (x_1' + x_2)$ respectively, referred to the secondary. Finally, it is assumed that $R \gg r$ and $X \gg x$. It is useful to express all these parameters as per-unit quantities. Equations (7.5) and (7.6) can be combined to yield:

$$\text{Voltage regulation} \approx \left[\frac{r}{R} + \frac{x}{X} \right] 100\% \tag{7.7}$$

Further, since in many transformers the X/R is ratio lies between 10 and 50, the resistive term can sometimes be ignored, leading to the approximate result:

$$\text{Voltage regulation} \approx \left[\frac{x}{X} \right] 100\% \tag{7.7a}$$

Equation (7.7) is significant, since it predicts approximately how the voltage regulation varies as the load changes. For small loads, both R and X will be large, and the regulation will be low, i.e. the output voltage will scarcely change. However, as the load increases, both R and X will decrease, leading to an increase in both the resistive and reactive voltage drops, and hence a fall in the output voltage. This will also be accompanied by a slight phase shift across the transformer's impedance.

Figure 7.10 shows how the regulation varies as a function of power factor for a constant VA load. In this example $r = 0.01$ pu and $x = 0.06$ pu, and the load impedance has been set at 1 pu, independent of the power factor, i.e. $Z_{pu} = 1 / \sqrt{\frac{1}{R^2} + \frac{1}{X^2}} = 1\text{pu}$. The load power factor is equal to $\cos(\arctan(R/X))$.

Figure 7.10 includes both positive and negative values for X, corresponding to inductive and capacitive loads, or lagging and leading power factors respectively. Although for phase angles less than ±90° the power factor is always positive, in Figure 7.10 lagging power factors are defined as positive while leading power factors are negative. (This terminology is frequently used in energy metering applications to clearly distinguish between lagging and leading loads.)

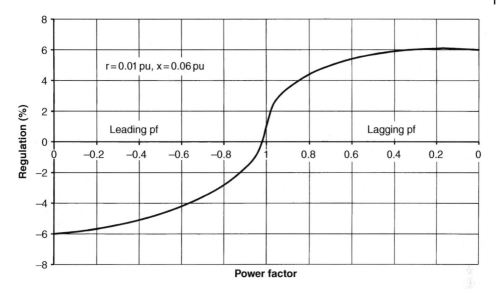

Figure 7.10 Transformer regulation as a function of power factor, for a 1 pu load.

The figure shows that at unity PF the regulation is small and positive (0.01 pu). This is simply due to the r/R term. As the load becomes inductive, the regulation increases rapidly and the output voltage falls. This is due to the fact that $x > r$ and therefore as the inductive component increases so does the regulation. It flattens as the PF decreases further, peaking when the voltage dropped across $r + jx$ lies in phase with the output voltage V_2. This occurs when $PF = \cos(\arctan(x/r))$. As the power factor continues to decrease, the regulation improves slightly.

The regulation curve becomes particularly interesting when the power factor is leading, i.e. as the load becomes capacitive. In these circumstances the x/X term is negative and the regulation decreases as a result and the terminal voltage will exceed that in the winding. When the load power factor reaches $-\cos(\arctan(r/x))$, the reactive term will cancel the resistive term and the regulation will equal zero. This means that the magnitude of the winding EMF and the transformer terminal voltage will then be equal.

As the power factor becomes progressively more capacitive, the reactive term begins to dominate and the regulation becomes negative, and therefore the output voltage rises beyond the secondary winding potential. In this case it reaches a minimum of -0.06 pu when the load is entirely capacitive; the regulation then being equal to $-x/X$. This behaviour is typical of the situation where a voltage support capacitor connected to an inductive source generates a small voltage rise.

7.4.1 Voltage Regulation in Yy Transformers without Flux Linkage

One disadvantage of Yy transformers where there is no flux linkage between phases lies in their ability to support unbalanced loads. Consider the case of a three-phase shell form transformer comprising a five-limb magnetic circuit. Because of its construction this type of transformer provides a low reluctance path for the production of zero-sequence flux and therefore it presents a high zero-sequence impedance. This presents

Table 7.1 Measured phase voltages on an asymmetrically loaded Yy transformer without flux linkage.

Phase	No load voltage (pu)	Phase to neutral load (pu)	Loaded phase voltage (pu)
A	1.0	0	1.56
B	1.0	0	1.15
C	1.0	0.03	0.73

two problems: the first as mentioned above is the inability to supply unbalanced loads, while the second relates to the low value of earth fault or zero-sequence currents that will flow in the transformer under fault conditions. This restriction may make the detection of earth faults difficult.

The voltage regulation of an asymmetrically loaded Yy transformer can be most easily understood in the case where a load is applied to one secondary phase only and where no primary neutral connection exists (which is frequently the case). In such a situation the primary current required by the loaded phase must be supplied via the other two, but since these are only carrying magnetising current, they appear as large inductive impedances. Accordingly the loaded phase voltage falls while the other two rise, generally asymmetrically.

This situation is illustrated in the following example. A 6 kVA three-phase Yy transformer consists of three 2 kVA single-phase units, and when loaded asymmetrically the loaded phase voltage falls while the others rise, as shown by the tabulated results below.

Table 7.1 shows that for a small asymmetrical load (0.03 pu, resistive) a Yy transformer exhibits a substantial change in phase voltages. This has been caused by a shift in the star point of the secondary winding. Such a situation occurs because of the generation of a zero-sequence flux within each leg and thus a zero-sequence voltage within each winding. The phasor diagram corresponding to this example appears in Figure 7.11.

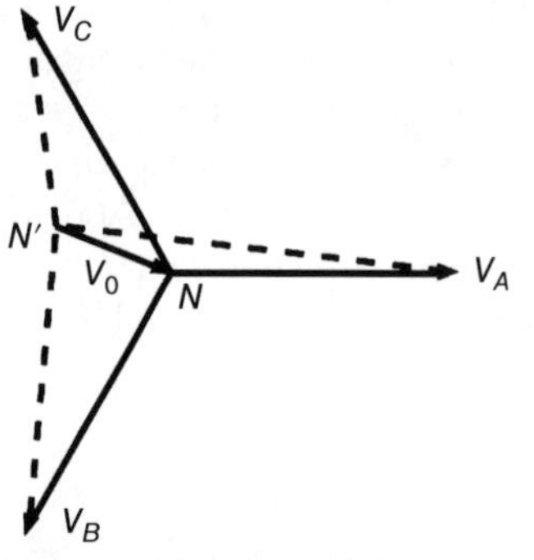

The zero-sequence voltage (V_0) can be shown to be equal to the shift in the star point voltage (i.e. $V_{NN'}$).

The situation can be remedied by the provision of a set of delta connected windings that generate a zero-sequence flux sufficient to cancel nearly all of the offending voltage V_0.

Table 7.2 shows the effect of including a closed delta winding on the transformer used in the previous example. Here we observe very small changes in phase voltage on the unloaded phases, as a result of a substantial load increase on the C phase. Clearly the delta winding has cancelled most of the zero-sequence voltage and has restored the regulation. For these reasons delta connected tertiary windings are usually fitted to Yy shell form transformers.

Figure 7.11 Presence of a zero-sequence voltage (V_0), shifts the star point from N to N'.

Table 7.2 Measured phase voltages on a Yy transformer without flux linkage, incorporating a closed delta winding.

Phase	No load voltage (pu)	Phase to neutral load (pu)	Loaded phase voltage (pu)
A	1.0	0	1.01
B	1.0	0	1.00
C	1.0	0.32	0.97

7.4.2 Voltage Regulation in Yy Transformers with Flux Linkage

How would this situation have changed had a core form transformer been used instead, i.e. one where a flux linkage exists between phases? In this case, the flux created by the secondary current on the loaded phase must split and return via the other two legs. By so doing, each of those phases would draw an equivalent primary balance current which collectively would be exactly sufficient to match that from the loaded phase. As a result, the star point voltage will not see a shift and therefore there will be no zero-sequence voltage developed. The data in Table 7.3 illustrates such an instance for a similarly sized core form transformer, under a heavy asymmetrical load.

Finally, Table 7.4 shows the effect of including a *delta tertiary winding* on this transformer. From the foregoing discussion we might not expect a closed delta winding to provide much of a phase voltage improvement on a transformer with good flux linkage between phases, and from Table 7.4 this would appear to be the case; however, there is

Table 7.3 Measured phase voltages on an asymmetrically loaded core type Yy transformer.

Phase	No load voltage (pu)	Phase to neutral load (pu)	Loaded phase voltage (pu)
A	1.0	0	0.93
B	1.0	0	1.05
C	1.0	0.46	0.94

Table 7.4 Measured phase voltages on a core form Yy transformer incorporating a closed delta winding.

Phase	No load voltage (pu)	Phase to neutral load (pu)	Loaded phase voltage (pu)
A	1.0	0	1.00
B	1.0	0	1.00
C	1.0	0.46	0.94

a slight improvement. This is likely to be due to the fact that even when a flux linkage exists between phases, it is still possible for a small zero-sequence flux to exist in the core. A closed delta winding may therefore result in a small improvement in regulation and a reduction in the unbalance between phase voltages.

7.5 Magnetising Current Harmonics

Figure 7.12 shows the magnetising current and flux waveforms for a single-phase transformer to which a sinusoidal voltage has been applied. Since the flux density is proportional to the integral of this voltage, then it too must have a sinusoidal waveform. However, in order to generate this, the magnetising current must take on high peak values, as demanded by the B–H loop. As a result the magnetising current is rich in low order odd harmonics, in particular the third, which assists the core in achieving its peak flux density.

As we will see in Chapter 13, the harmonics inherent in a balanced set of non-sinusoidal currents tend themselves to be balanced. This means, for example, that the 5th harmonic component of such a set of currents will exist as an individually balanced system, where each phase current has the same magnitude and is separated from its neighbours by 120°. Harmonic currents also have similar properties to those of the fundamental; in particular, they can be shown to have positive, negative or zero-sequence characteristics. Table 7.5 shows the sequence classification of some low order odd harmonics. Even harmonics tend to cancel in symmetrical current waveforms and, except during start-up, these generally do not appear in a transformer's magnetising current.

Harmonic currents that are a multiple of three times the fundamental frequency are zero-sequence harmonics. They are *co-phasal* and are often referred to as *triplen harmonics*, and as a result they cannot flow towards an isolated star point, as can positive or negative sequence currents. Similarly, they cannot flow towards a set of delta

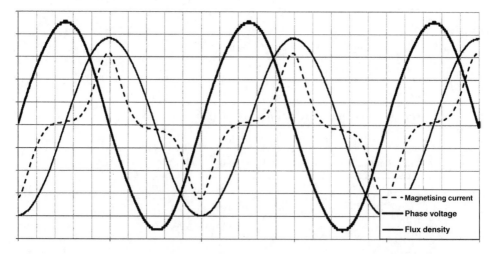

Figure 7.12 Magnetising current, flux density and voltage waveforms in a single-phase transformer.

Table 7.5 Characteristics of different harmonics.

Harmonic number	Characteristics
3rd	Zero sequence
5th	Negative sequence
7th	Positive sequence
9th	Zero sequence
11th	Negative sequence

connected windings, and therefore transformers that present a high zero-sequence impedance also exclude the triplen harmonic component from their magnetising current. This can have an unexpected effect on the resulting winding voltages.

7.5.1 Magnetising Current in Star Connected Shell Form Transformers

Consider the case of a three-phase shell form transformer (or three single-phase transformers) with star connected primary windings and no neutral connection. In this case there is no path for triplen currents to flow, since the star point is isolated from ground. The magnetising current therefore consists of a component at the fundamental as well as the low order odd non-triplen harmonics (i.e. 5th, 7th, 11th etc.). As a result, the flux waveform is flat topped and passes through zero at a slightly faster rate than in the case of a sinusoidal flux waveform, as shown in Figure 7.13.

Each phase voltage contains a significant third harmonic (triplen) voltage in addition to one at the fundamental frequency, as illustrated in Figure 7.14. This causes an 'oscillation' as the waveform passes through zero, as well as a slightly peaky maximum.

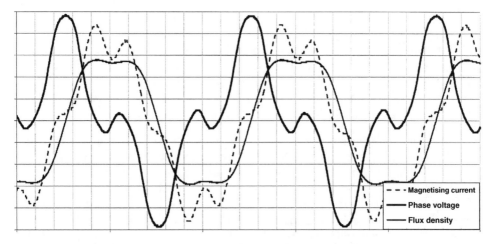

Figure 7.13 Magnetising waveforms for a shell form star connected transformer.

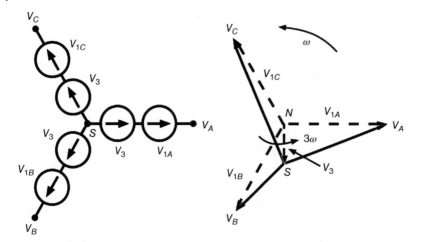

Figure 7.14 Presence of a third harmonic voltage in each phase winding causes the star point to oscillate at the third harmonic.

This zero-sequence triplen voltage creates an oscillation in the star point (S) with respect to the supply neutral, as shown in Figure 7.14. The triplen potential cancels in the line voltages which, as expected, remain sinusoidal.

Effect of a Tertiary Delta Connected Winding

If a closed delta winding is incorporated into such a transformer (a tertiary delta), the situation alters significantly. Zero-sequence triplen voltages induced into each of the delta windings cause a third harmonic current to circulate within the delta. The combined effect of these ampere-turns is to restore the flux waveform in each leg to very near sinusoidal and, as a result, the phase voltages become sinusoidal, as shown in Figure 7.15, and the star point voltage oscillation ceases.

The flux restoration process that leads to almost sinusoidal phase voltages can be understood from Figure 7.15, which shows the magnetising current flowing in the star connected primary winding, as well as the triplen current flowing in the delta. If these ampere-turns are added, the resulting waveform resembles the magnetising current shown in Figure 7.12 for a single-phase transformer, and thus a sinusoidal flux waveform is generated. This beneficial effect represents another reason why a tertiary delta winding is usually included in star connected shell form transformers.

7.5.2 Magnetising Current in Star Connected Core Form Transformers

Since core form transformers cannot support a zero-sequence flux (either at the fundamental or at a triplen harmonic) neither can they support the oscillating neutral potential described above. The flux waveform will therefore be sinusoidal, generated by a magnetising current rich in odd harmonics. Whether triplen harmonics are present depends on the neutral termination. A grounded neutral will provide a path for triplen currents to flow, and the magnetising current will look similar to that shown in Figure 7.12. An ungrounded neutral will block this path and the magnetising current will contain higher levels of the 5th and 7th, but in both cases the flux waveform will

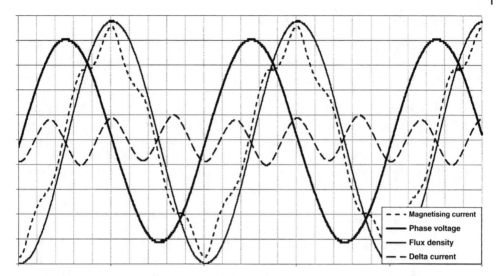

Figure 7.15 Magnetising waveforms in shell type transformer incorporating a delta connected winding.

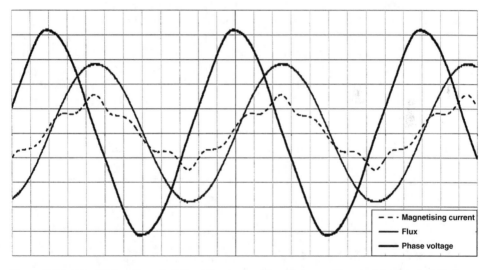

Figure 7.16 Magnetising characteristics of a core type transformer with an isolated neutral. (Note that the magnetising current contains significant levels of the 5th and very little 3rd harmonic).

remain nearly sinusoidal, and the neutral terminal will therefore lie close to ground potential. As a result, the phase voltages will generally be close to sinusoidal.

Figure 7.16 shows the magnetising characteristics of a core form transformer with an isolated neutral. The magnetising current is rich in odd non-triplen harmonics, and the peak flux density is achieved through the presence of elevated levels of the 5th and 7th. The resulting phase voltage is very nearly sinusoidal.

7.5.3 Magnetising Current in a Delta Connected Winding

When the excited winding is delta connected, a sinusoidal flux must be induced within the core capable of supporting the applied line voltage. Since third harmonic currents cannot flow from the supply towards a delta winding, the magnetising current contains components at both the fundamental plus odd non-triplen harmonics. This generates a slightly flat-topped flux waveform similar to that shown in Figure 7.13, which induces a small third harmonic potential in each winding. In a similar fashion to the tertiary delta winding discussed above, this voltage sets up a small third harmonic circulating current, which restores the flux waveform to a sinusoidal shape as shown in Figure 7.17. The small third harmonic voltage is dropped entirely across the winding impedance and therefore it does not appear in the line voltage.

7.5.4 Transformer Inrush Current

The magnetising current waveforms shown in the previous illustrations correspond to steady-state transformer operation, by which time any start-up transients will have died away. At all times, the applied voltages must be matched by the potential induced within each winding, and depending on the point on the voltage waveform when the transformer is energised, the magnetising current required to ensure this may become very large and asymmetrical. This phenomenon is known as *inrush current*, and it can exist in a large transformer for many seconds before finally achieving its steady state, by which time the magnetising current is again small and symmetrical.

Figure 7.17 shows that during each positive half cycle of the voltage waveform, the flux in the core must change from $-\phi_{max}$ to $+\phi_{max}$, a change of $+2\phi_{max}$, with a similar situation occurring during a negative half cycle. Therefore if a demagnetised transformer is energised at a voltage zero, during the next half cycle the flux must increase from zero to $2\phi_{max}$. This flux change may be further increased should a residual flux be present in the core prior to switch on.

Because the initial flux change is uni-polar, the resulting magnetising current will become very large and asymmetrical and may temporarily exceed the rating of the

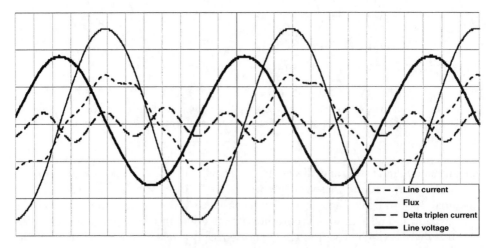

Figure 7.17 Magnetising characteristics for a delta connected winding.

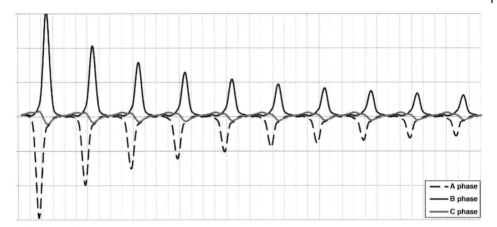

Figure 7.18 Measured inrush current waveforms.

transformer. A flux swing of this magnitude will usually not be achieved in practice, since a portion of the applied voltage will be absorbed by the source impedance as well as the impedance of the primary winding itself, due to the large current flow. The effect of these losses is to progressively reduce amplitude of both the flux change and the magnetising current, and to restore their symmetry about the time axis. In a large transformer this process can take tens of seconds, during which time the transformer can appear to have an internal fault, since primary current is consumed that is unmatched by a proportional secondary current.

On the other hand, if a transformer is energised at a voltage maximum, the flux waveform increases from zero towards ϕ_{max} before returning to zero 90° later. In this case, the steady state is achieved instantly without the need for a transient inrush current. Unfortunately, it is not possible to simultaneously energise all three phases of a transformer at a voltage maximum, and therefore there will always be some degree of inrush current on energisation.

Figure 7.18 shows the inrush currents for a three-phase transformer energised when the phase A and B voltages were approximately equal, and the phase C voltage was close to a voltage maximum. The A and B phase magnetising currents are uni-polar, each containing a decaying DC component. After a period of time all three phases will achieve steady state operation, characterised by small and symmetrical magnetising currents.

7.6 Tap-changing Techniques

Since system voltages tend to vary with diurnal changes in loading, steps must be taken to maintain them within acceptable limits. This is frequently done by switching shunt capacitors banks into the network as the load increases and also by changing taps on both transmission and distribution transformers. We have seen in Section 7.4.1 that as the load on a transformer increases the transformer regulation generally causes the secondary voltage to decrease. Therefore in order to maintain a constant secondary potential the turns ratio must be increased accordingly. This can be done either by increasing the number of secondary turns slightly or by decreasing the number of primary turns.

Equation 4.13 shows that the voltage-per-turn of the primary winding is proportional to the peak flux density $\hat{B}$ to which the core is excited:

$$E_{RMS}/N = 4.44\,\hat{B}\,Af$$

Therefore provided that the ratio of the applied voltage to the number of turns remains constant, the excitation level will not alter. Selecting taps to maintain a constant secondary voltage as the primary potential varies will therefore not change the excitation level of a transformer. This is known as a *constant flux voltage variation* (CFVV).

It is often necessary, however, to further increase the secondary voltage of a distribution transformer in order to compensate for the regulation losses within the transformer itself, as well as those in the downstream network. To achieve this, the turns ratio must be reduced slightly independent of the applied voltage. This will lead to an increase in the flux density in the core and a proportional increase in the secondary voltage as well. Operation of the tap-changer in this mode is called *variable flux voltage variation* (VFVV) and both the voltage-per-turn and the inter-tap potential vary with the tap selected.

Most network transformers are required to maintain one winding voltage within pre-scribed limits, regardless of the applied load, and therefore their tap-changer operations will usually include a mix of these categories. This is referred to as *combined voltage variation* (CbVV), and sufficient headroom must be provided in the transformer's design to accommodate the additional flux density required if saturation is to be avoided.

Tap changers may be installed on either the HV or the LV winding, but they are usually installed in the HV winding, where the current is lower and the number of turns higher.

7.6.1 Off-Load Tap Changers

Small pole- or pad-mounted distribution transformers usually have off-load tap-changers, sometimes called *off-circuit tap-changers* or *de-energised tap-changers*. These usually have between five and seven tappings included in the HV winding that can be used to adjust the number of HV turns in order to appropriately set the transformer's LV potential. These tap-changers are based on a simple selector switch and cannot be adjusted under load, and therefore the transformer must be de-energised whenever a tap adjustment is made. Figure 7.19 shows three arrangements commonly used to accommodate off-circuit tap-changers. The first is a *line end* arrangement for delta windings, where the tap-changer is located at a line terminal of each winding. The second is a *mid winding* arrangement which reduces the voltage that must be accommodated by the tap-changer to half the phase voltage. The third, places the tap-changer at the neutral end of a star connected winding, thereby placing the tapping voltages close to zero.

7.6.2 On-Load Tap Changers

Larger transmission and distribution transformers must be capable of tapping under load and therefore are fitted with *on-load tap-changers* or simply *load tap-changers* (LTC). These must be able to maintain supply during tap-changing, while providing an

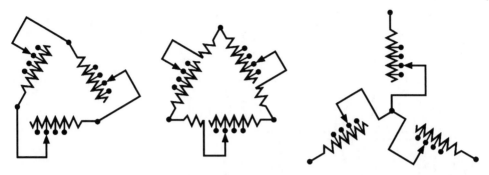

Figure 7.19 Off-load tap-change arrangements. (Adapted and reprinted with permission from IEEE. Copyright IEEE C57.131 (2012). All rights reserved. Permission for further use of this material must be obtained from IEEE. Requests may be sent to: stds-ipr@ieee.org).

almost indiscernible voltage change. This requirement generally means that a large number of taps must be available. In contrast to an off-load tap-changer, a load tap-changer is quite complex electromechanical device and occupies a substantial portion of a transformer's volume.

There are two basic types of tap-changer, both of which operate on a similar principle, as illustrated in Figure 7.20. In order to maintain the flow of load current, it is necessary to draw current from two tappings simultaneously throughout part of the tap-change. This process obviously cannot involve short-circuiting the tappings concerned. Therefore an impedance must be provided between them, one that will permit load current to flow from either tap without creating an appreciable load voltage drop and without permitting an excessively large circulating current to flow between taps. These schemes are shown in Figure 7.20, and both include inter-tap impedances.

The arrangement shown in Figure 7.20a uses a *bridging resistor* (or *transition resistor*) between taps. This type is known as the *high speed resistive tap-changer*, and because of the limited capacity of the resistive elements to dissipate heat, the tap-change is designed to occur in about 50 milliseconds. The second arrangement, shown in Figure 7.20b, is known as a *reactive tap-changer*, and uses an autotransformer to support the inter-tap voltage (also known as a *preventive autotransformer*). This has the advantage that it can carry load current continuously in the bridging position (shown below), and can thus be used to provide an intermediate voltage between adjacent taps and can therefore reduce the number of taps required on the tapping winding.

7.6.3 High Speed Resistive Tap Changer

The high speed resistive tap-changer is probably the most widely used type today, largely because of its compact size. Resistors are used to limit the circulating current during the tap-change process, and since these must be present in the circuit for a very brief period, the operating mechanism is designed to rapidly switch between taps.

In addition to the bridging resistors, there are two types of switch generally associated with this tap-changer; these are shown in Figure 7.21. The first are the *selector switches*, used to select the required tapping. Two sets of these are provided, one for the current tap and one for the next tap to be selected. These switches are not opened under load and hence do not suffer arcing damage and are therefore located within the main

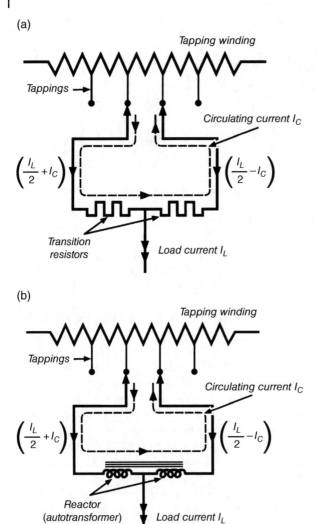

Figure 7.20 (a) Resistive tap-changer (b) Reactive tap-changer (both shown in the bridging position).

transformer tank. The second are the *diverter switches* (also called the *arcing switches*), designed to carry the arc created when the load circuit is partially interrupted.

The diverter switch has four contacts operated in rapid succession, usually by a spring-powered mechanism, so that the transition resistors are only briefly energised. Because these switches operate under load, they are prone to arcing and create carbon and combustible gasses in the surrounding oil. To avoid contaminating the oil in the main transformer tank, they are located in a separate *diverter tank* with its own oil reservoir. Contact wear must also be monitored and replacement contacts fitted as required.

Numerous schemes have been derived for switching between taps. We will only consider the *flag cycle* tap-change, so called because of the appearance of the phasor

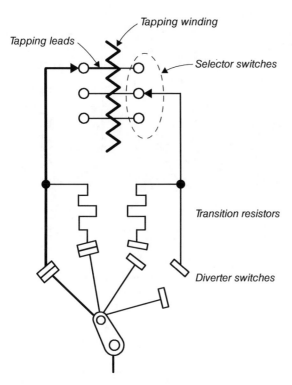

Figure 7.21 Tap changer selector and diverter switches.

diagram corresponding to this tap-change process. This cycle is characterised by the fact that load current is diverted to one transition resistor before the resistive bridge between the taps is completed and circulating current begins to flow.

Figure 7.22 illustrates the flag sequence for a tap-change between taps 1 and 2. Initially tap position 1 is selected and current flows from this tap to the load, via selector switch S1 and the left-hand diverter switch, Figure 7.22A. The tap-change commences when the diverter switch rotates to include resistor R1. The load current path initially remains unchanged. In Figure 7.22C the diverter switch has moved, and now the entire load current briefly flows via R1. By Figure 7.22D both R1 and R2 are available to carry the load and current is shared between taps 1 and 2, half flowing from each. At this time, the terminal voltage lies midway between taps 1 and 2, less the IR drop across the resistors. In Figure 7.22E, R1 has been switched out of circuit, and the entire load current now flows via R2, from selector switch S2. As the diverter switch rotates further (Figure 7.22F), this current is switched out of R2, and by Figure 7.22G all the current flows directly from S2 and the tap-change is complete.

At this point, selector switch S1 may move to tap 3 should a further increase in winding potential be required. The complete operation occurs in about 50–100 ms, thus the energy that must be dissipated by the resistors is relatively small. The diverter switch has only two stable states, corresponding to Figures 7.22A and 7.22G and once triggered the tap-change cycle proceeds rapidly from one to the other.

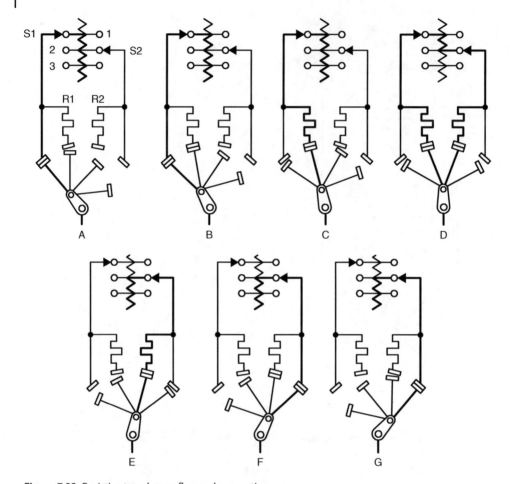

Figure 7.22 Resistive tap-change flag cycle operating sequence.

Flag Cycle Phasor Diagram

The phasor diagram corresponding to this sequence appears in Figure 7.23. We choose the winding voltage as the reference phasor and identify the present tap voltage V_1 and the subsequent tap voltage V_2. An inductive load current I_L is assumed to be flowing from the winding and a resistive circulating current I_C will flow between the taps when bridged.

Beginning again at Figure 7.22A, the load voltage is equal to V_1 and load current flows via selector switch S1. In Figure 7.22C load current is supplied by the left hand transition resistor; its voltage advances in phase and falls slightly in magnitude to $(V_1 - I_L R)$, point C in Figure 7.23. By Figure 7.22D bridging is complete, and $I_L/2$ flows in each transition resistor and a circulating current I_C flows between the taps. The load voltage in Figure 7.22D becomes $1/2(V_1 + V_2 - I_L R)$ (point D in Figure 7.23), rising to $(V_2 - I_L R)$ by E. Finally at F and G the load current is entirely supplied from tap 2 and its voltage is V_2. The load voltage therefore progresses from V_1 to V_2 via points C, D and E on the phasor diagram in Figure 7.23.

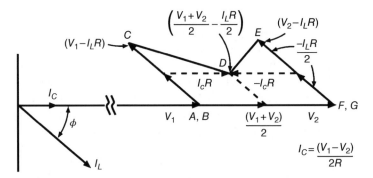

Figure 7.23 Phasor diagram corresponding to the flag cycle tap-change sequence.

When oriented vertically, Figure 7.23 resembles a flag and it is from this that the flag cycle sequence gets its name. There are numerous other tap-change schemes described in the IEEE standard C57.131 '*IEEE Standard Requirements for Tap-changers*', each also named after the appearance of its phasor diagram.

7.6.4 Reactive Tap-Changer

A similar tap selection scheme can be used with the reactive tap-changer shown in Figure 7.20b, but this design also lends itself to a slightly different mode of operation. Because the autotransformer can be designed to continuously carry full load current, it can be used to generate an intermediate voltage between two adjacent taps. In this configuration half the load current flows through each winding, thereby generating opposing fluxes in the core. This cancels most of the inductive impedance associated with the device, at the expense of a magnetising current, circulating between the selected taps. Using the autotransformer to generate intermediate voltages enables the number of taps to be reduced. This represents a significant saving in the size of the main transformer, since tapping leads tend to increase the size of both the windings and the core.

Reactive tap-changers are often purchased as an accessory by the transformer manufacturer and also contain the selector and diverter switches described above. Occasionally these switches are combined into one *arcing switch* which is located in the diverter tank. Since the autotransformer can continuously carry load current so there is no requirement for the diverter switches to operate as rapidly as in the case of the resistive tap-changer. We will discuss an example of a reactive tap-changer later in this chapter when we consider the operation of step regulators in Section 7.9.

7.6.5 Vacuum Interrupter Tap Changers

Since the early 1990s, vacuum interrupter technology of the kind used in vacuum circuit breakers has progressively found application in tap-changers in place of oil-insulated diverter switches. Vacuum interrupters give very good service and have obviated the need for tap-changer maintenance altogether in some transformers. In high tapping duty applications such as rectifier, phase shifting and arc furnace transformers, the maintenance interval can often be extended to between five and seven years.

Vacuum interrupters are hermetically sealed canisters in which the current is rapidly broken in a vacuum, and therefore the arcing products created are separated from the surrounding media. They offer a low contact resistance throughout the life of the interrupter, which can exceed 500,000 operations, and do not create the arcing by-products inherent in oil-insulated switches. Many manufacturers offer retro-fittable vacuum based diverter assemblies that can be easily installed in their existing oil based tap-changers. Vacuum interrupters are also quite capable of operating in free air when required, and are also now finding application in resin cast and small distribution transformers.

Figure 7.24 shows the tapping sequence for a reactive vacuum tap-changer, which in addition to one vacuum interrupter, also requires two *bypass switches* in each phase.

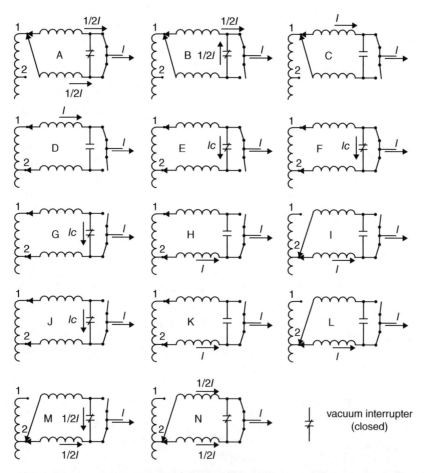

Figure 7.24 Tapping sequence vacuum interrupter based diverter switch. (Adapted and reprinted with permission from IEEE. Copyright IEEE C57.131 (2012) All rights reserved. Permission for further use of this material must be obtained from IEEE. Requests may be sent to: stds-ipr@ieee.org).

7.6.6 Typical Winding Arrangements

There are many different winding arrangements used in tap-changer circuits, some of which are shown in Figure 7.25. We will analyse the single reversing changeover configuration frequently used in the phase windings of star connected transformers.

A tapping winding is generally wound independent of the phase winding with which it is associated. This permits it to be located at either the line end, the centre or the neutral end of this winding. Of these, the neutral end option minimises the insulation requirements of the tap-changer, making it particularly attractive in star connected transformers. This particular arrangement is shown in more detail in Figure 7.26, where three windings are provided, all of which are on the same core leg: the *main phase winding*, across which most of the applied voltage is dropped, the *tapping winding*, in this case having eight taps, and an *auxiliary or tickler winding*, the purpose of which will become apparent shortly. A reversing switch is provided, enabling the phase of the tapping winding to be reversed with respect to that of the main winding.

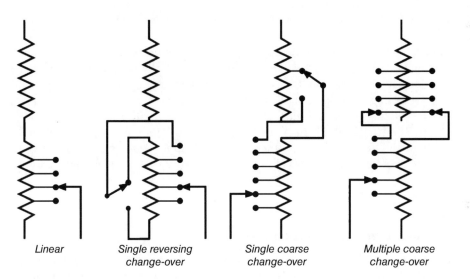

| Linear | Single reversing change-over | Single coarse change-over | Multiple coarse change-over |

Figure 7.25 Common tap winding connections.

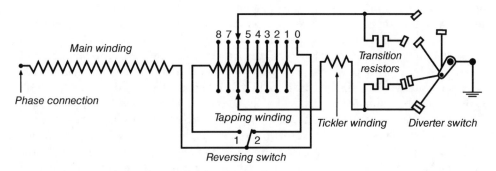

Figure 7.26 Typical tapping winding configuration.

Consider initially the situation when tap 0 is selected. This is the neutral tap, since the tapping winding is not in circuit, thus the phase voltage is applied across the main winding only. If this tap also corresponds to the nominal primary and secondary voltage ratings of the transformer it is also referred to as the principal tap.

Since the reversing switch resides in the main tank, the tap-change mechanism will only permit it to change state when tap 0 is selected, thus avoiding arc damage and oil contamination. If this switch is initially in position 2, then selecting tap 1 will introduce a small segment of the tapping winding in series with the main winding. However, because of the position of the reversing switch, the voltage across the tapping winding opposes that in the main winding. Since the applied phase voltage is fixed, this now must be supported by effectively fewer turns than previously was the case. Accordingly the transformer's turns ratio (N_s/N_p) has increased slightly, and with it the secondary voltage. Tap 2 will see this effect enhanced, and thus the secondary voltage will progressively increase until tap 8 is reached, at which point the secondary voltage will have achieved its maximum value.

When the reversing switch is in position 1 the situation changes. Tap 8 now introduces a few additional in-phase turns in series with the main winding, and thus the effective number of turns available to support the phase voltage increases. Accordingly, this will lead to a decrease in the effective turns ratio and with it a decrease in the secondary voltage. Tapping down through the winding will progressively reduce the secondary voltage until tap 1, by which point the lowest secondary voltage will have been reached. The use of a reversing switch therefore effectively doubles the number of available taps.

The tickler winding provides a further doubling effect. It will be noticed that this winding is included in the circuit on every alternate tap, and since the voltage developed across it is exactly half the inter-tap potential on the tapping winding, it doubles the effective number of tap positions available. This tap-changer therefore produces a total of thirty-three tap positions (including the neutral tap), from a total of only eight leads on the tapping winding!

7.6.7 Typical Tap Changer Hardware Components

As mentioned earlier, the diverter switch contacts require maintenance from time to time, and for this reason these are usually easy to withdraw from the diverter tank. Figure 7.27 shows a diverter switch assembly from a rectifier transformer built in the 1970s. In such applications the tap-changer operates often, usually under high current, and thus there is a need to frequently maintain the diverter switches and replace their insulating oil. In this case, the diverter assemblies are mounted inside a diverter drum consisting of three identical sections each containing the transition resistors and the diverter switch contacts for one phase. The drive shaft, located at the centre of the drum, is operated by a spring-powered mechanism that drives all three phases simultaneously.

The transition resistors each consist of two series connected wafers, immersed in the diverter switch oil supply to assist in dissipating the heat generated by each tap-change. The diverter switch contacts terminate on the surface of the drum, in the form of a spring-loaded contact. These mate with fixed contacts on the inside of the diverter tank, and allow the drum to be withdrawn for maintenance. The diverter switch contacts can be seen in Figure 7.28, where each copper-tungsten contact can be individually replaced as required.

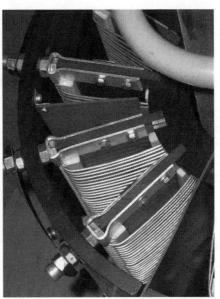

Figure 7.27 Left: diverter switch drum, Right: transition resistors.

Figure 7.28 Diverter switch contacts (open) inside the diverter drum.

Many transformer manufacturers purchase tap-changer mechanisms from third-party suppliers and design their transformer tanks to accommodate tap-change assemblies like those shown in Figure 7.29. These can be inserted through a circular opening in the tank lid and are suspended from the circular flange sealing the opening. The diverter switches and transition resistors are contained in a fibre glass drum immediately below the transformer lid, thus facilitating their maintenance.

The selector switches reside beneath the diverter tank, within the main oil supply, which allows easy connection of the tapping leads, and the reversing switch is mounted beside the selector assembly, as shown in Figure 7.29.

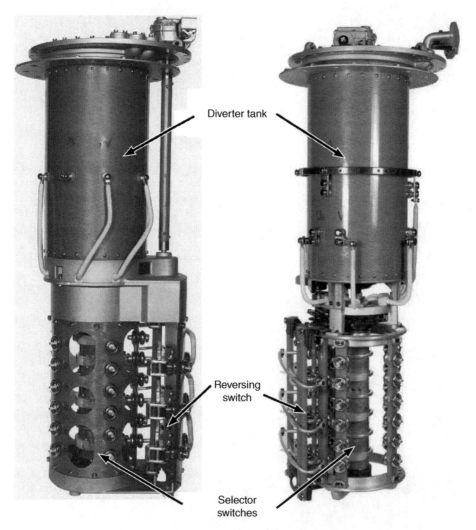

Figure 7.29 High speed resistive tap-changers. (Photo courtesy of Guizhou Changzheng Electric Co, PRC).

7.7 Parallel Connection of Transformers

It is frequently necessary to operate transformers in parallel, to supply a common load. In such situations it is necessary to ensure that paralleled transformers present secondary voltages *that have the same phase relationship and are almost equal.* Transformers that are of the *same vector group* and voltage rating will satisfy this requirement. While this represents the norm, it is also possible for certain other vector groups to operate satisfactorily in parallel. For example, a parallel combination of Yy and Dd transformers would operate satisfactorily (if their turns ratios were set appropriately), as would a Yd and Dy pair.

Two transformers having the same impedance and voltage rating will share a common load equally, provided that they operate at the same tap. Differences in tap setting will give rise to a circulating current flowing between the two secondaries, and a bus voltage that lies in between the two secondary potentials.

This situation is illustrated in Figure 7.30, where E_1 and E_2 are the individual winding voltages and V_b is the resulting bus voltage and Z_1 and Z_2 are the respective transformer impedances. Z is the load impedance and currents I_1 and I_2 flow in each winding.

The bus voltage V_b can be determined using Millman's Theorem as:

$$V_b = \frac{E_1 Y_1}{Y_1 + Y_2 + Y} + \frac{E_2 Y_2}{Y_1 + Y_2 + Y}$$

or

$$V_b = \frac{E_1 Z_2 Z}{Z_1 Z + Z_2 Z + Z_1 Z_2} + \frac{E_2 Z_1 Z}{Z_1 Z + Z_2 Z + Z_1 Z_2} \tag{7.8}$$

where the admittances Y_1, Y_2 and Y correspond to the impedances Z_1, Z_2 and Z respectively.

We also know that $I_1 = \dfrac{E_1 - V_b}{Z_1}$ and $I_2 = \dfrac{E_2 - V_b}{Z_2}$. Substituting Equation (7.8) into these equations yields:

$$I_1 = \frac{E_1 Z_2 + (E_1 - E_2) Z}{Z_1 Z_2 + (Z_1 + Z_2) Z} \tag{7.9}$$

Figure 7.30 Parallel transformer equivalent circuit.

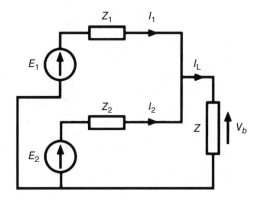

and

$$I_2 = \frac{E_2 Z_1 - (E_1 - E_2) Z}{Z_1 Z_2 + (Z_1 + Z_2) Z} \tag{7.9a}$$

The second term in the above equations is interesting. When $E_1 = E_2$ this term becomes zero, and therefore:

$$I_1 = \frac{E_1}{(Z_1 \| Z_2 + Z)} \cdot \frac{Z_2}{(Z_1 + Z_2)} = I_{load} \cdot \frac{Z_2}{(Z_1 + Z_2)} \tag{7.10}$$

and

$$I_2 = \frac{E_2}{(Z_1 \| Z_2 + Z)} \cdot \frac{Z_1}{(Z_1 + Z_2)} = I_{load} \cdot \frac{Z_1}{(Z_1 + Z_2)} \tag{7.10a}$$

Thus the winding currents are determined by the relative transformer impedances. Multiplying Equations (7.10) and (7.10a) by the bus voltage gives the VA loading seen by each transformer:

$$S_1 = I_1 V_b = S_{load} \frac{Z_2}{(Z_1 + Z_2)} \text{ and } S_2 = I_2 V_b = S_{load} \frac{Z_1}{(Z_1 + Z_2)} \text{ or } S_1/S_2 = Z_2/Z_1 \tag{7.11}$$

Therefore the fractional load carried by one transformer depends on the impedance of the other. It is possible that with dissimilar transformers, the heavier load may ironically be carried by the smaller transformer.

When $E_1 \neq E_2$, the second term in Equation (7.9) becomes approximately $\frac{(E_1 - E_2)}{(Z_1 + Z_2)}$, which can become very large, even for moderate values of $(E_1 - E_2)$, due to the generally small values of the transformer impedances, Z_1 and Z_2. This highlights the need to ensure that the winding voltages are similar in both magnitude and in phase.

Consider the case where the paralleled transformers have the same *impedance* (Z_t) but are not operating from the same tap, i.e. $(E_1 - E_2) \neq 0$. Ignoring the small effect of the load impedance, the bus voltage in this case will lie roughly midway between E_1 and E_2. A circulating reactive current will exist between them, of a magnitude sufficient to support this voltage difference. The higher voltage transformer will deliver VArs to the load as well as to its partner. Depending on the VAr flow demanded by the load and the voltage difference, the lower voltage transformer may even consume reactive power. It will certainly deliver less reactive power to the load and will therefore operate at a higher power factor than its partner. (This effect can be used to advantage by deliberately operating dissimilar transformers out of step so as to ensure that they both operate close to the same power factor.) This situation is depicted in Figure 7.31.

The currents I_1 and I_2, defined in Figure 7.31, can be shown each to contain a load current component, $(\approx I_L/2)$ together with the circulating current (I_C) flowing between the taps. For small voltage differences, roughly half the active power will be supplied by each transformer, therefore we can write $I_1 \approx \left(\frac{I_L}{2} + I_C \right)$ and $I_2 \approx \left(\frac{I_L}{2} - I_C \right)$, yielding:

Figure 7.31 Parallel transformer circulating current.

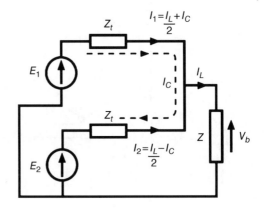

$$I_1 - I_2 \approx 2I_C \approx \frac{(E_1 - E_2)}{Z_t} \tag{7.12}$$

Assuming that $Z \gg Z_t$, and:

$$I_1 + I_2 = I_L = \frac{(E_1 + E_2)}{2(Z_t / 2 + Z)} = \frac{V_b}{Z} \tag{7.12a}$$

Equations (7.12 and 7.12a) suggest that the reactive power flow between the transformers is very sensitive to the tap position, whereas the active power flowing in the load is only slightly influenced by the tap position of either transformer, since changing one tap will only mildly influence the load voltage. Where both the impedances and the tap settings are different the approximation in Equation (7.12) is unreliable and a precise analysis using Equations 7.10 and 7.10a is required.

7.7.1 Tap Change Control Strategies

Having seen a little of how paralleled transformers behave, it is instructive to consider some of the tap control strategies used in practice. The principal function of the *voltage regulator* is to select transformer taps so as to maintain the bus voltage within acceptable limits. This requires the specification of the *voltage set point*, together with the *bandwidth* (also called *voltage dead-band*), about the set point, within which no tap-change will be initiated. This is illustrated in Figure 7.32, where V_s is the set-point voltage, V_{Bus} is the bus voltage and BW is the voltage bandwidth. A time delay TD is applied before a tap-change is initiated, so as to prevent instantaneous tap-changing in response to short-term voltage excursions.

Figure 7.32 shows the case where the bus voltage falls below the lower bandwidth limit. After the delay time TD has expired, the tap-change is initiated, restoring the voltage to within the acceptable range. In the case of a master-slave control strategy the master controller initiates the tap-change on the master transformer and the others are then commanded to follow. In this way, all transformers remain on the same tap, i.e. they remain in step.

If each transformer had an independent controller, the first to initiate the tap-change may well restore the bus voltage to within the desired dead-band. This would result in the

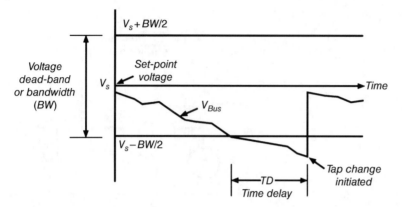

Figure 7.32 Voltage set-point and bandwidth. (Adapted with the kind permission of the Beckwith Electric Co).

other transformers remaining in their original tap positions, with an undesirable circulating current now flowing between them. Subsequent voltage disturbances would again see the fastest transformer make a corrective tap-change, thus driving the others further out of step. Operating similar transformers from different tappings will result in considerable reactive flows between them causing additional I^2R losses. The master-slave control strategy is generally implemented when transformers must be run in parallel.

Power Factor Paralleling Method
The ability to interchange VArs between paralleled transformers permits several other methods to be used for choosing the optimal tap positions of paralleled transformers. One such method is to choose tap settings that will ensure that paralleled transformers operate at about the same power factor while maintaining the bus voltage within limits. This requires the measurement of the current phase angle of each transformer; the controller blocks tap-changes that would drive these angles further apart, while maintaining the desired bus voltage. This ensures that both transformers operate at similar power factors, thus preventing asymmetrical transformer loading. It should be noted that this control strategy is rarely implemented in industrial power systems.

Circulating Current Method
The circulating current approach aims to minimise the magnitude of any circulating current flowing between transformers. To achieve this, equipment is required that will isolate the circulating current from the load current. With this information the controllers can bias the tap-change on the high tap transformer so as to reduce its voltage, while doing the opposite for the low tap device. In this way the circulating current is minimised while still maintaining the desired bus voltage.

VAr Balancing Method
This approach operates so as to equalise the VAr contribution from each transformer. In the case where transformers of different ratings are paralleled, the CT ratios should be appropriately chosen so as to reflect these ratings. Thus the sharing can be appropriately apportioned between transformers.

7.8 Transformer Nameplate

The nameplate contains an essential record of a transformer's voltage, current and VA ratings, its impedance, how its windings are configured, what tappings are available as well as data on accessories such as the cooling equipment and any associated current transformers. As such, it is second in importance only to the factory test report, and therefore it should always remain attached to the transformer. An example of a typical nameplate appears in Figure 7.33, in this case for a 110 kV:33 kV, 60 MVA, YyNn0 distribution transformer. There is no standard layout used by all manufacturers, but all nameplates generally include the following items:

Transformer ratings: Because of their importance, these usually appear somewhere near the top of the nameplate. In this case, the voltage ratings appear in the top right-hand table, the power ratings are in the column opposite, while the current ratings appear in the tapping chart. It is not uncommon for a transformer of this size to have two power ratings (in this case 42 MVA and 60 MVA) depending on whether fan forced cooling is applied to the radiators. This is indicated as either ONAN (oil natural air natural) or ONAF (oil natural air forced). (Had pumps been provided to circulate oil through the radiators, the latter abbreviation would have read OFAF.)

The upper table also includes information on the withstand capability of the winding insulation, and is characterised by two parameters. The first is the *basic insulation level* (BIL) or the *standard lightning impulse withstand voltage* that the transformer is capable of supporting (550 kV in this example). This is the peak value of a unidirectional surge voltage applied between the transformer terminals and ground, having a rise time of the order of 1.2 μs and a time to half-value of 50 μs. The second is the RMS *short duration power frequency withstand voltage* (230 kV). Collectively these parameters define the insulation level of the transformer.

Transformer impedances: The positive (negative) and zero-sequence impedances appear in the second table on the right-hand side, for taps 1, 7 and 23. The positive sequence impedance associated with each tap position also appears in the tapping table. The impedance varies slightly with each tap, roughly as the square of the effective number of turns on the HV winding, though not necessarily monotonically. In this case, the impedance is relatively high, having been chosen to set the LV fault current at a level suitable for the downstream switchgear. The zero-sequence impedances are of similar values, indicating that for this core form transformer sufficient leakage flux is established, probably via the tank walls, to permit a zero-sequence voltage to be generated in each phase as the result of zero-sequence current flowing in the windings.

Winding configuration and vector grouping: The winding configuration and vector group are shown in the lower right-hand corner of this nameplate. The HV terminals are always identified by capital letters and the LV by lowercase letters. Windings identified by the same letter are wound on the same limb, therefore the main HV winding A15-A14, the tapping winding A13-A1 and the LV winding a2-a1 are all concentric on the 'A' phase limb. The LV winding generally lies closest to the core, and the tapping winding is usually the outermost. The star connected HV winding has the on-load tap-changer located at its neutral end where the insulation requirements are less severe.

The HV winding has a current transformer on the 'B' phase; this is used to excite the winding temperature indicator with a current proportional to the instantaneous load on

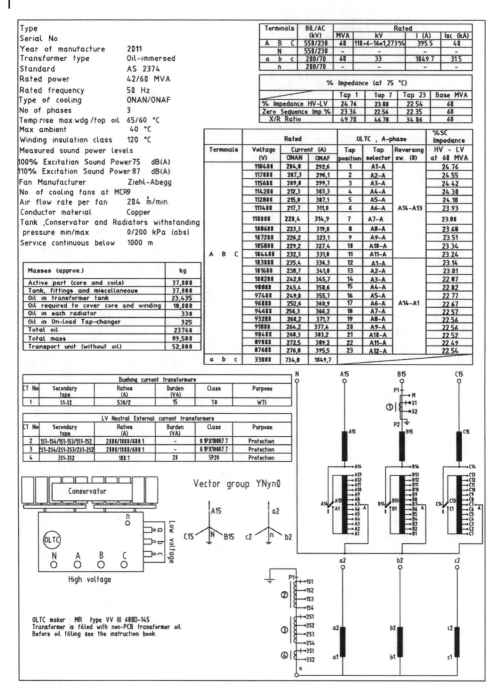

Figure 7.33 Example of a transformer nameplate.

the transformer. This information, coupled with the top oil temperature (usually shown on a separate indicator), is used to synthesise a temperature representative of the hotter parts of the windings. The star connected LV windings are fitted with protection class current transformers, located between the neutral terminal and ground, external to the tank.

Temperature limitations: The nameplate also indicates the permitted maximum rise above ambient (40°C) for the top oil and winding temperatures (60/65°C respectively), as well as the maximum temperature permitted for the hottest part of the winding (120°C). These temperatures are determined by the class of insulation used to insulate the windings. The maximum average winding temperature for this transformer would therefore be $40 + 65 = 105°C$, leaving a 15°C margin for the 'hot spot' temperature.

Tappings: The tapping chart shows the connections required on the HV winding for each of the 23 tap positions. The stated HV voltage on each tap is the no load voltage required to produce the rated voltage (33 kV) at the secondary terminals. On its principal tap, this transformer has current ratings of 220.4 and 314.9 A corresponding to each of its MVA ratings. The associated current ratings vary with the applied HV voltage, in such a way that the overall MVA rating of the transformer remains constant.

Transformer plan view: A plan view of the transformer is always included on the nameplate, showing the HV and LV terminal locations, as well as the location of the radiators and the tap-changer. Above the radiators is mounted the cylindrical conservator tank shown in Figure 7.1a, which provides a reservoir of oil above the main tank to accommodate the changes in oil volume that occur with diurnal temperature and load changes.

Current transformer specifications: The specifications of any current transformers supplied with or internal to the transformer are also always included on the nameplate. Internal CTs are generally not a good idea as they can be very difficult to test or replace, therefore apart from the temperature indicator CT mentioned above, it is usual to associate all protection and metering CTs with the HV and LV switchgear.

Component masses: The mass of the main components, both with and without oil is included on the nameplate so that, in the future, should the transformer need to be moved (possibly for repair) the transport weights are known.

7.9 Step Voltage Regulator

We consider as an example the Eaton Cooper Power series step regulators, frequently found in distribution networks. These are autotransformer based voltage regulators, inserted at regular intervals on MV distribution lines and feeders, for the purpose of providing voltage support along the length of the line. They are made in Wisconsin, USA, and are used in many countries worldwide. Although these regulators find application on both three- and four-wire feeders, they are usually manufactured as single-phase units, thereby making them easy to transport, maintain and, when necessary, to replace. It is instructive to analyse the operation of the step regulator since it provides an insight into the operation of autotransformers as well as the application of a reactive tap-changer.

Figure 7.34 shows a regulator together with its nameplate. It uses an autotransformer to either *buck* or *boost* the line voltage by up to 10%, and therefore the kVA rating

(a)

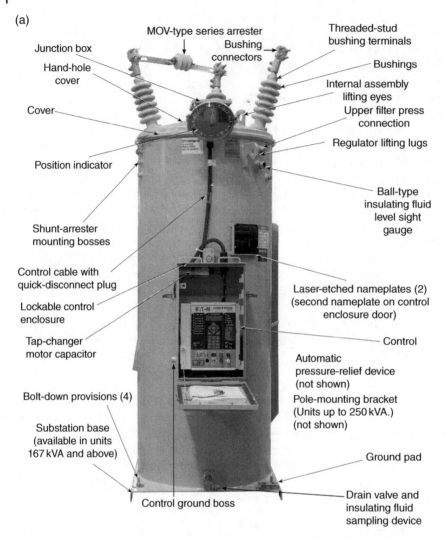

Figure 7.34 (a) Eaton Cooper Power series VR-32 step voltage regulator. (Illustration courtesy of Eaton).

required of the regulator is relatively low. For example, the device whose nameplate appears above has a nominal rating of 100 amps on a 22 kV line voltage. The single-phase VA capacity of this line is $22,000 V \times 100 A / \sqrt{3} = 1.27$ MVA, but the volt-amp rating required of the regulator is considerably less, being the regulator's *boost voltage* multiplied by its rated current, i.e. $2200 V \times 100 A = 220$ kVA.

Type 'A' and 'B' Regulators

There are two main variants of the basic step regulator, defined in the American National Standard ANSI C57.15 as Type 'A' and Type 'B' regulators. They both operate on the same principle and represent years of product refinement. We will investigate a 32-step Type 'B' regulator which provides a regulation range of ±10% of the system line

(b)

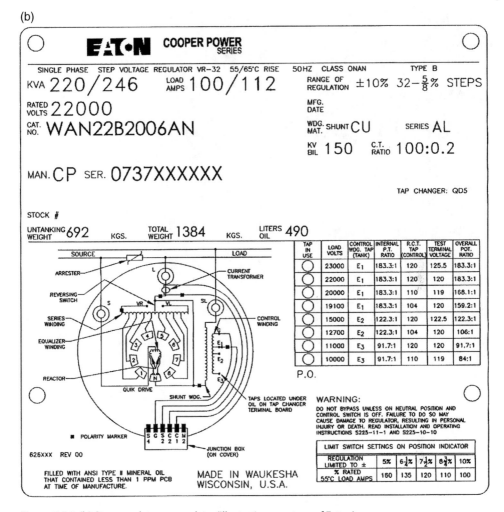

E:T·N COOPER POWER SERIES

SINGLE PHASE STEP VOLTAGE REGULATOR VR–32 55/65°C RISE 50HZ CLASS ONAN TYPE B

KVA 220/246   LOAD AMPS 100/112   RANGE OF REGULATION ±10% 32–$\frac{5}{8}$% STEPS

RATED VOLTS 22000

CAT. NO. WAN22B2006AN

MAN. CP SER. 0737XXXXXX

MFG. DATE

WDG. MAT. SHUNT CU   SERIES AL

KV BIL 150   C.T. RATIO 100:0.2

TAP CHANGER: QD5

STOCK #

UNTANKING WEIGHT 692 KGS.   TOTAL WEIGHT 1384 KGS.   LITERS OIL 490

TAP IN USE	LOAD VOLTS	CONTROL WDG. TAP (TANK)	INTERNAL P.T. RATIO	R.C.T. TAP (CONTROL)	TEST TERMINAL VOLTAGE	OVERALL POT. RATIO
◯	23000	E_1	183.3:1	120	125.5	183.3:1
◯	22000	E_1	183.3:1	120	120	183.3:1
◯	20000	E_1	183.3:1	110	119	168.1:1
◯	19100	E_1	183.3:1	104	120	159.2:1
◯	15000	E_2	122.3:1	120	122.5	122.3:1
◯	12700	E_2	122.3:1	104	120	106:1
◯	11000	E_3	91.7:1	120	120	91.7:1
◯	10000	E_3	91.7:1	110	119	84:1

SOURCE — LOAD

ARRESTER — CURRENT TRANSFORMER

REVERSING SWITCH

SERIES WINDING

EQUALIZER WINDING — CONTROL WINDING

REACTOR

QUIK DRIVE

SHUNT WDG.

■ POLARITY MARKER

626XXX REV 00

FILLED WITH ANSI TYPE II MINERAL OIL THAT CONTAINED LESS THAN 1 PPM PCB AT TIME OF MANUFACTURE.

P.O.

TAPS LOCATED UNDER OIL ON TAP CHANGER TERMINAL BOARD

JUNCTION BOX (ON COVER)

MADE IN WAUKESHA WISCONSIN, U.S.A.

WARNING:
DO NOT BYPASS UNLESS ON NEUTRAL POSITION AND CONTROL SWITCH IS OFF. FAILURE TO DO SO MAY CAUSE DAMAGE TO REGULATOR, RESULTING IN PERSONAL INJURY OR DEATH. READ INSTALLATION AND OPERATING INSTRUCTIONS S225–11–1 AND S225–10–10

LIMIT SWITCH SETTINGS ON POSITION INDICATOR					
REGULATION LIMITED TO ±	5%	6$\frac{1}{4}$%	7$\frac{1}{2}$%	8$\frac{3}{4}$%	10%
% RATED 55°C LOAD AMPS	160	135	120	110	100

Figure 7.34 (b) Step regulator nameplate. (Illustration courtesy of Eaton).

voltage, in 16 steps up of 0.625% and 16 down, in addition to a neutral tap (tap 0), on which the regulator has an effective voltage gain of 1. The fine step size means that regulation action is unlikely to be noticed by customers. Two 22 kV pole-mounted regulators in an open delta configuration are shown in Figure 7.35.

The regulator's operating principle is illustrated in Figure 7.36a where a Type 'A' regulator is shown. Here the supply voltage V_S, is connected across the shunt winding, and the load voltage V_L, is obtained across both the series and shunt windings in combination. The series winding is excited by the shunt winding, since they are both wound on the same transformer core. Depending on the position of the reversing switch, the regulator either boosts or bucks the source voltage, yielding an effective turns ratio of $V_L/V_S = (N_p \pm N_s)/N_p$, where N_p is the number turns on the shunt winding, and N_s is the number of selected turns on the series winding.

Figure 7.35 Typical pole-mounted installation (open delta connection). (Note the pole top bypass and isolation switches).

(a)

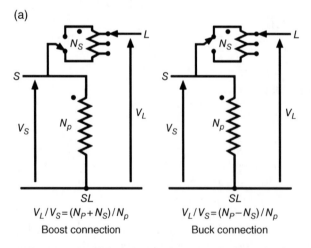

$$V_L/V_S = (N_P + N_S)/N_p$$

Boost connection

$$V_L/V_S = (N_P - N_S)/N_p$$

Buck connection

Figure 7.36a Type 'A' regulator. (Adapted and reprinted with permission from IEEE. Copyright IEEE C57.15 (1986) All rights reserved. Permission for further use of this material must be obtained from IEEE. Requests may be sent to: stds-ipr@ieee.org).

(b)

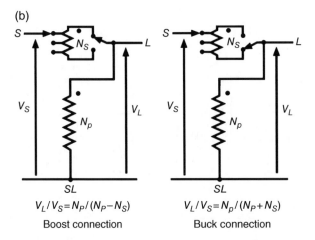

$$V_L/V_S = N_P/(N_P - N_S)$$

Boost connection

$$V_L/V_S = N_P/(N_P + N_S)$$

Buck connection

Figure 7.36b Type 'B' regulator. (Adapted and reprinted with permission from IEEE. Copyright IEEE C57.15 (1986) All rights reserved. Permission for further use of this material must be obtained from IEEE. Requests may be sent to: stds-ipr@ieee.org).

Type 'B' regulator appears in Figure 7.36b. Here the supply voltage is connected through the tap-changer and is applied to both the series and shunt windings in combination. The load voltage V_L appears across the shunt winding and, depending on the position of the reversing switch, it is either boosted above V_S or bucked below it. The turns ratio is given by $V_L/V_S = N_p/(N_p \pm N_s)$.

As a result of this connection, the shunt winding sees a regulated voltage, and hence its excitation is less variable that of the Type 'A' regulator. To raise the load voltage in a Type 'A' device, the reversing switch must be in the lower position (Figure 7.36a), while in the Type 'B' regulator must be in the upper position (Figure 7.36b).

Thirty-Two-Step Voltage Regulator

The schematic diagram in Figure 7.37a provides a simplified view of a thirty-two-step voltage regulator. It has three main terminals, S (source), L (load) and SL (source-load reference terminal). Figure 7.37b shows the regulator applied to a single-phase line. Here the reference terminal (SL) is connected to the neutral, the source (S) is connected to the incoming phase terminal, and the load terminal (L) supplies potential to the line downstream of the regulator. Bypass and isolation switches (disconnectors) are incorporated into the installation to facilitate removal of the regulator without the need to take the line out of service.

Between the reference (SL) and the load terminal (L) is the shunt winding of the series transformer. On the same magnetic circuit are the series and control windings. The series winding has eight tapping leads and a total voltage equal to 10% of that seen by the shunt winding. A portion of this voltage will be either added or subtracted – through the action of the tap-changer – to the source potential and delivered to the load terminal. The control winding serves two functions: firstly it acts as a voltage transformer, providing the controller with a representation of the downstream voltage for regulation purposes, and secondly it supplies power for both the tap-change motor and the control circuitry. Finally, a current transformer provides load current information to the

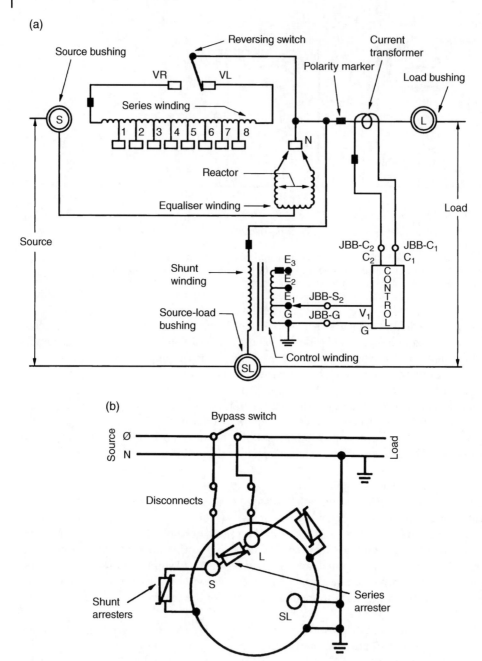

Figure 7.37 (a) Simplified schematic of an Eaton 32 step regulator (b) Single-phase two-wire regulator connection. (Illustrations courtesy of Eaton).

controller. This is used for both power flow and energy measurement, as well as for line drop compensation purposes.

Tap-Changer

The heart of the regulator is the tap-changer, and from the schematic diagram in Figure 7.37a it can be seen that there are no separate diverter switches. Instead a *combined diverter-selector switch* (also called an *arcing switch*) is used. This makes the tap-changer mechanism simpler and more compact, but steps must be taken to minimise any arcing of these contacts. This is especially important because they reside in the general volume of insulating oil, and not in the equivalent of a diverter tank. The phase of the series winding is changed using a conventional reversing switch, which only operates when the tap-changer is in its neutral position.

Figure 7.38 shows the diverter-selector switch bridging two adjacent tapping contacts. Notice that an insert has been silver soldered to both the leading and trailing edges of each fixed tapping contact. This material is usually sintered copper and tungsten; as such it presents a hard wearing surface with good resistance to arc erosion, and it effectively performs the function of a diverter switch by being able to cope with minor arcing without sustaining serious damage. The wiping contacts are made from a similar material.

In contrast, the reversing switch, shown in Figure 7.39, is much simpler. Since it only switches when the neutral tap is selected and therefore does not interrupt current, it requires no special materials in its construction.

The schematic of Figure 7.37a shows the tap-changer in its neutral position, where the S and L terminals are interconnected. This position *must* be used when the regulator is to be taken out of service, since this is the only one in which it is safe to close the bypass switch (shown in Figure 7.37). Once the bypass has been closed and the isolation

Figure 7.38 Combined diverter-selector switch. (Illustration courtesy of Eaton).

Figure 7.39 Reversing switch. (Illustration courtesy of Eaton).

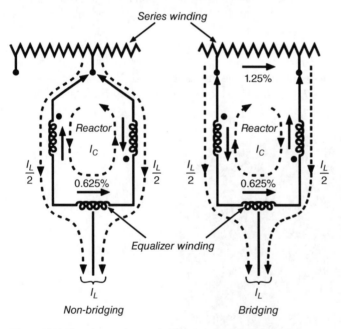

Figure 7.40 Tap changer in non-bridging and bridging positions.

switches have been opened, the regulator can be earthed, and maintenance activities can begin, leaving the line operational.

The schematic also shows that there are eight tapping leads. A centre-tapped equaliser winding and a split phase reactor are inserted between the two diverter-selector switches (see Figure 7.40). This configuration suggests that the tap-changer operates in both bridging and non-bridging modes, which is indeed the case. Thus there are effectively 16 boost steps and 16 buck steps, in addition to the neutral tap. A tap position indicator displaying the current tap is mounted at the top of the tank. The voltage between adjacent tapping leads on the series winding is 1.25% of the primary voltage and, courtesy of the equaliser winding, the load voltage steps are half this value (0.625%).

The equaliser winding is a little like the autotransformer discussed earlier, but in this case it is wound on the same core as the series winding. It therefore has a constant potential induced within it, which is set to half the inter-tap voltage, or 0.625% of the applied primary voltage. The midpoint of this winding is connected to the source terminal, S.

When both diverter-selector switches are connected to the same tapping lead, the voltage induced in the equaliser winding is dropped across the reactor windings. These are interleaved on two sides of a rectangular magnetic circuit (Figure 7.41), and their tight magnetic coupling ensures that a very low leakage reactance exists between them. The reactor is dimensioned so that a reactive current I_c, equal in magnitude to about half the rated regulator current, circulates through the reactor and equaliser windings in non-bridging tap positions. Conversely, when the selector switches are bridging two adjacent tapping leads, a current of this magnitude also circulates, but now in the opposite direction.

The reactor also ensures that half of the load current ($I_L/2$), flows through each side of the tapping circuit. Therefore the *total* current flowing in one reactor winding is ($I_L/2 + I_c$), while that in the other is ($I_L/2 - I_c$). For this current sharing to be effective both reactor windings must see the same flux, so they must be tightly coupled and, as previously mentioned, this is an essential element of the reactor's design. In addition, because the reactor operates close to the line potential, it is mounted on three white support

Figure 7.41 The tap-changer reactor. (The copper strap connects turns interleaved on one leg with those on the other).

insulators which permit its core to approach this potential as well (Figure 7.41). This substantially reduces its insulation requirements.

It must be remembered that the circulating current is largely reactive, lagging behind the inter-tap voltage by close to 90°. It therefore also lags the load current by about 60–90°. The load current components flow in opposing directions in the reactor windings as well as in each half of the equaliser winding. Because of this there is no associated flux generated in either of these components and therefore they present no reactive impedance to the flow of load current.

On the other hand, the circulating current is limited by the inductance of the reactor itself, and to achieve the desired impedance, an air gap is incorporated within its core. When either diverter-selector switch opens, the entire load current commutates to the other, while the circulating current is interrupted. This current change is not as dramatic as interrupting the current flow entirely, and while it leads to some arcing of the contacts, the equaliser winding assists by reducing the voltage that would otherwise be developed across the opening diverter-selector switch (known as the *recovery voltage*) by an amount equal to 0.625% of the primary potential. Arcing is further minimised by the swift operation of the tap-changer. The tap-changers fitted to 32-step regulators operate under oil, which assists in quenching the arc, and are generally capable of making a complete tap-change in about 250 ms, thus minimising the time during which arcing can take place.

Regulator Connections

Step regulators can be installed on two-, three- and four-wire distribution feeders. We have already seen the single-phase two-wire connection. The four-wire arrangement, shown in Figure 7.42, requires three regulators and a grounded neutral conductor. In this configuration each regulator can operate independently if required, although they are usually operated in a leader-follower arrangement where two regulators follow the tap-changes of the third (the lead regulator). The load voltages so produced are in phase with source voltages and any residual current flows in the neutral.

Two regulators can also be installed in an open delta configuration on a three-wire line as shown in Figure 7.43. Here the B phase has been used as the common phase. It can be seen from the phasor diagram that as the A′B and C′B voltages increase, due to the action of the regulator, so the A′C′ voltage increases as well, gaining half its increase from each regulator. For this reason it is important to operate open delta regulators in the leader-follower mode, ensuring that all line voltages are maintained in step.

One possible disadvantage with the open delta arrangement is the neutral shift voltage that occurs as a result. The neutral voltage lies at the centroid of the line voltage triangle, and since the source and load line phasor triangles are not concentric, there will be a slight shift in the neutral potential of the load side with respect to that of the source. This can cause nuisance tripping of some earth fault protection schemes when lines are paralleled.

On long feeders it may be necessary to install several sets of regulators at intervals along the line. In such cases the neutral shift problem can be partially alleviated if three open delta regulators are installed, each using a different reference phase. The effect is to alter the neutral shift voltage by 120° at each installation so that by the third, the neutral point has returned to its original location. The effectiveness of this technique depends upon the load applied to each regulator installation and the voltage drop along the line between them.

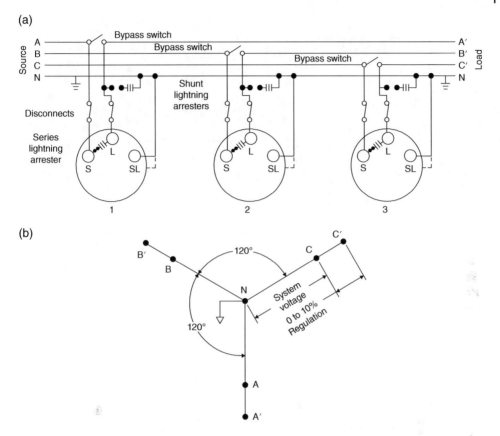

Figure 7.42 (a) Four-wire wye connection. (b) Phasor diagram. (Illustrations courtesy of Eaton).

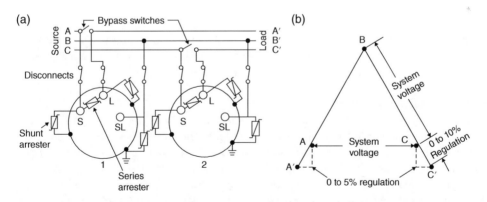

Figure 7.43 (a) Two regulator open delta connection (b) Phasor diagram (Illustrations courtesy of Eaton).

The final arrangement frequently used is the three regulator closed delta connection, illustrated in Figure 7.44. It has the advantage that both the source and load line phasor triangles are concentric and therefore no neutral shift potential exists. It also offers an additional 5% regulation range, so that ±15% of the nominal line voltage can be accommodated.

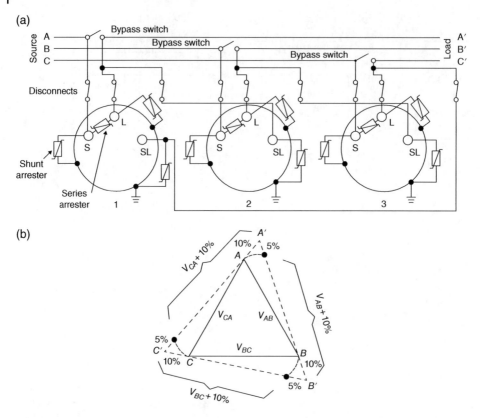

Figure 7.44 (a) Three regulator, closed delta connection (b) Phasor diagram (Illustrations courtesy of Eaton).

Although the source and load phasor triangles are concentric, there is a slight phase shift between corresponding source and load voltages. At maximum regulation, this shift is about 5°, as shown in Figure 7.44, and therefore care must be exercised when paralleling feeders downstream of different regulator installations.

Paralleling Autotransformer Based Regulators

Autotransformer based regulators of any type present an impedance that varies roughly as the square of the boost or buck turns inserted into the circuit. Therefore when operating on the neutral tap, a regulator's impedance is effectively zero. It generally rises from zero to a relatively low value at the maximum boost (or buck) tap available. For example, the maximum impedance presented by an Eaton regulator is only about 0.5%, when the entire series winding is used to either boost or buck the supply voltage. Therefore large and damaging circulating currents will flow between paralleled regulators when operating from different taps, or between paralleled regulators where an inherent phase shift exists between them. For this reason, regulators may only be paralleled if they are each operated in the leader-follower mode, and are used in conjunction with an external current limiting device. For example, they may be supplied from different transformers, or alternatively they may each have a series current limiting reactor fitted, in order to reduce the magnitude of the circulating currents that will arise during tapping operations.

Sometimes, it is necessary to parallel MV feeders during switching operations. The neutral shift potential created by open delta connections can cause nuisance tripping of earth fault protection schemes. This problem occurs when such feeders are paralleled with others without open delta regulators or with feeders having a different number of open delta regulators. In these cases it is the residual current resulting from the neutral shift voltage between the two lines that is detected by the protection relay.

In order to avoid this problem, it is common practice to place the tap-changers involved on the neutral tap and leave the regulator in its manual mode, so that no auto-tapping can occur, before the lines are paralleled. This approach is generally successful, but the line voltage may drift out of limits during switching operations as a result. It is also worth mentioning the need to ensure that the neutral tap is selected before the bypass switch is closed. Failure to do so will result in very high fault currents and severe damage to the unit.

Line Drop Compensation

One useful feature incorporated into the regulator controller is the ability of the device to regulate the voltage at a remote point downstream of the installation; this feature is known as *line drop compensation* (LDC). This is achieved by modifying the information provided by the voltage transformer at the regulator, to account for the line drop between the regulator and the downstream load centre. Figure 7.45 represents the traditional LDC arrangement.

In addition to the voltage transformer that senses the downstream potential, the regulator is also equipped with a current transformer in series with the load terminal. This measures the load current flowing, and its secondary current can be made to flow through a miniature replica of the downstream line, located within the controller (shown in Figure 7.45). If the controller's X and R values are set so that they reflect those of the line, then the voltage sensing circuit will effectively see a potential corresponding to that at the desired downstream location. The regulator will therefore control the potential there, at the expense of a slightly higher line voltage at the regulator itself.

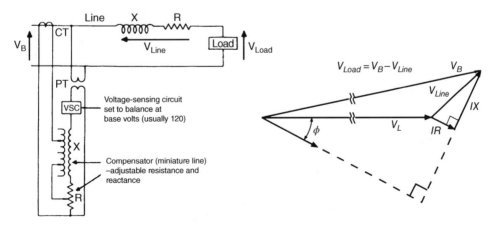

Figure 7.45 Line drop compensation. (Illustrations courtesy of Eaton).

Modern regulator controllers usually perform the LDC calculations digitally, although they still require the user to specify the X and R settings corresponding to that portion of the feeder downstream of the regulator.

Reverse Modes of Operation

In most applications the regulator will be installed on a line where the power flow is unidirectional, and in these cases regulation will always be applied towards the load side. However, on some lines the power flow may reverse from time to time, and in such cases it is desirable that the direction of regulation does so as well. (This requirement occurs more frequently nowadays with the increased penetration of renewable power sources in the distribution network, including wind and photovoltaic distributed generation.) This requires that the controller be able to determine the power flow direction, and apply the regulation to the source side rather than the load side. The latter requires that the source voltage be used as the reference for regulation; however, since the voltage transformer is located on the load side, steps must be taken to enable the regulator to obtain source voltage information. This may be done by installing as second voltage transformer on the source side, or in the case of some controllers, the source voltage can be obtained from the load voltage by knowing the tap selected, the current flowing and the impedance of the series winding.

There are six modes of reverse operation provided in Eaton regulator controllers. In the bidirectional mode the controller measures the power flow along the line and thus determines the magnitude and direction of the current. Should the current exceed the reverse direction threshold, the controller will then use the source voltage as its reference, and the regulator's operation will switch towards the source, regulating this potential instead.

7.10 Problems

1 Consider Equations (7.3) and (7.4). What options are available to a designer wishing to build a *high impedance* transformer, i.e. one with a high X/R ratio?

2 Phase shifting transformer or phase angle regulator (PAR). Figure 6.12 shows a 132 kV:132 kV, 200 MVA phase shifting transformer, whose nameplate appears in Figures 7.46 and 7.47. It was installed on an Australian transmission line running between Armidale and Kempsey. Armidale is one of the substations that receive a strong in-feed when energy is imported into New South Wales from Queensland. During such times, this line frequently becomes overloaded and in order to prevent this, the power imported must be reduced accordingly. In order to remove this constraint a phase shifting transformer was installed.

By regulating the phase shift across this transmission line, the power flowing through it can be controlled. This transformer can vary the phase at one end of the line and therefore influence the power flowing in the line without appreciably altering the local Armidale bus voltage.

The transformer in Figure 6.12 is actually a composite transformer, consisting of two tanks each containing a transformer and tap-changer. The main transformer is

PHASE-SHIFTING AND TAP-CHANGING TRANSFORMER

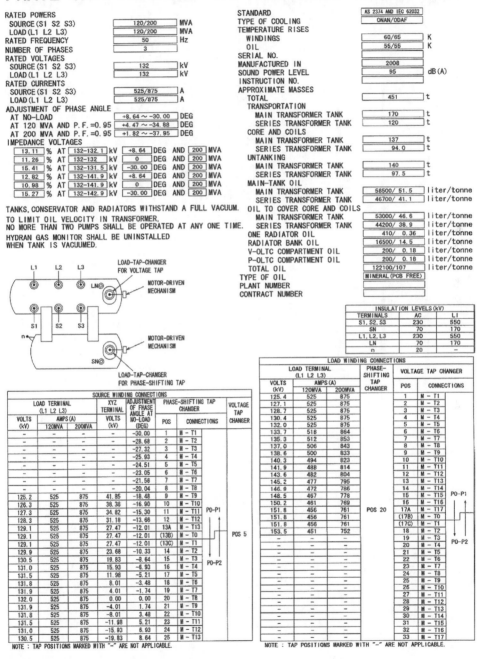

Figure 7.46 Phase shift transformer nameplate.

PHASE-SHIFTING AND TAP-CHANGING TRANSFORMER

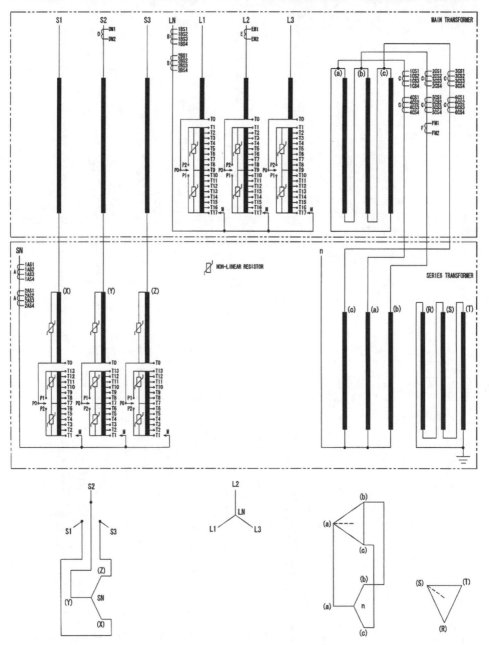

Figure 7.47 Phase shift transformer nameplate, continued.

in one tank and the series transformer in the other. Between them are two cabling ducts, as shown in plan view in Figure 7.46.

The main transformer is a star-star connected 132 kV:132 kV transformer with a tertiary delta winding, as shown in Figure 7.47. Its primary (source) terminals are S1, S2 and S3 and its tap-changer is capable of regulating the voltage at the load terminals L1, L2 and L3. The tertiary winding associated with this also provides excitation to the series transformer, located in the other tank. The local bus is connected to terminals S1, S2 and S3 and the transmission line to be controlled is fed by terminals L1, L2 and L3.

The series transformer is also a star-star connected device with a tertiary delta winding (see Figure 7.47). Its secondary and tap-changer windings are connected in series with the source windings of the main transformer, but because of the interconnections between the two transformers, the series voltages are in quadrature with the source voltages of the main transformer.

Phase regulation: The series transformer's tap-changer has the ability to regulate both the sign and magnitude of the voltage at terminals X, Y and Z in Figure 7.47. As a result the voltage applied to the main transformer primary windings can be varied in phase with respect to that of the local bus. This phase shift is achieved by adding a small quadrature voltage to each of the main transformer's primary windings, as shown in Figure 7.48.

The quadrature voltages are applied between neutral end of the main primary windings and the star point. They are generated by the series transformer, and appear at terminals X, Y and Z.

As shown, the quadrature voltages can be altered in magnitude (and sign) by the tap-changer associated with the series transformer. By using a tap-changer reversing winding and changeover switch, shown in Figure 7.47, it is possible for the tap-changer voltage to either buck or boost that in the fixed windings X, Y and Z. For example, if the reversing switch is in position P0-P2 then the tapping winding bucks the fixed winding, and thus it is possible that at one particular tap position, for the voltages at Z, Y and Z to become zero. An inspection of the tapping chart shows that this occurs on tap 20. When the quadrature voltages are zero there will be no induced phase shift between the supply and the load voltages. The phase shift tapping chart confirms this. Tap 20 is therefore the null tap.

In the buck condition described above, should the tapping winding voltage exceed that of the fixed winding, then the quadrature voltage will change sign, and with it the sign of the phase shift angle as well. In this case the load voltage will lead that at

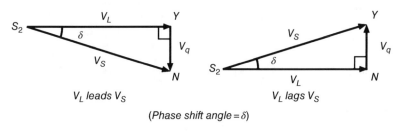

V_L leads V_S V_L lags V_S

(*Phase shift angle* = δ)

Figure 7.48 Quadrature phase shifting.

the source terminals. The tapping chart also demonstrates this. (In this case, lagging angles are defined as negative and leading angles are positive.)

On the other hand, when the tap-changer boosts the voltage in windings X, Y and Z the quadrature voltage will become larger, and with it the induced phase shift as well. As shown in Figure 7.48, this will result in a lagging phase shift between the primary voltage of the main transformer and the applied bus voltage. Because the load windings are also wound on the main transformer, the load voltage will also lag the applied voltage by the angle δ.

Voltage regulation: Figure 7.48 shows that, as the phase of the primary winding voltage is changed, unfortunately so is its magnitude. Changes in magnitude of the load voltage will affect the VAr flow and to a lesser extent the power flow as well. In order to prevent this, the load voltage must generally be held constant. This is the task of the voltage tap-changer associated with the load windings L1, L2 and L3. Thus in normal operation, the phase shift through the transformer will be set using the phase tap-changer – to achieve the desired power flow – and the resulting load voltage will be held constant by the voltage tap-changer. Tap 5 is the null tap on the voltage tap-changer, since the load voltage here is 132 kV.

Transformer under load: The tapping charts assume that a source voltage of 132 kV is applied and that the transformer is not loaded. The adjustment of phase angle table in the upper part of Figure 7.46 shows that the induced phase shift also becomes more lagging when the transformer is inductively loaded. The equivalent circuit in Figure 7.49 can be used to demonstrate this effect, which shows that the load voltage under zero current conditions (V_L^*) is given by the sum of the supply and quadrature components (V_S and V_q), as described already.

When the transformer is inductively loaded at an angle ϕ, the terminal voltage V_L, lags V_L^* by an additional angle δ', as shown in Figure 7.50.

Finally, Figure 7.51 shows the combination of a leading phase shift between V_L and V_S, induced by the quadrature voltage and the lagging effect produced by the combined effect of the load current and the transformer's impedance, X.

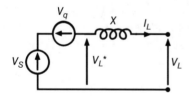

Figure 7.49 Equivalent circuit of phase shifting transformer. (V_L^* = Load voltage when I_L = 0, V_S = source voltage and V_q = quadrature voltage).

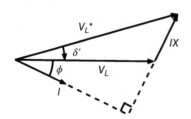

Figure 7.50 Effect of a lagging load current *I* on the terminal voltage V_L.

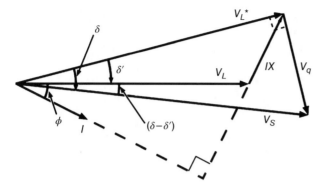

Figure 7.51 Overall phase shift becomes more lagging as a result of load current, I.

Questions:

A Why will phase control regulate the power flow on a transmission line? Explain.

B By using of phasor diagrams, show how the quadrature voltages discussed above are generated for each phase. (You will need to consider the main and series transformer interconnections shown in Figure 7.47.)

C From the voltage information in the tapping table, estimate the voltage induced in the phase shift windings X, Y and Z, when the applied bus voltage is 132 kV. (The tapping charts always assume that the nominal bus voltage is applied, i.e. 132 kV.) (Answer: 27.47 kV, each phase tap corresponds approximately to 3.9 kV.)

D Calculate the phase shift generated when the tap-changer is in position 9. Compare this result with the tapping chart. (Answer: 18.48°)

E Calculate the resulting load winding voltages for this condition. Assume that the transformer is unloaded. Compare your result with the tapping chart. (Neglect any compensation from the voltage tap-changer.) (Answer: 125.2 kV)

F Which tap should be selected from the voltage tap-changer in order to restore the load voltage to 132 kV? (Answer: Either tap 9 or 10)

G The phase shift table in Figure 7.46 shows that a maximum leading phase shift of 8.64° can be achieved when the transformer is unloaded. Using the equivalent circuit of Figure 7.51, calculate the maximum leading phase shift that you expect from the transformer when it is supplying its rated load at a power factor of 0.95. You will need to know the transformer's impedance. Compare your result with that given in Figure 7.46. (Answer: Max leading angle = 1.82°)

3 *Out-of-step transformer operation.* In this question we consider the out-of-step operation of a pair of transformers, T1 and T2. To simplify our analysis we will neglect the resistive component of each transformer's impedance, and we will assume that the magnitude of the reactive component is independent of the tap position. The single-phase equivalent circuit of these transformers appears in Figure 7.52. The load impedance is assumed to be 1.0 pu, with a lagging power factor of 0.966 and it is connected to a bus whose voltage, V_B is 1 pu. These high impedance transformers are identical, each with a reactance $X = 0.33$ pu.

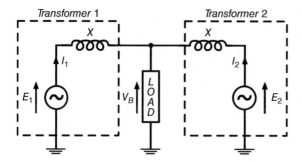

Figure 7.52 Single-phase equivalent circuit.

A We begin with the transformers operating from the same tap. Draw the phasor diagram for the bus voltage, V_B the reactive drop (V_x) and obtain from it the open circuit secondary voltages, E_1 and E_2, and the power angle δ. Assume that each transformer supplies half the load current. Hint: Make V_B the reference phasor, and use a scale of 1.0 pu = 300 mm and calculate the power angle from the dimensions of your drawing, don't measure it directly.
(Answer: $E_1 = E_2 = 1.055$ pu, $\delta \approx 8.2°$.)

B Assume that we now adjust the taps on the transformers so that E_1 is increased by 5% and E_2 is reduced by 5%. Since these changes are symmetrical with respect to V_B we might assume that V_B remains largely unaltered. In addition, the voltages E_1 and E_2 must remain in phase. (Why?) Modify your phasor diagram to show this, and from it evaluate the new transformer currents I_1 and I_2.
(Answer: $I_1 = 0.586$ pu, $I_2 = 0.454$ pu.)

C From your phasor diagram, calculate the phase angles between the bus voltage and I_1 and I_2 respectively. Calculate also the values of P and Q delivered by each transformer to the load. Show that these quantities are consistent with the power delivered to the load. Why has the tap-change affected Q more than P? (Answer: $\theta_1 \approx 29.7°$, $\theta_2 \approx -3.7°$, $P_1 \approx 0.46$ pu, $P_2 \approx 0.45$ pu, $Q_1 \approx 0.291$ pu, $Q_2 \approx -0.034$ pu) The unbalance is due to a significant circulating current that leads to high reactive power losses. Calculate the total reactive power each primary winding must provide under the tap settings of part 2. Where is the extra reactive power consumed?
(Answer: $Q_{1pri} = 0.40$ pu, $Q_{2pri} = 0.035$ pu)

D Modify your phasor diagram to include the circulating current, I_c, shown in Figure 7.53. Does this provide a partial justification of our initial assertion that E_1 and E_2 should lie in phase? Explain. What magnitude do you expect for I_c? Why must this current exist? Hint: What phase relationship do you expect between I_c and $(E_1 - E_2)$?
(Answer: $I_c = 0.162$ pu.)

E If T2 is now taken out of service, with its present tap settings, what value would the bus voltage adopt? (Assume that the load impedance remains unchanged.) If T1 is taken out of service instead, how would V_B change? In the light of this information, devise an operational procedure that you might use as a system operator, for the purpose of taking one of a pair of transformers out of service.
(Answer: 0.98 pu (T2 off), 0.88 pu (T1 off).)

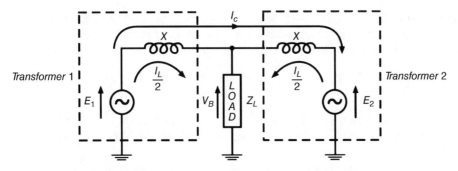

Figure 7.53 Inclusion of the circulating current, i_c.

4 A new 11,000 V:433 V Dyn11 transformer has been installed to share an LV load with an existing transformer of the same vector group, but initially the pair were not operating from the same tap. The original transformer was on the 11550 V:433 V tap and the new transformer was installed on its principal tap (11,000 V:433 V). The data relating to these transformers, measured at the connection to the bus, appears in the table below. The load voltage in this configuration was 422 volts.

	Rating	Impedance	Supply voltage	P (kW)	Q (kVAr)
Existing transformer	1 MVA	11.56%	11 kV	230	−220
New transformer	1 MVA	8.75%	11 kV	320	344

Tappings: 10,450 V:433 V, 10,725 V:433 V, 11,000 V:433 V, 11,275 V:433 V, 11,550 V:433 V 11,825 V:433 V 12,100 V:433 V.

A By making any reasonable assumptions, and assuming that the impedance values quoted are wholly inductive and include the transformer, the LV cable and bus connections, calculate the power angle δ.
  (Answer $\delta \approx 1.4°$)

B Calculate the magnitude of the reactive current circulating between these transformers.
  (Answer: ≈ 300 A.)

C Calculate the load impedance, assuming it to be represented by a resistive element in parallel with an inductive one.
  (Answer: $0.324 + j1.63$ ohms)

D Assuming that the load impedance remains constant, determine the tap setting for the new transformer which will see the load shared most evenly between the two. You may assume that the transformer impedances do not vary appreciably with tap setting.
  (Answer: 11,825 V:433 V)

5 Most MV-LV distribution transformers are of the Dyn11 vector grouping. Under what conditions would a Zyn11 transformer be an acceptable substitute and why?

6 Show by constructing a suitable phasor diagram, that a phase shift of about 5° arises between corresponding source and load voltages in the three regulator closed delta configuration shown in Figure 7.44.

7.11 Sources

1 Say MG. '*Alternating current machines*', 2nd ed, 1976, Pitman Press, London.
2 IEEE Std C57.131 '*IEEE standard requirements for tap-changers*', 2012, IEEE Power Engineering Society, New York.
3 IEC 60076-1 '*Power transformers – Part 1 General*', 2011, IEC International Electrotechnical Commission, Geneva.
4 IEEE Std C57.15. *Requirements, terminology and test code for step voltage and induction-voltage regulators*', 1988, IEEE Power Engineering Society, New York.
5 AS 60214.1 '*Tap changers part 1; Performance requirements and test methods*', 2005, Standards Australia, Sydney.
6 Harlow JH. '*Load tap-changing control*' Beckwith Electric Co, Largo FL 34649.
7 Sankar V. '*Preparation of transformer specifications*' Power Engineering Society Industry Application Seminar 2012; IEEE Power Engineering Society, New York.
8 Jonsson L, Sundqvist D. '*Vacuum interrupters as an alternative to traditional arc quenching on-load tap-changers*' ABB Sweden, TechCon Asia Pacific; 2007.
9 Harlow JH. '*Transformer tap-changing under load a review of concepts and standards*', 1993, Beckwith Electric Co, Largo FL 34649.
10 Beckwith Electric Co '*Introduction to paralleling of LTC transformers by the circulating current method*' Application note 11, Beckwith Electric Co, Largo FL 34649.
11 Jauch E.T. '*Advanced transformer paralleling*' 2001, IEEE Power Engineering Society, New York.
12 Areva T&D '*Network protection & automation guide*', 2005 Edition, Areva T&D, Paris.
13 Kloss A. '*A guide to power electronics*' 1984, John Wiley & Sons, Chichester.
14 Schaffer J. '*Rectifier circuits, theory and design*' 1965, John Wiley & Sons, New York.

8

Voltage Transformers

Voltage transformers (VTs), or potential transformers (PTs) as they are also called, are used to provide a scaled representation of the potentials on MV or HV busses, one that may conveniently be handled by protection relays, energy meters and general instrumentation equipment. Depending on the intended application, VTs may either be connected to system phase voltages or across line voltages. Therefore a VT's primary rating must be appropriate to both the bus potential and its intended method of connection.

Because a VT's secondary voltage is ideally a scaled replica of the bus voltage to which it is connected, it is important that the device does not introduce errors of its own into the measurement. Accordingly, manufacturers limit the burden that may be connected to the secondary winding(s), so as to reduce the potential dropped across the transformer's short-circuit impedance. Therefore the *rated* burden current is generally well below the *thermal current limit* of a VT. Despite this, magnitude and phase errors are introduced into the secondary voltages, but provided they are kept sufficiently small, their effects can be accepted. These errors are more of a problem in metering than in protection applications, since they directly affect the accuracy of the metered data. Limitations on the errors introduced are imposed by the various *metering class* designations.

Voltage transformer secondary potentials generally take one of several standardised voltages, as outlined in Table 8.1. With the exception of line connected VTs used in two element metering applications, modern voltage transformers tend to be phase connected, i.e. between phase and ground potential. In addition, they are generally single-phase devices, and are therefore capable of supporting a zero-sequence voltage, should one exist. This feature is important in protection applications. (Three-phase, three-limb VTs generally do not have this ability.) Despite being phase connected the voltage ratio is usually specified in terms of the rated line voltage of the associated bus, thus a 132 kV VT will have its ratio specified as: 132/√3 kV:110/√3 V.

8.1   Inductive and Capacitive Voltage Transformers

Voltage transformers broadly fall into two categories, single- and three-phase *inductive (iron cored)* VTs and single-phase *capacitive* VTs. Medium voltage transformers usually fall into the former category, and are built in a similar fashion to small single- or three-phase distribution transformers. Three-phase devices are often oil immersed, while

AC Circuits and Power Systems in Practice, First Edition. Graeme Vertigan.
© 2018 John Wiley & Sons Ltd. Published 2018 by John Wiley & Sons Ltd.

Table 8.1 Voltage Transformer standard secondary voltages.

Location	Secondary line voltages	Secondary phase voltages
Europe	100 V and 110 V	100/√3 V and 110/√3 V
Europe (extended secondary circuits)	200 V	200/√3 V
China	100 V	100/√3 V
USA & Canada (distribution)	120 V	120/√3 V
USA & Canada (transmission)	115 V	115/√3 V
USA & Canada (extended secondary circuits)	230 V	230/√3 V

(IEC 61869-3 ed.1.0 Copyright © 2011 IEC Geneva, Switzerland. www.iec.ch)

11000/V3
110/V3 110/V3 110/V3 V

VOLTAGE TRANSFORMER

TYPE VEN 12-02	F-Nr. 02/ 10302234		
	AS 1243-1982	IL:	12/28/75 kV
1.5xUn 30 s		Isol. E	50 Hz

No.	A2 - A1	Voltage Ratio / V	Class/	Rated Output	Ith
1a2-1a1	11000/V3	110/V3	1P	10VA	2A
2a2-2a1	11000/V3	110/V3	0. 2M	10VA	2A
3a2-3a1	11000/V3	110/V3	0. 2M	10VA	2A

WANDLER- UND TRANSFORMATOREN-WERK
WIRGES GMBH
Made in Germany   D-56419 Wirges

Figure 8.1 Three-winding 11 kV VT together with its nameplate.

single-phase VTs designed to be incorporated into metal clad switchgear, are usually resin cast. High voltage VTs on the other hand are increasingly of the single-phase capacitive type. These use a capacitive voltage divider to reduce the potential supplied to an iron-cored transformer, and by so doing make its design less expensive.

The voltage transformer in Figure 8.1 is a resin cast, single-phase, three-winding inductive device. The 11 kV HV terminal can be seen on top, and the secondary terminals are at the lower front. The low voltage end of the primary winding also frequently appears with these terminals, and it is very important that it remains grounded while the VT is in service. Its nameplate shows that this VT is designed to be connected to a primary potential of 11000/√3 V (6350 V). Each of its secondary windings delivers a phase voltage of 110/√3 V (63.5 V) and has a VA rating of 10 VA, which means that the minimum burden impedance that may be connected to any winding is $(63.5)^2/10 = 403$ ohms. One of these windings is a *protection class* (1P) winding while the other two are *metering class* (0.2 M). The thermal rating for each secondary winding is 2 A, beyond which overheating may occur. While it is common for modern VTs to be equipped with

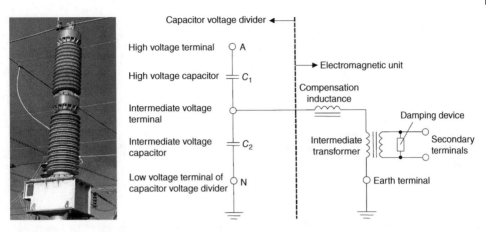

Figure 8.2 (a) A 220 kV capacitive voltage transformer (CVT) (b) CVT equivalent circuit. (IEC 61869-5 ed.1.0 Copyright © 2012 IEC Geneva, Switzerland. www.iec.ch).

two secondary windings, it is unusual for a VT to have three. This particular device was built to an old Australian standard (AS1243, *Voltage Transformers for Measurement and Protection*, 1982) and was one of several provided to a large customer who required independent revenue and check metering class windings, in addition to a protection class winding.

Figure 8.2 shows a single-phase capacitive voltage transformer (CVT), designed to operate on a phase voltage of $220/\sqrt{3}$ kV. Its HV terminal is internally connected to a capacitive voltage divider (C_1, C_2), the output of which supplies an oil immersed medium voltage inductive transformer (the intermediate transformer). The Thévenin equivalent impedance of this divider is equal to the reactance of the parallel capacitor combination (C_1, C_2). These capacitors are relatively small, and were this impedance not compensated for, the transformer would exhibit particularly poor voltage regulation.

The Thévenin equivalent impedance forms a series circuit, resonant at the supply frequency, with the compensation inductance, shown in Figure 8.2. This may be a discrete inductance or it may comprise the leakage inductance of the intermediate transformer. By creating this resonance, the effects of the Thévenin impedance can be annulled, and the iron-cored transformer effectively becomes fed from a low impedance source.

Because of the difficulty in achieving tight tolerances in the values of C_1 and C_2, tappings are often included on the intermediate transformer to provide a degree of ratio adjustment. The compensation inductance must similarly be adjustable.

For frequencies close to the system frequency a capacitive VT performs well, its magnitude and phase errors depending largely on the short-circuit impedance of the inductive transformer. However, at frequencies remote from the system frequency, the voltages reported by capacitive VTs will contain errors and cannot be relied upon; these devices therefore find limited application in harmonic studies.

8.1.1 Winding Arrangements

Because the secondary voltages must be a scaled representation of the associated primary potentials, voltage transformers are usually connected in a star-star (wye-wye) configuration, which allows both phase and line voltages to be measured. Occasionally

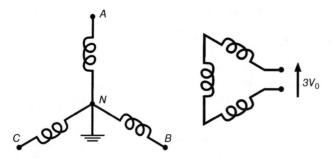

Figure 8.3 Open delta connection for the detection of residual voltages.

a delta-delta connection is used in situations where line voltages alone are sufficient for the intended application, particularly in ungrounded or high impedance grounded networks. In both these arrangements there is no significant phase shift introduced between the primary and secondary potentials.

There is also a third connection that is used in protection applications. This relates to the detection of residual or zero-sequence voltages, arising during fault conditions. Residual voltage transformers have a grounded star connected primary; their secondary windings are connected in an open delta arrangement, across which the residual voltage appears (Figure 8.3). A residual VT may be constructed using three single-phase transformers or alternatively it may be based on a three-phase transformer. In the latter case a five-limb core (i.e. a shell type core) must be used, so that a zero-sequence flux can be supported, as required to generate the zero-sequence voltage in each winding. The residual voltage developed across the open delta will thus be equal to three times the zero-sequence voltage.

8.2 Voltage Transformer Errors

All voltage transformers introduce small errors into their representations of the applied primary voltage. These are generally not critical in protection applications, but any introduced error will directly affect the accuracy of metering circuits. As a result, voltage transformers used in metering applications fall into various classes according to the *size* of these errors, as defined in the appropriate standard.

Magnitude error, as its name suggests, relates to the magnitude difference between the actual secondary voltage of a given voltage transformer and that which an ideal transformer would produce. Similarly, the phase error represents the angular difference between the actual secondary voltage and that produced by an ideal device. The magnitude and phase errors associated with both the voltage and current transformers used in energy metering circuits introduce errors into the energy measured. In order to achieve a required accuracy of measurement, these transformers must therefore be of an appropriate accuracy class. For example, in order to correctly meter energy to an accuracy of ±0.5%, class 0.2 voltage and current transformers will generally be required.

Magnitude Error Definition

The magnitude error e (sometimes called the ratio error) is defined as:

$$e = \frac{\left(V_{sec}.N - V_{pri}\right)}{V_{pri}}.100\% = \left[\frac{V_{sec}}{V_{pri}/N} - 1\right].100\% \tag{8.1}$$

where V_{pri} is the actual primary voltage, V_{sec} is the actual secondary voltage, and N is the nominal turns ratio ($V_{rated\,pri}/V_{rated\,sec}$)

Therefore a VT with a magnitude error of +0.2% presents a secondary voltage which is in excess of the actual voltage by this amount.

Phase Error Definition

The phase error γ, is the difference in phase between the primary voltage phasor and the secondary voltage phasor. The phase error is defined as positive when the secondary voltage phasor leads that of the primary.

Phase errors are usually measured in either minutes of arc or centiradians. (1 Crad = 34.4 minutes of arc.)

(*Radian measure* is based on the arc length subtended by a radius, R. An angle expressed in radians is defined as the length of the enclosed arc divided by the radius. For example, a semicircle of radius R subtends an arc of length πR and therefore the subtended angle of 180° is equal to π radians. Thus one radian is equal to 57.3°. A centiradian is one hundredth of a radian, and equals 0.573° or 34.4 minutes of arc.)

The magnitude and phase errors, as defined under IEC standards, are independent of one another, and as we shall see, they can be related to the short-circuit impedance of the voltage transformer concerned. To ensure compliance with these standards, the magnitude and phase errors must lie within the appropriate rectangle shown in Figure 8.4.

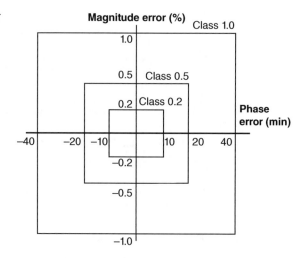

Figure 8.4 Magnitude and phase error limits for classes 0.2, 0.5 and 1.0. (IEC 61869-3 ed.1.0 Copyright © 2011 IEC Geneva, Switzerland. www.iec.ch).

8.2.1 Voltage Transformer Standards

We shall confine our interest in voltage transformer standards to three documents, the first two of which are European standards: IEC61869-3 *Additional Requirements for Inductive Voltage Transformers* and IEC61869-5 *Additional Requirements for Capacitor Voltage Transformers*. The third is the American standard IEEE C57.13 *IEEE Standard Requirements for Instrument Transformers*, including its supplement, IEEE C57.13.6 *IEEE Standard for High-Accuracy Instrument Transformers*. The last two documents apply to *both* voltage and current transformers. Both the European and American standards define the requirements of voltage transformers for both metering and protection applications. We will initially consider the requirements for metering class VTs as defined by each standard.

8.2.2 Magnitude and Phase Errors Defined under IEC 61869-3 and IEC 61869-5

IEC 61869-3 and IEC61869-5 define magnitude and phase error limits for VTs used in metering applications according to Table 8.2. Five classes of voltage transformer are defined. Class 0.1 applies to inductive laboratory grade instruments, and will not generally be found in metering applications. Classes 0.2 and 0.5 and 1.0 are prescribed for revenue metering applications, while class 3.0 is generally only used for panel instrumentation.

Table 8.3 defines the standard burden ratings for voltage transformers. Two ranges are prescribed: range I applies to very lightly loaded transformers, typical of modern metering or protection equipment: these burdens are assumed to be resistive. Range II applies to more traditional VTs where a heavier burden must be supported, one that is partially inductive. The error limits outlined in Table 8.2 apply for 0–100% of the rated

Table 8.2 Accuracy class limits for metering classes 0.1 – 3.0.

Metering class	Magnitude error limits	Phase error limits	
0.1*	±0.1%	±5 minutes	±0.15 Crad
0.2	±0.2%	±10 minutes	±0.30 Crad
0.5	±0.5%	±20 minutes	±0.60 Crad
1.0	±1.0%	±40 minutes	±0.1.2 Crad
3.0	±3.0%	Not specified	Not specified

(IEC 61869-3 ed.1.0 Copyright © 2011 IEC Geneva, Switzerland. www.iec.ch)

Table 8.3 Voltage transformer standard burdens as defined in IEC 61869-3.

Burden range	Rated burden (VA)	Burden power factor	Compliance range (% of rated burden)
I	1.0, 2.5, 5, 10	1.0	0–100
II	10, 25, 50, 100	0.8	25–100

(IEC 61869-3 ed.1.0 Copyright © 2011 IEC Geneva, Switzerland. www.iec.ch)

burden for range I, and 25–100% of rated burden for range II. Burden ratings specified under IEC 61869-3 apply to each winding on a voltage transformer.

Finally, the prescribed error limits must be maintained for voltages within the range 0.8–1.2 pu, and in the case of capacitive voltage transformers, over a frequency range of 99% to 101% of the nominal device frequency (50 or 60 Hz.)

8.2.3 Ratio and Phase Errors Defined under IEEE C57.13

The American standard IEEE C57.13 defines three standard classes of metering transformer (0.3, 0.6, 1.2) and its supplement IEEE C57-13.6, defines a further high accuracy class, 0.15. This standard does not directly prescribe magnitude and phase errors, but rather places limits on a related quantity known as the *transformer correction factor* (TCF).

As mentioned above, any magnitude or phase errors inherent in the associated VT will introduce a small error into the measurement of the power or energy flowing. This error may be corrected using the *transformer correction factor*, which may be calculated from the *ratio correction factor* (RCF), the phase error γ of the associated VT, and the phase angle of the metered load ϕ, using:

$$TCF_{VT} = \frac{\cos(\phi)RCF_v}{\cos(\phi+\gamma)} = \frac{P_{correct}}{P_{measured}} \tag{8.2}$$

(Note that this equation does not include the errors associated with the current transformers (CTs) used in the metering installation.)

The RCF is defined as the factor by which the stated ratio of an instrument transformer must be multiplied to obtain the actual ratio. Therefore the RCF can be used to correct the measured voltage presented by a VT for the effects of magnitude error, as follows:

$$RCF = \frac{|V_{correct}|}{|V_{measured}|}$$

where $V_{measured}$, is the voltage measured by the VT in question, and $V_{correct}$ is this quantity corrected for the magnitude error introduced by the VT. The RCF is related to the magnitude error e, according to:

$$RCF = \frac{1}{\left(1+\dfrac{e}{100}\right)}$$

For example, if the magnitude error, $e = +0.2\%$, then the associated RCF will be 0.998.

IEEE C57.13 prescribes limits on the TCF, which appear in Table 8.4. These must be maintained from 0% to 100% of the transformer's rated burden, for metered loads with lagging power factors between 0.6 and 1.0. They must also be maintained over a range of applied voltages between 0.9 and 1.1 pu.

Prescribing limits applicable to the TCF indirectly places limits on both the RCF and the phase error of a VT. The TCF is largest when the power factor of the metered load the lies at the lower end of the permitted range, i.e. when it equals 0.6. Under these conditions Equation (8.3) defines the relationship between these parameters. (This is fully discussed in Chapter 10.)

$$TCF = RCF + \gamma/2600 \tag{8.3}$$

Table 8.4 Accuracy class limits for metering classes 0.15 to 1.2.

Metering Class	Transformer correction factor (TCF)	
	Minimum	Maximum
0.15	0.9985	1.0015
0.3	0.997	1.003
0.6	0.994	1.006
1.2	0.988	1.012

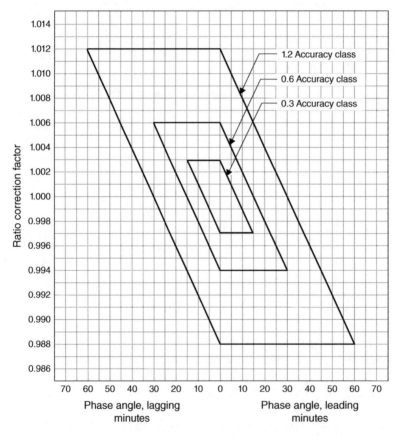

Figure 8.5 Voltage transformer ratio correction factor and phase error limits. (Adapted and reprinted with permission from IEEE. Copyright IEEE C57.13 (2008) All rights reserved. Permission for further use of this material must be obtained from IEEE. Requests may be sent to stds-ipr@ieee.org).

As a result, the ratio correction factor and the phase error are no longer independent of each other, and instead are constrained to lie within the parallelogram appropriate to the class of the VT, as shown in Figure 8.5.

Table 8.5 Standard voltage transformer burdens.

Burden designation	VA rating	Burden power factor	Impedance at 120V (line voltage) ohms	Impedance at 69.3V (phase voltage) ohms
W	12.5	0.1	1152	384
X	25	0.7	576	192
M	35	0.2	411	137
Y	75	0.85	192	64
Z	200	0.85	72	24
ZZ	400	0.85	36	12

IEEE C57.13 defines a list of standard burdens for voltage transformer accuracy testing, as shown in Table 8.5. For voltages different from those listed in the table, the burden impedance can be calculated according to $Z_{Burden} = V_{sec}^2 / VA\ Rating$. The accuracy class for a particular voltage transformer is specified in conjunction with a standard burden designation. For example, class 0.3ZZ implies that the transformer satisfies the 0.3 class accuracy requirements with a 400 VA burden of 0.85 pf. It will also be compliant with lower designations of the same power factor; however 0.3 class accuracy is not necessarily maintained for lower burden designations of a different power factor.

8.3 Voltage Transformer Equivalent Circuit

The equivalent circuit of a voltage transformer can be used to derive expressions for both the magnitude and phase errors. A simplified equivalent circuit appears in Figure 8.6, where all circuit elements have been referred to the low voltage (secondary) winding. The resistive component of the short-circuit impedance r relates to the winding resistances, while the inductive reactance x arises due to the leakage

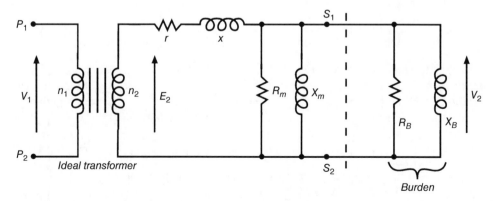

Figure 8.6 Simplified voltage transformer equivalent circuit, including inductive burden.

flux generated by each winding that does not link the other. The magnitude error is a function of the turns ratio, the short-circuit impedance and the burden applied to the VT.

X_M represents the inductance through which the magnetising current flows in order to establish the required flux in the core, while R_M reflects the associated hysteresis and eddy current losses within the core. The magnetising impedance will depend to a small extent on the magnitude of the applied primary voltage; however, these components generally only have a very small influence on the device errors over the normal range of operating voltages.

In general, the applied burden will be slightly inductive, although in recent years the trend towards electronic energy meters and numerical protection relays has seen many voltage transformers become resistively loaded, and frequently very lightly. Burden range I in Table 8.3 is designed to reflect this. The parallel combination of R_B and X_B therefore reflects a slightly inductive burden. Except for the special case of zero burden, we find that $R_B \ll R_M$ and $X_B \ll X_M$, so the magnetising impedance can usually be neglected.

8.3.1 Turns Compensation in Voltage Transformers

Because the secondary voltage of any VT falls slightly when it is burdened, manufacturers generally apply some additional turns to the secondary winding, in order to slightly increase the magnitude of the secondary voltage. This permits an increase in the burden that can be applied before the magnitude error falls outside the lower class limit, at the expense of creating positive magnitude errors when small burdens are applied. This technique, called *turns compensation*, usually results in an actual turns ratio between 0.2% and 1.0% greater than the nominal ratio, depending on the class of the VT. Turns compensation has a marked effect on the magnitude error of the device, but makes no difference to the phase error.

8.3.2 VT Magnitude Error Equation

The voltage across the secondary terminals (S1, S2) depends on the actual turns ratio used (N_{act}), the short-circuit impedance ($r + jx$) and the burden connected between S1 and S2. We will show that provided $r \ll R_B$ and $x \ll X_B$, then the magnitude error can be expressed to a very good approximation as:

$$e = \left[\frac{N}{N_{act}} \frac{R_B}{(R_B + r)} \frac{X_B}{(X_B + x)} - 1 \right] . 100\% \qquad (8.4)$$

where e = the magnitude error in %, N = the *nominal* turns ratio, $E_{1\,rated}/E_{2\,rated}$, N_{act} = the *actual* turns ratio, n_1/n_2, R_B = the resistive component of the burden, X_B = the inductive component of the burden, r = the resistive component of the short-circuit impedance and x = the inductive component of the short-circuit impedance.

The ratio $\dfrac{N}{N_{act}}$ is the *normalised turns ratio*, and typically lies between 1.002 and 1.01. Equation (8.4) may be derived from the secondary circuit phasor diagram shown in Figure 8.7. The burden voltage V_2 has been chosen as the reference phasor, and the

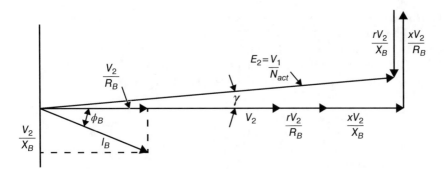

Figure 8.7 VT secondary circuit phasor diagram.

inductive burden current I_B lags this by an angle ϕ_B. We resolve this current into an in-phase component V_2/R_B and a quadrature component V_2/X_B.

The voltage drops across r and x due to each of these currents, when added to V_2, will yield the voltage developed across the secondary winding which, allowing for the application of turns compensation, will equal $|V_1/N_{act}|$, as shown in Figure 8.7. It is therefore apparent that:

$$\frac{|V_1|}{N_{act}} = |V_2| + |V_2|\frac{r}{R_B} + |V_2|\frac{x}{X_B}$$

Thus:
$$\frac{|V_1|}{|V_2|N_{act}} \approx \left[1 + \frac{r}{R_B} + \frac{x}{X_B}\right] \approx \left[\frac{(R_B+r)(X_B+x)}{R_B \quad X_B}\right]$$

Equation (8.1) can be re written as:

$$e = \left[\frac{|V_2|N}{|V_1|} - 1\right].100\% = \left[\frac{|V_2|N_{act}}{|V_1|}\left\{\frac{N}{N_{act}}\right\} - 1\right].100\%$$

So:

$$e = \left[\frac{N}{N_{act}}\frac{R_B}{(R_B+r)}\frac{X_B}{(X_B+x)} - 1\right].100\%$$

which is the same as Equation (8.4).

It is interesting to consider the case where no burden is applied to the VT (i.e. when both R_B and $X_B \rightarrow \infty$), the magnitude error then becomes:

$$e_0 = \left[\frac{N}{N_{act}} - 1\right].100\%$$

In the unusual case where *no* turns compensation is applied, the zero burden magnitude error will be zero. However, in most cases, turns compensation is employed, and therefore e_0 generally lies between +0.2 and +1.0%. Note that the zero burden error is independent of the burden and therefore also of its power factor.

The RCF can also be expressed in terms of the equivalent circuit parameters as:

$$RCF = \left[\frac{N_{act}}{N} \right] \frac{(R_B + r)(X_B + x)}{R_B X_B} \tag{8.4a}$$

8.3.3 VT Phase Error Equation

Figure 8.7 can also be used to obtain an expression for the phase error γ, in terms of the elements in the equivalent circuit, from which we may write:

$$\tan(\gamma) \approx -\frac{\left[\dfrac{xV_2}{R_B} - \dfrac{rV_2}{X_B} \right]}{V_2} = \left[\frac{r}{X_B} - \frac{x}{R_B} \right]$$

And since $\tan(\gamma) \approx \gamma$ for small γ (when expressed in radians), we may write:

$$\gamma \approx \left[\frac{r}{X_B} - \frac{x}{R_B} \right].100 \, \text{Crad} \tag{8.5}$$

Because $r \ll R_B$ and $x \ll X_B$, Equation (8.5) generally provides a very good approximation for γ. However, in the case when zero burden is applied, Equation (8.5) suggests that the phase error will be zero. While this is nearly so, we often find that γ takes small non-zero values, determined largely by the small burden imposed by the magnetising impedance, R_M and X_M. We may therefore modify Equation (8.5) accordingly:

$$\gamma \approx \left[\frac{r}{X_B} - \frac{x}{R_B} \right].100 + \gamma_0 \, \text{Crad} \tag{8.5a}$$

where γ_0 is the zero burden phase error, expressed in Crad.

8.4 Voltage Transformer 'Error Lines'

Figure 8.8 shows a plot of the magnitude and phase errors of a class 1.0 voltage transformer, as a function of the applied burden, in VA. To a very good approximation, both these quantities are linear functions of the applied burden, and therefore we might expect that when plotted against each other, the resulting graph will also be linear.

Figure 8.9 demonstrates that this is indeed the case. We call the line in this figure an *error line* and it is useful for finding both magnitude and phase errors at any fractional burden applied to the VT. Magnitude and phase error information is frequently provided only at two fractions of the rated burden. However, this small amount of information is sufficient to enable the evaluation of errors at any other burden, even one of a different power factor.

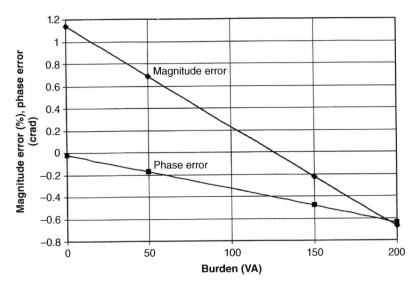

Figure 8.8 Class 1.0 voltage transformer magnitude and phase errors (UPF burden).

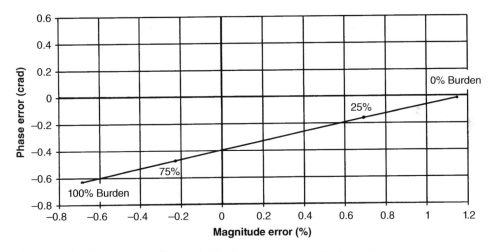

Figure 8.9 Error line corresponding to the VT whose errors appear in Figure 8.8.

8.4.1 Properties of Error Lines

Before we can make full use of error lines it is necessary to understand some of their properties. We begin by plotting an error line for a VT from the magnitude and phase errors measured at two burden points. These are usually either 25% and 100%, or perhaps 0% and 100% of the applicable burden. Let us assume that we have data for the former. We plot the error line from this data, as shown in Figure 8.9.

At this stage, a summary of the important properties of error lines is useful.

1) For a particular VT, all the errors corresponding to fractions of a particular burden lie on the same error line.

2) The error at any fractional burden is distributed along the error line in direct proportion to the fraction concerned. For example, the errors corresponding to 50% of the rated burden lie half way between the zero burden point and the 100% point.

3) The zero burden point can therefore be found by extrapolating the error line to the right of the 25% point, by one third of the distance between the 25% and 100% points. Once the zero burden point (e_0, γ_0) has been found, the value of the normalised turns ratio N/N_{act} can be determined.

4) A unique error line exists for each particular burden power factor. The error line corresponding to a unity power factor burden is known as the base line.

5) All error lines for a particular VT must pass through the zero burden point, since this is independent of the burden power factor.

6) All error lines for a particular voltage transformer are the same length. Therefore the fractional burden errors from one error line can be easily transferred to another, by scribing arcs of appropriate radii, centred on the zero burden point.

7) The error line corresponding to a lagging burden of power factor pf, lags the base line by an angle of arcos(pf). Thus the error line corresponding to a 0.8 pf burden lags the base line by 36.9°. Therefore once one error line has been plotted, others can be quickly constructed.

The following example will clarify these concepts.

8.4.2 Error Conversion between Burdens of Differing Power Factors

Occasionally it may become necessary to evaluate the magnitude and phase errors at a burden whose power factor is different from that at which it was originally tested. For example, if the error measurements were made with a 0.8 pf burden for a range II, 10 VA VT, which is subsequently to be used as a range I VT, it will be necessary to demonstrate compliance with the IEC standard when burdened at unity power factor. The steps outlined below demonstrate how the unity power factor errors can be graphically obtained from the 0.8 pf results.

Example: We will use the 0.8 pf test data from a 44 kV, Class 0.5, 100 VA VT presented in Table 8.6, to evaluate the errors at unity power factor. (Note that in this instance the

Table 8.6 Measured error results from a 44 kV Class 0.5, 100 VA VT (graphed in Figures 8.9 and 8.10).

Volts (pu)	Burden (pu)	Magnitude error (%)		Phase angle (Crad)	
		0.8 pf	1.0 pf	0.8 pf	1.0 pf
0.8	0.25	0.21	0.23	0.029	−0.07
0.8	1.00	−0.23	−0.14	0.058	−0.30
1.0	0.25	0.21	0.23	0.029	−0.07
1.0	1.00	−0.23	−0.14	0.058	−0.30
1.2	0.25	0.21	0.23	0.029	−0.07
1.2	1.00	−0.23	−0.15	0.058	−0.03

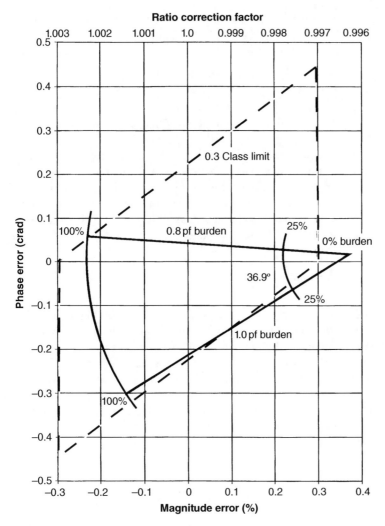

Figure 8.10 Error lines for the VT whose error results are shown in Table 8.6. (This illustration is also known as a circle diagram or a fan diagram).

test results are independent of the applied voltage over the range 0.8–1.2 pu. This is usually the case.)

We proceed as follows:

Step 1: Plot the error line corresponding to the 0.8 power factor data for 25% and 100% burdens. This is the upper line in Figure 8.10.

Step 2: Determine the zero burden point as described above. In this case, the zero burden error (e_0, γ_0) is (0.36%, 0.02 Crad), thus the normalised turns ratio for this VT is 1.0036.

Step 3: Since the unity pf line leads the 0.8 pf line by an angle equal to arcos(0.8) or 36.9°, this error line can be constructed by simply drawing a line that leads the 0.8 pf line by this angle, passing through the zero burden point (the lower line in Figure 8.10).

Step 4: All error lines for a given VT have the same length, therefore the 100% error point can be scaled directly from the 0.8 pf line. Errors for any other fractional burden may also be similarly derived.

Since the unity pf error line was used constructed from the 0.8 pf data, the unity pf errors obtained from the graph should be compared with those in the Table 8.6. It can be seen from Figure 8.10 that this IEC 61869-3 Class 0.5 VT does not quite satisfy the IEEE C57.13 Class 0.3 limits, particularly when a unity power factor burden is applied.

In summary, because of the linear nature of voltage transformers, if two sets of magnitude and phase errors are known for a particular burden, the magnitude and phase errors may be derived for any other burden. This simple graphical process does not require knowledge of any of the VT's internal parameters, nor does it require complex computations. Annex ZA of IEC 61869-3 includes a description of this process as well as an equivalent numerical approach to error conversion.

8.5 Re-rating Voltage Transformers

Occasionally, it may be necessary to re-rate a VT from its rated class to a higher one. For example, a Class 1 VT may be used in the Class 0.5 *over a restricted range of burdens*. The error line can be useful in determining the required burden range.

Figure 8.11 shows an example of a VT error line, drawn for a unity pf burden. The IEC 61869-3 error limits for Class 0.5 instruments are also plotted on this graph. The error line lies within these for burdens between 35% and 79% of the 200 VA rating. Therefore for Class 0.5 operation, the maximum burden permitted is 79% of 200 VA (158 VA) and the minimum burden required is 35% of 200 VA (55.6 VA).

This technique can be very useful when a superior class of VT is required. It is necessary, however, to ensure that the minimum burden calculated for the new class is actually applied to the device.

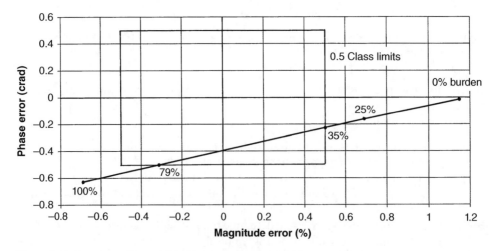

Figure 8.11 Re-rating a Class 1 VT to Class 0.5 over a restricted range of burdens.

8.6   Accuracy Classes for Protective Voltage Transformers

The same voltage transformer is frequently used in both metering and protection applications. While the accuracy required for protection need not be as high as that for metering, a protection VT must operate satisfactorily over a considerably wider range of voltages. IEC 61869-3 defines two accuracy classes for protection VTs, 3P and 6P, shown in Table 8.7. These limits apply from 5% of rated voltage, to the rated voltage multiplied by the applicable *voltage factor* F_v, over the range of burdens shown in Table 8.3.

It is usual for a VT with a single secondary winding to be assigned both a metering and a protection rating, for example 0.5/3P.

8.6.1   Voltage Factor (F_v)

The *voltage factor* depends on the magnitude of the over-voltages that the VT may encounter as a result of faults on or near the bus to which it is connected. The maximum primary voltage that can be supported by a VT can be found by multiplying its rated voltage by the voltage factor F_v. Table 8.8 shows voltage factors for single and three-phase VTs prescribed by IEC 61869-3, for both *effectively earthed* and *non-effectively earthed* systems. Together with the F_v values, the table also specifies the duration for which the over-voltage may exist.

Voltage factors vary between 1.2 and 1.9 and are determined largely by the performance of the earthing system at the location of the VT in question. A voltage factor of 1.9 requires that the VT be capable of supporting almost twice its rated voltage. It must therefore be conservatively designed to operate at a relatively low flux density at its nominal voltage and be equipped with sufficient insulation to cope with the required over-voltage.

8.6.2   Earth Fault Factor and Effective Earthing

Put simply, the earth fault factor is the ratio of the highest phase to earth voltage on a healthy phase during an earth fault on a system, to the magnitude of the same voltage in the absence of such a fault.

The earth fault factor has a direct impact on the potential that voltage transformers must support near to a fault. It quantifies the maximum in potential experienced by the healthy phase(s) during fault conditions. The earthing system at the fault location is classified as either effectively earthed or non-effectively earthed, according to the magnitude of the earth fault factor.

Table 8.7 Protective voltage transformer magnitude and phase error limits (IEC 61869-3).

Protection accuracy class	Magnitude error (%)	Phase error	
		±Minutes	±Crad
3P	3	120	3.5
6P	6	240	7.0

(IEC 61869-3 ed.1.0 Copyright © 2011 IEC Geneva, Switzerland. www.iec.ch)

Table 8.8 Rated voltage factors (F_v) for single-phase inductive VTs.

Rated voltage factor	Rated time	Method of connecting the primary winding and system earthing conditions
1.2	Continuous	Between phases in any network
		Between transformer star-point and earth in any network
1.2	Continuous	Between phase and earth in an effectively earthed neutral system
1.5	30 s	(IEC 61869–1: 2007, 3.2.7a)
1.2	Continuous	Between phase and earth in a non-effectively earthed neutral system
1.9	30 s	(IEC 61869–1: 2007, 3.2.7b)
		with automatic earth fault tripping
1.2	Continuous	Between phase and earth in an isolated neutral system (IEC 61869–1:
1.9	8 h	2007, 3.2.4) without automatic earth fault tripping or in a resonant earthed system (IEC 61869–1: 2007, 3.2.5) without automatic earth fault tripping

(IEC 61869-3 ed.1.0 Copyright © 2011 IEC Geneva, Switzerland. www.iec.ch)

Note 1. The highest continuous operating voltage of an inductive voltage transformer is equal to the highest voltage for the equipment (divided by $\sqrt{3}$ for transformers connected between a phase of a three-phase system and earth) or the rated primary voltage divided by the factor 1.2, whichever is the lower.

Note 2. Reduced rated times are permissible by agreement between manufacturer and purchaser.

A three-phase earthed neutral system is said to be effectively earthed at a given location, if the earth fault factor at that location does not exceed 1.4. On the other hand, a system is said to be non-effectively earthed if the earth fault factor there exceeds 1.4.

The earth fault factor depends on the system impedances at the fault, and it can be shown that a system can be considered effectively earthed if the ratio of resistive component of the zero-sequence impedance R_0, to the positive sequence reactance X_1, is less than one, and the ratio of the zero-sequence reactance X_0, to the positive sequence reactance X_1, is less than three.

The over-voltage problem occurs during faults to earth, when a potential difference arises between the neutral terminal of the supply transformer and the star point of the VT in question. This *neutral shift voltage* occurs as a result of the fault current flowing through an impedance between these points. This may be the resistance between the ground connections of the supply transformer and the star point of the VT, the impedance of the neutral conductor in MV systems when the neutral is distributed, or the impedance of the neutral connection where impedance earthing is used.

The potential difference between these terminals is generally small, and can usually be neglected. However, during earth faults, where substantial ground currents can flow, this potential may become significant. If the VT in question is near a phase-to-ground fault, the resulting potential rise appears as a neutral shift voltage, causing a reduction in the potential seen by the faulted phase, and an increase in the potential seen by at least one other.

This is illustrated in Figure 8.12, where a phase-to-ground fault near the VT causes a substantial rise in the local earth potential. In this case, it is assumed that the earth resistance between the remote earth and that of the supply transformer is dominant,

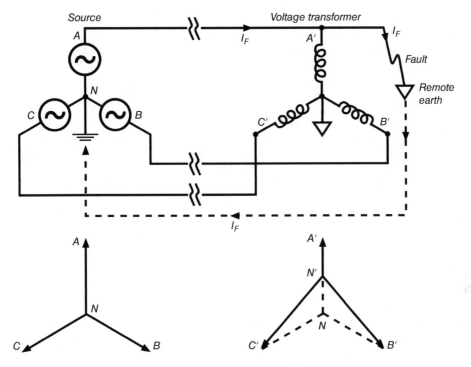

Figure 8.12 Neutral shift voltage resulting from an earth fault.

and therefore most of the faulted phase potential is dropped between the two ground connections. As a result, the neutral potential at the VT rises, reducing the A phase voltage in proportion, and increasing both the B and C phase voltages seen by the VT.

If the earth fault factor at a particular location is 1.4 or less, then allowing for bus voltages of up to 1.1 pu, the maximum phase potential seen by a VT at that location under fault conditions, will be equal to $1.4 \times 1.1\text{pu} = 1.5\text{pu}$. Therefore a voltage factor of 1.5 is appropriate.

The largest earth fault factor possible is $\sqrt{3}$, occurring when a phase-to-ground fault occurs *at* the VT and faulted phase potential is therefore zero. If the maximum bus voltage is 1.1 pu, then the voltage factor will be $\sqrt{3} \times 1.1 = 1.9$.

The duration associated with the voltage factor depends on the operating time of the earth fault protection equipment. In some cases, such as resonant or impedance earthing, the fault may exist for some time before being cleared, and therefore the VT may need to have a continuous over-voltage capability. In other cases, the fault will be cleared relatively quickly, and the VT need only be short term over-voltage rated.

8.6.3   IEEE C57.13 Over-Voltage Ratings

IEEE C57.13 defines five groups of voltage transformers each with over-voltage and exposure durations ratings, shown in Table 8.9.

Table 8.9 IEEE Over-voltage limits.

Group	Intended primary connection	Over-voltage factor	Duration
1	Line to line or line to ground	1.25	8 hours (@ 64% of the thermal burden)
2	Line to line	1.1	Continuous
3	Line to ground	$\sqrt{3}$	1 minute
4a	Line to ground	1.25	8 hours (@ 64% of the thermal burden)
5	Line to ground	1.4	1 minute

(Adapted and reprinted with permission from IEEE. Copyright IEEE C57.13 (2008) All rights reserved. Permission for further use of this material must be obtained from IEEE. Requests may be sent to stds-ipr@ieee.org).

8.7 Dual-Wound Voltage Transformers

Recent years have seen a move towards the use of single-phase VTs with two secondary windings. Occasionally one winding is designated for metering and the other for protection, but more commonly, both are rated for either metering or protection duties (for example a dual 0.5/3P rating). However, this does not mean that all windings are so rated, and care must be taken to ensure that metering circuits are supplied appropriately. Dual-wound VTs provide a degree of flexibility where both revenue and check meters are required to be supplied from separate VT windings. It also allows for the segregation of protection and metering circuits should this be required.

The presence of two secondary windings fed from a common primary, means that the burden on one secondary will affect the errors associated with the other. Provided that both windings are equally burdened, a dual-wound VT can be treated as a single winding device from the point of view of magnitude and phase errors. While this is rarely the case, modern protection and metering systems generally impose very light burdens on VTs, and a small imbalance in burden between windings may be insignificant.

Under IEC 61869-3 the rated burden applies to *each* secondary winding. Error testing is therefore usually carried out when both windings are burdened at 25% and 100% of their rating. On the other hand, the burden defined under IEEE C57.13 is the *total* burden that can be applied to *both* windings. Therefore VTs must be class compliant when this burden is apportioned in *any* ratio between the secondary windings.

8.8 Earthing and Protection of Voltage Transformers

Star-star connected VTs are generally only used in solidly grounded networks where a low zero-sequence impedance exists between the VT star point and the supply neutral. This permits the third harmonic component of the VT's magnetising current (a zero-sequence current) to flow, and thus allows the flux in the VT to achieve its correct value. Without this current, the flux waveform will be a little flat-topped, and an erroneous third harmonic component will be present in each phase voltage, sufficient to corrupt energy metering and power quality measurements, both of which require an accurate representation of the phase voltages. This makes the use of conventional three-element metering impractical in ungrounded three wire circuits.

Fortunately, the line voltages reported by star connected VTs are not similarly corrupted (since the zero-sequence voltage component cancels in the subtraction of phase potentials), and therefore this configuration can be used in applications where only line voltage information is required. However, it is more likely that line connected VTs will be used in ungrounded networks.

Where single-phase star connected VTs are applied to ungrounded networks, it is necessary to ensure that the voltage rating of each is equal to the maximum rated line voltage of the network. This is because a phase-to-ground fault may exist for some time before being cleared.

8.8.1 Ferro-Resonance

There is a danger of exciting a ferro-resonant oscillation when VTs are connected between line and ground on ungrounded networks or networks grounded through a high impedance. As explained in Chapter 11, ungrounded networks tend to be loosely tied to ground through the action of the parasitic capacitances that exist between each phase and ground, as depicted in Figure 8.13. Because there is no firm ground connection, the source phase voltages can vary with respect to ground in response to a disturbance such as ferro-resonance.

Ferro-resonance is a highly non-linear oscillation that arises between the magnetising inductance of a transformer and the stray system capacitance existing in parallel with it. It can be triggered by a temporary over-voltage, possibly as a result of a switching transient, which drives the transformer deep into the saturated region of its magnetising characteristic. The large magnetising currents that result can generate very high oscillatory voltages, across a range of frequencies, both above and below the system frequency, due to the non-linear nature of the transformer's magnetising inductance. This may result in the destruction of the VT and possible damage to nearby equipment.

As with any resonant condition, the amplitude of a ferro-resonant oscillation is influenced by the degree of damping present, and therefore a resistive burden on the VT can play an important role in suppressing ferro-resonance. In cases where star connected VTs are used on ungrounded networks, a sufficiently resistive burden must be applied to avoid the possibility of ferro-resonance. This may impose a significant thermal

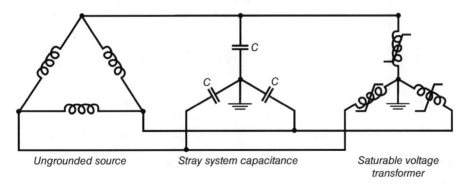

Ungrounded source   Stray system capacitance   Saturable voltage transformer

Figure 8.13 Ungrounded network.

burden on a VT, beyond that permitted for metering purposes. Therefore in addition to the reasons cited above, this connection will generally only be used for residual voltage (ground fault) detection.

8.8.2 Ferro-Resonance in Capacitive VTs

While all VTs are subject to ferro-resonance in adverse circumstances, capacitive VTs are slightly more so, given that they already contain both capacitive and inductive elements. In order to prevent sustained ferro-resonance, damping components are generally included in parallel with the secondary winding, as shown in Figure 8.14. The damping system chosen should not load the VT excessively at the fundamental frequency, since this will reduce its available VA capacity. Circuits like that shown in Figure 8.14a are often used in the secondary circuit of the inductive voltage transformer. Here a low impedance resistive path exists across the secondary winding at frequencies away from the system frequency. Since ferro-resonance usually generates energy at many harmonic frequencies, the damping so provided will help to reduce the oscillation. Figure 8.14b shows an alternative arrangement where a spark gap connected resistor is connected across the VT secondary. This will consume no energy under normal circumstances, but will provide damping when the gap breaks down under high voltage conditions. This circuit also includes a saturable inductor L which will present a low impedance when saturated by large secondary voltages.

In addition to these techniques, non-linear devices such as metal oxide varistors (MOVs) are also used as damping elements. These devices are made from a crystalline zinc oxide structure, the crystal boundaries of which form p-n junctions that pass very little current under low voltage conditions. At high applied voltages these junctions break down in an avalanche mode, and the device impedance falls to a low value. However, they can only dissipate a certain amount of energy before permanent damage occurs, and therefore the energy rating should be carefully matched to the intended application.

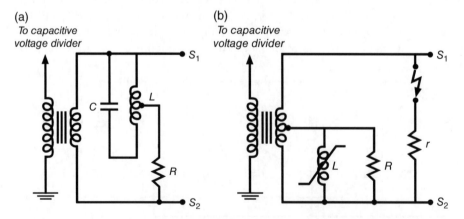

Figure 8.14 (a) CVT off frequency damping (b) CVT spark gap, saturable inductor damping.

IEC 61869-5 deals with capacitive voltage transformers, and specifies that for any burden up to the device rating that once triggered, a ferro resonance condition shall not be sustained. This is generally satisfied by the provision of adequate damping of the types described above. This standard defines a maximum instantaneous error $\hat{\varepsilon}_F$, defined by:

$$
\hat{\varepsilon}_F = \frac{\hat{U}_S T_F - \dfrac{\sqrt{2}U_P}{k_R}}{\dfrac{\sqrt{2}U_P}{k_R}} = \frac{k_R \hat{U}_S T_F - \sqrt{2}U_P}{\sqrt{2}U_P}
\tag{8.6}
$$

where $\hat{\varepsilon}_F$ is the maximum instantaneous error, $\hat{U}_S$ is the peak secondary voltage, U_P is the primary voltage (r.m.s), U_{Pr} is the rated primary voltage (r.m.s), k_R is the transformation ratio and T_F is the duration of ferro-resonance (seconds).

The maximum instantaneous error $\hat{\varepsilon}_F$, shall fall to a value less than 10% after the duration T_F, as defined in Tables 8.10 and 8.11. Note that it is possible that the effects of ferro-resonance may have entirely disappeared by time T_F, or that the oscillation may still be decaying. In either event the value of $\hat{\varepsilon}_F$ shall be less that 10%.

Table 8.10 Ferro-resonance requirements for effectively earthed CVTs.

Primary voltage U_P	Ferro-resonance oscillation duration T_F (Seconds)	Error $\hat{\varepsilon}_F$ (%) after duration T_F
0.8 U_{Pr}	≤0.5	≤10
1.0 U_{Pr}	≤0.5	≤10
1.2 U_{Pr}	≤0.5	≤10
1.5 U_{Pr}	≤2.0	≤10

(IEC 61869-5 ed.1.0 Copyright © 2012 IEC Geneva, Switzerland. www.iec.ch)

Table 8.11 Ferro-resonance requirements for non-effectively earthed CVTs (isolated neutral).

Primary voltage U_P	Ferro-resonance oscillation duration T_F (Seconds)	Error $\hat{\varepsilon}_F$ (%) after duration T_F
0.8 U_{Pr}	≤0.5	≤10
1.0 U_{Pr}	≤0.5	≤10
1.2 U_{Pr}	≤0.5	≤10
1.9 U_{Pr}	≤2.0	≤10

(IEC 61869-5 ed.1.0 Copyright © 2012 IEC Geneva, Switzerland. www.iec.ch)

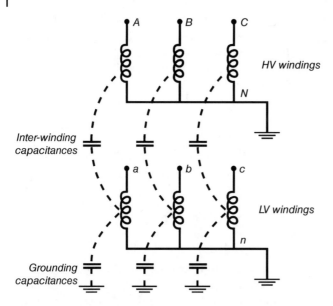

Figure 8.15 Capacitive coupling between windings and ground in a voltage transformer.

8.8.3 Earthing the LV Star Point

The LV star point should always be earthed, although this is more for reasons of safety. If the secondary star point was permitted to float, then it is possible that a substantial voltage could develop between the secondary star point and ground. This occurs because of the capacitive coupling between each HV winding and the associated LV winding, and between the each LV winding and ground, as is illustrated in Figure 8.15.

Provided that corresponding capacitances are equal, the secondary star point will remain close to ground potential, but if there is a mismatch between these capacitances then a neutral shift voltage will be generated which may pose a safety risk to both personnel and equipment. For this reason, the LV star point must be earthed so that such a potential cannot exist.

In the case of line connected VTs, where the secondary windings have no star point, it is customary to ground the white phase terminal, for the same reasons.

8.8.4 VT Protection

Voltage transformers are expensive and deserve some degree of protection, on both the primary side and the secondary side. Primary fusing is normally selected to protect the VT in the event of internal faults or short circuits close to the secondary terminals. It should be capable of carrying the inrush current of the device without causing nuisance tripping. Similarly, secondary fusing or magnetic circuit breakers should also be provided to protect the secondary from downstream faults and to provide a convenient method of isolating the VT. It is always preferable for the secondary protection to operate before the primary fuse, and therefore the secondary protection should be chosen accordingly.

Figure 8.16 VT failure – note the lack of HV fusing, the thermal damage to the left-hand VT and the carbon condensate damage throughout the cubicle.

Figure 8.16 shows what can go wrong when a VT fails internally, where the fault is not cleared by a primary fuse. The right-hand VT in the figure shows a solid copper connection to the HV terminal; this link would normally be replaced with an HV fuse. In this case the left-hand VT failed internally, and since it was not fuse protected the fault progressed to the stage where carbon condensate damage occurred throughout the cubicle in which it was housed. Where a VT is connected to an incoming feeder, such a fault can only be cleared by tripping the associated circuit breaker or by a bus zone protection operation, neither of which is desirable.

8.9 Non-Conventional Voltage Transformers

Both inductive and capacitive voltage transformers exhibit a non-linear frequency response which severely limits their usefulness in measuring harmonic voltages. Capacitive VTs are tuned to the fundamental frequency and therefore their performance is only guaranteed over a very narrow bandwidth. Inductive VTs also suffer from non-linear effects in the frequency domain due to the existence of parasitic winding and inter-winding capacitances. These create resonances in the frequency response, leading to large errors in the apparent turns ratio. Because HV VTs require more turns on the primary winding than do MV VTs, they are subject to larger parasitic capacitances and, as a result, the first resonance usually occurs at lower frequencies, further restricting their use for harmonic measurement.

Non-conventional voltage transformers have been developed for power quality applications as well as for use in gas insulated switchgear (GIS). These devices are generally either based on the principle of a voltage divider or on a fibre optic material sensitive to the electric field associated with the potential to be measured. Resistive divider networks may be used to produce a scaled version of the measured voltage, but the effects of stray capacitance limit the device response at high frequencies.

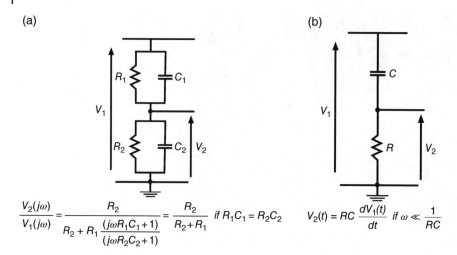

Figure 8.17 (a) RC voltage divider sensor (b) RC differential voltage sensor.

The potential divider shown in Figure 8.17a contains both resistive and capacitive elements connected in parallel and provided that $R_1C_1 = R_2C_2$ then linear performance can be achieved to frequencies in excess of 10 kHz, making it applicable to broadband harmonic measurements.

The simple CR divider circuit shown in Figure 8.17b is used in some GIS switchgear and provides an output voltage proportional to the derivative of the input voltage, which must be amplified and integrated to reconstruct the input voltage. Since these circuits are incapable of driving any appreciable burden, the associated amplifier circuitry must also perform this function. Modern electronic protection relays and control equipment present very little burden and therefore it is quite practical to provide a low power analogue replica of the measured voltage. However, these instruments may also be configured to provide digital information to the substation secondary equipment over a fibre optic link. This approach is usually adopted in the case of fibre optic voltage sensors.

The IEC standard IEC 61850-6 describes a digital interface between field based current and voltage transformers and the associated control room instrumentation. Under this standard, current and voltage information from either conventional or non-conventional instrument transformers (NCITs) is digitised and time synchronised in the field by a *merging unit*, and conveyed optically to the associated protection and control equipment within the substation. This approach has three significant advantages: firstly, it greatly simplifies the wiring required and entirely removes the effects of cabling and equipment burdens; secondly multiple items of secondary equipment may be connected to the same instrument transformers, and where non-conventional instrument transformers are fitted, these may be generic and not application specific, thus simplifying the design and specification of the primary equipment. Finally, non-conventional voltage transformers do not suffer from the effects of ferro-resonance. For all these reasons, it is likely that many more NCITs will find application in the future.

8.10 Problems

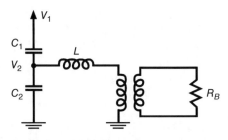

Figure 8.18 Diagram for question 1. ($C_1 = 0.0192\,\mu F$, $C_2 = 0.078\,\mu F$, $V_1 = 110\,kV/\sqrt{3}$, $f = 50\,Hz$, the output voltage is $110v/\sqrt{3}$ and rated burden of the CVT is 150 VA).

1 Apply Equation (6.2) to the CVT circuit shown in Figure 8.18, and show that the power delivered by a CVT is given by $P = 2\pi f C_1 V_1 V_2 Sin(\delta)$. Hence determine the phase shift (δ) that will occur in V_2, with respect to the voltage at V_1. Will V_2 lag or lead V_1? Assume that the intermediate transformer is lossless and that all the power is therefore consumed by the burden resistance.
(Answer: +1.8°)

2 Calculate the value of the tuning inductance L in question 1. Assuming that the transformer is ideal, find the maximum steady state current that this component must carry.
(Answer: 104.35H, 12 mA)

3 Calculate the phase shift that you expect between V_2 and the transformer output terminals in the schematic above, and hence determine the overall phase shift between V_1 and the output, under the assumptions made above.
(Answer: –1.8°, 0°)

4 The intermediate transformer in question 1 has a short-circuit impedance of $0.12 + j0.2$ ohms and a normalised turns ratio of 1.004. The tuning inductance has a Q value of 20 at the fundamental frequency. Determine the magnitude and phase errors of this CVT when fully burdened: (a) at 50 Hz at 0.8 pf and (b) at 150 Hz.
(Answers: 50 Hz: $e = -0.76\%$, $\gamma = -0.23$ Crad; 150 Hz: $e = -2.13\%$ $\gamma = 8.3$ Crad.)

5 An $11\,kV/\sqrt{3}{:}110v/\sqrt{3}$, 50 VA inductive VT has the magnitude and phase errors shown in Table 8.12 when burdened at 0.8 power factor. Find the short-circuit impedance of this device and its corresponding short-circuit current, assuming negligible impedance on the primary side. What size primary and secondary fuses would you consider using to protect this VT, and why?
(Answers: $r = 0.39\,\Omega$, $x = 0.16$ ohms. $I_{sc} = 151\,A$, HV fuse rating $\approx 3\,A$, LV fuse rating $\approx 4-6A$)

6 Does the VT in question 5 comply with the 0.3 Class limits as defined under C57.13? If not, at what burden would it comply?
(Answers: Not with a 0.8 pf burden, 43 VA)

Table 8.12 VT error information.

0.8 pf burden (%)	Ratio correction factor	Phase error (Crad)
0	0.997	0.02
100	1.002	0.15

7 A 22 kV VT is installed at a location remote from its associated supply transformer. If the positive, negative and zero-sequence impedances at this location are shown in Table 8.13, calculate the earth fault factor at this location for a phase-to-ground fault close to the VT. Is this VT effectively earthed or non-effectively earthed? How would this change if $R_g \approx X_1$?
(Answer: ≈ 1.4)

8 A Class 0.3 VT has the magnitude and phase errors given in Table 8.14, for a *resistive* burden. Plot these errors for this VT at 100% burden as a function of the burden power factor, and hence show that the phase error is slightly more sensitive to the

Table 8.13 Impedance information.

Positive and negative sequence impedance ($R_1 = R_2 \approx 0$)	Total zero-sequence impedance	Ground resistance from VT location to the supply transformer neutral
$Z_1 = Z_2 = jX_1$	$R_0 + jX_0 = X_1 + j3X_1$ $R_g \approx 0.7X_1$	

Table 8.14 VT error information.

% burden	Magnitude error (%)	Phase error (Crad)
0	0.272	0.00
100	−0.016	−0.16

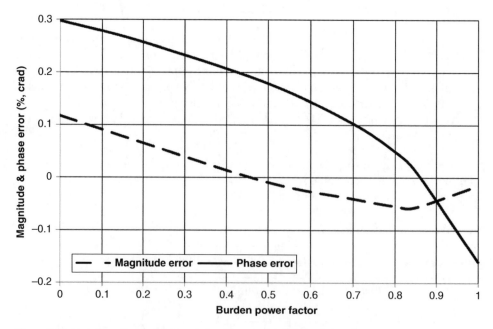

Figure 8.19 Variation of magnitude and phase errors with burden power factor.

power factor than the magnitude error. Your graphs should resemble those in Figure 8.19. (Note that the shape of these graphs depends to a large extent on the actual characteristics of the VT in question.)

How would you expect the errors to behave if this VT were to be burdened with a progressively more *capacitive* load?

8.11 Sources

1 IEC International Standard 61869-3, 2011. *Instrument transformers Part 3: Additional requirements for inductive voltage transformers.* International Electrotechnical Commission, Geneva.

2 IEC International Standard 61869-5, 2012. *Instrument transformers Part 5: Additional requirements for capacitor voltage transformers.* International Electrotechnical Commission, Geneva.

3 IEEE Standard C57.13, 2008. *IEEE standard requirements for instrument transformers.* IEEE Power Engineering Society, New York.

4 IEEE Standard C57.13.6, 2005. *IEEE standard for high accuracy instrument transformers.* IEEE Power Engineering Society, New York.

5 Australian Standard AS1243, 1982. *Voltage transformers for measurement and protection.* Standards Australia: Sydney.

6 ABB Inc. *Instrument transformers, technical information and application guide.* Pinetops, NC 27864, USA.

7 General Electric Company. '*Manual of instrument transformers*'. Meter & Instrument Business Dept, Somersworth, N.H. 03878, USA.

8 Areva. *Network Protection and Application Guide*, 2005. Areva, Paris.

9 IEC/TR 61850–1 Ed. 2.0: *Communication networks and systems for power utility automation – Part 1: Introduction and overview*, 2013, International Electrotechnical Commission, Geneva.

10 Reza Tajali PE. *Line to ground voltage monitoring on ungrounded and impedance grounded power systems.* Square D Company, Power Systems Engineering Group LaVergne. Tennessee 37086.

11 Fuchsle D, Stanek M. *Experiences with non-conventional instrument transformers (NCITs)*, ABB High Voltage Products, Switzerland.

9

Current Transformers

Current transformers (CTs) enable the measurement of currents in situations where either the system voltage or the magnitude of the current is too large to permit direct measurement. The current to be measured flows in the CT primary winding (usually just a single turn) and a scaled replica flows in its secondary. Current transformers are essential components in both protection and metering circuits, and while they are rugged devices they do have some limitations.

For example, like voltage transformers, CTs introduce both magnitude and phase errors into the secondary current. While these can be made arbitrarily small for metering applications, protection CTs must operate correctly during system faults, when primary currents can reach many times their normal value. Under these conditions a CT may approach saturation, and as a result its error will inevitably increase. In addition, in this region the non-linear magnetising characteristic will introduce a significant harmonic content into the magnetising current, making phasor analysis inapplicable.

To minimise the likelihood of saturation and to reduce the size of the magnitude and phase errors, current transformers are designed to normally operate at very low flux densities. This is achieved by imposing a restriction on the size of the secondary voltage that a CT must generate, which in turn is achieved by limiting the *burden impedance* that can be connected across its secondary winding. As a consequence, the VA rating of current transformers tends to be small. For example, metering class CTs rarely have ratings in excess of 30 VA, and usually considerably less than this.

The *rated burden* is the maximum load (expressed in either VA or ohms) that can be supported by a CT while maintaining its accuracy requirements. Any burden less than the rated can be safely applied, but *overburdening* a CT will generally result in an additional measurement error. The VA rating of a CT is given by the square of the rated secondary current, multiplied by the rated burden, expressed in ohms.

Because of the significant difference in performance required for protection duties as compared to metering, current transformers are usually designed either for one purpose or the other. Protection cores tend to be larger, and they therefore contain more transformer steel than their metering counterparts. This enables them to carry substantial fault currents and to support high secondary voltages, without becoming heavily saturated. Metering cores, on the other hand, are smaller and cheaper and operate over a much narrower range of currents. During fault conditions they become saturated, thereby limiting the exposure of the attached metering equipment to damaging currents.

Figure 9.1 shows a 220 kV pedestal CT, used in aerial switchyards. This type of CT can be fitted with several independent cores, either for protection or metering purposes.

AC Circuits and Power Systems in Practice, First Edition. Graeme Vertigan.
© 2018 John Wiley & Sons Ltd. Published 2018 by John Wiley & Sons Ltd.

Figure 9.1 220 kV pedestal CT.

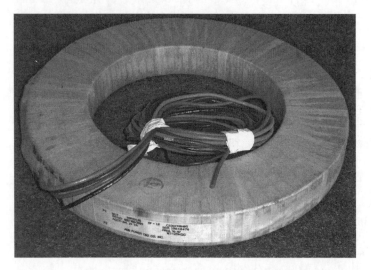

Figure 9.2 A 15 VA, Class 1, 600,400,200:5 toroidal bushing CT.

The CT depicted in Figure 9.2 is one of several cores that can be fitted around the bushing of a *dead tank circuit breaker*. It consists of a tape wound toroidal core upon which is wound the secondary winding, in this case one with three tappings.

9.1 CT Secondary Currents and Ratios

Current transformers are generally designed to deliver secondary currents of either 1 A or 5 A, when operating at rated primary current, the choice of which depends to a large extent on the distance between the CT and its associated protection relay or energy meter. Where this distance is short, 5 A secondary currents are usually chosen, as is usually the case in LV metering applications. This helps to reduce the cost of the CT, albeit at the expense of requiring a lower turns ratio, which can adversely affect the magnitude and phase errors. On the other hand, for long cabling runs, the fraction of the available burden consumed by the voltage drop in the interconnecting cables can become significant, and in such cases 1 A CTs are generally used instead. Protection CTs, particularly in aerial switchyards, are therefore usually 1 A devices, although in the United States there is a preference for 5 A CTs in protection applications.

Current transformer ratios are expressed as the ratio of rated primary to secondary current, e.g. 1000:1 for a 1 A CT or 1000:5 for a 5 A device. Frequently, tappings are included on the secondary winding so as to provide multiple current ratios from the same CT. For example, the ratio of a three tap CT will usually be expressed in the form 1000, 800, 400:5. Tappings enable the ratio to be altered as load growth occurs, and provide a choice of ratio in protection applications, without the need to physically change the CT.

9.1.1 CT Terminal Labelling Convention and Connection Arrangements

The terminals of a current transformer are always clearly marked since the correct phasing of current transformers is critical in both protection and metering applications. The CT primary terminals are marked P1 and P2, usually on the case of the device, as are the corresponding secondary terminals, S1 and S2. Figure 9.3a shows that a primary current flowing from P1 to P2 generates a proportional secondary current flowing from S1 to S2. In the case of multi-tap CTs, the secondary terminals are labelled progressively S1, S2, S3 ... Sn, from the common end of the secondary winding (S1), to the highest tapping (Sn).

In order to further reduce the cabling burden imposed on a CT, it is common to star connect the secondary terminals at the CT marshalling enclosure, and again at the burdening device, as shown in Figure 9.3a. This permits one side of each secondary winding to be earthed, and requires only four wires to deliver the current information to the associated instrumentation. In the case of three-wire MV or HV circuits, the primary phase currents will sum to zero and thus the current in the neutral wire will also be zero.

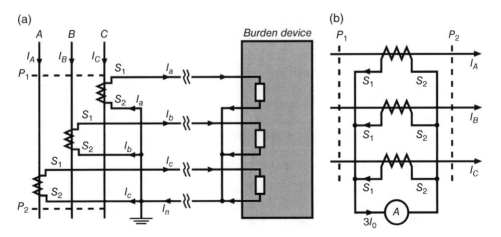

Figure 9.3 (a) Four wire connection (b) Residual current connection.

This roughly halves the cabling burden seen by each CT. This approach is generally adopted where long cable runs are necessary.

In four-wire MV or LV circuits, where the load is balanced or approximately so, a similar benefit will also exist, since the neutral wire will only carry the residual current. For unbalanced loads the burden seen by the CTs will generally be larger, since the voltage drop in the neutral must also be supported by each CT.

The residual connection of Figure 9.3b is frequently used to isolate the zero-sequence current from a set of three-phase currents. Each CT generates all three sequence components in its secondary current, but being balanced, the positive and negative sequence currents add to zero in the summation, leaving three times the zero-sequence component flowing in the burden.

9.2 Current Transformer Errors and Standards

Like voltage transformers, current transformers also introduce small magnitude and phase errors into their secondary currents. These arise due to the need for a magnetising current to generate sufficient flux density to produce the voltage demanded by the burden. Since the magnetising current is consumed internally, the current seen by the burden differs slightly in both magnitude and phase from that which might otherwise be expected. In addition, the magnetising impedance is not constant, but instead depends on the flux density that the CT is called upon to support in order to circulate secondary current through the burden. Therefore both the applied burden and the actual secondary current, influence the CT errors.

As mentioned above, current transformers may saturate during fault conditions when large currents flow. Metering CTs are designed to saturate under such conditions, but protection CTs must be dimensioned to operate correctly throughout network faults, without becoming heavily saturated. As a result, a significant performance difference exists between the two, and their characteristics are separately defined in current transformer standards.

We will again confine our general interest in current transformer standards to two documents. The first is the European standard IEC61869-2 *Additional Requirements for Current Transformers* and the second is the same American standard that also applies to voltage transformers, IEEE C57.13 *IEEE Standard Requirements for Instrument Transformers*, together with its supplement, IEEE C57.13.6 *IEEE Standard for High-Accuracy Instrument Transformers*.

However, there are two additional standards to consider in the specification of current transformers associated with protective relaying. The first is the European standard IEC61869-1 *General Requirements for Instrument Transformers* and the second is the American standard IEEE C37.110 *IEEE Guide for the Application of Current Transformers used for Protective Relaying Purposes*.

9.2.1 Current Transformer Magnitude and Phase Errors

Current transformers introduce errors into their representation of the primary current, in much the same way as voltage transformers introduce errors into their representation of primary voltage. In the case of current transformers, magnitude and phase errors arise as a result of the magnetising current required to generate the flux required in the core.

The *magnitude (or ratio) error* (*e*) of a current transformer is "the error in the magnitude of the fundamental component of the actual secondary current when a sinusoidal current flows in the primary" (AS1675, 1986). It can be expressed as:

$$e = \frac{\left(NI_s - I_p\right)}{I_p}.100\% \tag{9.1}$$

where N is the rated transformation ratio, I_p is the actual primary current and I_s is the actual secondary current.

The phase error β, is the difference in phase between the primary current phasor and the secondary current phasor. The phase error is defined as positive when the secondary current phasor leads that of the primary. It is measured in minutes of arc (Min) or in centiradians (Crad) where 1 Crad = 0.01 radian = 34.4 Min.

9.2.2 IEC 61869-2 Metering Class Magnitude and Phase Errors

IEC 61869-2 prescribes eight standard accuracy classes for measuring current transformers: 0.1, 0.2, 0.2S, 0.5, 0.5S, 1, 3 and 5. Class 0.1 applies to laboratory grade instruments, while classes 0.2–1 are used in revenue metering applications, and classes 3 and 5 are for non-critical applications such as panel instrumentation.

Table 9.1 outlines the error limits for classes 0.1–1, as prescribed under IEC 61869-2. Since current transformer errors are a function of the secondary voltage, both the secondary current and the applied burden must be specified when defining the error limits. The standard specifies a burden power factor (pf) of 0.8, which means that the rated burden includes both resistive and inductive elements. This is not to say that a compliant CT must always be so burdened, but rather that error testing must be done with a 0.8 pf burden.

The standard also requires that a CT be compliant between 25% to 100% of its rated burden, at 5, 20, 100 and 120% of rated current. Manufacturers generally error test their CTs at 25% and 100% of the rated burden, at each specified percentage of rated primary current. It can be seen from the table that the error limits are considerably more relaxed at small primary currents, i.e. where both the secondary current and voltage are also small. The reason for this will become apparent in Section 9.5.

Table 9.1 Metering class error limits.

| Accuracy class | Magnitude error (±%) | | | | Phase error (±Crad) | | | |
| | Primary current (%) | | | | Primary current (%) | | | |
	5	20	100	120	5	20	100	120
0.1	0.40	0.20	0.10	0.10	0.45	0.24	0.15	0.15
0.2	0.75	0.35	0.20	0.20	0.90	0.45	0.30	0.30
0.5	1.50	0.75	0.50	0.50	2.70	1.35	0.90	0.90
1	3.00	1.50	1.00	1.00	5.40	2.70	1.80	1.80

(IEC 61869-2 ed.1.0 Copyright © 2012 IEC Geneva, Switzerland. www.iec.ch)

Table 9.2 Metering class error limits for class 0.2S and 0.5S CTs.

Accuracy class	Magnitude error (±%)					Phase error (±Crad)				
	Primary current (%)					Primary current (%)				
	1	5	20	100	120	5	10	20	100	120
0.2S	0.75	0.35	0.20	0.20	0.20	0.90	0.45	0.30	0.30	0.30
0.5S	1.5	0.75	0.50	0.50	0.50	2.70	1.35	0.90	0.90	0.90

(IEC 61869-2 ed.1.0 Copyright © 2012 IEC Geneva, Switzerland. www.iec.ch)

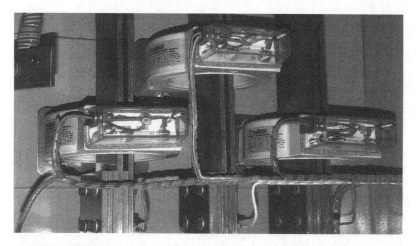

Figure 9.4 Busbar mounted multi-tap metering CTs (3000, 2000, 1000:5, 15 VA).

As shown in Table 9.2, class 0.2S and 0.5S transformers (sometimes called *special class* CTs), must maintain the prescribed error for currents as small as 1% of the primary rating. They must also achieve their ultimate error limits by 20% of rated current.

In the case of multi-tap CTs, the higher ratios will generally offer improved accuracy, since the core excitation is proportional to the ampere-turns applied. With more secondary turns available, less magnetising current is required to achieve a given flux density, and therefore the errors introduced into the secondary current reduce accordingly. For example, it is possible that the multi-tap metering CTs in Figure 9.4 may satisfy the 0.5 class requirements on the 1000:1 tap and the 0.2 class on the 3000:1 tap. The appropriate accuracy class will always be stated for the lowest tap, and may sometimes be stated for higher taps as well.

9.2.3 IEC 61869-2 Metering Class Burdens and Extended Range Operation

The rated burden is the load, usually expressed in VA or in ohms, which a CT can supply while maintaining its prescribed accuracy, at rated primary current. For example, a 1000:1 15 VA CT can support a maximum burden impedance of $15/1^2 = 15$ ohms, whereas for a 1000:5, 15 VA transformer, the burden is limited to $15/5^2 = 0.6$ ohms.

IEC 61869-2 defines a range of standard burdens for metering class CTs, specifically 2.5, 5, 10, 15 and 30 VA. In many LV metering applications the current transformers are located close to the energy meter, and in such cases CTs with a relatively small rated burden can be chosen, 5 VA being typical. However, in HV protection applications, particularly in aerial switchyards where the route length of the secondary cables is long, larger CT burdens will generally be specified. This will avoid the need to install long lengths of heavy secondary cable.

9.2.4 Extended Range CTs

IEC 61869-2 also permits the use of *extended current ratings* for metering class CTs with accuracy classes between 0.1 and 1. Tables 9.1 and 9.2 show that a standard metering CT is class compliant up to 120% of its rated primary current. This means that the magnitude and phase errors prescribed for 100% of rated current must also apply at 120%. *Extended range CTs* are similarly class compliant, but to a higher multiple of rated current, possibly as high as 200% or 250%. It is also a requirement that the continuous *thermal rating current* for the CT equals or exceeds the extended range primary current.

Extended range CTs are frequently used in LV metering applications, where they enable a wide range of currents to be accommodated within a particular accuracy class. For example, an 800:5 class 0.5 CT, with an extended range of 250% is class compliant from 40 A through to 2000 A. The use of such CTs avoids the need to change tappings as load growth occurs, and enables a considerably smaller range of CTs to be held in stock for LV applications.

9.3 IEEE C57.13 Metering Class Magnitude and Phase Errors

The American standard IEEE C57.13 defines three standard accuracy classes of metering transformer (0.3, 0.6 and 1.2) and its supplement, IEEE C57-13.6, defines two additional high accuracy classes, 0.15 and 0.15S. As with voltage transformers, this standard does not directly prescribe magnitude and phase errors, instead it places limits on the *transformer correction factor* (TCF).

As is the case with voltage transformers, any magnitude or phase errors inherent in a current transformer will introduce a small error into the measurement of the power or energy by a metering installation. This error may be corrected using the TCF, which relates the measured power ($P_{measured}$) to the correct power ($P_{correct}$), according to:

$$TCF_{CT} = \frac{P_{correct}}{P_{measured}} \tag{9.2}$$

The TCF may be calculated from the ratio correction factor (RCF), the phase error β and the phase angle of the metered load ϕ, according to:

$$TCF_{CT} = \frac{\cos(\phi) RCF_i}{\cos(\phi - \beta)} \tag{9.2a}$$

(Note that this equation does not include the errors associated with the voltage transformers also used in the metering installation.)

Table 9.3 Accuracy class limits for metering classes 0.3–1.2.

	Transformer correction factor (TCF)			
	At 100% rated current		At 10% rated current	
Accuracy class	Minimum	Maximum	Minimum	Maximum
0.3	0.997	1.003	0.994	1.006
0.6	0.994	1.006	0.988	1.012
1.2	0.988	1.012	0.976	1.024

(Adapted and reprinted with permission from IEEE. Copyright IEEE C57.13 (2008) All rights reserved. Permission for further use of this material must be obtained from IEEE. Requests may be sent to stds-ipr@ieee.org)

Table 9.4 Accuracy class limits for metering classes 0.15 and 0.15S.

	Transformer correction factor (TCF)			
	At 100% rated current		At 5% rated current	
Accuracy class	Minimum	Maximum	Minimum	Maximum
0.15	0.9985	1.0015	0.9970	1.0030
0.15S	0.9985	1.0015	0.9985	1.0015

(Adapted and reprinted with permission from IEEE. Copyright IEEE C57.13.6 (2010) All rights reserved. Permission for further use of this material must be obtained from IEEE. Requests may be sent to stds-ipr@ieee.org)

The RCF for a current transformer is defined in a similar way to that for a voltage transformer. It is the factor by which the stated ratio of an instrument transformer must be multiplied to obtain the actual ratio. Therefore the RCF can be used to correct the current measured by a CT for the effects of magnitude error:

$$RCF_{CT} = \frac{|I_{correct}|}{|I_{measured}|}$$

where $|I_{measured}|$ is the magnitude of current measured by the CT in question, and $|I_{correct}|$ is this quantity corrected for the ratio error introduced by the CT.

IEEE C57.13 prescribes limits on the TCF for accuracy classes 0.3–1.2, which appear in Table 9.3. Those prescribed by IEEE C57.13.6 apply to classes 0.15 and 0.15S and appear in Table 9.4. These limits must be maintained from 0% to 100% of the transformer's rated burden and for metered loads with lagging power factors between 0.6 and 1.0. The limits applicable to rated current must also be maintained up to the continuous thermal current rating of the device, which effectively gives the CT an extended operating range.

Prescribing limits applicable to the TCF, indirectly places limits on both the RCF and the phase error of a CT. The TCF is largest when the power factor of the metered load the lies at the lower end of the permitted range, i.e. when it equals 0.6. Under these

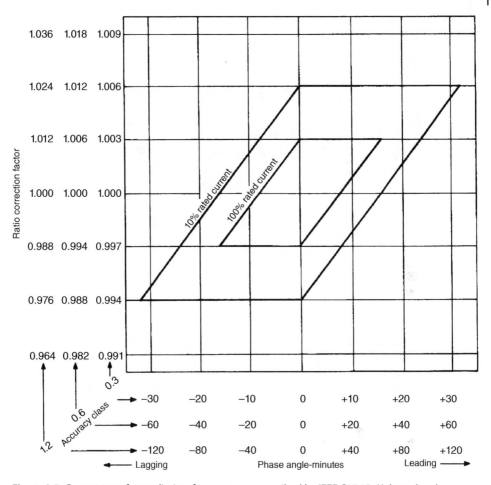

Figure 9.5 Current transformer limits of accuracy as prescribed by IEEE C57.13. (Adapted and reprinted with permission from IEEE. Copyright IEEE C57.13 (2008) All rights reserved. Permission for further use of this material must be obtained from IEEE. Requests may be sent to stds-ipr@ieee.org).

conditions Equation (9.3) defines the relationship between these parameters. (This equation is fully discussed in Chapter 10.)

$$TCF = RCF - \beta/2600 \qquad (9.3)$$

As a result, the ratio correction factor and the phase error are no longer independent of each other, and instead are constrained to lie within the parallelogram appropriate to the class of the CT, shown in Figure 9.5.

9.3.1 Error Comparison between IEC 61869-2 and IEE C57.13

The *total error* (or *overall error*) is a scalar figure of merit for a CT, which embodies both magnitude and phase error information in the measurement of the power flow to a load. It is related to the TCF according to:

$$Total\ Error = \frac{(P_{measured} - P_{correct})}{P_{correct}} \times 100\% = (1/TCF - 1) \times 100\%$$

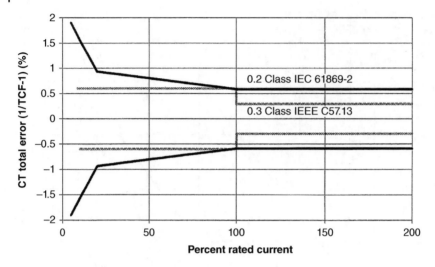

Figure 9.6 Comparison of CT specifications (primary circuit pf = 0.6).

So if the TCF is 0.995, the total error will be +0.5%.

A graphical comparison between the total error of two current transformers as a function the percentage of primary current appears in Figure 9.6. One of these is a 0.2 class device, compliant with the European Standard IEC 61869-2, while the other is a 0.3 class CT, compliant with IEEE C57.13. Although the nomenclature suggests otherwise, an inspection reveals that the performance of the 0.3 class device is superior to that of the 0.2 class. This is because both the magnitude and phase errors are interdependent in the case of the 0.3 class CT, whereas they are not so constrained in the specification of the 0.2 class CT.

9.4   Current Transformer Equivalent Circuit

Current transformer magnitude and phase errors can be shown to be related to components of the CT equivalent circuit. In particular, both these errors depend to a large extent on the value of the magnetising current I_m.

Consider the equivalent circuit shown in Figure 9.7. The CT on the left is an ideal current transformer, i.e. one requiring no magnetising current. In parallel with its secondary winding is an admittance $(G_m - jB_m)$, through which flows the magnetising current I_m demanded by a real CT. Provided the CT is not operated near saturation, this admittance is considerably smaller than that of the burden, and therefore the CT errors will be quite small.

Note that N_{act} in the figure above is the *actual* turns ratio used, which may be slightly different from the nominal ratio N in the case where turns compensation (see Section 9.5.2) has been applied. (In cases where n_1 equals one turn, N_{act} equals n_2.)

Unlike voltage transformers, the leakage inductance associated with a toroidal CT is generally negligible. This is particularly so if the primary consists of a single turn, centred in a toroidal core with a uniformly distributed secondary winding, and the

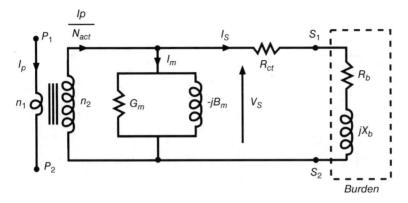

Figure 9.7 CT equivalent circuit.

magnetic effects of the return primary conductor, or those of the other phases, are negligible. Therefore the only series element to be considered is the secondary winding resistance R_{ct}.

In general, the burden connected to the CT can be represented by a series connected resistance and inductance ($R_b + jX_b$). The secondary voltage generated by the CT, V_s, is that developed across the magnetising admittance. Since this is also the voltage applied to the series connection of the winding resistance and the burden, it is convenient to associate these elements, therefore we write $R'_b = R_b + R_{ct}$.

We use an admittance approach in the evaluation of CT errors, since it provides both a simpler analysis and error equations closely analogous to Equations 8.4 and 8.5, previously derived for voltage transformer errors. The equivalent circuit of a CT may be considered as the reciprocal of that used for voltage transformers. This is because in the case of a VT, the short-circuit impedance is connected in series with the burden, whereas in a CT, the magnetising admittance and the burden are parallel connected. Further, the burden connected to a VT is represented by a parallel R and L combination, whereas for a CT a series R and L combination is used.

9.4.1 CT Magnitude Error

We analyse the error performance of the CT using the phasor diagram depicted in Figure 9.8, where the burden current I_S has been chosen as the reference phasor. The secondary voltage leads this current by an angle ϕ, determined by the burden impedance. We resolve this voltage into two components, one in-phase with I_S, the other in quadrature with it.

According to Kirchhoff's current law, the current flowing from the secondary winding of the ideal CT (I_p/N_{act}) is equal to the sum of the burden current I_S and the magnetising current I_m. We resolve the latter into contributions from both voltage components, each of which gives rise to a current component lying in-phase with I_S as well as one in quadrature with it. Note from Figure 9.8 that the in-phase components principally affect the magnitude error, while those in quadrature with I_S, determine the phase error. An inspection of this figure reveals that:

$$\left| \frac{I_P}{N_{act}} \right| \approx |I_S| + |I_S| R'_b G_m + |I_S| X_b B_m$$

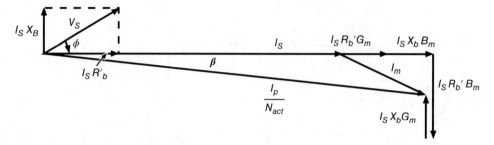

Figure 9.8 CT phasor diagram. (Note: The magnetising current components have been considerably enlarged for clarity).

Thus:

$$\frac{\left|I_p / N_{act}\right|}{\left|I_s\right|} \approx \left(1 + R'_b G_m + X_b B_m\right) = \left(1 + \frac{G_m}{G'_b} + \frac{B_m}{B_b}\right) \approx \left(\frac{(G_m + G'_b)(B_m + B_b)}{G'_b B_b}\right)$$

where $G'_b = \dfrac{1}{R'_b}$, $B_b = \dfrac{1}{X_b}$ and $G_m \ll G_b$ and $B_m \ll B_b$.

Therefore: $\dfrac{\left|I_s\right| N}{\left|I_p\right|} \approx \left(\dfrac{N}{N_{act}} \dfrac{G'_b B_b}{(G_m + G'_b)(B_m + B_b)}\right)$

We rewrite Equation (9.1) in the form:

$$e = \left[\frac{\left|I_s\right| N}{\left|I_p\right|} - 1\right] 100\% \tag{9.1a}$$

Substituting the above equation into (9.1a) yields:

$$e = \left[\left(\frac{N}{N_{act}} \frac{G'_b}{(G_m + G'_b)} \frac{B_b}{(B_m + B_b)}\right) - 1\right] 100\% \tag{9.4}$$

Equation (9.4) bears a striking resemblance to Equation (8.4), derived previously for voltage transformer magnitude errors, with the exception that the resistances and reactances have been replaced by conductances and susceptances. The ratio correction factor may also be expressed in terms of these equivalent circuit parameters:

$$RCF = \left[\frac{N_{act}}{N}\right] \frac{(G'_b + G_m)(B_b + B_m)}{G'_b B_b}$$

9.4.2 CT Phase Error

We can also derive an expression analogous to Equation (8.5) for the phase error for a CT, using Figure 9.8, from which we observe that:

$$\tan(\beta) \approx \frac{\left[I_s R'_b B_m - I_s X_b G_m\right]}{I_s} = \left[\frac{B_m}{G'_b} - \frac{G_m}{B_b}\right]$$

Figure 9.9 Simplified CT phasor diagram.

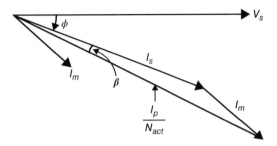

And since $\tan(\beta) \approx \beta$ for small β, when β is expressed in centiradians, we may write:

$$\beta \approx \left[\frac{B_m}{G'_b} - \frac{G_m}{B_b} \right].100\,\text{Crad} \tag{9.5}$$

The phase error β, is defined as positive when I_S leads I_P. Equation (9.5) suggests that when the burden power factor is greater than that of the magnetising admittance, the phase error will be positive, as illustrated in Figure 9.9. When the power factor of the burden equals that of the magnetising admittance, the phase error will be zero, otherwise it will be negative.

9.5 Magnetising Admittance Variation and CT Compensation Techniques

The magnetising conductance and susceptance have a major influence on CT magnitude and phase errors, but unlike the short-circuit impedance of a VT, they are generally not constant, but vary with the flux density in the core. A typical variation is depicted in Figure 9.10, for a current transformer constructed from grain oriented silicon steel (GOSS).

At very low flux densities, a GOSS CT core exhibits a low incremental permeability and therefore requires a disproportionately large excitation current to produce the required flux waveform. As a result, the magnetising susceptance B_m increases in magnitude as the flux density decreases.

At slightly higher flux densities, however, the core becomes progressively easier to magnetise, and the susceptance decreases in magnitude, and adopting a roughly constant value throughout the linear region of the magnetising characteristic. As the core approaches saturation, the susceptance (and to a lesser extent the conductance) increase substantially as the magnetising current becomes very large and rich in harmonics.

Since metering class CTs operate at quite low flux densities, the relatively large magnetising susceptance there leads to an increase in both the magnitude and phase errors, as an inspection of Equations (9.4) and (9.5) will reveal. In practice, this occurs when either the burden or the secondary current is small (or both), and it is for this reason that current transformer standards permit an increase in magnitude and phase errors at small secondary currents.

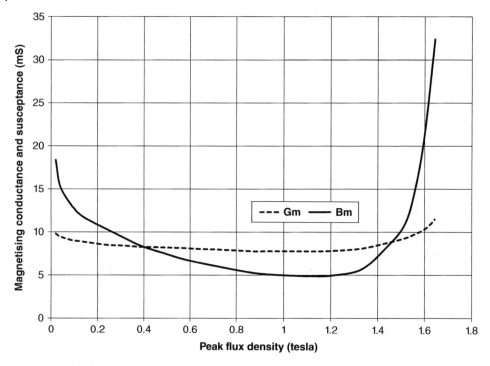

Figure 9.10 Magnetising admittance of a 200:5 CT (GOSS core).

9.5.1 Error Reduction Techniques: Composite Magnetic Circuits

One technique used by designers to reduce these errors is the use of a *composite magnetic circuit*. Figure 9.12a shows a 200:5 CT wound on a composite core, one consisting of a parallel combination of GOSS and *Mu-metal* laminations. Mu-metal is a nickel-iron alloy comprising 77% nickel and 15% iron, and it is highly permeable (easy to magnetise) at low flux densities, but it saturates at a much lower flux density than GOSS. (Mu-metal is also known under the trade name of Permalloy, and is frequently also used in magnetic shielding applications.)

In a composite core of this type, the Mu-metal carries the flux at very low flux densities providing acceptably low G_m and B_m values. At flux densities where Mu-metal saturates, the GOSS carries the flux quite easily, thus ensuring that G_m and B_m remain low. This has beneficial effects on both magnitude and phase errors and generally removes the need to provide turns compensation (see Section 9.5.2).

The admittance graph of Figure 9.11 demonstrates the beneficial effect of the use of Mu-metal on B_m and to a lesser extent on G_m. It also shows the smooth transition from Mu-metal to GOSS at around 0.3 tesla.

9.5.2 Turns Compensation

Turns compensation involves the use of a turns ratio that is slightly less than nominal. An inspection of Equation (9.4) will show that the magnitude error e is a function of the normalised turns ratio N/N_{act}. It turns out that e is particularly sensitive to this

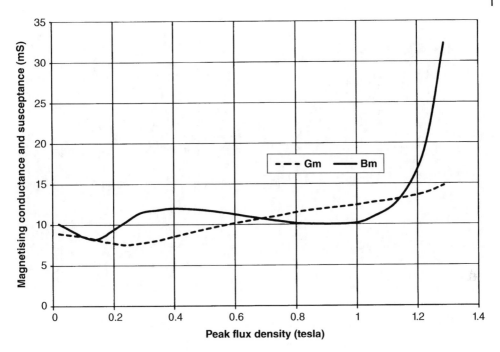

Figure 9.11 Magnetising conductance and susceptance for the composite 200:5 CT shown in Figure 9.12a.

parameter and CT manufacturers take advantage of this fact and use turns compensation to reduce the magnitude error in low ratio CTs. This approach is used to good effect in the manufacture of voltage transformers, where a few additional turns are added to the secondary winding to partially compensate for the voltage dropped across the short-circuit impedance.

In the case of current transformers, it is desirable to reduce the number of turns on the secondary slightly, thereby slightly increasing the normalised turns ratio. This slightly increases the current flowing from the secondary winding of the ideal transformer (I_p/N_{act}) in Figure 9.7, permitting some to be consumed by the magnetising branch while maintaining that in the burden close its desired level. As is evident from Equation (9.5), turns compensation has no effect on the phase error.

Low ratio CTs benefit most from turns compensation. This is because low ratio devices have relatively few secondary turns, and therefore require a substantial magnetising current to excite the core to the flux density demanded by the burden.

High ratio CTs, by virtue of the larger number of turns available, consume quite small magnetising currents and therefore do not require turns compensation. As a rule of thumb, GOSS CTs with more than about 400–500 secondary turns will generally not require compensation.

Dropping a whole turn in the case of a low ratio CT will generally lead to the device becoming *over compensated*, resulting in large and positive magnitude errors. For example, in the manufacture of a 200:5 A CT, it is usually only necessary to drop around 1/3rd to 1/5th of a turn to provide a useful improvement. While this sounds impossible,

(a) (b)

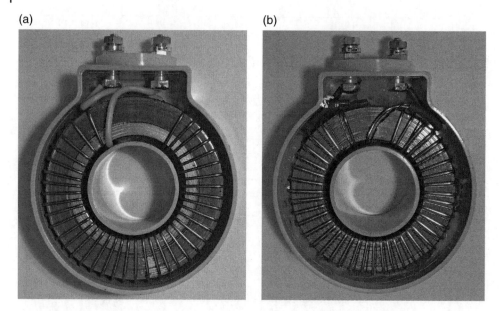

Figure 9.12 (a) Composite core construction, 200:5 CT (b) Turns compensated 200:5 CT.

manufacturers have devised a method of *split turn compensation* to drop between 20% and 80% of a single turn.

An example of this appears in Figure 9.12b, where a 200:5 turns compensated metering CT is shown. This device has two full span windings; the heavier winding has 40 turns and the lighter *compensation winding* has 39. Assuming that 80% of the secondary current flows in the heavy winding and 20% in the compensation winding, the effective number of turns will be $0.8 \times 40 + 0.2 \times 39 = 39.8$ turns. So by including a parallel compensation winding, the normalised turns ratio can be increased from unity to 1.005, offsetting the magnitude error by +0.5%.

In higher ratio CTs it may not be necessary to provide a *full span* compensation winding as in the example above. Instead this may only span a fraction of the main winding, in which case it will always be located at the S1 end of the main winding. Therefore unless it is known that no turns compensation has been applied, when using multi-tap CTs the S1 secondary connection must always be included in the circuit.

9.5.3 Multiple Primary Turns

Finally, in some particularly low ratio metering CTs, magnitude and phase errors can be improved by applying multiple primary turns. This permits a proportional increase in the number of turns on the secondary winding, and with it a reduction in the magnetising current demanded by the CT.

The CT in Figure 9.13 is a Class 0.5 instrument used in a two-element, pole-mounted MV metering transformer, containing two CTs and two line-connected VTs. The CTs in this case are multi-tapped, having ratios of 100, 50, 25:5 and a rated burden of 15 VA. On the lowest tap and with a single primary turn, the number of secondary turns would be limited to 5, which would require quite a large magnetising current to support its

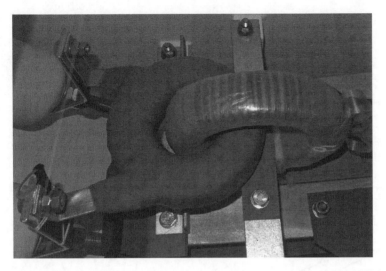

Figure 9.13 A 100/50/25:5, 15 VA wound primary metering CT (photographed through oil).

15 VA burden. As a result the magnitude and phase errors would become excessive. Instead, however, this CT is provided with a 12 turn wound primary, thereby enabling 60 secondary turns to be used for the 25:5 tap, resulting in a corresponding reduction in both magnetising current and CT errors. Multi-turn primary windings are also used in some protection applications.

9.6 Composite Error

The magnitude and phase errors as defined above are meaningful only so long as the magnetising (or excitation) current is sinusoidal, i.e. when the magnetic circuit can be considered linear. When a CT is operated within or near saturation, the non-linear nature of the magnetising characteristic introduces a significant harmonic content into this current, as shown in Figure 9.14. As a result, phasor analysis is no longer appropriate, since more than one frequency is involved. Instead, in this region the concept of a *composite error* is used to quantify the error inherent in a CT.

Composite error, defined by Equation (9.6), is essentially the ratio of the *RMS value of the magnetising current to the primary current, when the latter is reflected into the secondary circuit, according to the rated transformation ratio.* It is expressed in percent.

$$\varepsilon_c = \frac{100 I_{mRMS}}{I_p/N}\% = \frac{100}{I_p}\sqrt{\frac{1}{T}\int_0^T \left(N i_s(t) - i_P(t)\right)^2 dt}\% = \frac{100}{I_p/N}\sqrt{\frac{1}{T}\int_0^T \left(i_s(t) - i_P(t)/N\right)^2 dt}\%$$

(9.6)

where I_{mRMS} = the RMS value of the secondary magnetising current, I_p = the RMS value of the primary current, N = the *rated* turns ratio, $i_P(t)$ = the instantaneous primary current, $i_s(t)$ = the instantaneous secondary current and T = the period.

This definition is useful for quantifying the current error introduced by a CT when operating in the non-linear region of its magnetising characteristic. Protection CTs

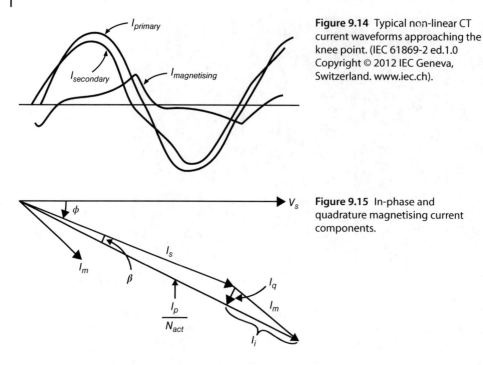

Figure 9.14 Typical non-linear CT current waveforms approaching the knee point. (IEC 61869-2 ed.1.0 Copyright © 2012 IEC Geneva, Switzerland. www.iec.ch).

Figure 9.15 In-phase and quadrature magnetising current components.

operate in this region during fault conditions and the composite error is thus used in the specification of IEC class P and PR protection cores.

A simple expression for the composite error of a CT may be derived using the phasor diagram of Figure 9.15, in terms of the magnitude and phase errors occurring at low flux densities, where the magnetising current is sinusoidal. Here the magnetising current is resolved into a component in-phase I_i with the reflected primary current I_p/N and one in quadrature with it, I_q. The in-phase component is proportional to the magnitude error e, while the quadrature component is proportional to the phase error β. It can be shown that:

$$\text{Composite Error } \varepsilon_c = \sqrt{e^2 + \beta^2} \qquad (9.7)$$

(This will be left as a tutorial exercise.)

In the case where the magnetising current is non-linear, the composite error will exceed the value suggested by Equation (9.7), and it can only be determined by a process of measurement. As a consequence, the composite error will always exceed the highest possible value of either ratio error or phase error. By ensuring that the *knee point voltage* of a CT is not exceeded (see Section 9.8.1), the composite error can be kept at or below 10%. This fact can be used to advantage in the application of overcurrent relays, since these are sensitive to ratio error, and in directional relays, which are sensitive to phase errors.

9.6.1 Determination of Composite Error

From the foregoing discussion it is apparent that an analytical determination of the composite error is not easy. In practice, the composite error is measured against a reference CT. The test arrangement depicted in Figure 9.16 can be used to determine the

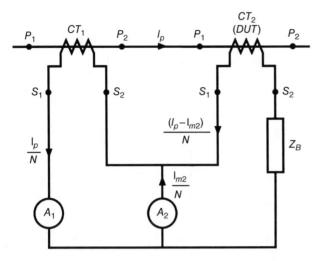

Figure 9.16 Evaluation of composite error using a reference CT of the same ratio as the DUT. (IEC 61869-2 ed.1.0 Copyright © 2012 IEC Geneva, Switzerland. www.iec.ch).

composite error of a current transformer when a reference transformer (CT_1) of the same ratio (N) is available. The *device under test* (DUT), CT_2, is burdened with impedance Z_B. It is assumed that the reference CT has negligible composite error, and that both CTs are supplied with the same sinusoidal primary current. Thus the current flowing in ammeter A_2 is the magnetising current of CT_2 and therefore the composite error of CT_2 is the ratio of the RMS current indicated by A_2, to that measured by A_1, i.e.:

$$\varepsilon_c = \frac{A_2}{A_1} \times 100\% = \frac{I_{m2}}{I_p} \times 100\%$$

One limitation with this arrangement is the difficulty in finding a reference CT with negligible composite error at the current at which the measurement is to be made. The composite error is usually measured at the rated accuracy limit primary current, which may be many times the rated current. As a result, the reference CT must have negligible composite error at this current, while having the same turns ratio as the DUT.

A more convenient test arrangement is shown in Figure 9.17. Here the reference transformer (CT_1) is considerably larger than the DUT, and therefore an interposing CT (CT_3) is introduced to compensate.

Assuming that the turns ratio of CT_1 is N_1, that of CT_2 is N_2 and that of CT_3 is N_3, then for this arrangement to operate correctly, it is necessary that:

$$N_1 = N_2 N_3$$

For example, if the CT under test has a ratio of 400:1 and an *accuracy limit factor* (see Section 9.8.2.) of 20, then the test will normally be carried out at a primary current of about 8000 A. A 10,000:1 CT would be suitable as the reference device, since 8000 A is well within its rated current, and its magnetising current may therefore be considered negligible. Given this, then according to the equation above, CT_3 must have a ratio of 25:1, and will only be called upon to carry about 20 A.

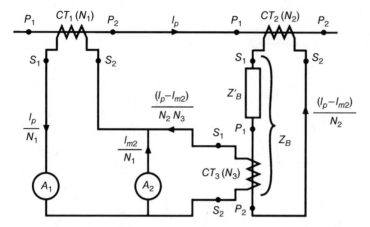

Figure 9.17 Improved composite error test arrangement. (IEC 61869-2 ed.1.0 Copyright © 2012 IEC Geneva, Switzerland. www.iec.ch).

In this event the current measured by ammeter A_2 in Figure 9.17 will be equal to the RMS magnetising current of CT_2, and therefore the composite error of CT_2 will again be given by the ratio of the ammeter readings, A_2/A_1.

9.7   Instrument Security Factor for Metering CTs

Metering CTs generally saturate during fault conditions and thereby limit the fault current that the connected instrumentation must tolerate until the fault is cleared. The *instrument security factor FS* is useful for determining the likely magnitude of this current in metering CTs. It is defined as the ratio between the rated instrument limit primary current and the rated primary current of the CT, when the applied burden is equal to the rated burden.

The rated instrument limit primary current is the minimum value of primary current at which the composite error becomes equal to or greater than 10%. Beyond this value the core may be considered saturated, and in this state, throughout most of the AC cycle, the flux change is seen by the CT will be zero and therefore the secondary voltage will be as well. Only near the primary current zero crossings will a rapid flux change occur, as the flux density swings from one extreme to the other. This gives rise to a short duration, high amplitude voltage pulse, which in turn generates quite a large, though short duration, current pulse in the burden.

The *secondary limiting EMF E_{FS}* for a metering CT is that developed across the secondary terminals of a fully burdened transformer, operating at the onset of saturation. It may be expressed in terms of the security factor according to Equation (9.8).

$$E_{FS} = FS.I_S \sqrt{\left(R_{ct} + R_{b\,Rated}\right)^2 + X_{b\,Rated}^2} \qquad (9.8)$$

where E_{FS} is the *secondary limiting EMF*, I_S is the rated secondary current, FS is the instrument security factor (usually 5 or 10), $R_{b\,Rated}$ and $X_{b\,Rated}$ are the rated burden resistance and reactance respectively and R_{ct} is the CT winding resistance, corrected to a temperature of 75 °C.

IEC61869-2 prescribes two values for FS, namely 5 and 10, and where this parameter is specified on the nameplate of a measurement CT, it appears after the class designation in the CT specification, e.g. 2000:1, 15 VA, Class 0.2, FS 10.

In practice, the instrument security factor depends on the actual burden applied. Since a given CT saturates at a particular secondary voltage, as the connected burden impedance reduces, a larger primary current will be required to achieve saturation. Therefore where the connected burden is less than the rated burden, the effective security factor becomes:

$$FS = FS_{Rated} \frac{\left(\sqrt{\left(R_{ct} + R_{b\,Rated} \right)^2 + X_{b\,Rated}^2} \right)}{\left(\sqrt{\left(R_{ct} + R_b \right)^2 + X_b^2} \right)}$$

where $R_{b\,Rated}$ and $X_{b\,Rated}$ represent the rated resistive and reactive burden elements respectively.

For example, in the case of the 2000:1, 15 VA CT mentioned above, the rated 0.8 pf burden is $12 + j9$ ohms. If this CT has a winding resistance of 8 ohms, then when fully burdened, its secondary limiting EMF becomes:

$$E_{FS} = 10 \times 1 \times \sqrt{(8+12)^2 + 9^2} = 219V$$

When *resistively* burdened with 4 ohms, this CT will still have a secondary limiting EMF of 219 V, and in order to generate this, a secondary current of 18.2 A will be required. The security factor will therefore become:

$$FS = FS_{Rated} \frac{\left(\sqrt{(8+12)^2 + 9^2} \right)}{\left(\sqrt{(8+4)^2 + 0^2} \right)} = 18.2$$

Therefore at this reduced burden the connected instrumentation will be required to support up to 18.2 A during fault conditions (assuming that the primary fault level is greater than $2000 \times 18.2 \approx 36\,kA$). If this secondary current cannot be safely supported without damage, then the burden on the CT should be artificially increased. Figure 9.18 shows the variation in security factor for this CT as a function of its connected burden.

The secondary limiting EMF (E_{FS}) can be estimated for a given CT by exciting the secondary winding with a sinusoidal voltage at rated frequency with the primary winding open circuit. The secondary voltage is increased until the RMS value of the magnetising current reaches 10% of the rated secondary current multiplied by the security factor FS (since at this potential the CT has a 10% composite error). The resulting potential will in practice, be a little smaller than E_{FS}, since the effect of the burden current flowing in the winding resistance is not included.

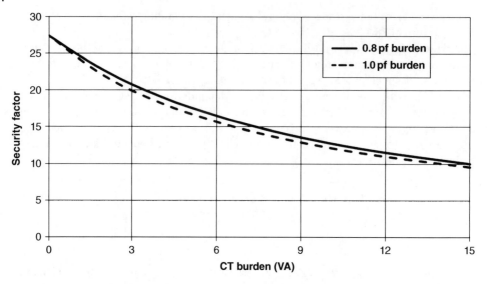

Figure 9.18 Security factor as a function of the applied burden for the CT in the example above.

9.8 Protection Current Transformers

Metering CTs are generally only required to operate within their error class limits, provided that the maximum primary current remains less than 120% of its rated value (or perhaps 200–250% in the case of an extended range CT). Metering errors incurred during fault conditions are assumed to be negligible, since faults will be cleared within a very short time by the associated protection equipment. It is more important that a metering CT does saturate during a fault, thereby protecting the connected meter.

On the other hand, protection CTs must function correctly during faults, when the primary current may exceed the nominal device rating by a factor of 20 or more. As a result, they must be specifically designed to cope with significant transient overcurrents, well beyond their nominal ratings.

The limiting factor to CT accuracy occurs when the flux density within the core approaches saturation. At such densities the magnetising current rises well beyond its normal value, and because of the shape of the B–H loop, the magnetising impedance becomes highly non-linear and therefore the magnetising current also contains a significant harmonic content, as shown in Figure 9.14. As a result, a considerable difference exists between the primary and secondary current wave shapes. Protection CTs therefore generally contain considerably more core material than metering CTs of the same ratio.

9.8.1 Knee Point Voltage

Once heavily saturated the secondary winding of a CT appears almost as a short circuit. This is because in this condition effectively no flux change[1] occurs over most of the cycle, despite a sinusoidally changing primary current. Therefore, except for a short

---

1 The magnetising inductance has therefore effectively fallen to zero.

duration, high amplitude pulse, as the flux density rapidly swings from its saturation value in one direction to the corresponding one in the other, little voltage is induced within the secondary winding.

The onset of saturation is somewhat more gradual however, and it is useful to define a maximum voltage that can be usefully generated by a CT prior to saturation. This is known as the knee point voltage and it is derived from a plot of the applied secondary voltage against the resulting RMS magnetising current, with the primary winding open. It may be regarded as the boundary between the linear region of the magnetisation characteristic and the saturation region.

From the foregoing discussion of the secondary limiting EMF defined for metering class CTs (E_{FS}), we also expect that the composite error of a protection core will be close to 10% at the knee point voltage as well, and this is generally the case.

The CT standards we are considering each define this potential slightly differently, and therefore two different knee point voltages can be obtained for the same CT, depending upon which definition is applied.

IEC 61869-2 defines the knee point voltage as that which, when increased by 10%, causes the RMS magnetising current to increase by 50%. This definition is illustrated in the upper graph in Figure 9.19, where a knee point voltage of 66 V is observed.

IEEE C57.13 on the other hand, requires that this graph be plotted on logarithmic axes, each having the same decade length. This standard defines the knee point as the voltage corresponding to *intersection of the graph with a tangent inclined at 45° to the horizontal axis*. For the same CT this definition yields a knee point voltage of 43 V, as shown in the lower graph in Figure 9.19. (The knee point voltage defined in IEC 61869-2 also lies close to the intersection of the tangents to the linear portions of this logarithmic plot; hence it will always be the greater of the two values.)

9.8.2 Accuracy Limit Factor and Secondary Limiting EMF for Protection CTs

The accuracy limit factor for a protection CT is analogous to the instrument security factor for a metering CT. It is defined in IEC 61869-2 as *the ratio of the rated accuracy limit primary current to the rated primary current*.

The rated accuracy limit primary current is the *value of primary current up to which the current transformer will comply with the requirements for composite error* (IEC 61869-2).

The secondary limiting EMF E_{ALF} for protection CTs is given by an equation very similar to (9.8), specifically:

$$E_{ALF} = ALF.I_S \sqrt{\left(R_{ct} + R_{b\,Rated}\right)^2 + X_{b\,Rated}{}^2} \tag{9.9}$$

where *ALF* is the accuracy limit factor for the CT, standard values for which are 5, 10, 15, 20 and 30. Since the accuracy limit factor for protection CTs is specified in terms of the rated burden, it will increase when smaller burdens are applied and therefore it will vary in exactly the same way with changes in burden as does the security factor for metering class CTs.

So long as the secondary voltage of a CT does not exceed the secondary limiting EMF during fault conditions, the CT will remain unsaturated and will therefore operate correctly.

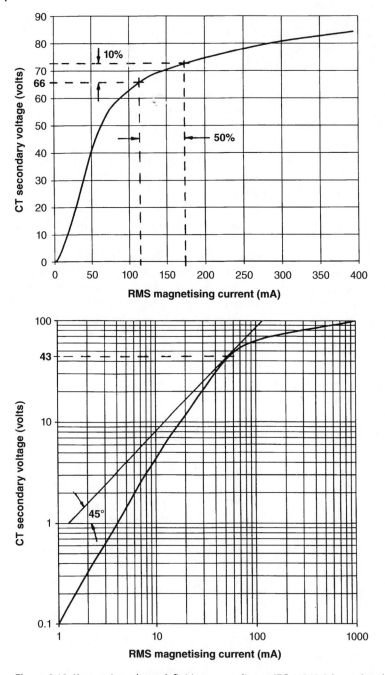

Figure 9.19 Knee point voltage definitions according to IEC 61869-2 (upper) and IEEE C57.13 (lower).

9.8.3   Protection Current Transformer Classes as defined in IEC 61869-2

IEC 61869-2 defines several classes of protection CTs; these include class P, class PX, and class TP, each of which will be discussed in turn.

Class P (and PR)

Class P are low leakage reactance CTs, used for general purpose applications where transient performance is not important. The performance of these CTs is specified in terms of the secondary voltage at the accuracy current limit (the secondary limiting EMF) E_{ALF}. There is no way of determining the transient performance of class P or PR devices from nameplate data, and therefore they should not be used in differential applications where the saturation of one CT might prevent the correct operation of the associated protection relay. They are designed to operate in the presence of symmetrical fault currents, without a significant transient current component.

Class P CTs have no limit to the remnant flux that may exist in the core upon the sudden interruption of the primary current. Class PR cores on the other hand, are built so that the remnant flux is limited to 10% of the saturation flux. This is usually done by including several tiny air gaps evenly distributed around the core. The magnitude, phase and composite errors prescribed for class P and PR CTs appear in Table 9.5.

These CTs are designated according to their rated composite error at the applicable accuracy limit factor, thus 10PR20 indicates a PR class CT, with a 10% composite error, at 20 times rated primary current. The full designation will also include the turns ratio and the VA rating of the device, for example: 800:1, 30 VA, 10PR20. The error limits above apply at rated burden of 0.8 power factor, unless the VA rating is less than 5 VA, in which case a unity power factor shall be used. This does not mean that a 0.8 power factor burden must be used in protection applications, but this is a requirement for error testing.

A check on the composite error of a P or PR class CT may obtained by sinusoidally exciting the secondary winding at the secondary limiting EMF E_{ALF} (with the primary open circuit), while simultaneously measuring the RMS magnetising current. The latter, when expressed as a percentage of $I_s \times ALF$, should be similar to the composite error limits specified in Table 9.5.

Class PX (and PXR)

Class PX and PXR CTs are specified by defining the magnetising characteristic through the knee point voltage and the magnetising current corresponding to it, in addition to the winding resistance and the rated resistive burden. The knee point voltage is related to the winding and burden resistances by an equation similar to (9.9), including the *dimensioning factor* K_x according to:

$$E_k = K_x \left(R_{ct} + R_b \right) I_s \tag{9.10}$$

Table 9.5 Error limits for class P and PR CTs.

Accuracy class	Magnitude error at rated primary current (%)	Phase error at rated primary current		Composite error at the accuracy limit primary current (%)
		(Crad)	(Min)	
5P or 5PR	±1	±1.8	±60	5
10P or 10PR	±3	–	–	10

(IEC 61869-2 ed.1.0 Copyright © 2012 IEC Geneva, Switzerland. www.iec.ch)

Table 9.6 Error limits for class PX and PXR CTs as prescribed in IEC 61869-2.

Accuracy class	Magnitude error at rated primary current (%)	Phase error at rated primary current	Remanence factor
PX	±0.25	Not specified	None
PXR	±1	Not specified	<10%

(IEC 61869-2 ed.1.0 Copyright © 2012 IEC Geneva, Switzerland. www.iec.ch)

Thus from a knowledge of the knee point voltage, the winding resistance and the dimensioning factor, the maximum permissible burden resistance for a given fault current can be calculated.

PXR class CTs, like the PR class, are equipped with tiny air gaps distributed around the core, to reduce any remanent flux to less than 10% of the saturation value. These CTs must operate within the error limits prescribed in Table 9.6 at rated primary current.

The nameplate of a PX or PXR CT includes the turns ratio, the knee point voltage E_k, the magnetising (excitation) current I_e corresponding to E_k, the winding resistance of the CT R_{ct} (corrected to 75 °C), the dimensioning factor K_x, and burden resistance R_b.

For example: 800:1, $E_k = 300V$, $I_e = 0.25A$, $R_{ct} \leq 3\Omega$, $K_x = 40$, $R_b = 4.5\Omega$

Class TP

Class TP CTs are designed to operate under conditions where a significant *transient current* occurs in addition to the symmetrical fault current, as shown by Equation (9.11).

When a fault occurs on a power system the inductive nature of the system impedance generates a fault primary current of the form:

$$i_p(t) = \frac{E_p}{\sqrt{R^2 + \omega^2 L^2}} \left[\sin(\omega t + \gamma - \phi) + \sin(\phi - \gamma)e^{\frac{-tR}{L}} \right] \tag{9.11}$$

where $i_p(t)$ is the CT primary fault current, E_p is the peak system voltage prior to the fault, R is the resistive component of the system impedance at the fault, L is the inductive component of the system impedance at the fault, γ is the fault inception angle and ϕ is the system power factor angle ($= \tan^{-1}(\omega L/R)$).

The term $\dfrac{E_p}{\sqrt{R^2 + \omega^2 L^2}} \sin(\omega t + \gamma - \phi)$ represents the steady state peak fault current;

while the term $\dfrac{E_p}{\sqrt{R^2 + \omega^2 L^2}} \left[\sin(\phi - \gamma)e^{\frac{-tR}{L}} \right]$ represents the transient component (sometimes called the DC component of the fault current); this decreases exponentially with the primary time constant L/R.

Figure 9.20a shows the form of the resulting fault current, including both AC and DC components. The resulting flux that a CT must support may be considered as comprising two parts, an *AC flux*, generated by the symmetrical component of the fault current, and a *transient flux* (or DC flux), produced by the transient component, shown in Figure 9.20b.

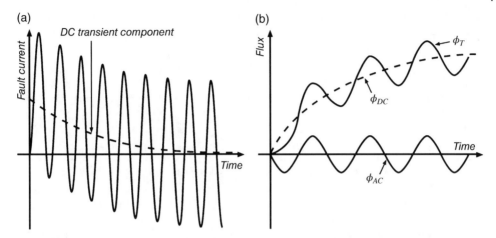

Figure 9.20 (a) Transient fault current waveform (b) Flux waveform AC and DC components.

Depending on the fault inception angle γ, the system X/R ratio at the fault, and the operation of any auto-reclosing circuit breakers, the transient current may substantially increase the peak flux that the CT must support. As an approximation, the total flux ϕ_T can be expressed in terms of the alternating flux ϕ_{AC}, the transient or DC flux ϕ_{DC}, and the system X/R ratio:

$$\phi_T = \phi_{DC} + \phi_{AC} = \phi_{AC}\left(X/R + 1\right)$$

Therefore at locations in a network where the X/R ratio is high, substantial transient currents may be generated during fault conditions. As a result, TP class protection CTs must be considerably increased in size in order to avoid saturation and the mal-operation of the associated protection relays. TP class CTs therefore include a transient dimensioning factor K_{td} in the evaluation of the equivalent secondary limiting EMF E_{al} applied to cater for this transient flux:

$$E_{al} = K_{SSC}\,K_{td}\left(R_{ct} + R_b\right)I_s \tag{9.12}$$

where E_{al} is the secondary limiting EMF for class TP CTs, K_{SSC} is the symmetrical short-circuit current factor = (primary symmetrical short circuit current)/(rated primary current) and K_{td} is the transient dimensioning factor (typically 10–25).

The transient dimensioning factor depends on several parameters, including both primary and secondary time constants, the fault inception angle and the on-off duty cycle of any auto-reclosing circuit breakers interrupting the fault current. The latter is important since the DC flux in the core may not have decayed significantly following the first fault interruption, when the primary circuit is re-energised. The residual flux, plus that arising from a subsequent re-energisation, is very likely to drive the core into saturation unless due allowance has been made for this contingency in the dimensioning of the core. Therefore TP class CTs must be designed with the specific application in mind.

In cases where the X/R ratio is extremely large, it is impractical to provide a CT of sufficient dimension to entirely avoid saturation, and in such cases the associated

Table 9.7 Error limits for class TPX, TPY and TPZ CTs.

CT class	At rated primary current		Peak instantaneous error	Remanence limit
	Ratio error (%)	Phase error (min)		
TPX	±0.5	±30	$\hat{\varepsilon}=10\%$	None
TPY	±1.0	±60	$\hat{\varepsilon}=10\%$	<10%
TPZ	±1.0	180±18	$\hat{\varepsilon}_{ac}=10\%$	<10%

(IEC 61869-2 ed.1.0 Copyright © 2012 IEC Geneva, Switzerland. www.iec.ch)

protection relay is designed to operate before the CT saturates. A full discussion of the evaluation of the transient dimensioning factor is beyond the scope of this text, but the interested reader is referred to IEC 61869-2 and to Technical Report, IEC 61869-100 TR.

The saturation behaviour of class TP CTs is specified in terms of the peak value of the instantaneous error $\hat{\varepsilon}$, for class TPX and TPY CTs, or in the case of class TPZ CTs, the peak value of the alternating error $\hat{\varepsilon}_{ac}$, as outlined in Table 9.7. In each case the rated burden is resistive. The instantaneous error current $i_\varepsilon(t)$, is expressed in terms of the instantaneous difference between the primary and secondary currents, referred to the primary winding, thus:

$$i_\varepsilon(t) = \left(Ni_s(t) - i_p(t) \right)$$

where both AC and DC components exist, i_ε may be expressed as:

$$i_\varepsilon(t) = i_{\varepsilon AC}(t) + i_{\varepsilon DC}(t) = \left(Ni_{sAC}(t) - i_{pAC}(t) \right) + \left(Ni_{sDC}(t) - i_{pDC}(t) \right)$$

The peak instantaneous error $\hat{\varepsilon}$, is defined as:

$$\hat{\varepsilon} = \frac{\hat{i}_\varepsilon}{\sqrt{2}I_{psc}} \times 100\%$$

And the peak alternating error $\hat{\varepsilon}_{AC}$, is given by:

$$\hat{\varepsilon}_{AC} = \frac{\hat{i}_{\varepsilon AC}}{\sqrt{2}I_{psc}} \times 100\%$$

where $\hat{i}_\varepsilon$ is the peak instantaneous error current, $\hat{i}_{\varepsilon AC}$ is the peak alternating component of the instantaneous error current, I_{psc} is the RMS value of the AC component of the primary current and N is the transformation ratio.

TPX class CTs have no limit imposed on the level of residual flux remaining in the core following the sudden interruption of primary current, and therefore these cores contain no air gaps. The magnetising inductance is therefore large and the secondary time constant[2] long, varying between 5 and 20 seconds. The resulting magnetising current is small and these cores offer good magnitude and phase error performance.

---

2 The secondary time constant is the ratio of the magnetising inductance to the secondary loop resistance, i.e. that of the CT plus the applied burden. It determines the rate at which the DC flux decays following the interruption of primary current.

Both TPY and TPZ CTs each have a limit of 10% imposed on the remanent flux within the core, and as a result they both contain air gaps. TPY cores are generally fitted with the minimum gap required to achieve this, and as a result the magnetising inductance is smaller, and the secondary time constant proportionally shorter, typically 0.2–2 seconds.

TPZ cores contain considerably larger air gaps, and in practice retain negligible remanent flux. As a result their B–H relationship is substantially linear, since the magnetising inductance is determined largely by the cumulative dimension of the distributed air gap. The TPZ class magnitude and phase errors are correspondingly larger than those of TPX or TPY cores. They are also much less prone to saturation than either of these classes, and the secondary time constant is considerably shorter than either of the above, typically 50–70 ms. TPZ CTs therefore cannot accurately reproduce the DC current component of fault current and it is for this reason that their error performance is based on the peak alternating error $\hat{\varepsilon}_{AC}$. Gapped cores like these generally do not find extensive application in the United States.

The designation of class TP transformers is outlined in the following examples:

Class TPX 20×12.5 ($K_{SSC} = 20$, $K_{td} = 12.5$), $R_{ct} \leq 3\Omega$, $R_b = 5\Omega$,
Class TPY 20×12.5 ($K_{SSC} = 20$, $K_{td} = 12.5$), $R_{ct} \leq 3\Omega$, $R_b = 5\Omega$,
Class TPZ 20×12.5 ($K_{SSC} = 20$, $K_{td} = 12.5$), $R_{ct} \leq 3\Omega$, $R_b = 5\Omega$, $T_s = 60$ms.

9.8.4 Protection Current Transformer Classes as defined by IEEE C57.13

Unlike European practice where there is a preference for 1 A protection CTs, those used in the United States, defined by IEEE C57.13 almost always have a 5 A rated secondary winding. They frequently also have multiple tappings on the secondary so that different ratios may conveniently be selected. IEEE C57.13 defines two basic classes of protection CTs, class C and T, for which the ratio error limits appear in Table 9.8.

Class C CTs have a uniformly distributed winding around a toroidal core, with single primary conductor, centrally located inside it. They therefore have negligible leakage reactance and therefore their performance can be calculated from the magnetisation characteristic like the one shown in Figure 9.21.

The family of curves shown in Figure 9.21 relate to a multi-tapped non-gapped CT, each curve corresponding to a different tap. The higher taps, having more turns, consume less magnetising current and have a correspondingly higher knee point voltage. This varies in direct proportion to the number of turns, as shown by Equation (4.13). These curves can be used to calculate an *overcurrent ratio curve* for any particular burden, applied to any tapping of the CT, similar to those depicted in Figure 9.22b.

Table 9.8 Limits of ratio error for class C and T protection CTs defined by IEEE C57.13.

Limits of ratio error: Classes C and T		
@ rated current	@ 20×rated current	Phase error
3%	10%	Not specified

(Adapted and reprinted with permission from IEEE. Copyright IEEE C57.13 (2008) All rights reserved. Permission for further use of this material must be obtained from IEEE. Requests may be sent to stds-ipr@ieee.org)

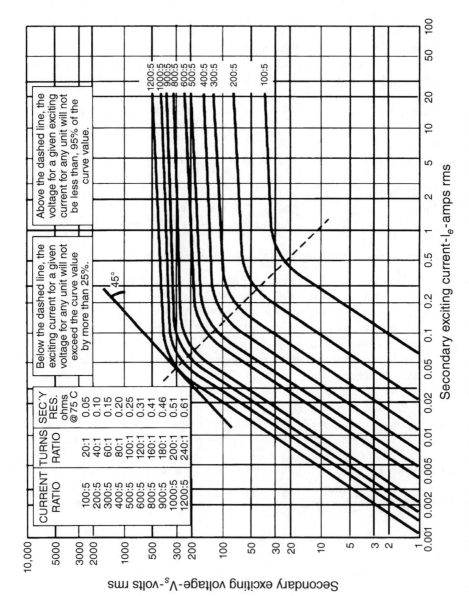

The following text appears within the figure:

Above the dashed line, the voltage for a given exciting current for any unit will not be less than, 95% of the curve value.

Below the dashed line, the exciting current for a given voltage for any unit will not exceed the curve value by more than 25%.

45°

CURRENT RATIO	TURNS RATIO	SEC'Y RES. ohms @75 C
100:5	20:1	0.05
200:5	40:1	0.10
300:5	60:1	0.15
400:5	80:1	0.20
500:5	100:1	0.25
600:5	120:1	0.31
800:5	160:1	0.41
900:5	180:1	0.46
1000:5	200:1	0.51
1200:5	240:1	0.61

Curve labels: 1200:5, 1000:5, 900:5, 800:5, 600:5, 500:5, 400:5, 300:5, 200:5, 100:5

Y-axis: Secondary exciting voltage-V_s-volts rms

X-axis: Secondary exciting current-I_e-amps rms

Figure 9.21 Typical magnetising characteristics for a multi-tapped, non-gapped, class C transformer. (Adapted and reprinted with permission from IEEE. Copyright IEEE C57.13 (2008) All rights reserved. Permission for further use of this material must be obtained from IEEE. Requests may be sent to stds-ipr@ieee.org).

(a)

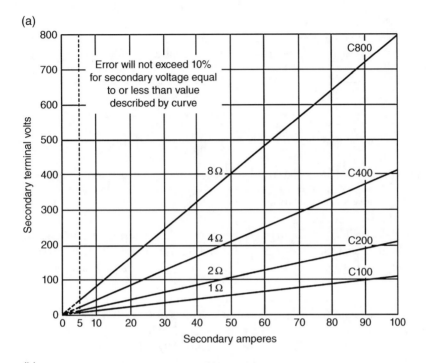

(b)

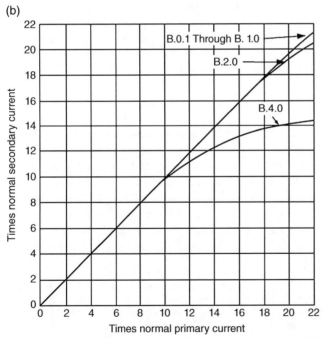

Figure 9.22 (a) Class C CT voltage classifications (b) Typical Class T overcurrent ratio curves. (Adapted and reprinted with permission from IEEE. Copyright IEEE C57.13 (2008) All rights reserved. Permission for further use of this material must be obtained from IEEE. Requests may be sent to stds-ipr@ieee.org).

Table 9.9 Standard burdens for relaying.

Burden designation	Impedance (Ω)	Resistance (Ω)	Inductance (mH)	Secondary voltage @ 20×5 A
B-1.0	1.0	0.50	2.3	100 V
B-2.0	2.0	1.00	4.6	200 V
B-4.0	4.0	2.00	9.2	400 V
B-8.0	8.0	4.00	18.4	800 V

(5 amp secondary winding, 60 Hz, burden $PF = 0.5$) (Adapted and reprinted with permission from IEEE. Copyright IEEE C57.13 (2008) All rights reserved. Permission for further use of this material must be obtained from IEEE. Requests may be sent to stds-ipr@ieee.org)

Class T CTs usually have a wound primary (similar to the metering core shown in Figure 9.13), and hence they lack the symmetry and the close coupling of class C devices. As a result, the leakage reactance of class T CTs is not negligible and it significantly increases the secondary voltage that the CT must generate, especially at larger multiples of rated current. This in turn increases the magnetising current to the point where it is not possible to calculate the performance of these cores to within the 10% ratio error limit called for in Table 9.8. It is therefore necessary for the manufacturer to provide a set of overcurrent ratio curves, like those in Figure 9.22b. These are obtained by ratio testing each class T transformer, and they can be used to infer the performance of the CT at any particular burden.

Class C and T current transformers are classified according to the secondary terminal voltage (secondary limiting voltage) generated at 20 times rated current, while maintaining a ratio error of no more than 10%, when loaded with the appropriate standard burden, defined in Table 9.9.

The standard burdens have a power factor of 0.5 and generate the rated secondary terminal voltage at 20 times rated current. Therefore, for example, burden B-8.0 will generate $20 \times 5\,A \times 8$ ohms $= 800$ volts. The voltage–current relationship for each standard burden is plotted in Figure 9.22a.

The overcurrent ratio curves in Figure 9.22b apply to a class T200 transformer, since for burdens equal to or less than B-2.0, the error is less than 10% at 20 times rated current. These curves are plotted for all standard burdens up to that which causes a ratio correction of about 50%, in this case the B-4.0 curve. At some particular value of secondary current, this burden will be sufficient to cause the CT to begin to saturate, and the secondary current will no longer match that in the primary, due to the rapidly increasing magnetising current.

Multi-Tapped CTs

Where C or T class CTs are multi-tapped, the voltage classification applies to the largest tapping, i.e. to the entire secondary winding. Therefore for example, a 1200, 800, 600, 400, 100:5, class C200 CT will only support 100 V on the 600 A tap. Further, the maximum burden impedance that can be applied to a lower tapping is equal to the standard burden *scaled according to the fraction of the winding in use*. So in the case of the 600 A tapping, the maximum burden applicable is 2 ohms × 50% = 1 ohm, while that for the 400 A tapping is only 0.66 ohms. Finally, the voltage classification relates

Table 9.10 Characteristics of a C200, 1200:5 multi-tap CT.

CT ratio	Maximum burden impedance (class C200)	Secondary terminal voltage (class C200)	VA rating
1200:5	2.00 Ω	200 V	50
800:5	1.33 Ω	133 V	33
600:5	1.00 Ω	100 V	25
400:5	0.67 Ω	67 V	17
100:5	0.17 Ω	17 V	4

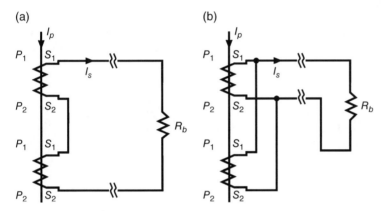

Figure 9.23 (a) Series CT connection (b) Parallel CT connection.

only to the potential developed across the external burden, it does not consider the potential dropped internally across the winding resistance. Therefore as long as the secondary current is less than or equal to 100 A, the CT winding resistance need not be considered.

Table 9.10 summarises these results, and in particular demonstrates the limited performance of multi-tap CTs at lower transformation ratios.

For example, consider the overcurrent protection of a 20 MVA, 22 kV transformer for which the rated current is 525 A, using the above CT. The 600:5 ratio would at first glance seem appropriate since this would yield a tripping current around 5 A for a 1.14 pu (600 A) primary current. If the external burden on this CT, including cabling and that of the protection relay, is 1.2 ohms, and the three-phase fault level is 12 kA, then under fault conditions we might expect a terminal voltage of 120 volts. However, Table 9.10 shows that on the 600 A tapping the CT can only generate 100 V, and therefore at this current it will partially saturated, and thus the ratio error will be greater than 10%.

The 800:5 tapping is a suitable alternative, since the secondary current will be 75 A during a fault, generating CT terminal voltage of only 90 V (However, this would require a tripping current of 3.75 A). Instead however, two (or more) CTs may be used with their primaries and secondaries series connected, as shown in Figure 9.23a. In this arrangement, each CT sees half the total burden voltage (assuming that they each have identical ratios and magnetising characteristics). Therefore if two 600:5 CTs are series connected in the example above, each will see a terminal voltage of 60 V, well within its capability, and as a result the ratio error will be less than 10%.

It is also possible to parallel connect the secondaries of two CTs, as shown in Figure 9.23b. However, this configuration doubles the burden seen by each CT, but it also effectively halves the overall transformation ratio. It is advantageous in this case to parallel connect two 1200:5 CTs, since each is capable of supporting 200 V and under this fault condition the secondary terminal voltage will only be 120 V. In this respect, parallel connected CTs are the equivalent of using 600:5 CT of a higher terminal voltage rating, but the ratio and phase errors generated by 1200:5 CTs will be considerably lower than those that would be obtained from a single 600:5 device.

The above calculations are based on the secondary terminal voltage rating, and while they are usually quite sufficient, they cannot specify the actual ratio error that will occur under fault conditions (other than to determine whether it is greater than or less than 10%). It is possible however to use the magnetisation characteristic to *calculate* the CT ratio that applies during fault conditions and hence to determine the actual ratio error that applies for a selected burden and fault current.

Consider the magnetisation curves of Figure 9.21. Assume that the same 20 MVA transformer is to be overcurrent protected using the 600:5 tap. Assume that the cabling burden is 0.7 ohms, the relay burden is 0.5 ohms with a power factor of 0.8, and the CT resistance is 1 ohm. From the associated magnetisation curve we can calculate the ratio error that would apply in this case as follows:

1) If the tripping current is to remain 600 A and the three-phase fault current is 12 kA, then under fault conditions 100 A will notionally flow in the CT secondary.
2) The total burden on the CT is approximately $0.5 + 0.7 + 1.0 = 2.2$ ohms. Ordinarily we would add these impedances vectorially, but algebraic addition is simpler and provides a slightly larger burden and therefore a more pessimistic result.
3) The total voltage generated by the CT will be $100 \times 2.2 = 220$ V, and from the 600:5 magnetisation curve the corresponding magnetising current will be about 5 A. Therefore the net current flowing in the CT burden will be $100 - 5 = 95\,\mathrm{A}^3$, and therefore the effective turns ratio will be $12{,}000 / 95 = 126.3 : 1$ or 631:5, which represents a ratio error of 5%. It will be apparent that if the CT burden were to increase slightly, the magnetising current needed would grow rapidly, and with it the ratio error as well.

Transient Fault Currents
Where the primary fault current contains a substantial transient component, class C and T CTs may be used successfully so long as the core dimensions are adequately increased to cope with the DC flux. As discussed above, this is achieved by including the transient dimensioning factor K_{td} in Equation (9.12), typical values for which lie in the range 10–25. As a result, these CTs are substantially increased in size, since the

---

3 This subtraction should also be performed vectorially, but the angle by which these currents lag the CT secondary voltage is generally not known. By performing an algebraic subtraction we are tentatively assuming that they are in phase.

ability to support a high terminal voltage effectively increases the cross-sectional area of the core, as suggested by comparing Equations (9.12) and (4.13):

$$E_k = K_{SSC}\,K_{td}\left(R_{ct} + R_b\right)I_s$$

$$E_k = 4.44\,\hat{B}N_2 A f$$

Short-Term Mechanical and Thermal Current Ratings

Part of the specification of class C or T transformers are the short-term mechanical and thermal current ratings. The short-term mechanical rating relates to the AC component of an asymmetrical primary current that the device is capable of carrying for one second, without sustaining damage, while its secondary is short circuited.

The short-term thermal current rating is the RMS symmetrical primary current that can be carried by a CT for one second, with its secondary winding shorted, without exceeding the winding temperature limits (250 °C for copper windings and 200 °C for aluminium).

9.9 Inter-Turn Voltage Ratings

The dangers of open circuiting the secondary winding of a current transformer were discussed in Chapter 4, where the possible generation of very high secondary voltages was explained. While it is never intended to operate a CT with its secondary winding open, from time to time this does occur, usually as the result of CT links being left open or not properly tightened. In such cases high ratio CTs will usually generate sufficient potential to cause an arc to form across the open link, the sound of which is an indication of the open circuit.

So as to prevent permanent damage from occurring in such situations, both standards require that compliant CTs be able to operate under 'emergency conditions' with the secondary winding open while the accuracy limit current flows in the primary, for a period of at least one minute. Such operation will certainly result in complete saturation of the CT and, as a result, a very peaky waveform will appear across the open secondary winding. IEC 61869-2 specifies that under these conditions the crest voltage be less than 4500 V, while IEEE C57.13 specifies crest voltages not exceeding 3500 V.

In particularly high ratio CTs, these limits may be exceeded, and in such cases the CT must be provided with some form of voltage limiting device, either a spark gap or a suitably rated varistor, capable of clamping the secondary voltage to a safe level during open circuit operation. In the case of a varistor, the energy dissipated may ultimately result in the device's destruction and its replacement might therefore be necessary after an open circuit event. While this is inconvenient, it is much preferable to damaging the CT.

Current transformers that suffer a prolonged period of open circuit operation may also overheat, depending on the magnitude of the hysteresis loss in the core under heavily saturated conditions. In this event, permanent damage may well occur to the device.

9.10   Non-Conventional Current Transformers

While magnetic current transformers have the advantage of being simple and robust, they do have some disadvantages, such as the effects of magnetic saturation and limited burden capacity. However, despite this, magnetic CTs generally exhibit a linear frequency response to about the 50th harmonic, providing that the burden is small and its power factor is close to unity. (Inductive burdens lead to increase in secondary voltage with frequency, thus demanding more magnetising current, leading to an increase in CT error.) Beyond this frequency, winding capacitances and leakage inductance effects become significant, making magnetic current transformers less suitable for higher harmonic measurement.

Magnetic CTs still make up the vast majority of current measuring devices in use, but there are several other technologies that also find application in power circuits. These generally operate by sensing the magnetic field created by the current to be measured and therefore they all require some signal processing in order to obtain the required current information. These measuring principles are linear in both amplitude and frequency, and are not limited by the saturation of a magnetic circuit. The associated electronics on the other hand cannot operate over an unlimited dynamic range and must be chosen with the specific application in mind.

IEC standard IEC 60044-8 *Electronic current transformers* defines the metering class error limits shown in Table 9.11 for electronic current transformers up to the 13th harmonic.

Rogowski Coils

A Rogowski coil is shown in Figure 9.24; it is similar to a conventional CT except that it operates without a magnetic circuit and couples into the magnetic field surrounding the current carrying conductor, generating a small voltage proportional to $d\phi/dt$. Since the flux is directly proportional to the current flowing, the winding voltage is also proportional to current derivative, specifically:

$$v(t) \approx \mu_0 N A \, di \, / \, dt$$

Table 9.11 Harmonic error limits for electronic current transformers.

| | | | | | Phase error | | | | | | | |
| | Ratio error (%) | | | | Degrees | | | | Centiradians | | | |
Accuracy class	2nd – 4th	5th, 6th	7th – 9th	10th – 13th	2nd – 4th	5th, 6th	7th – 9th	10th – 13th	2nd – 4th	5th, 6th	7th – 9th	10th – 13th
0.1	1	2	4	8	1	2	4	8	1.8	3.5	7	14
0.2	2	4	8	16	2	4	8	16	3.5	7	14	28
0.5	5	10	20	20	5	10	20	20	9	18	35	35
1.0	10	20	20	20	10	20	20	20	18	35	35	35

(IEC 60044-8 ed.1.0 Copyright © 2002 IEC Geneva, Switzerland. www.iec.ch)

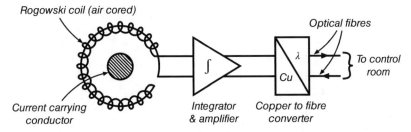

Figure 9.24 Rogowski coil current measurement.

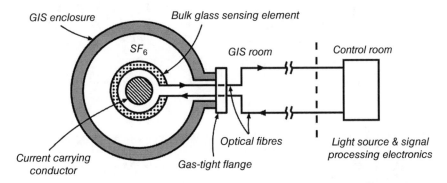

Figure 9.25 Optical sensor applied to a GIS busbar.

where N is the turn density (turns/m) and A is the area of each turn. The coil voltage must therefore be integrated and amplified to provide a replica of the primary current.

It is interesting to note that the end of the winding in Figure 9.24 returns *inside* the coil. This is to ensure that it is insensitive to external magnetic fields, for which the contributions from the coil and the return conductor cancel one another out. Rogowski coils are finding application in gas insulated switchgear (GIS) and aerial switchyard protection circuits, but the homogeneity of the coil, its position and temperature drifts all contribute to the error in the device, which may be beyond that required for Class 0.2 metering applications, unless drift compensation is applied.

Optical Current Transformers (Faraday Effect Sensors)

Optical current sensors are a common choice for alternative current transformers (Figure 9.25). They operate on the Faraday effect, whereby a polarised light wave in an optical fibre or glass block will have its direction of polarisation altered by the presence of a magnetic field. Therefore an optical fibre wrapped around a current carrying conductor will experience a change in polarisation in direct proportion to the current flowing. Through the use of suitable electronics, a digital representation of the current signal can be obtained. Optical sensors are linear in both current and frequency, but the signal processing package does have a limited dynamic range.

Hall Effect Current Sensors

Hall effect current sensors operate on the principle that moving charges, such as electrons, experience a force in the presence of a magnetic field, and are thus deflected in their

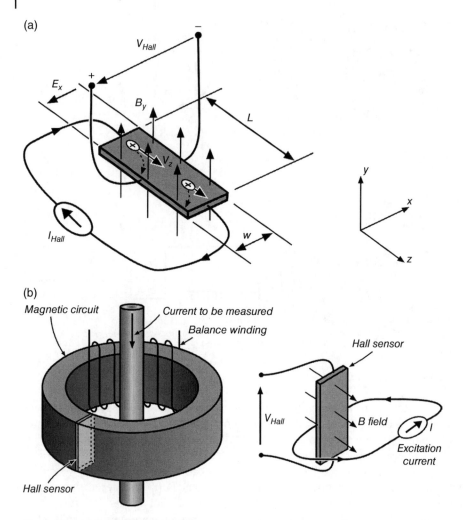

Figure 9.26 (a) Hall effect (b) Zero flux current measurement.

direction of travel. This is illustrated in Figure 9.26a, where a current is set up in a semi-conductor material by an external potential. The magnetic field created by the current to be measured, is arranged to pass orthogonally through the plane of the material and the electron flow is deflected as a result. This creates an accumulation of positive charge on the left and negative on the right, which gives rise to an electric field which establishes an equal and opposite force to that from the magnetic field. Numerically the total force F can be expressed in terms of the difference between the electric and magnetic forces:

$$F = \left(E_x q - B_y q v_z \right) = 0 \text{ and } E_x = V_H / w$$

where E_x is the electric field strength in the x direction created by the charge displacement, V_H is the Hall voltage, q is the charge on an electron, B_y is the flux density set up by the current being measured, v_z is the charge velocity in the z direction and w is the width of the semiconductor material.

The Hall voltage can therefore be expressed as $V_H = B_y v_z w$ and is directly proportional to the magnetic field surrounding the current carrying conductor. Practical current measurement is made using the arrangement in Figure 9.26b, where a Hall sensor is inserted in a gapped core surrounding the current carrying conductor. An electronics package excites a multi-turn balance winding with an AC current in such a way that both the Hall voltage and the magnetic field are driven to zero. The balance current then becomes a scaled replica of that in the primary conductor, since the primary and balance ampere-turns are equal.

The zero flux technique is also applicable to direct currents and is frequently used in the manufacture of current transformers for the measurement of large DC currents used in metal refining industries. Because the flux in the core is always held at or very close to zero, magnetic saturation presents no limitation to the amplitude of the current that can be measured.

Summary
Non-conventional current transformers are linear devices unconstrained by magnetic saturation; they have a wide dynamic range limited only by the associated electronics. They generally do not deliver an analogue replica of the measured current over a copper circuit in the same way as do magnetic CTs. Instead the derived current signal is usually digitally encoded and transmitted optically over a fibre link to suitable numerical protection relays. This communication technique means that there are no burden limitations and one sensor can serve numerous secondary devices.

9.11 Problems

1   A star connected set of metering CTs is to be connected to a remote energy meter. The CT specification ratio is: 1000:1, 10 VA, 0.5 class, FS10. If the route length is 250 m, using the cable data in Table 9.12, determine the minimum cable size that will ensure that the CTs are not overburdened. Assume that the burden presented by the meter is 1.0 VA at rated secondary current.
   (Answer: $2.5\,mm^2$)

2   Repeat question 1 if the CT ratio is changed to 1000:5. (You may assume that a 5 A meter has been provided and that it presents the same burden to the CT.)
   (Answer: $16\,mm^2$!)

Table 9.12 Cable resistance/km.

Cable size (mm^2)	Resistance per km (Ω @ 75 °C)
2.5	9.01
4	5.61
6	3.75
10	2.23
16	1.39

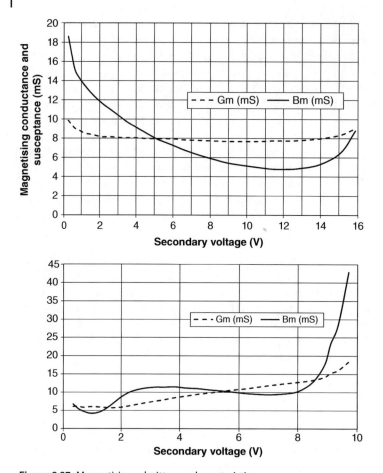

Figure 9.27 Magnetising admittance characteristics.

3 The magnetising admittance characteristics for two 200:5, 5 VA, 0.5 class metering CTs appear in Figure 9.27. The upper graph is that of an un-compensated CT with a large GOSS core and a winding resistance of 47 milliohms, while the lower is for composite core, consisting of a smaller GOSS core together with a Mu-metal inner section, like that pictured in Figure 9.12(a). It has a winding resistance of 50 milliohms.

For each of these CTs, calculate the magnitude and phase errors that might be expected when fully burdened according to IEC 61869-2, and delivering:

A full rated current

B half rated current.

(The manufacturer's test results for these devices are: CT1: magnitude error = -0.335%, phase error = 0.149 Crad; CT2: magnitude error = -0.174%, phase error = 0.015 Crad)

4 Repeat the calculations in question 3 for the first CT, assuming that it has been fitted with turns compensation, providing an effective turns ratio of 39.8:1. (Answer: Magnitude error ≈ +0.153%, phase error remains unchanged).

Table 9.13 CT magnetising data.

Magnetising current (mA)	Exciting voltage (V)	Magnetising current (mA)	Exciting voltage (V)	Magnetising current (mA)	Exciting voltage (V)
1	0.45	27	81.81	196.8	338.23
2	1.50	31.6	101.24	254.9	355.39
4	4.54	37.3	128.48	307.4	368.38
7	11.58	42.45	152.09	393.4	382.77
9	16.80	50.84	188.86	500	399.52
10.5	21.34	62.5	227.00	700	422.22
13.3	30.65	79.4	262.41	1000	454.00
18.07	46.22	120.2	303.91		
22.2	61.43	154.2	321.43		

5 Show that when a CT is operating far from saturation, and its magnetising current is sinusoidal, its composite error can be expressed in the form: $\varepsilon_c = \sqrt{e^2 + \beta^2}$.

6 Plot the CT magnetisation characteristic of the protection core described by the data in Table 9.13. From your graph determine the knee point voltages according both IEC 61869-2 and IEEE C57.13.
(Answer: $E_K \approx 300\,\text{V}$ (IEC 61869-2), $E_K \approx 215\,\text{V}$ (IEEE C57.13))

7 If the protection CT described in question 6 has a turns ratio of 2000:1, a winding resistance of 5 ohms and an accuracy limit factor of 15, determine:
   A the largest permitted burden resistance that may be connected
   B the VA rating applicable
   C the maximum symmetrical fault current that the CT is capable of accurately reporting.
   (Answer: (a) 15 ohms (b) 15 VA (c) 30 kA.)

8 If the CT in question 6 was used in a metering application instead, with an applied burden of $(8 + j2)$ ohms, determine the applicable security factor and the maximum current that the meter would see during the fault conditions described.
   (Answer: $FS = 22.8$; $I_{max} = 22.8\,\text{A}$, assuming that the primary fault level is equal to or greater than 45.6 kA.)

9 It is often necessary to *accurately* determine the turns ratio of a current transformer to which turns compensation has been applied. Simultaneously measuring the primary and secondary currents seldom provides the degree of precision required. The method shown in Figure 9.28 uses a reference CT whose turns ratio is known, against which the device under test (DUT) is compared. This method has the advantage that only one instrument is required, and that the resistance need not be

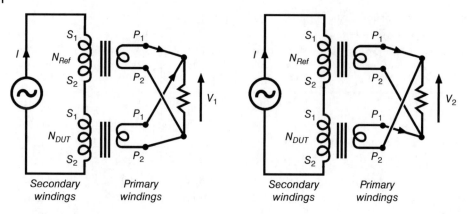

Figure 9.28 Precise measurement of CT turns ratio.

calibrated, although it is convenient to use an ammeter shunt for this purpose. Sufficiently heavy conductors must be provided for the primary circuit, since large currents will flow there. This requirement can be eased, however, by reducing the magnitude of the secondary current, but it is important that this be held constant during change in the CT burden.

A Show that the turns ratio of the device under test N_{DUT} is given by:

$$N_{DUT} = N_{ref} \frac{(1-a)}{(1+a)}$$

where $a = \dfrac{V_2}{V_1}$.

B It is important that neither CT is permitted to saturate during testing. In which part of this test is saturation most likely and why?

C If the CTs in Figure 9.28 are 300:5, 15 VA, class 0.5 metering cores (the reference CT being un-compensated), and the shunt generates 100 mV @ 500 A, determine the size of the primary cabling necessary if the CTs are to see rated burden. Assume that each cable is 450 mm long.

(Answer: 44 mm² cable. In practice, a smaller shunt and cables could be used so long as the secondary current is kept well below 5 A.)

D What additional measurements would you make to ensure that these CTs do not saturate during this test?

10 *CT error determination.* The measurement of CT phase and magnitude errors can be done either by comparing the unknown CT with a reference (or a standard) CT of the same ratio and whose errors are known (a process known as *primary injection testing*), or by measuring the turns ratio, the winding resistance and the magnetising admittance, so that the CT's equivalent circuit can be obtained, allowing its device errors to be calculated from first principles. This method is generally known as *secondary injection testing*.

Primary injection involves passing a known primary current through both CTs and comparing their secondary currents. Today this comparison is accomplished

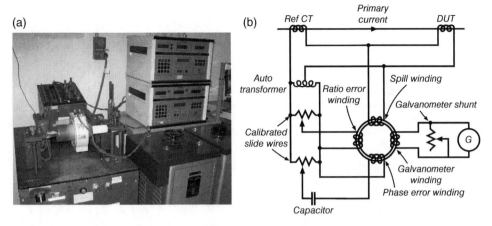

Figure 9.29 (a) Primary injection testing of a metering class current transformer. (LHS: CT under test; background transformer: multi-tapped reference CT. RHS: digital comparators.) (b) Simplified equivalent circuit of the Petch–Elliott current transformer test set.

with the use of a digital comparator like those shown in Figure 9.29(a), where a production metering class CT is being error tested. The comparators are provided with the secondary current from each CT, as well as the errors relating to the reference CT, and from this information they can provide the magnitude and phase errors of the CT under test.

In 1939 an English inventor patented a simple but very effective current transformer test set, which was a considerable improvement on the testing techniques in use at the time. It was based on the principle of primary injection and thus required a reference CT as well as an external current source. Its simplified equivalent circuit is shown in Figure 9.29(b), and although now obsolete, it is instructive to appreciate the simplicity and elegance of its design.

The external source provides primary current to both the reference CT and the DUT, both of which must have the *same nominal turns ratio*. We will initially assume that the reference CT has no magnitude or phase errors. The secondary windings of each CT are connected in series so that most of the secondary current circulates between them. Any small current difference that arises is due to the magnitude and phase errors within the DUT. This current flows in the spill winding of the toroidal transformer, shown in the schematic. As a result, a small flux is generated within its core, one which can be detected by a galvanometer sensing the potential developed across a second winding on the same core.

The phasor diagram in Figure 9.30 shows that the spill current (enlarged here for clarity) can be resolved into a component in-phase with the current in the reference CT (proportional to the magnitude error of the DUT) and one in quadrature with it (proportional to the phase error of the DUT).

The secondary current from the reference CT also flows through a small auto-transformer, chosen for its low leakage reactance. This in turn provides current to a pair of calibrated potentiometers, see Figure 9.29(b). The upper one is connected to a ratio or magnitude error winding on the same toroidal core, and is adjusted to

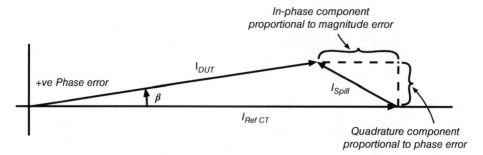

Figure 9.30 In-phase and quadrature components of spill current.

create an equal but opposite flux to that created by the in-phase component of the spill current.

In a similar way, the lower potentiometer in the figure and its associated phase error winding, generate an equal but opposite flux to that generated by the quadrature component of the spill current. This is achieved by injecting a leading current into the phase error winding, created as a consequence of the capacitor in the circuit.

The nulling potentiometers are each calibrated according to the degree of compensation they provide, the ratio error potentiometer in per cent and the phase error potentiometer in minutes of arc. The instrument is nulled by adjustment of each potentiometer until the galvanometer reading is minimised. With careful adjustment, a sharp null will be observed, and the flux in the toroid will then be near zero, at which time the ratio and phase errors of the DUT can be read from the potentiometers.

The elegance of this design lies in the fact that when the errors are determined there is no net flux in the toroidal core, and thus this core does not contribute any error to the measurement. The spill winding thus presents a very low *non-inductive* impedance, and as a result the DUT sees virtually zero burden. (If magnitude and phase errors are required at a particular burden, then this can be inserted in series with the DUT, and the instrument has provision for this.) There are therefore no stringent requirements for the toroidal core, other than that it should be easily permeable, so that should even a small flux be present, its presence can be readily detected, and thus Mu-metal would be a good core material. Many similar comparison circuits have since been used for CT error measurement, several of which appear in IEEE C57.13.

On its nominal range this instrument measures magnitude and phase errors relative to those of the reference CT, to within 0.01% and better than 1 minute of arc. Therefore in order to obtain the absolute error for the DUT, the errors of the reference CT must be added to those obtained from the instrument.

A This instrument must be able to cater for both lagging and leading phase errors, depending for example, on the power factor of connected burden. The description above outlines its operation when the DUT has a leading (positive) phase error. How would you modify the design to permit it to cater for negative phase errors?

B This instrument must also be able to cater for turns compensated CTs, which will generate positive magnitude errors. How would you modify its design to cater for these?

C The phase correction winding must inject leading ampere-turns in order to null the quadrature component in the spill winding. Since the compensation current is supplied from a potentiometer with a small but resistive Thévenin impedance, the capacitive ampere-turns will not lead the potentiometer voltage by exactly 90°. How might this defect be corrected?

Hint: Consider the use of an additional winding, suitably connected in parallel with the existing phase error winding, in which the current is resistively controlled. Will this modification completely correct the problem?

11 *DC transient flux.* The general expression for primary fault current, given by Equation (9.11), can be simplified in the worst case by letting $(\alpha - \gamma) = \pi/2$. This makes the DC transient largest, maximising the asymmetry, and in this case $i_p(t)$ can be written:

$$i_p(t) = \frac{E_p}{\sqrt{R^2 + \omega^2 L^2}} \left[\sin(\omega t - \pi/2) + e^{\frac{-tR}{L}} \right]$$

The flux in the CT depends on the voltage developed in the secondary winding. This will be equal to the secondary current multiplied by the burden resistance R_b. In general, the flux in the core is given by:

$$\phi = \frac{1}{N_s} \int E_s dt = \frac{1}{N_s} \int I_s R_b dt$$

where I_s is the secondary current, R_b is the applied burden and n_s is the number of secondary turns.

It is useful to examine the *AC flux* ϕ_{AC}, and the transient or *DC flux* ϕ_{DC}, when considering how the CT will react to the fault current.

A Show that the peak steady state flux $\hat{\phi}_{AC}$ during the fault is given by:

$$\hat{\phi}_{AC} = \frac{\hat{I}_s R_b}{N_s \omega} \text{ webers}$$

B Obtain an equation for time variation of the DC flux. Show also that the peak DC flux $\hat{\phi}_{DC}$, is given by:

$$\hat{\phi}_{DC} = \frac{\hat{I}_s R_b L}{N_s R} \text{ webers}$$

C The ratio $\hat{\phi}_{DC} / \hat{\phi}_{AC}$ is thus equal to $\omega L/R = X/R$, where X and R are the system reactance and resistance at the fault respectively. Since the core must support both flux components, we can write:

$$\phi_T = (\phi_{AC} + \phi_{DC}) = \phi_{AC}(1 + X/R) = \phi_{AC}(1 + \omega L/R) \tag{9.13}$$

The term $(1 + X/R)$, sometimes called the transient factor, quantifies the amount by which the flux in the core is increased by the transient component of the fault current. X/R ratios of 10–20 are common, and thus the CT core must ultimately be able to support a flux 10–20 times as large as the steady state fault flux, without saturating.

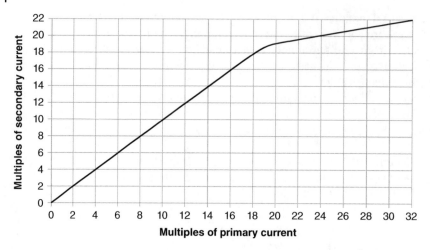

Figure 9.31 Currents in primary and secondary. (Magnitude error at 20 times primary current ≈ 5%).

(a) (b)

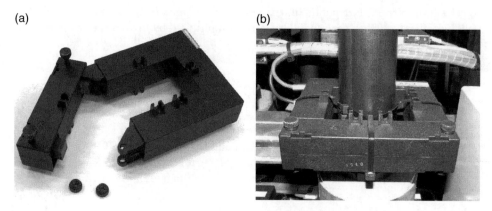

Figure 9.32 (a) Split core current transformer (b) Cable mounted split core CT.

In view of Equation (9.13) and the equation for the DC flux obtained in part (B), is the total flux in the core always likely to reach the value suggested in this equation, and if not why not? If the transient dimensioning factor were based on this transient factor, might the CT not then be slightly overdesigned?

12 Using a similar process to that outlined for class C CTs above, derive and graph the overcurrent ratio curve for the 1200:5 CT whose magnetising characteristic appears in Figure 9.21. Repeat this analysis for the 600:5 tapping. Assume in each case that the external burden on the CT is 4.4 ohms and that it has a winding resistance of 0.6 ohms. Your results should look similar to Figure 9.31.

13 Occasionally it is necessary to install an energy metering installation in a circuit that contains no suitable current transformers – in the primary circuit of a cable-fed distribution transformer, for example. In some cases, it is possible to install cable mounted split core CTs, as shown in Figure 9.32. These are current transformers

with a magnetic circuit that can be opened, allowing them to be placed around a cable in-situ, without requiring an outage.

MV cables usually include an earthed screen, outside the primary insulation layers and just inside the cable's outer covering. This may be either a solid sheath of lead, steel wire armouring or a multi-stranded copper screen. Cable-mounted CTs function very well in this situation provided that the screen is not earthed at both ends of the cable.

A What precautions would you take when fitting a split core CT around a cable carrying current?

B What effect will earthing the cable screen at both ends have on the performance of these current transformers, and why?

C Suggest two methods by which this effect may be avoided.

9.12 Sources

1 IEC 61969-2 *Instrument transformers – Part 2: Additional requirements for current transformers*, 2012, International Electrotechnical Commission, Geneva.

2 IEEE Std C57.13 *IEEE Standard requirements for instrument transformers*, 2008, IEEE Power Engineering Society, New York.

3 IEC 61869-1 *Instrument transformers – Part 1 General requirements Edition 1.0*, 2007-10, International Electrotechnical Commission, Geneva.

4 IEEE Std C37.110 *IEEE Guide for the application of current transformers used for protective relaying purposes*, 2007, IEEE Power Engineering Society, New York.

5 AS1675 *Current transformers – measurement & protection* 1986, Standards Australia, Homebush NSW.

6 IEC 60044-8 *Instrument transformers: Part 8 Electronic current transformers* 2002-07, International Electrotechnical Commission, Geneva.

7 ABB High Voltage Products Department, *Instrument transformers application guide*, ABB Instrument Transformers.

8 Blackburn JL, Domin TJ, *Protective relaying principles and applications*, Third Edition 2007, CRC Press, Boca Raton FL.

9 Elmore WA, *Protective relaying: Theory and applications*, CRC Press, Boca Raton FL.

10 Relay Work Group, *White paper on relaying current transformer application*, 2014, Western Electricity Coordinating Council, Salt Lake City, Utah.

11 General Electric Power Management, *Dimensioning of current transformers for protection application*, GE Publication No: GER-3973B, 2006, GE Multilin.

12 General Electric Company, *Manual of instrument transformers*, Meter & Instrument Business Dept, Somersworth, NH 03878.

13 Cigre Study Committee A3 *State of the art of instrument transformers*, 2009, CIGRE.

14 IEC 60044-8 *Instrument transformers – Part 8: electronic current transformers*, 2002–07, International Electrotechnical Commission, Geneva.

15 Cataliotti A, Di Cara D, Di Franco PA, Emanuel AE, Nuccio S, *Frequency response of measurement current transformers* 2008, IEEE International Instrumentation Measurement Technology Conference, Victoria, Canada.

10

Energy Metering

Once practical uses for electricity were discovered, its value as a saleable item was immediately realised, and the need for accurate measurement of electrical energy arose. Early energy meters were induction disc instruments, and their design has changed little for over 100 years. These instruments are based on a small induction motor, driving a gear train which registers the energy used on a series of decadal dials, visible from the face of the meter. The speed of rotation of the disc and thus of the gear train, depends on the applied voltage and current, as well as the angle between them. Induction disc meters are designed to produce maximum torque at unity power factor where, for a given voltage and current, the power flow is greatest. The speed of the disc is then proportional to the load power factor, and thus to the power flowing to the customer. Temperature compensation is usually provided, as well as damping magnets for precise speed calibration. Depending on the class of the instrument, accuracies can lie in a range from 0.2% to 3%, though being a mechanical device, periodic calibration is necessary to ensure that this accuracy is maintained.

An example of an early single-phase residential meter appears in Figures 10.1 and 10.2. This is a Shallenberger instrument made in the USA by the Westinghouse Electric Manufacturing Company, about 1900. It was owned by the Melbourne City Council (Australia), which built a power station in 1894 to provide street lighting and to supply some private customers.

This instrument is the subject of patents granted between August 1888 and April 1890 (*prior* to Andre Blondel's metering theorem), yet it is not dissimilar to domestic single-phase induction disc meters in use today. It displays the accumulated energy consumed in kilowatt-hours on a register consisting of a series of four gear driven dials. One interesting feature of this particular instrument is the lack of an iron magnetic circuit found in modern induction disc meters. It has two voltage windings insulated with black tape and within these is the heavier current coil, positioned at an angle with respect to them. Inside this is the induction disc which rotates on a vertical shaft. Damping is achieved by vanes attached to the lower portion of the shaft, which presumably were a close fit with the inside of the instrument's cover. By suitably bending these, one could adjust the damping of the instrument and thus calibrate its speed of rotation. The upper portion of the shaft drives the gearbox, mounted behind the register plate.

AC Circuits and Power Systems in Practice, First Edition. Graeme Vertigan.
© 2018 John Wiley & Sons Ltd. Published 2018 by John Wiley & Sons Ltd.

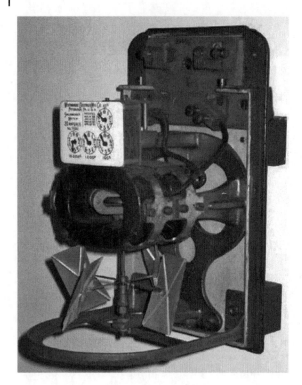

Figure 10.1 Single-phase electricity meter, circa 1900 (minus its case).

Modern Energy Meters

Although there are many induction disc meters still in service, energy meters installed today are almost exclusively electronic devices, generally known as *static meters* (to indicate that there are no moving parts). Static meters are cheaper to mass produce and offer the advantage of consistently greater accuracy than can be obtained from electro-mechanical instruments, as well as providing a time based profile of the energy consumed. Kilowatt-hours, kiloVAr-hours, kVA-hours (both import and export), harmonic information and various other power quality measurements are available from most static meters.

The establishment of electricity markets in many countries and the need for prompt market settlements has led to a need to interrogate participating customer metering much more frequently than once per billing period, without the need to visit each customer's premises. Static meters provide a convenient means of communication since they can be interfaced to a modem, and in recent years many induction disc devices have been replaced with static meters for just this purpose.

In this chapter we will mainly confine our interest in energy metering to that of three-phase commercial and industrial customers. Smaller, single-phase or three-phase LV customers are generally provided with whole current meters (see section 10.1.1), capable of directly accepting currents of 100–200 A per phase. These meters require no current or voltage transformers and their connection is relatively straightforward.

Figure 10.2 Meter registers – note the patent dates. (Courtesy of the Victorian Museum, Australia).

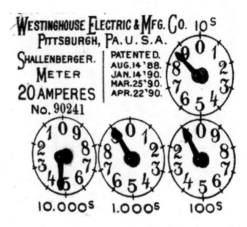

WESTINGHOUSE ELECTRIC & MFG. CO. 10 S
PITTSBURGH, PA. U. S. A.

SHALLENBERGER. | PATENTED.
METER | AUG.14 '88.
| JAN. 14 '90.
| MAR.25 '90.
20 AMPERES | APR.22 '90.
No. 90241

10.000S   1.000S   100S

10.1 Metering Intervals

We saw in Chapter 3 that the general expression for the active and reactive power consumed by any three-phase load can be expressed as:

$$P = |V_a||I_a|\cos(\phi_a) + |V_b||I_b|\cos(\phi_b) + |V_c||I_c|\cos(\phi_c) \tag{10.1}$$

$$Q = |V_a||I_a|\sin(\phi_a) + |V_b||I_b|\sin(\phi_b) + |V_c||I_c|\sin(\phi_c) \tag{10.2}$$

where $|V_a|$ is the magnitude of the voltage phasor V_a, and $|I_a|$ is the magnitude of the current phasor I_a and similar definitions apply to the other phases. The angles ϕ_a, ϕ_b and ϕ_c relate to the impedance of the connected load, and therefore are positive when the current lags the applied voltage.

Induction disc meters accumulate the active energy (kWh) consumed by a customer during the billing period (which may be between 1 and 3 months), on a mechanical register like that shown in Figure 10.1b. Such an accumulation provides no information as to when the energy was used or indeed how much was used at any particular time.

Modern static energy meters, on the other hand, usually record as a minimum the active and reactive energy (kWh and kVArh) consumed during each *metering interval* throughout the billing period. Metering intervals are generally either 15 or 30 minutes, although they can sometimes be as short as 5 minutes. The energy recorded represents the time integral of P and Q, in Equations (10.1) and (10.2) over this interval. For customers with *co-generation equipment*, the metered energy is segregated into *import* and *export* registers within the meter. In this way, the retailer knows how much energy is imported from, or exported to, the network and when the exchange occurred. This allows both power and time-of-use profiles to be obtained for the load, in addition to the total energy consumed.

The power, averaged over each metering interval, can be found by dividing the accumulated energy by the metering interval expressed in hours. This information is used by the network owner to determine the *maximum demand charge* applicable to the customer concerned. The *maximum demand* is the largest power demand (or alternatively the VA demand), recorded in any metering interval throughout either the billing period or the calendar year. A maximum demand charge is levied on all commercial and industrial customers by the network operator to cover the cost of the distribution and transmission networks as well as any specific connection assets used in the delivery of energy to the customer concerned. The reactive power demanded during each metering interval may also be the subject of a maximum demand limits under some connection agreements.

The total active energy accumulated during each metering interval throughout the billing period is paid for by the customer. Early static meters delivered active and reactive energy information at the end of every metering interval to a separate data logger. Today, however, the logging function is carried out within the meter itself. Depending on how many parameters are logged, most modern meters have sufficient data capacity to store between one and two months of 15 minute data.

From a practical point of view there are two *Blondel compliant*[1] metering connections that are commonly used to measure the energy consumed by three-phase customers, these are known as three-wire and four-wire metering connections, and an analysis of each one follows. There are also many metering configurations that have approximate Blondel compliance; these are generally only correct when the load is balanced, but they are attractive because most offer a saving in the hardware required.

10.1.1 Three-Element Metering (Four-Wire Metering)

Equations (10.1) and (10.2) suggest that three *metering elements* are required when metering a three-phase load, each of which is contained within the same meter. Each element receives a representative voltage and current from one phase and computes the associated energy by integrating the VI product over a metering interval.

In the case of LV customers the voltages and currents may be directly connected to the meter, provided that the phase currents are not excessive. Such metering is called *whole current* or *direct connect*, and meters are capable of accepting phase currents of the order of 100 A, although some can accommodate considerably more. In the USA for example, the standard *ANSI meter sockets* are widely used for both single- and three-phase LV applications. These accept a wide range of energy meters, and have current jaws capable of accepting currents as high as 200 A per phase.

Where these current limits are exceeded, metering class current transformers are required to provide the meter with a scaled representation of the load current. The metering of HV loads requires both current and voltage transformers, since energy meters clearly cannot accommodate HV potentials.

A three-element (or four-wire) HV metering installation is shown in Figure 10.3, where the energy delivered by a 110 kV:11 kV distribution transformer is metered. This arrangement requires three current transformers (CTs) and three voltage transformers (VTs). Four wires are taken from the star connected VTs to the meter (hence the name), the three phase voltages and the neutral. The CTs are also star connected, with the star point generally located near the CTs, so as to minimise the cabling burden. Four wires

1 The meaning of this term is fully discussed in Section 10.1.2. Put simply, Blondel compliant metering correctly records the energy metered under all conditions, regardless of the degree of unbalance, either in the supply voltages or in the currents consumed by the load. Not all metering configurations in use are strictly Blondel compliant.

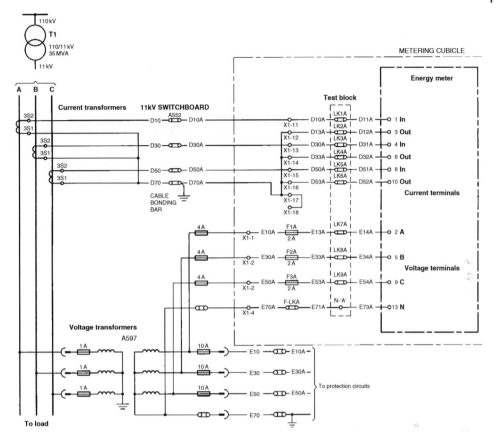

Figure 10.3 Four-wire, three element HV metering connection, including the test block.

also run from the CTs to the meter. Each of the three metering elements is thus supplied with a representation of one phase voltage and its associated phase current. The meter aggregates the energy contribution from each phase, and the sum is logged against the appropriate metering interval.

When a meter is read, the energy logged must be rescaled according to a meter multiplier, equal to the product of the VT and the CT ratios. For example, if a metering installation is supplied from a $22\text{kV}/\sqrt{3}:110\text{V}/\sqrt{3}$ VT and an $800:5\,\text{A}$ CT, then the meter multiplier will be $200 \times 160 = 32{,}000$. The meter multiplier can be programmed into the meter in the form of voltage and current transformer ratios, in which case the meter will report its energy in terms of the CT and VT primary quantities, or alternatively it can be applied when the energy data is downloaded.

10.1.2 Two-Element Metering (Three-Wire Metering)

The two-element or three-wire metering installation is used to measure energy in MV and HV distribution and transmission circuits, where only the phase conductors are distributed. It finds application in metering customers who are directly connected to the MV distribution system, and relies on Blondel's theorem, as derived below.

Blondel's Theorem

In 1893, French engineer André Blondel devised a method of metering the energy in a polyphase system, supplied via N conductors using $N-1$ metering elements. This means that in a three-wire system where the load currents are supplied by the three-phase conductors (i.e. no neutral), the total energy can be correctly measured using two metering elements only.

This technique is therefore referred to as three-wire metering or two-element metering. In 1893, this theorem must have been quite a breakthrough, since it saved one metering element, one CT and one VT, at a time when these items were particularly expensive. Blondel's theorem can be demonstrated by initially considering the total power supplied to a three-phase load measured using three metering elements, as given by Equation (10.1):

$$P = |V_{an}||I_a|\cos(\phi_a) + |V_{bn}||I_b|\cos(\phi_b) + |V_{cn}||I_c|\cos(\phi_c) \tag{10.3}$$

We can rewrite Equation (10.3) using the dot product of two vectors, since this returns the product of their magnitudes multiplied by the cosine of the angle between them, thus:

$$V_{an} \bullet I_a = |V_{an}||I_a|\cos(\phi_a)$$

so $\quad P = V_{an} \bullet I_a + V_{bn} \bullet I_b + V_{cn} \bullet I_c \tag{10.4}$

where V_{an} is the complex a phase voltage and I_a is the complex a phase current. Similar definitions apply for the b and c phases. We introduce an arbitrary voltage node x as shown in Figure 10.4, such that:

$$V_{an} = V_{ax} + V_{xn} \tag{10.5}$$

$$V_{bn} = V_{bx} + V_{xn} \tag{10.6}$$

$$V_{cn} = V_{cx} + V_{xn} \tag{10.7}$$

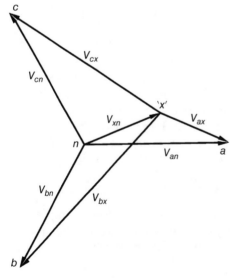

Figure 10.4 Definition of voltages with respect to the arbitrary node 'x'.

Where V_{xn} is the voltage between the new node x and the neutral terminal. (Note that we have imposed no limitations on the location of the node x.) Substituting Equations (10.5) (10.6) and (10.7) into Equation (10.4) yields:

$$P = \left(V_{ax} + V_{xn}\right) \bullet I_a + \left(V_{bx} + V_{xn}\right) \bullet I_b + \left(V_{cx} + V_{xn}\right) \bullet I_c$$
$$P = V_{ax} \bullet I_a + V_{bx} \bullet I_b + V_{cx} \bullet I_c + V_{xn} \bullet I_a + V_{xn} \bullet I_b + V_{xn} \bullet I_c$$

or $\quad P = V_{ax} \bullet I_a + V_{bx} \bullet I_b + V_{cx} \bullet I_c + V_{xn} \bullet \left(I_a + I_b + I_c\right)$ (10.8)

Without a neutral conductor, Kirchhoff's current law requires that:

$$I_a + I_b + I_c = 0$$

(It should be emphasised that the phase currents need **not** be balanced; they must simply add to zero.) Thus Equation (10.8) becomes:

$$P = V_{ax} \bullet I_a + V_{bx} \bullet I_b + V_{cx} \bullet I_c$$ (10.9)

Node x was arbitrarily chosen, and therefore it can be placed anywhere on the complex plane. It is usual in three-wire metering installations, to let x lie on the b phase voltage, whence $V_{bx} = V_{bb} = 0$ and therefore Equation (10.9) becomes:

$$P = V_{ab} \bullet I_a + V_{cb} \bullet I_c$$ (10.10)

Equation (10.10) can be expressed alternatively as:

$$P = |V_{ab}||I_a|\cos(\phi_{ab}) + |V_{cb}||I_c|\cos(\phi_{cb})$$ (10.11)

or alternatively $\quad P = |V_{ab}||I_a|\cos(\phi_a + 30) + |V_{cb}||I_c|\cos(\phi_c - 30)$ (10.12)

Equation (10.10) shows that the total power delivered by all three-phases can be correctly calculated using two metering elements, provided that the VTs are connected across the ab and bc line voltages and that the a and c line currents are associated with their respective line voltage. This result is particularly convenient in ungrounded or impedance grounded systems, since it permits the use of line connected VTs instead of the star-star connection used in four-wire circuits. The latter suffers from induced voltage errors in these situations, since the third harmonic component of the magnetising current is constrained by the high zero-sequence impedance.

By a similar argument, the reactive power, Q can be expressed as:

$$Q = |V_{ab}||I_a|\sin(\phi_{ab}) + |V_{cb}||I_c|\sin(\phi_{cb})$$ (10.13)

or alternatively $\quad Q = |V_{ab}||I_a|\sin(\phi_a + 30) + |V_{cb}|I_c\sin(\phi_c - 30)$ (10.14)

where ϕ_a is the angle between the a phase current and the a phase voltage, ϕ_c is the angle between the c phase current and the c phase voltage, ϕ_{ab} is the angle between the a phase current and the ab line voltage, and ϕ_{cb} is the angle between the c phase current and the cb line voltage, as shown in Figure 10.5.

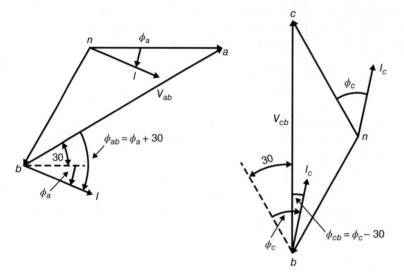

Figure 10.5 Three-wire metering angle definitions.

The left-hand phasor diagram in Figure 10.5 shows that the angle between the *a* phase current and the *ab* line voltage is $(\phi_a + 30)$ degrees. The right-hand diagram shows that angle between the *c* phase current and the *cb* line voltage is $(\phi_b - 30)$ degrees. Thus:

$$\phi_{ab} = (\phi_a + 30°) \text{ and } \phi_{cb} = (\phi_c - 30°)$$

Note: Phase angles are defined as positive if lagging.

This situation is depicted in a simpler fashion in Figure 10.6, where the phase voltages have been omitted for clarity. Most meter manufacturers use a vector diagram like this to provide the user with a pictorial representation of the line currents and voltages.

Figure 10.7 shows a schematic diagram of a three-wire metering installation using a pole-mounted CT/VT metering transformer, similar to the one shown in Figure 10.8. In this case three phase-connected VTs are used to supply the meter with a representation of the line voltages V_{ab} and V_{cb}.

10.1.3 Blondel Compliant Metering

A metering installation is said to be Blondel Compliant if it correctly records the energy metered under all conditions, regardless of the degree of unbalance, either in the supply voltages or in the currents consumed by the load. Some supply authorities use non-Blondel compliant metering connections in order to save on metering elements, CTs and sometimes VTs. In nearly all of these, correct metering depends upon a balanced set of supply voltages and sometimes balanced load currents as well. An example of this is the Form 2S metering connection used in the United States for metering domestic single-phase 120/240 volt supplies.

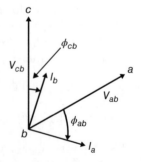

Figure 10.6 Simplified three-wire connection phasor diagram.

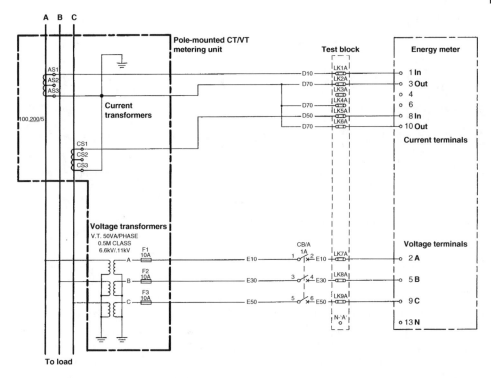

Figure 10.7 Three-wire (two-element) metering connection.

Figure 10.8 Pole-mounted current and voltage metering transformer.

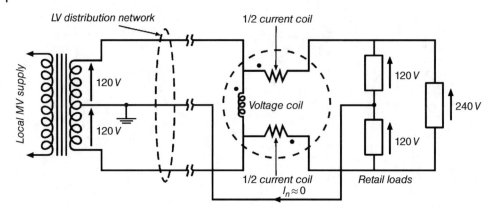

Figure 10.9 Form 2S Metering Configuration (USA).

This ingenious metering connection uses a single element induction disc meter to record the total energy consumed from both the 120 and 240 volt potentials, and is shown in Figure 10.9. It has two current coils, one in each phase conductor, each dimensioned to excite the metering element with a flux proportional to half the current flowing, and a single voltage winding connected across the 240 volt potential. The instrument registers correctly when metering the 240 volt loads, since both current windings contribute a flux proportional to the full current flow. Loads connected across either 120 volt potential, consume current that flows through one half current winding only, but since the voltage winding is excited by twice the load potential, the energy is generally measured correctly. Errors arise when the supply voltages are not balanced and when the neutral current is not close to zero. Since the torque on the disc contains components proportional to the power in all three loads, the meter aggregates the total energy consumed.

Unless the load is heavily unbalanced and the voltages are skewed, this connection achieves acceptable accuracy; however, because the metered energy cannot be guaranteed correct under all load scenarios, this connection is deemed to be non-Blondel compliant.

Another non-Blondel compliant installation is the Form 14S configuration sometimes used to meter three-phase, four-wire circuits in the USA. This arrangement, also known as 2½ *element metering*, uses an instrument with two voltage windings, connected across the *a* and *c* phase voltages respectively, and four current windings. These are arranged so that the *a* phase element effectively sees the difference between the *a* and phase *b* currents, and the *c* phase element sees the difference between the *c* and *b* phase currents. This can either be achieved using two current windings per element, as shown in Figure 10.10, or by arranging the current difference to flow through a single current coil associated with each element.

As a result of its current connections, the 2½ element installation evaluates the power flowing according to:

$$P = |V_a||I_a|\cos(\phi_a) + |V_a||I_b|\cos(60-\phi_b) + |V_c||I_c|\cos(\phi_c) + |V_c||I_b|\cos(60+\phi_b)$$
$$P = |V_a||I_a|\cos(\phi_a) + |V_c||I_c|\cos(\phi_c) + \left[|V_a||I_b|\cos(60-\phi_b) + |V_c||I_b|\cos(60+\phi_b)\right]$$

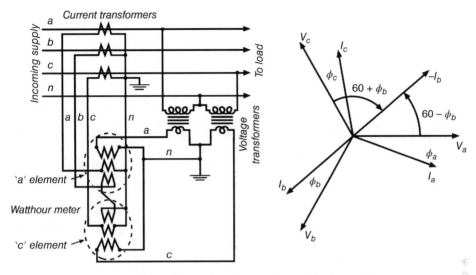

Figure 10.10 Metering installation for 2½ element metering, and its phasor diagram.

A three-element installation would correctly measure P according to:

$$P = |V_a||I_a|\cos(\phi_a) + |V_b||I_b|\cos(\phi_b) + |V_c||I_c|\cos(\phi_C)$$

Therefore if the 2½ element installation is to be correct, we require that:

$$|V_b||I_b|\cos(\phi_b) = |V_a||I_b|\cos(60 - \phi_b) + |V_c||I_b|\cos(60 + \phi_b)$$

or: $$|V_b||I_b|\cos(\phi_b) = 1/2 \left[|V_a||I_b|\cos(\phi_b) + |V_c||I_b|\cos(\phi_b) \right]$$ (10.15)

Equation (10.15) can only be true if the voltages are balanced in magnitude and phase, i.e. $|V_a| = |V_b| = |V_c|$ and adjacent phase voltages are separated by 120°. Note that there is no requirement that the currents be balanced, since $|I_b|$ appears on both sides of Equation (10.15).

Since there is a regulatory limit to the degree of unbalance in a three-phase system, there will also be a limit to the error associated with this metering connection. The a and c phases are metered correctly, so any voltage unbalance present will only affect the b element. The b phase metering error is thus given by one-third of the percentage voltage unbalance.

10.2   General Metering Analysis using Symmetrical Components

In the example above since the b phase voltage is not measured, it is not possible to determine the magnitude of any negative or zero-sequence components present in the supply. It is useful to investigate how these affect the metered energy.

Consider a three-element installation metering a three-phase four-wire circuit. Assume that phase voltages and currents contain small negative and zero-sequence components in addition to the positive sequence component. In this event, the metered power over all three-phases can be written:

$$P = \left(V_{1a} + V_{2a} + V_0\right) \bullet \left(I_{1a} + I_{2a} + I_0\right) + \left(V_{1b} + V_{2b} + V_0\right) \bullet \left(I_{1b} + I_{2b} + I_0\right)$$
$$+ \left(V_{1c} + V_{2c} + V_0\right) \bullet \left(I_{1c} + I_{2c} + I_0\right)$$

When summed over all three-phases this equation reduces to:

$$P = 3|V_1||I_1|\cos(\phi_1) + 3|V_2||I_2|\cos(\phi_2) + 3|V_0||I_0|\cos(\phi_0) \qquad (10.16)$$

(The derivation of this equation will be the subject of a tutorial exercise.)

Here the sequence symbols have their usual meanings and ϕ_1, ϕ_2 and ϕ_0 are the angles between the sequence currents and their respective sequence voltages.

Equation (10.16) effectively states that, when summed over all three phases, the power contribution generated from the product of one sequence component with another will be zero, i.e.:

$$\sum_{i \neq j}(V_i \bullet I_j) = 0$$

Therefore unless there are negative or zero-sequence voltage components present, any negative or zero-sequence currents will not contribute to the total power flow.

A similar analysis can be applied to the VAr flow, which will yield the equation:

$$Q = 3|V_1||I_1|\sin(\phi_1) + 3|V_2||I_2|\sin(\phi_2) + 3|V_0||I_0|\sin(\phi_0)$$

In the 2½ element metering example above, it has been tacitly assumed that the voltage supply is balanced, that it does not contain negative or zero-sequence components. However, when these components do arise, in both the voltage and current, then errors occur. For example, consider the effect of a zero-sequence current flowing in the neutral conductor, perhaps as the result of unbalanced single-phase loads. Because each element in this metering configuration sees the difference between two phase currents, any zero-sequence current will be cancelled, and therefore the zero-sequence power flow cannot be metered.

Similarly, since the b phase voltage is not measured, the presence of negative sequence voltages and currents will also lead to metering errors, since the meter is incapable of resolving the negative sequence potential. For these reasons, this metering configuration is classed as non-Blondel compliant; it meters correctly when the supply voltages contain positive sequence components only, i.e. when they are balanced.

On the other hand, the three-element (four-wire) metering installation is Blondel compliant, and therefore it will operate correctly in the presence of any degree of unbalance. This is because all phase voltages and all phase currents are measured. Consider the operation of this installation with balanced phase voltages and unbalanced currents, as shown in the phasor diagram of Figure 10.11.

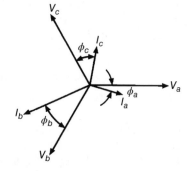

Figure 10.11 Balanced voltages and unbalanced phase currents.

According to Equation (10.16) the metered power should then be:

$$P = 3|V_1||I_1|\cos(\phi_1) \tag{10.17}$$

where $|V_1|$ is the magnitude of the positive sequence voltage, $|I_1|$ is the magnitude of the positive sequence current and ϕ_1 is the angle between the two.

Recall that I_1 is given by: $I_1 = (I_a + \alpha I_b + \alpha^2 I_c)/3$

Therefore: $3I_1 = |I_a|\angle - \phi_a + 1\angle 120°|I_b|\angle - (120° + \phi_b) + 1\angle - 120°|I_c|\angle(120 - \phi_c)$

or: $3I_1 = |I_a|\angle - \phi_a + |I_b|\angle - \phi_b + |I_c|\angle - \phi_c$

Then I_1 can be written in Cartesian form as:

$$3I_1 = |I_a|\cos(\phi_a) + |I_b|\cos(\phi_b) + |I_c|\cos(\phi_c) - j\{|I_a|\sin(\phi_a) + |I_b|\sin(\phi_b) + |I_c|\sin(\phi_c)\}$$

From which we can derive both the magnitude and the phase of I_1:

$$|I_1| = \frac{\sqrt{\{|I_a|\cos(\phi_a) + |I_b|\cos(\phi_b) + |I_c|\cos(\phi_c)\}^2 + \{|I_a|\sin(\phi_a) + |I_b|\sin(\phi_b) + |I_c|\sin(\phi_c)\}^2}}{3}$$

and:

$$\angle I_1 = -\arccos\left[\frac{|I_a|\cos(\phi_a) + |I_b|\cos(\phi_b) + |I_c|\cos(\phi_c)}{\sqrt{\{|I_a|\cos(\phi_a) + |I_b|\cos(\phi_b) + |I_c|\cos(\phi_c)\}^2 + \{|I_a|\sin(\phi_a) + |I_b|\sin(\phi_b) + |I_c|\sin(\phi_c)\}^2}}\right]$$

If we substitute these values into Equation (10.17) we find:

$$P = 3|V_1| \frac{\sqrt{\{|I_a|\cos(\phi_a) + |I_b|\cos(\phi_b) + |I_c|\cos(\phi_c)\}^2 + \{|I_a|\sin(\phi_a) + |I_b|\sin(\phi_b) + |I_c|\sin(\phi_c)\}^2}}{3}$$

$$x \frac{|I_a|\cos(\phi_a) + |I_b|\cos(\phi_b) + |I_c|\cos(\phi_c)}{\sqrt{\{|I_a|\cos(\phi_a) + |I_b|\cos(\phi_b) + |I_c|\cos(\phi_c)\}^2 + \{|I_a|\sin(\phi_a) + |I_b|\sin(\phi_b) + |I_c|\sin(\phi_c)\}^2}}$$

So $P = |V_1|\{|I_a|\cos(\phi_a) + |I_b|\cos(\phi_b) + |I_c|\cos(\phi_c)\}$

Since the supply voltages are balanced, we obtain the familiar result:

$$P = |V_a||I_a|\cos(\phi_a) + |V_b||I_b|\cos(\phi_b) + |V_c||I_c|\cos(\phi_c)$$

When the voltage supply is not balanced, and negative and zero-sequence components also exist, then negative and zero-sequence power components will appear in the summation, as predicted by Equation (10.16). These can each be found using a similar analysis to that above.

10.2.1 Two-Element (Three-Wire) Metering Installation

In the light of this discussion, it is useful to review the operation of the two-element metering installation with imbalances in both the supply voltages and phase currents. While this installation *is* Blondel compliant, it has the restriction that $I_a + I_b + I_c = 0$, therefore zero-sequence currents cannot exist. Thus the only possible imbalance is due to the presence of negative sequence components. This installation will respond correctly to both positive and negative sequence currents and voltages, as shown below.

We express the measured power using the dot product notation introduced above, thus:

$$P = V_{ab} \bullet I_a + V_{cb} \bullet I_c$$

therefore:
$$P = \left(V_{an} - V_{bn}\right) \bullet I_a + \left(V_{cn} - V_{bn}\right) \bullet I_c$$
$$P = V_{an} \bullet I_a + V_{cn} \bullet I_c - V_{bn} \bullet \left(I_a + I_c\right)$$

and since $(I_a + I_c) = -I_b$, we can write:

$$P = V_{an} \bullet I_a + V_{cn} \bullet I_c + V_{bn} \bullet I_b$$

whence: $P = |V_a||I_a|\cos(\phi_a) + |V_b||I_b|\cos(\phi_b) + |V_c||I_c|\cos(\phi_c)$

This result is similar to the three-element metering configuration and therefore the two-element metering configuration will also operate correctly for both positive and negative sequence components.

10.2.2 Metering in the Presence of Harmonics

Non-linear loads within distribution networks consume currents rich in harmonics, and because the network impedance is non-zero, these harmonics generate small harmonic voltages throughout the distribution network. Thus the non-linear load of one customer creates harmonic disturbances for others.

Since the loads generating harmonic currents are usually balanced, the harmonic voltage sources so created will also be roughly balanced as well. These voltage sources will cause linear loads to consume small amounts of harmonic current, since they exist in addition to the fundamental voltage throughout the network. Therefore all customers consume tiny amounts of harmonic power, in addition to that supplied at the fundamental.

Static energy meters are generally capable of measuring the harmonic components in the total energy supplied, to within the bandwidth constraints of the meter. Meters are usually provided with a bandwidth of about 4–5 kHz and therefore they can include harmonic energy contributions up to the 80th or perhaps the 100th harmonic.

It should be stressed that harmonic power can only arise as the result of *both* voltage and current *at the same harmonic frequency*. Harmonic currents flowing in a network where there is little or no voltage distortion will not significantly contribute to the net power flow, but they will contribute to the harmonic VAr flow and thus to a reduction in the total power factor, which is measured as follows:

$$\text{Total Power Factor} = \frac{V_{RMS} I_{RMS}}{\text{Active Power}} = \frac{S}{P}$$

where $S = V_{RMS} I_{RMS}$ are the volt-amps consumed, including all harmonic components present.

The total power P per phase in the presence of harmonics can be written as:

$$P = P_1 + P_h$$

$$P = V_1 I_1 \cos(\phi_1) + \text{Average}\left[\sum_{n=2}^{\infty} \hat{v}_n \cos(n\omega t).\sum_{n=2}^{\infty} \hat{i}_n \cos(n\omega t + \phi_n)\right]$$

$$P = V_1 I_1 \cos(\phi_1) + \sum_{2}^{\infty} V_n I_n \cos(\phi_n)$$

where P_1 is the fundamental power, P_h is the total harmonic power and V_n, I_n and ϕ_n are the nth order RMS harmonic voltage, current and phase angle, respectively.

Because the harmonic currents generally decrease as $1/n$ or sometimes as $1/n^2$, the power from higher-order harmonics will generally contribute very little to the total energy consumed. In addition, it is generally only the odd harmonics that are present in significant quantities.

10.2.3 Directional Power Flow

We saw in Chapter 3 that *active* and *reactive* power flows are independent of each other, and they can flow in different directions simultaneously. Energy meters segregate the active and reactive energies into separate import and export registers, so that imported energy is recorded separately to that exported. For example, residential customers with solar generation capacity, can import active power from the bus during some metering intervals and export it during others. By separating import and export flows, the retailer can pay the customer for exported energy and charge for energy imported. Different rates may also be applied to these power flows.

Since the active power delivered to a load is given by $P = 3VI\cos(\phi)$, the power flow depends on the sign of the power factor. It will be positive when ϕ lies between +90 and −90°. Thus when the load current lies in quadrants 1 or 4, positive power flow to the load occurs. In this situation, the load can be described as importing power from the bus, or alternatively the bus is exporting power to the load.

When ϕ is greater than 90° or less than −90°, cos (ϕ) changes sign and the power flow becomes negative. So when the current lies in quadrants 2 or 3 the load delivers power to the bus, i.e. it generates energy. In this situation the load is exporting power to the bus or alternatively, the bus is importing power from the load.

The reactive flow is likewise proportional to $\sin(\phi)$, and since most industrial loads operate at lagging phase angles, where ϕ is positive, they import reactive power from the bus. So for currents in quadrants 3 or 4, the bus exports reactive power to the load. This flow is usually defined as being positive. Alternatively, when the current leads the voltage, as in quadrants 1 and 2, the load generates or exports reactive power to the bus. Flow in this direction is usually defined as being negative. Figure 10.12 illustrates these power flows from the perspective of the load.

The definitions of import and export depend upon the perspective of either the bus or the load. Unfortunately, some meter manufacturers define power flows from the point of view of the bus while others do so from the point of view of the load and

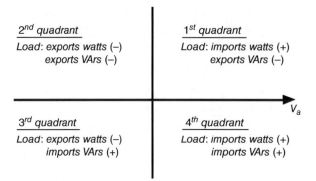

Figure 10.12 Power flow definitions from the perspective of the load.

therefore depending on the context, the terms import and export can appear to take on opposite meanings.

10.2.4 Measurement of Apparent Energy and Reactive Energy

The *apparent energy* (kVAh) is generally calculated from the integral of the RMS voltage and RMS current evaluated over a metering interval. Most static meters effectively carry out a true RMS calculation, so the effects of any harmonics present are included, up to the bandwidth of the meter concerned. This method of calculation is well suited, since the $V_{RMS}I_{RMS}$ product has no sign.

Reactive energy on the other hand is usually measured using the same metering element used to determine active energy, but either the voltage or the current must be phase shifted by 90°. Of these it is generally simpler to phase shift the voltage applied to each metering element. This can be done by applying a scaled version of the line voltage opposite the phase in question. For example consider the *a* phase element; if $1/\sqrt{3}$ times the *bc* line voltage is applied to this element as depicted in Figure 10.13, we find that the reactive power, Q_a is metered:

$$Q_a = \frac{|V_{bc}||I_a|\cos(90-\phi_a)}{\sqrt{3}} = |V_a||I_a|\sin(\phi_a)$$

Therefore by supplying each metering element with a scaled line voltage, the same element configuration used to calculate the active energy can also be used to evaluate the reactive energy consumed throughout each metering interval. This method of calculation also makes it possible for the meter to record the sign of the VAr flow. Accordingly, imported VArh are recorded in a different register from exported VArh, just as in the case of the active energy.

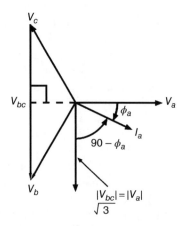

Figure 10.13 Technique of 90° phase shifting for VAr measurement.

Reactive energy could also be evaluated from the power triangle relationship $Q = \sqrt{S^2 - P^2}$, and while this equation would provide a useful numerical result, including the effect of any harmonics present, it cannot provide the important sign information for Q. Therefore most meters use the former method to evaluate VArh.

10.3   Metering Errors

Non-Blondel compliant metering requires that we accept that under some circumstances, errors will be introduced in the energy measured. Blondel compliant metering is therefore usually preferred, since the energy measured should be correct, regardless of the degree of unbalance, either in the supply voltages or in the load currents. There are however also other sources of error that must be considered.

As it happens, all metering installations include inherent errors, and these errors are independent of the metering configuration and are introduced by imperfections in the equipment used. The main sources of these include the current and voltage transformers used to provide the meter with a replica of what is happening in the primary circuit, as well as the energy meter itself. The instrument transformers do not provide a perfect representation of the primary quantities. As we have seen in earlier chapters they introduce both magnitude and phase errors into the secondary currents and voltages supplied to the meter. In the case of voltage transformers, the wiring running from the VT cubicle to the meter panel can also introduce additional magnitude and phase errors in the potentials seen by the meter.

Tables 10.1 and 10.2 outline the magnitude and phase error limits of metering class current and voltage transformers, as defined by the 2012 IEC standards IEC 61869-2 and IEC 61869-3 for inductive current and voltage transformers respectively. IEC 61869-2 prescribes six accuracy classes for CTs (class 0.1, 0.2, 0.5, 1.0, 3.0 and 5.0), in addition to sub-classes 0.2S & 0.5S, while IEC 61689-3 prescribes five classes for VTs. The magnitude error of each is expressed in per cent and the phase error is expressed either in minutes of arc or sometimes in centiradians. (1 Crad = 34.4 minutes).

Table 10.1 CT error limits.

| Accuracy class | Magnitude error limits (%) (At percentage of rated current) | | | | | Phase error limits (min) (At percentage of rated current) | | | | |
	1%	5%	20%	100%	120%	1%	5%	20%	100%	120%
0.2S	0.75	0.35	0.2	0.2	0.2	30	15	10	10	10
0.5S	1.5	0.75	0.5	0.5	0.5	90	45	30	30	30
0.1	Not specified	0.4	0.2	0.1	0.1	Not specified	15	8	5	5
0.2	Not specified	0.75	0.35	0.2	0.2	Not specified	30	15	10	10
0.5	Not specified	1.5	0.75	0.5	0.5	Not specified	90	45	30	30
1.0	Not specified	3.0	1.5	1.0	1.0	Not specified	180	90	60	60

(IEC 61869-2 Copyright © 2012 IEC Geneva, Switzerland. www.iec.ch)

Table 10.2 VT error limits.

Accuracy class	Magnitude error limits (per cent)	Phase error limits (minutes)
0.1	±0.1	±5
0.2	±0.2	±10
0.5	±0.5	±20
1.0	±1.0	±40
3.0	±3.0	Not specified

(IEC 61869-3 Copyright © 2011 IEC Geneva, Switzerland. www.iec.ch)

Of these, class 0.1 devices are generally laboratory grade instruments, class 0.2, 0.5 and 1.0 are routinely used for energy metering, while classes 3.0 and 5.0 are normally used for panel instrumentation.

The 0.2S and 0.5S sub-classes are for precision metering applications, and these require the CT to perform from 1% of rated current up to a minimum of 120% of rated current, or in the case of an *extended current rating*, up to the specified accuracy limit, which may be 200–250% of the nominal device rating. IEC 61869-2 also provides for an *extended burden range* to be specified by the manufacturer. In this event, the CT errors must lie within the limits specified over a burden range from 1 VA to the rated burden of the CT. When an extended burden range is specified, the standard requires that the CT be tested with a resistive burden (i.e. unity power factor). This requirement is particularly useful in static energy metering applications, which impose very small and generally resistive burdens on metering CTs.

Under the IEC standards, there is no inter-relationship between magnitude and phase error limits, but this in not true of the IEEE instrument transformer standard applicable in the USA, as we shall see later in this chapter.

The quantity of energy to be metered determines the degree to which metering errors can be tolerated and hence the class of instrument transformer required. Large electrical loads must generally be metered with little error, while larger metering errors can be accepted for smaller loads. For example, under the Australian *National Electricity Rules*, metering installations with a throughput of more than 1000 GWh per annum must be metered to within an accuracy of ±0.5%. Such installations require the use of high quality class 0.2 CTs, VTs and meters. However, where the energy throughput is less than 100 GWh per annum, the total metering error need only be measured to within ±1.5% and here standard class 0.5 devices can be used instead.

The meter itself is also slightly imperfect; it introduces small errors into the energy measured during each metering interval, although these are often less than those accruing from the instrument transformers. Unlike the current and voltage transformers, the meter error can be considered a scalar quantity, which reflects the difference between the metered energy and the true energy, as a percentage of the true energy measured:

$$Meter\ Error = \frac{(Metered\ Energy - True\ Energy)}{True\ Energy} \times 100\%$$

Table 10.3 IEC62053.22 Percentage error limits for polyphase meters with balanced loads.

Value of current	Power factor	Percentage error limits for meters of class	
		0.2S	0.5S
$0.01I_n \leq I \leq 0.05I_{max}$	1	±0.4	±1.0
$0.05I_n \leq I \leq I_{max}$	1	±0.2	±0.5
$0.02I_n \leq I \leq 0.1I_{max}$	0.5 inductive	±0.5	±1.0
	0.8 capacitive	±0.5	±1.0
$0.1I_n \leq I \leq I_{max}$	0.5 inductive	±0.3	±0.6
	0.8 capacitive	±0.3	±0.6
When specially requested by the user from: $0.1I_n \leq I \leq I_{max}$	0.25 inductive	±0.5	±1.0
	0.5 capacitive	±0.5	±1.0

(IEC 62053–22 Copyright © 2003 IEC Geneva, Switzerland. www.iec.ch)

Here the true energy is that read from a calibration instrument which we assume to be ideal. So a meter with a +0.1% error over-reads an ideal instrument by 0.1%. The IEC meter standard IEC62053.22 defines the error limits for electronic special class energy meters, which are summarised in Table 10.3.

The class 0.2S and 0.5S meter accuracies are limited to ±0.2% and ±0.5% respectively for resistive loads over most of the applicable current range of these meters (see Table 10.3 row 2). As the load power factor decreases, the permitted meter error increases to reflect the increasing sensitivity of the power measurement to the phase angle, ϕ. When ϕ approaches 90°, for example, a small error in determining ϕ results in a large error in the energy metered, and the meter accuracy diminishes markedly. This effect is not obvious from the table since it does not specify power factors less than 0.5.

Electronic meters generally function correctly at up to three or four times the rated meter current I_n, therefore I_{max} in Table 10.3 is typically $4I_n$. This capability is particularly useful when a CT is able to be used up to its thermal limit.

10.3.1 Overall Error Equation

The error associated with a metering installation is a function of *all* the errors associated with the CTs and VTs, as well as those of the meter itself. The following analysis leads to a simple result that can be used to evaluate the overall error of a metering installation, as defined by Equation (10.18).

$$Overall\ Error = \frac{(Measured\ Energy - Correct\ Energy)}{Correct\ Energy} \times 100\% \tag{10.18}$$

We begin our analysis by writing the actual voltage and current for the *a* phase as:

$$V_{actual} = \hat{v} \sin(\omega t)$$

$$I_{actual} = \hat{i} \sin(\omega t - \phi)$$

where ϕ is the phase angle between the current and voltage, and is positive if the current lags the voltage.

The measured voltage and current presented to the a phase meter element will include errors from associated voltage and current transformers, and therefore can be expressed as:

$$V_{measured} = \left(1 + \frac{e_v}{100}\right)\hat{v}\sin(\omega t + \gamma)$$

$$I_{measured} = \left(1 + \frac{e_i}{100}\right)\hat{i}\sin(\omega t - \phi + \beta)$$

where β is the phase error associated with the current transformer and γ is the phase error associated with the voltage transformer (expressed in minutes), and e_v and e_i are the associated magnitude errors respectively (expressed in per cent). Note that e_v and γ should include the effects of any VT wiring errors that may be significant.

An *ideal* meter element (i.e. one with zero internal error), will compute the measured power for this phase as the product of the RMS voltage, current and the cosine of the angle between them. Therefore the power measured by this metering element is:

$$P_m = \left(1 + \frac{e_v}{100}\right)\left(1 + \frac{e_i}{100}\right)|V||I|\left[\cos(-\phi + \beta - \gamma)\right]$$

$$= \left(1 + \frac{e_v}{100}\right)\left(1 + \frac{e_i}{100}\right)|V||I|\left[\cos(\phi)\cos(\beta - \gamma) + \sin(\phi)\sin(\beta - \gamma)\right]$$

$$\cong \left(1 + \frac{e_v}{100} + \frac{e_i}{100}\right)|V||I|\left[\cos(\phi)\cos(\beta - \gamma) + \sin(\phi)\sin(\beta - \gamma)\right]$$

$$= \left(|V||I| + |v||I| + |V||i|\right)\left(\cos(\phi)\cos(\beta - \gamma) + \sin(\phi)\sin(\beta - \gamma)\right)$$

where $|v| = \left(\frac{e_v}{100}\right)|V|$, is the absolute voltage error in *volts* and $|i| = \left(\frac{e_i}{100}\right)|I|$, is the absolute current error in *amps*.

The correct power recorded by a meter supplied from ideal instrument transformers, P_c is given by:

$$P_c = |V||I|\cos(\phi)$$

Since $e_i, e_v \ll 1$ and the phase errors β and γ are small, any second-order terms will be very small and can be neglected, thus the fractional error can be shown to be:

$$\frac{(P_m - P_c)}{P_c} = \frac{\Delta P}{P_c}$$

$$= \frac{|V||I|(\cos(\phi)\cos(\beta - \gamma) + |V||I|\sin(\phi)\sin(\beta - \gamma)) + |v||I|\cos(\phi) + |V||i|\cos(\phi) - |V||I|\cos(\phi)}{|V||I|\cos(\phi)}$$

Simplifying and removing any additional second order terms yields:

$$\frac{\Delta P}{P_c} = \tan(\phi)\sin(\beta - \gamma) + \frac{e_v}{100} + \frac{e_i}{100} = \varepsilon \tag{10.19}$$

where ε can be defined as the fractional error, $\dfrac{\Delta P}{P_c}$ for *one element* within the meter. Alternatively we can write:

$$P_m = P_c\left(1+\varepsilon\right)$$

Clearly ε relates only to the error associated with the instrument transformers of one phase, since the meter has thus far been assumed error free.

10.3.2 Effect of Meter Error

Equation 10.19 applies in the case where an ideal meter is used. In the general case, where the meter error is non-zero, we must include a new term to account for this:

$$P_m = P_c\left(1+\varepsilon\right)\left(1+e_m/100\right) \tag{10.20}$$

where e_m is the meter error in per cent.

Equation (10.20) can be simplified as follows, by neglecting the second order term $\varepsilon\,e_m/100$:

$$P_m = P_c\left(1+\varepsilon+\frac{e_m}{100}\right) \tag{10.21}$$

or,

$$\frac{\Delta P}{P_c} = \left(\varepsilon+\frac{e_m}{100}\right)$$

Equation (10.21) applies to the error associated with one element within a metering installation. How we combine the error results of all metering elements depends on the metering configuration used. We will demonstrate this for two Blondel compliant circuits: the three-element (four-wire) and the two-element (three-wire) configurations.

10.3.3 Overall Error as Applied to the Three-Element (Four-Wire) Metering Configuration

In the case of three-element metering, each element contributes approximately equally to the total power measured and therefore to the overall error as well, which is given by:

$$\frac{\Delta P}{P_c} = \frac{\left(\varepsilon_a+\varepsilon_b+\varepsilon_c\right)}{3}+\frac{e_m}{100} \tag{10.22}$$

where ε_a, ε_b and ε_c are the fractional errors for the a, b and c phases respectively, according to Equation (10.19). Here the meter error e_m, applies to the total energy recorded by the meter.

Finally, the overall error is equal to $\dfrac{\Delta P}{P_c}\times 100\%$,

$$Overall\ Error = \left[\frac{\left(\varepsilon_a+\varepsilon_b+\varepsilon_c\right)}{3}.100+e_m\right]\%$$

10.3.4 Overall Error as Applied to Two-Element (Three-Wire) Metering Configuration

In the case of two-element metering, the power measured is given by:

$$P_c = |V_{ab}||I_a|\cos(\phi_a + 30) + |V_{cb}||I_c|\cos(\phi_c - 30)$$

Assuming a reasonably balanced load, it is easy to see that the power measured by each element under this configuration is *not* equal. For example, if ϕ_a and ϕ_b are each equal to about 30°, then the power measured by the *ab* element will be only half that measured by the *cb* element. It is therefore necessary to weight the fractional error contribution from each element accordingly. This can be achieved as follows:

$$\frac{\Delta P}{P_c} = \frac{\cos(\phi_a + 30)}{\cos(\phi_a + 30) + \cos(\phi_c - 30)}\varepsilon_{ab} + \frac{\cos(\phi_c - 30)}{\cos(\phi_a + 30) + \cos(\phi_c - 30)}\varepsilon_{cb}$$

If we include the metering error e_m in per cent, the overall error becomes:

$$Overall\ Error = \left[\frac{\cos(\phi_a + 30)}{\cos(\phi_a + 30) + \cos(\phi_c - 30)}\varepsilon_{ab}\right.$$
$$\left. + \frac{\cos(\phi_c - 30)}{\cos(\phi_a + 30) + \cos(\phi_c - 30)}\varepsilon_{cb}\right].100 + e_m\ \% \tag{10.23}$$

The weighting factor applied to each of the element errors ε_{ab} and ε_{cb} in Equation (10.23), are plotted in Figure 10.14 over the normal range of load power factors. These weightings are only equal at unity PF, and as the PF falls we see that the weighting factor relating to the *cb* element increases, while that for the *ab* element decreases.

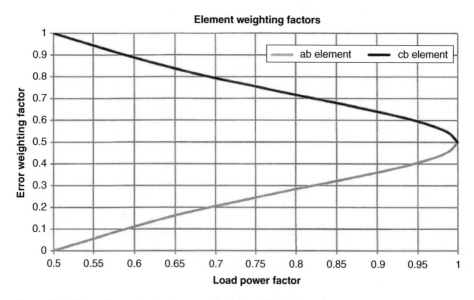

Figure 10.14 Two-element metering, overall error weighting factors.

This highlights one slight disadvantage with two-element metering. Because the power measured by each element is generally not equal, the *cb* element will usually contribute significantly more to the overall error than will the *ab* element. It is therefore important to ensure that the *cb* instrument transformers are well within their appropriate error class limits.

10.4 Ratio Correction Factors

In the USA the concept of a ratio correction factor is used to quantify magnitude errors in voltage and current transformers. We have seen that the magnitude error in both current and voltage transformers means that the actual ratio of the device does not exactly match its stated ratio. The ratio correction factor (RCF) is defined as the factor by which the stated ratio of an instrument transformer must be multiplied to obtain the actual ratio. Therefore the RCF can be used to correct the measured voltage in a VT for the effects of magnitude error as follows:

$$RCF_v = \frac{|V_c|}{|V_m|}$$

Similarly for a CT:

$$RCF_i = \frac{|I_c|}{|I_m|}$$

where V_m, is the voltage measured by the VT in question, and V_c, is this quantity corrected for the magnitude error introduced by the VT. Similar definitions apply for I_m and I_c.
From these equations it is easy to show that:

$$RCF = \frac{1}{\left(1 + \dfrac{e}{100}\right)}$$

Thus if the magnitude error, $e = +0.2\%$, then the associated RCF becomes 0.998.

10.4.1 Overall Error in Terms of Ratio Correction Factors

We now derive a result analogous to Equation (10.19) for overall error in terms of ratio correction factors of the associated CT and VT. Beginning with the measured power P_m in one element of an installation, we can write:

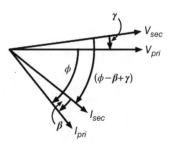

$$P_m = |V_m||I_m|\cos(\phi - \beta + \gamma)$$

where $|V_m|$ is the magnitude of the measured voltage and $|I_m|$ is the magnitude of the measured current, β is the phase error associated with the CT and γ is the phase error associated with the VT. Figure 10.15 shows

Figure 10.15 Angle definitions.

the load phase angle as seen by the meter. (For simplicity unity CT and VT ratios have been assumed.)

The correct power P_c is: therefore:

$$P_c = |V_m||I_m|\cos(\phi)RCF_v RCF_i$$

where RCF_v and RCF_i are the RCFs of the voltage and current transformers respectively. We can therefore write:

$$P_c = |V_m||I_m|\cos(\phi - \beta + \gamma)\frac{\cos(\phi)RCF_v RCF_i}{\cos(\phi - \beta + \gamma)}$$

$$= P_m \frac{\cos(\phi)RCF_v RCF_i}{\cos(\phi - \beta + \gamma)}$$

$$= P_m FCF$$

where FCF is the final correction factor which, when multiplied by the measured power, provides the correct power for the metering element under consideration.

$$FCF = \frac{\cos(\phi)RCF_v RCF_i}{\cos(\phi - \beta + \gamma)} \qquad (10.24)$$

It can easily be shown that the quantity $\frac{1}{1+\varepsilon}$ is also the factor by which the measured power must be multiplied to obtain the corrected power, therefore:

$$FCF = \frac{1}{1+\varepsilon}$$

Example

The CT and VT magnitude and phase errors and corresponding RCFs associated with the one phase of a particular three-phase load, are outlined in Table 10.4:

From this data the overall percentage error for the associated metering element can be independently calculated using Equations (10.19) and (10.24), as shown in Table 10.5.

Thus, regardless of the method used, the percentage error is virtually the same. It is interesting to note that *the overall error increases as the load power factor decreases.* (This is due to the $\tan(\phi)\sin(\phi_i - \phi_v)$ term in Equation ((10.19).))

Table 10.4 magnitude and phase errors and corresponding RCFs.

Parameter	CT	VT
magnitude error (%)	+0.2	+0.3
phase error (min)	+5	-7
RCF	0.998	0.997

Table 10.5 Calculation of overall percentage error.

	Percentage error	
Method	$\cos(\phi)=0.866$	$\cos(\phi)=0.5$
$\varepsilon.100\%$	0.701%	1.10%
$\left(\dfrac{1}{FCF}-1\right).100\%$	0.704%	1.11%

10.4.2 IEEE Standard C57.13: Standard Requirements for Instrument Transformers

American standard IEEE C57.13 defines three standard accuracy classes for metering current and voltage transformers. The accuracy requirements for these appear in Table 10.6, and rather than specifying magnitude and phase error limits, this standard uses the concept of a *transformer correction factor* (TCF), which combines the magnitude and phase errors into a single parameter. This is the factor which must be applied to a meter reading to correct for the magnitude and phase errors associated with the instrument transformer in question.

The TCF has a similar form to the final correction factor described above, and applies to a metering installation where *only the transformer in question contains errors*. Therefore the TCF for a current transformer is given by:

$$TCF_{CT} = \frac{\cos(\phi)RCF_i}{\cos(\phi-\beta)} \tag{10.25}$$

and for a voltage transformer:

$$TCF_{VT} = \frac{\cos(\phi)RCF_v}{\cos(\phi+\gamma)} \tag{10.26}$$

Table 10.6 IEEE C57.13 Transformer correction factors for voltage and current transformers.

	Voltage transformers		Current transformers				
	(90% to 110% rated voltage)		@ 100% rated current		@ 10% rated current		Metered load power factor
Accuracy class	Minimum	Maximum	Minimum	Maximum	Minimum	Maximum	limits
0.3	0.997	1.003	0.997	1.003	0.994	1.006	0.6 – 1.0
0.6	0.994	1.006	0.994	1.006	0.988	1.012	0.6 – 1.0
1.2	0.988	1.012	0.988	1.012	0.976	1.024	0.6 – 1.0

(Adapted and reprinted with permission from IEEE. Copyright IEEE C57.13 (2008) All rights reserved. Permission for further use of this material must be obtained from IEEE. Requests may be sent to stds-ipr@ieee.org)

The final correction factor for a particular phase can be expressed to a good approximation as the product of the current and voltage transformer correction factors:

$$FCF \approx TCF_{VT} . TCF_{CT}$$

In the light of these equations, IEEE C57.13 also defines a range of power factors applicable to the load being metered. It aims to limit the metering error introduced by an individual current or voltage transformer, to the magnitude of the specified TCF, for load power factors between 0.6 and 1.0. This implies that a relationship must also exist between a transformer's RCF and its corresponding phase error which, as we will show, is indeed the case.

As shown in the example above and in Equation (10.19), the overall error increases as the power factor reduces, therefore the largest error will occur when the load power factor is at its minimum of 0.6, i.e. when the angle between the current and voltage in the load circuit is 53.13°. If we substitute this value into Equations (10.25) and (10.26) we find that:

$$\frac{0.6RCF_i}{TCF_{CT}} = \cos(53.13° - \beta) \quad \text{and} \quad \frac{0.6RCF_v}{TCF_{VT}} = \cos(53.13° + \gamma)$$

These equations summarise the inter-relationship between RCF and the phase error and provide *limits* for these values corresponding to a particular TCF. They can be simplified with a slight approximation, into the two linear equations given in the standard, which relate the RCF and phase error of the transformer in question to its associated TCF. Thus:

$$\beta = 2600\left(RCF_i - TCF_{CT}\right) \tag{10.27}$$

And

$$\gamma = 2600\left(TCF_{VT} - RCF_v\right) \tag{10.28}$$

Equations (10.27) and (10.28) enable the TCF to be calculated when the RCF and phase error are known. The limits for this can be expressed graphically in the form of the parallelograms shown in Figure 10.16. Provided that the errors given on the transformer's test certificate lie within the appropriate parallelogram, the device is deemed compliant. However, if the RCF, the phase error or the primary circuit power factor lie outside the prescribed limits, the resulting transformer correction factor will not comply with the standard.

IEEE C57.13 also prescribes a range of *standard burdens* for both current and voltage transformers at which the associated TCFs are measured. Burdens for metering class current transformers range between 2.5 and 45 VA and are specified at a burden power factor of 0.9, while for voltage transformers standard burdens lie between 12.5 and 400 VA. Since the burden presented to voltage transformers by both electronic energy meters and protective relays tends to be quite low, modern VTs are seldom rated beyond 50 VA.

It is not necessary that the standard burden be applied when a current or voltage transformer is used in a metering application, and in practice the actual burden will generally be very much lower than this. The rated burden simply represents the maximum for which the transformer will remain compliant. At burdens lower that the specified maximum, the transformer will also generally remain class compliant.

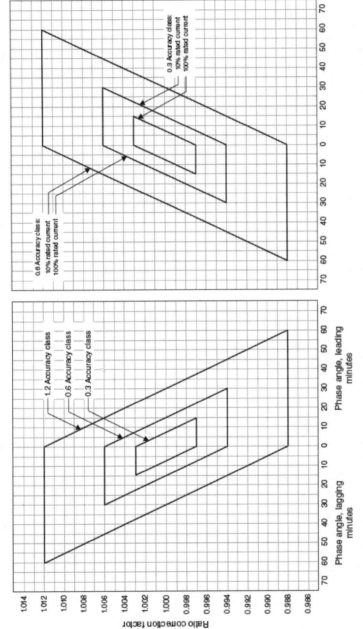

Figure 10.16 IEEE C57.13 metering class accuracy limits: VT (left) and CT (right). (Adapted and reprinted with permission from IEEE. Copyright IEEE C57.13 (2008) All rights reserved. Permission for further use of this material must be obtained from IEEE. Requests may be sent to stds-ipr@ieee.org).

Table 10.7 TCF limits for class 0.15 and 0.15S instrument transformers.

Accuracy class	Voltage transformers (90% to 110% rated voltage)		Current transformers				Metered load power factor limits
			@ 100% rated current		@ 5% rated current		
	Minimum	Maximum	Minimum	Maximum	Minimum	Maximum	
0.15	0.9985	1.0015	0.9985	1.0015	0.997	1.003	0.6–1.0
0.15S	N/A	N/A	0.9985	1.0015	0.9985	1.0015	0.6–1.0

(Adapted and reprinted with permission from IEEE. Copyright IEEE C57.13.6 (2010) All rights reserved. Permission for further use of this material must be obtained from IEEE. Requests may be sent to stds-ipr@ieee.org)

10.4.3 IEEE Standard C57.13.6 High Accuracy Instrument Transformers

In 2005, the IEEE released a supplement to IEEE C57.13 in the form of IEEE C57.13.6, which defines two new high accuracy classes of instrument transformer, 0.15 and 0.15S for current transformers and 0.15 for voltage transformers. These reflect the general move to static energy meters from induction disc types, as well as a need for higher accuracy when metering very large loads.

Because the burdens imposed by static meters tend to be small and resistive, this standard also defines new 5 VA and 1 VA resistive CT burdens, which must be applied when these transformers are tested. The operational burden applied to these CTs must be equal to or less than these, and since the secondary cabling usually constitutes a substantial fraction of the burden, care must be used in its specification to ensure that the CT is not overburdened.

Class 0.15 current transformers must also comply with the specified TCF requirements over an extended range from 5% to 100% rated current, as shown in Table 10.7, and up to the thermal rating current of the transformer where this is greater than the rated current. Further, the 0.15S class devices must maintain the same TCF from 5% rated current to 100% and possibly beyond.

Voltage transformers must comply with the TCF requirements from 0 VA up to the rated burden for the device, for voltages between 90 and 100% of nominal.

10.5 Reactive Power Measurement Error

The same analysis used to derive the error equations for active power can also be applied to the measurement of reactive power, and will produce the following results:

$$\frac{\Delta VAr}{VAr_c} = \left[\cot(\phi)\sin(\gamma - \beta) + \frac{e_i}{100} + \frac{e_v}{100} \right] = \varepsilon_{VAr} \tag{10.29}$$

In terms of ratio correction factors, this error can be expressed as:

$$FCF_{VAr} = \frac{\sin(\phi) RCF_v\, RCF_i}{\sin(\phi - \beta + \gamma)} \tag{10.30}$$

For three-element metering installations, the accumulated reactive error is given by:

$$\text{Overall Reactive Error} = \left[\frac{\left(\varepsilon_a + \varepsilon_b + \varepsilon_c \right)}{3} .100 + e_m \right] \%$$

And for *two-element* installations it can be expressed as:

$$\text{Overall Reactive Error} = \left[\frac{\sin\left(\phi_a + 30 \right)}{\sin\left(\phi_a + 30 \right) + \sin\left(\phi_c - 30 \right)} \left(\varepsilon_{ab} \right) \right.$$
$$\left. + \frac{\sin\left(\phi_c - 30 \right)}{\sin\left(\phi_a + 30 \right) + \sin\left(\phi_c - 30 \right)} \left(\varepsilon_{cb} \right) \right].100 + e_m \%$$

where ε_{ab} and ε_{cb} are the fractional errors associated with the *ab* and *cb* elements respectively. Note that the metering error for reactive energy will generally be different from that for active energy.

10.6   Evaluation of the Overall Error for an Installation

When evaluating the overall error for a metering installation it is necessary to choose a load current and voltage, together with the corresponding power factor which are representative of the majority of the energy flowing. For this scenario it is then necessary to compute both the CT and VT errors corresponding to the applied burdens, from previously obtained test results. VT errors can be simply linearly interpolated according to the applied burden and bus voltage. CT errors however are generally not linearly related to either the applied burden or to the secondary current flowing, largely because of the non-linear nature of the magnetising admittance. Despite this, because of the limited quantity of error information usually available, a linear interpolation of test results may be the best that can be achieved.

In some cases, where cyclic load fluctuations occur, it may be necessary to compute overall error figures corresponding to several different load scenarios. In such cases these figures must then be combined into a representative overall error figure. This can be done by weighting each according to its energy contribution.

For example, if a generator installation exhibits an overall error of 0.2% when operating at 95% of its rated load and 1.5% error when operating at 20% rated load *and* it operates at 95% of rated load for 75% of the time and the remainder at 20% load, then the overall error can be calculated as follows:

$$\textit{Energy Delivered @ 0.95 pu rated load} = 0.95 \times 0.75 = 0.712 \, \text{pu}$$
$$\textit{Energy Delivered @ 0.20 pu rated load} = 0.20 \times 0.25 = 0.05 \, \text{pu}$$

Thus the overall error becomes:

$$\text{Overall error} = 0.2\% \times \frac{0.712}{\left(0.712 + 0.05 \right)} + 1.5\% \times \frac{0.05}{\left(0.712 + 0.05 \right)} = 0.285\%$$

With this definition the overall error is 0.285%, therefore the total energy metered by the installation will exceed the true energy by 0.285%. From this calculation it can be seen that despite potentially large errors when operating at low output levels, the component of the total energy contributed by this scenario is low. As a result, so is its contribution to the overall error.

10.6.1 Error Correction within the Meter

Of the error equations derived above, those associated with active power are usually the most important simply because this is the quantity that customers pay for. In most jurisdictions there are rules limiting the size of the overall error applicable to metering installations and generally the use of the appropriate class of instrument transformers and energy meter will ensure that this is achieved. However, occasionally instrument errors accumulate adversely, and despite the use of the correct instrument class, overall error compliance may not be achieved. When this occurs it is possible to provide error correction within the meter itself. Most meter manufacturers provide meters with the ability to correct for VT and CT errors during computation. To achieve this, the user must provide the meter with a look-up table of the appropriate magnitude and phase correction factors. With this information the meter can compensate for known VT and CT errors, thereby significantly reducing the installation's overall error.

The ratio correction factor is used to correct magnitude errors, while phase errors are nulled by applying a phase correction equal to the opposite of the known phase error. It is necessary for the user to interpolate measured or calculated errors over a range of operational voltages at the burden applied to a VT, in order to evaluate these correction factors. Similarly, CT correction factors can be found by interpolating measured or calculated error data over a range of operational currents, at the metering burden applied to the CT. The accurate calculation of these correction factors requires a considerable amount of data about the VT or CT in question; however, once it is applied to the meter, the overall error for the installation will generally be reduced to about the error inherent in the meter itself.

10.6.2 Limits of Practical Energy Measurement

As discussed earlier, the active overall error tends to increase as the power factor of the metered load decreases. When taken to extremes, this effect can give rise to metering errors large enough to almost render the metered data useless. It may therefore not be possible to accurately measure a small quantity of active energy against a large quantity of reactive energy, and vice versa. The reason for this effect is the $\tan(\phi)\sin(\beta-\gamma)$ term in Equation (10.19), repeated here:

$$\frac{\Delta P}{P_c} = \tan(\phi)\sin(\beta-\gamma) + \frac{e_v}{100} + \frac{e_i}{100} = \varepsilon$$

As ϕ approaches 90° this term becomes very large, since the multiplicative $\sin(\beta-\gamma)$ term is generally non-zero. The resulting overall error for the installation then becomes large and greatly degrades the accuracy with which the energy can be measured. This is illustrated in Figure 10.17 where the overall error of a particular metering installation is plotted as a function of the power factor of the associated load.

A similar effect arises in the reactive overall error equation as the load power factor tends towards unity due to the $\cot(\phi)\sin(\gamma-\beta)$ term in Equation (10.29), repeated here:

$$\frac{\Delta VAr}{VAr_c} = \left[\cot(\phi)\sin(\gamma-\beta) + \frac{e_i}{100} + \frac{e_v}{100}\right] = \varepsilon_{VAr}$$

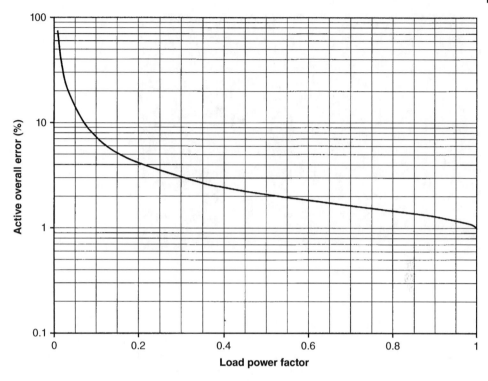

Figure 10.17 Typical variations in active overall error with load power factor.

This means that the measurement of a small quantity of reactive energy may be difficult against a large quantity of active energy.

10.7 Commissioning and Auditing of Metering Installations

The costs associated with lost or incorrect metering data can be high, either for the customer or the energy retailer. As a result, large electrical loads are frequently equipped with *check metering*. This is effectively a duplicate metering installation (ideally generated from independent current and voltage transformers), so that should a fault arise in the main circuit, metered data can still be obtained. Metering installations are usually audited every few years to verify that they continue to operate correctly.

Particular care must be taken when commissioning a new metering installation to ensure that both the instrument transformers and the meter have been configured correctly. Inadvertent wiring errors can result in the need for one party to repay considerable sums, particularly if an error has remained undetected for some time. For example, the reversal of the meter connections of just one CT can result in only one-third of the energy supplied being paid for by the customer.

To aid in the commissioning and auditing processes, most static meters are capable of being interrogated and can provide the metering technician with a phasor diagram of the load parameters, as seen by the meter. This makes the task of checking the

Figure 10.18 Typical metering test block, cover removed. (The CT links are on the left, and colour coded voltage terminals are on the right).

installation straightforward, since with a balanced or near balanced load, each phase current should have approximately the same phase relationship with its phase voltage. The reversal of a CT's connections at the meter, or the association of a CT with an incorrect phase voltage, will therefore immediately become apparent.

It is also desirable that the installer have some understanding of the load being metered. For example, does it have generation capacity? If so, the phase current might lie in the 2nd or 3rd quadrant (i.e. almost 180° out of phase with the voltage). Is PFC equipment provided? If so, the current may occasionally lead the voltage, particularly when the load is small. This type of information can be very useful when interpreting the phasor diagram.

Most metering installations also include the provision of a *metering test block*, usually adjacent to the meter, as shown in Figure 10.18. The test block is inserted into the metering circuit immediately prior to the meter. It is arranged with CT links on the left and colour-coded VT terminals on the right. There is a one-to-one correspondence between the currents and voltages on the test block and those supplied to each metering element.

The test block facilitates both meter testing and auditing of the metering installation. It permits the measurement of CT secondary currents or the phase angle between a voltage and its respective current. This is achieved by inserting either an ammeter or a *phase angle meter* (PAM) into the CT circuit and then opening the CT links. (In the case of phase angle measurements, the PAM must also be provided with the element voltage, as well as its current, in order to display the phase angle between them.)

The test block is also useful when the meter is to be tested, whereupon the CT secondaries are shorted at the test block and isolated from the meter, using the links provided. Currents from a *watt-hour standard* are injected into the meter. Most watt-hour standards synchronise their test currents to the local three-phase supply, which can be conveniently accessed from the potential terminals on the right-hand side of the test block. By comparing the data in the meter with that calculated by the watt-hour standard, the meter error can be evaluated.

10.8 Problems

1 When the AC supply is taken at medium voltage potentials, the load is always metered on the MV side of the associated MV/LV distribution transformer(s), rather than on the LV side, where the bulk of the load is usually connected, as shown in Figure 10.19. The LV load often consists of a mixture of balanced three-phase loads (such as induction motors) and single-phase loads of various sizes. As a result there will usually be some neutral current flowing on the LV side, i.e. a *zero-sequence* current exists there.

The delta-wye (Δ/Y) connected transformer supplying the load has a zero-sequence current flowing *within* its delta winding, but *not* in the three-wire circuit supplying it. We know that on the LV side, where the zero-sequence current flows, the energy consumed can be correctly metered with a three-element metering installation. However, since there is no zero-sequence current present on the MV side, the question naturally arises, will a three-element MV metering installation record the energy consumed correctly as well?

Consider an extreme case depicted in Figure 10.19, where a single-phase resistive load is applied to the *a* phase on LV side of the transformer, while the *b* and *c* phases remain unloaded.

A Derive the associated line currents flowing on the MV side of the transformer, in terms of the current flowing in the *A* phase on the LV side. (For simplicity you may assume that this transformer has a 1:1 turns ratio, as shown in Figure 10.19, and assume also that the phase voltages are balanced.)

B Show that the energy recorded by a three-element metering installation on the MV side equals that recorded by a similar installation on the LV side. You may assume that the transformer is lossless.

C Repeat part (b) for the case of a two-element installation on the MV side.

2 The results obtained in question 1 are really to be expected, otherwise MV metering would not be possible. We can demonstrate this for any unbalanced condition by considering the symmetrical components of the MV line currents and evaluating the power associated with each of them. We begin with a general expression for the

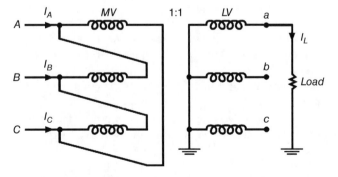

Figure 10.19 MV/LV distribution transformer.

three-phase power flow (assuming balanced voltages), in terms of the sequence components of the current.

$$P = Average\{V_a(I_{a1} + I_{a2} + I_{a0}) + V_b(I_{b1} + I_{b2} + I_{b0}) + V_c(I_{c1} + I_{c2} + I_{c0})\}$$

As we have seen, the three positive sequence currents in this expression sum to a total of $3V_aI_1 \cos(\phi_1)$ watts.

A Expand the negative and zero-sequence powers in the expression above and show that these sum to zero, provided the voltage contains positive sequence components only. Thus the total power measured is:

$$Total\ Power = 3V_aI_{a1} \cos(\phi_1)$$

B Modify your answer to part (A) to allow for the general case where the phase voltages are also unbalanced. Show that the only terms in this expression that are non-zero are those involving only positive sequence voltages and currents and negative sequence voltages and currents respectively, since in the case of a three-wire load, zero-sequence currents cannot exist, therefore:

$$Total\ Power = 3V_{a1}I_{a1} \cos(\phi_1) + 3V_{a2}I_{a2} \cos(\phi_2)$$

3 An underground mining site, supplied by a 44 kV MV supply is metered with a two-element installation. After a meter test the A phase CT was accidentally left short-circuited, meaning that the AB element was inoperative for a period of time. The meter did, however, continue to correctly record the CB element data, but it was only capable of recording the active and reactive energy imported from the bus throughout each 15 minute metering interval. The site consumes balanced currents and has power factor correction equipment installed and therefore its power factor tends to be quite high, typically lying between 0.92 and 0.95.

A Derive an expression for a correction factor k that can be applied to the measured data recorded for each metering interval, in order to obtain the correct energy consumed according to:

$$P_{correct} = k.P_{measured}$$

The correction factor k will be a function of the power factor of the load. In order to determine if k is a strong or a weak function of power factor, graph your expression for k over a range of power factors from 0.92 to 0.95. You should obtain a graph similar to that shown in Figure 10.20.

B If the median value of site power factor is chosen for the active energy correction, what is the maximum percentage error that can be expected in the corrected data?

C Can the active power maximum demand of the site be determined using this technique? If not, why not?

D Would k be as weak a function of power factor if the site's power factor was uncorrected and was in the range 0.65–0.75? What error would you expect in this case?

4 A Derive an equivalent correction factor, K for the *reactive* energy recorded during each metering interval in question 3. Show by plotting a graph, that K *is* a strong

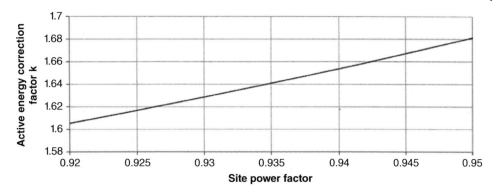

Figure 10.20 Active energy correction factor, k as a function of load power factor.

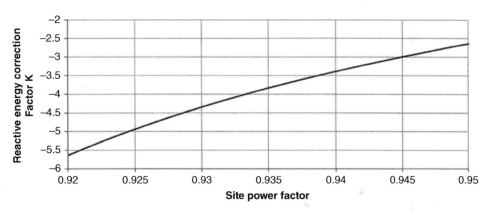

Figure 10.21 Reactive energy correction factor, K as a function of load power factor.

function of the load power factor. Your graph should look like the one in Figure 10.21. Why is K negative in this particular case? Will it always be negative?

B Why can't the reactive energy consumed by the site be recovered in the same way as the active? Explain.

C How might an estimate of the average reactive demand for the site be obtained assuming a median power factor? Explain.

5 Figure 10.22 shows a variation on the standard two-element metering installation, which uses only one CT and two line connected VTs. This arrangement, known as *negative B phase metering,* is often used to provide a check metering installation using the B phase CT when the other two CTs are already used in a Blondel compliant two-element arrangement.

A Sketch the phasor diagram for this installation, showing the relationship between the voltage and current as seen by each element. Assume that the phase currents lag their respective phase voltages by 10°.

B From your phasor diagram, determine the angle between the voltage and current as seen by each element, in terms of the load phase angles. In other words, derive expressions for ϕ_{AB} and ϕ_{CB}, the angles between the current and voltage applied to each meter element.

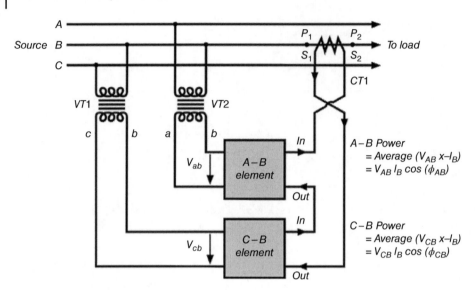

Figure 10.22 Negative B phase metering.

C Write an expression for the total power measured by the meter. Compare this with the expression for a standard two-element installation. Under what conditions will the two expressions be equal? Is this circuit Blondel compliant? Why might it be used?

6 The actual relationships between TCF, RCF and transformer phase errors γ and β appear below.

$$\frac{0.6 RCF_i}{TCF_{CT}} = \cos\left(53.13° - \beta\right) \text{ and } \frac{0.6 RCF_v}{TCF_{VT}} = \cos\left(53.13° + \gamma\right)$$

The linear Equations (10.27) and (10.28), provided in IEEE C57.13, are an approximation to these and make their interpretation simpler. If the ratio $\dfrac{RCF}{TCF}$ can be written in the form $\dfrac{1+x}{1+y}$ where $|x|,|y| \ll 1$ show that $\dfrac{RCF}{TCF} \approx RCF - TCF + 1$, and hence derive Equations (10.27) and (10.28), repeated here:

$$\beta = 2600\left(RCF_i - TCF_{CT}\right)$$

$$\gamma = 2600\left(TCF_{VT} - RCF_v\right)$$

7 The metering transformers on the site of a large HV connected industrial customer were upgraded and, at the customer's request, the old metering installation was retained for some time so that performance of the old and new could be compared. Power data was averaged over many metering intervals to remove any timing jitter and then corrected for the known CT, VT and meter errors in each case. Using the

Table 10.8 Comparative data of new and old metering. (You may assume that each set of current transformers share the same errors, as do each set of voltage transformers.)

	VT magnitude error (%)	CT magnitude error (%)	VT phase error (min)	CT phase error (min)	Meter error (%)	Load power factor	Average metered power (kW)
New installation	0.436	−0.082	−6	5	0.02	0.95	126234
Old installation	0.156	−0.075	−2	4	−0.36	0.95	125391

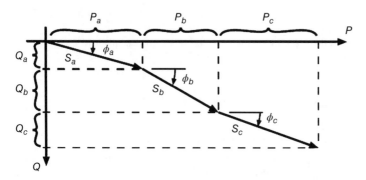

Figure 10.23 *S* vectors for a three-element metering installation.

data in Table 10.8, correct the measured power and show that, when corrected, both installations are effectively same.

Answer:

Corrected power (kW)	New installation	Old installation
	125634	125670

8   *S vectors (VA vectors)* – Equations 10.1 and 10.2 show that the active and reactive power measured by each metering element are related to the element volt-amp product and the angle by which the phase current lags the phase voltage. This suggests that it might be useful to consider an *S* vector (or a VA vector) for each metering element, whose magnitude is equal to the volt-amp product seen by that element, and whose angle is equal to that by which the element current lags the element voltage. These vectors can be plotted on the P-Q plane, as depicted in Figure 10.23. *S* vectors are not phasors in that they do not represent sinusoidal quantities, but they do possess similar properties and can be manipulated in a similar fashion.

Figure 10.23 shows *S* vectors for a three-element metering installation in which the meter sums the active and reactive energy contribution from each element. Figure 10.24 shows the *S* vectors for both three- and two-element metering installations, in this case for a balanced load. Provided the phase currents sum to zero, the total active and reactive energy metered by each installation will be equal.

Consider the vector plot in Figure 10.25, where the load currents are unbalanced. Show that the only possible location for the *b* phase vector S_b (shown dotted) requires that $I_a + I_b + I_c = 0$. What implication can be drawn when $I_a + I_b + I_c \neq 0$?

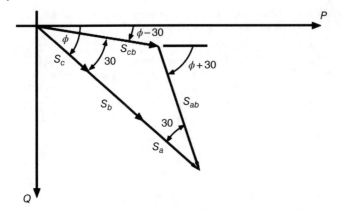

Figure 10.24 *S* vectors for three- and two-element metering installations for a balanced load.

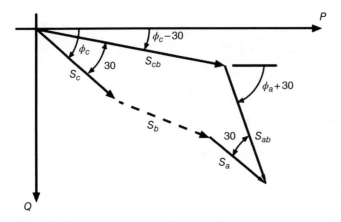

Figure 10.25 *S* vectors for three- and two-element metering installations with an unbalanced load.

10.9 Sources

1 Independent Electricity Market Operator *IMO market manual 3: metering Part 3.4; measurement error correction*, IMO, USA.

2 United States Department of the Interior, Bureau of Reclamation, *Watt-hour meter maintenance and testing* 2000, Facilities Engineering Branch, Colorado.

3 IEEE.C57.13, *IEEE standard requirements for instrument transformers*, 2008, IEEE Power Engineering Society, New York.

4 IEEE.C57.13.6, *IEEE standard for high-accuracy instrument transformers* 2005, IEEE Power Engineering Society, New York.

5 IEC 60044.1, *Instrument Transformers: Part 1 current transformers* 2007, International Electrotechnical Commission, Geneva.

6 IEC 60044-2, *Instrument transformers: Part2 voltage transformers* 2007, International Electrotechnical Commission, Geneva.

7 IEC 61869-2, *Instrument transformers – Part 2: additional requirements for current transformers* 2012, International Electrotechnical Commission, Geneva.

8 IEC 61869-3, *Instrument transformers – Part 2: additional requirements for voltage transformers* 2012, International Electrotechnical Commission, Geneva.

9 IEC 62053.22, *Electricity metering equipment (AC) – Part 22 static meters for active energy (classes 0.2S and 0.5S)*, 2003, International Electrotechnical Commission, Geneva.

10 GE, *Manual of instrument transformers, operation principles and application information* General Electric.

11 ESAA, *Manual of Australian metering practice*, 1971, Electricity Supply Association of Australia, Melbourne.

11

Earthing Systems

Low voltage distribution systems are generally based on a star connected set of windings where each phase voltage is referenced to the neutral terminal. The system of earthing the neutral determines the fault characteristics of the network, including the touch voltages to which people will be exposed in the event of a fault, and the magnitude of the current flowing. A solidly earthed neutral, for example, will result in large fault currents, but it will limit the extent to which over-voltages can exist. On the other hand, while earthing the neutral through an impedance reduces the magnitude of earth fault current, it does so at the expense of permitting considerable over-voltages to occur. Therefore electrical equipment must be capable of withstanding either high fault currents or large over-voltages, until such time as a fault can be detected and cleared.

LV Electrical systems are earthed for the following reasons:

- To provide a current path to earth in the event of a phase-to-ground fault
- To provide a limit on the voltage that exposed metallic parts of an installation may attain during a fault
- To permit the operation of protective devices that remove the risk of electric shock to personnel in the vicinity of a fault
- To limit the rise of the LV potentials due to a fault within the MV network

Earthing system design is based on mitigating the effects of electricity on the human body that can occur during an earth fault. Therefore before we begin a discussion of the various earthing systems in use, we need to understand a little of the adverse effects of electric current on the body.

11.1 Effects of Electricity on the Human Body

Considerable research into this topic was done in the late 1940s and early 1950s by an American professor by the name of Charles Dalziel at the University of California, Berkeley. He carried out rigorous electrical experiments on himself as well as on numerous volunteers, and gathered a large amount of data on the effects of both AC and DC on the body. The standard IEC 60479, *Effects of Current on Human Beings and Livestock* owes much to Dalziel's work, as well as that of others. Interestingly he also designed the first solid-state *residual current device* (RCD) in the USA.

AC Circuits and Power Systems in Practice, First Edition. Graeme Vertigan.
© 2018 John Wiley & Sons Ltd. Published 2018 by John Wiley & Sons Ltd.

The RCD is now common in both domestic and industrial applications, where it interrupts the AC supply when it detects the presence of a ground fault current.

11.1.1 Mild Electric Shocks

Electrical currents as low as 0.5 mA can provoke a reaction from a person touching a conductive surface, although this does depend on the force applied to the surface, whether the skin is wet or dry and other physiological characteristics of the individual concerned. Higher currents, of the order of 3–6 mA, though considerably more uncomfortable, are not debilitating and the person can release the conductor concerned in order to stop the current flow. However, if the current flowing through the body increases beyond this, then the person concerned may not be able to let go.

Dalziel conducted an experiment on a group of 28 women and 134 men to determine the current threshold beyond which letting go was not possible, the *let-go threshold*. The results of this work are summarised in the graph of Figure 11.1. He determined an average let-go threshold for men of 16 mA, and 10.5 mA for women. These values can be read from the graphs in Figure 11.1 at the 50th percentile. Half of the people in the sample tested were able to let go at currents above these values, and half were not. Today IEC 60479 assumes a let-go threshold of 10 mA for adult males. At higher shock currents ventricular fibrillation and electrocution become more likely.

11.1.2 Electrocution and Ventricular Fibrillation

Electrocution is defined as death occurring as a result of electric current flowing through the body. The most common electrocution effects relate either to the disruption of the heart's natural pumping rhythm, known as cardiac arrest, or to a

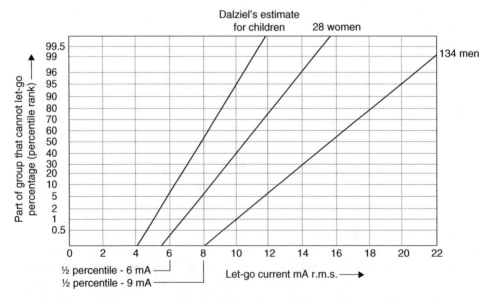

Figure 11.1 Let-go current distribution (60 Hz) (IEC 40679). (Originally published in: 'Deleterious Effects of Electric Shock', by Professor Charles Dalziel, 1961) (IEC/TS 60479-1 ed.4.0 Copyright © 2005 IEC Geneva, Switzerland. www.iec.ch).

cessation of the respiratory system, i.e. the breathing response. Ventricular fibrilla-tion is the most likely cause of cardiac arrest through electrocution. Both these effects prevent the body from receiving the oxygen it requires, either because (1) when in cardiac arrest, the heart fails to effectively pump blood, or (2) if breathing ceases insufficient oxygen enters the bloodstream. In either event the end result is the same.

Figure 11.2 shows a view of the heart and a typical electrocardiogram (ECG), which is a graph representative of the electrical potentials developed within the muscles of the heart. The numbers in Figure 11.2 designate the propagation of the muscle contraction with time throughout the heart.

Of particular interest is the *T wave* on the right-hand side of the ECG. The ventricular muscles are recovering during this period, and at this time in the cycle the heart is particularly vulnerable to ventricular fibrillation, triggered by an external shock cur-rent. Figure 11.3 shows the effect of fibrillation, arising from such a disturbance, intro-duced early in the T wave.

Following a shock of sufficient magnitude and duration, the regular ECG waveform is destroyed and as a result the ventricular muscle begins to contract erratically. The pumping action of the heart ceases and, as shown, the blood pressure falls. Unless the heart can be quickly restarted, death is very likely.

Through his research, Dalziel was able to show that the body responds to high ampli-tude short duration shocks in a similar way to low amplitude long duration ones. He summarised his results in an empirical equation that predicts the shock current flowing through the heart, at the threshold of ventricular fibrillation:

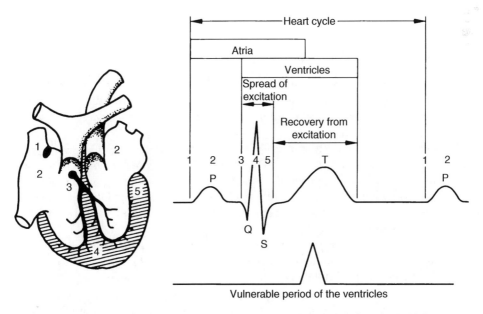

Figure 11.2 Electrocardiogram (IEC 60479). (IEC/TS 60479-1 ed.4.0 Copyright © 2005 IEC Geneva, Switzerland. www.iec.ch).

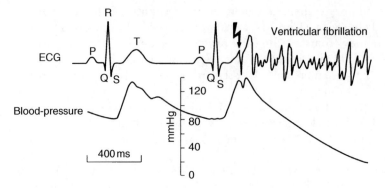

Figure 11.3 Ventricular fibrillation and loss of blood pressure. (IEC/TS 60479-1 ed.4.0 Copyright © 2005 IEC Geneva, Switzerland. www.iec.ch).

$$I = \frac{K}{\sqrt{t}} \tag{11.1}$$

where I = the threshold ventricular fibrillation AC current (mA), t = shock duration in seconds (0.03 < t < 3.0 sec), K is a constant relating to the body weight of the person concerned, (K = 116 for a 50 kg person and K = 157 for a 70 kg person).

The energy delivered as a result of any shock is proportional to the I^2t product, and to the body resistance R_b, thus we can write:

$$Shock\ Energy \propto I^2 t R_b \tag{11.2}$$

The shock energy required to trigger ventricular fibrillation is therefore proportional to:

$$Shock\ Energy \propto K^2 R_b \tag{11.2a}$$

Shock energies in excess of 50 joules can be sufficient to trigger ventricular fibrillation in individuals, the prevention of which is vital if injury or death is to be avoided. Therefore a well-designed earthing system must ensure that prospective shock energy is kept below the threshold of ventricular fibrillation.

While Equation (11.1) strictly applies for AC currents between 15 and 100 Hz, a similar result also applies for direct current. Direct current is less effective in causing fibrillation than is AC current. This is partly because the stimulation of nerves and muscles is associated with changes in the magnitude of the current flowing. Once established, direct current is roughly constant in magnitude, and as a result it is less likely to trigger ventricular fibrillation.

Experimental results suggest that the magnitude of the DC current required to trigger fibrillation is about 3.75 times greater than its AC equivalent. IEC 60479 defines a *DC/AC equivalence factor* (k) as follows:

$$k = \frac{I_{Fibrillation\ DC}}{I_{Fibrillation\ AC}} \approx 3.75 \tag{11.2b}$$

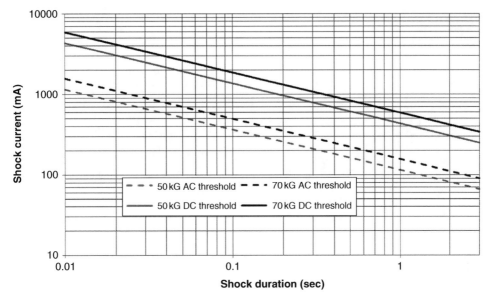

Figure 11.4 AC & DC ventricular fibrillation current threshold vs shock duration.

Equations (11.1) and (11.2b) can be combined to yield the graph in Figure 11.4, which shows the AC and DC ventricular fibrillation threshold currents as a function of shock duration. Clearly, as the duration of the shock increases, considerably less current is required to trigger ventricular fibrillation. It is interesting to note, however, that for shocks of 3 seconds duration, DC currents around 300 mA are required to trigger fibrillation, whereas AC currents of the order of only 80 mA will achieve the same result.

11.1.3   Electrical Resistance of the Human Body

The severity of a shock depends upon many things, including the magnitude of the shock current, its duration and timing with respect to the cardio rhythm, the points of entry and exit from the body, the associated body resistance between these points, and the physical characteristics of the person concerned.

Given all these variables, it is not surprising that people respond very differently to mild electric shocks. What may seem painful and unpleasant to one person may scarcely be felt by another. One parameter that is useful in estimating the shock current a person may receive is the *body's electrical resistance*. Figure 11.5 shows average values of body resistance for fractions of the general population as a function of the shock touch voltage. These are plotted for 5, 50 and 95 percentiles. The body resistance of 95% of the population lies below the upper curve, 50% of the population exhibit resistances below the middle curve, and only 5% of the population have body resistances below the lowest curve. These curves apply to either hand-to-hand or hand-to-foot current paths, both of which are possible shock scenarios.

The 5th percentile curve in Figure 11.5 is of interest in determining the worst case prospective AC or DC shock current, since it represents a body resistance lower than that possessed by 95% of the population. This curve suggests that an asymptotic body

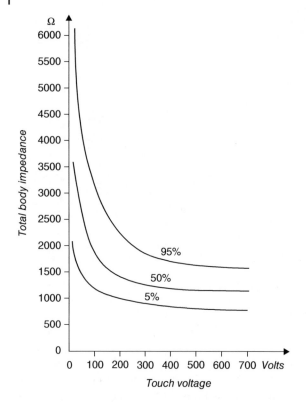

Figure 11.5 Body resistance (ohms) vs touch voltage for hand-to-hand or hand-to-foot current paths. (AS 3859, 1991) (© Standards Australia Limited. Copied by John Wiley & Sons with the permission of Standards Australia under Licence 1609-c119).

resistance of 1000 ohms is appropriate for AC or DC voltages up to 700 volts. Therefore the worst case prospective shock current can be estimated by measuring that drawn by a 1000 ohm resistance placed across any prospective shock potential.

The IEEE Standard 80 *Guide for Safety in AC Substation Grounding* assumes a value of 1000 ohms for body resistance. It also conservatively neglects the effects of shoe and glove resistances in the calculation of touch and step potentials.

11.1.4 Electrical Shock Current: (How Much is Too Much?)

With the ability to estimate shock currents in this way, it is reasonable to ask 'How much shock current is too much?' IEC 60479 provides guidance on this, in the form of a graph showing *time–current zones for AC* (Figure 11.6). This is somewhat similar to Figure 11.4 in that it considers the shock current, as a function of shock duration, but here the current spectrum is divided into four zones. The physiological effects of each are as follows:

Zone AC-1: There are usually no reaction effects, owing to the small shock current flowing.

Zone AC-2: Usually no harmful physiological effects.

Zone AC-3: Usually no organic damage to be expected. Likelihood of muscular contractions and difficulty in breathing, reversible disturbances of formation and conduction of impulses in the heart, including atrial fibrillation and transient cardiac arrest without ventricular fibrillation, increasing with current magnitude and time.

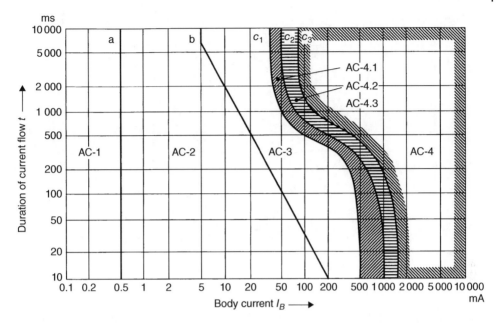

Figure 11.6 Time–current zones for AC. (IEC/TS 60479-1 ed.4.0 Copyright © 2005 IEC Geneva, Switzerland. www.iec.ch).

Zone AC-4: In addition to the effects of zone 3, probability of ventricular fibrillation increasing up to about 5% (zone AC-4.1), up to about 50% (zone AC-4.2) and above 50% beyond (zone AC-4.3). Increasing with magnitude and time, pathophysiological effects such as cardiac arrest, breathing arrest and heavy burns may occur.

The body can tolerate the effects of quite large AC currents (200–500 mA), provided the shock duration is kept very short. Unfortunately, as the shock current magnitude increases, it becomes more difficult for victims to extricate themselves from the source of the current, and the longer the duration of the shock the more likely is the onset of ventricular fibrillation.

The technical report IEC 61200-413 entitled *Explanatory Notes to Measures of Protection against Indirect Contact by Automatic Disconnection of Supply* defines permissible LV touch voltage exposure limits, illustrated in Figure 11.7. In order for the time–current zones defined in Figure 11.6 not to be violated as the prospective touch voltage rises, the duration of exposure must be reduced. In practice, this is achieved through the operation of fast-acting protection devices isolating the LV supply.

Figure 11.7 shows two permissible LV touch voltage curves, *L* and *Lp*. The former relates to normal operation in dry conditions, and corresponds to a sustained touch voltage of 50 volts, known as the safety voltage. The latter is designed to cater for wet areas in which the body's skin resistance is considerably lower. This corresponds to a sustained touch voltage of 25 volts.

These curves can be justified by assuming that the asymptotic body resistance of 1000 ohms in Figure 11.5 applies. The permissible touch voltage can then be plotted against duration on the time–current graph shown in Figure 11.8, since the touch voltage is approximately equal to the product of the shock current and the body resistance. Figure 11.8 shows that the resulting exposure curves avoid the likelihood of ventricular fibrillation.

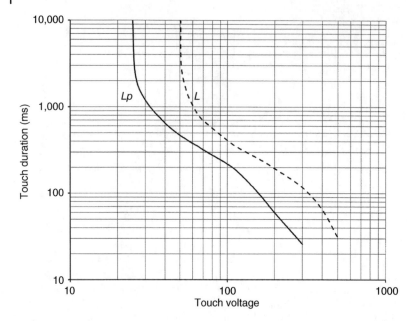

Figure 11.7 Maximum duration of permissible LV touch voltages. (Curve *L*: Corresponds to a maximum sustained touch voltage, $U_{Lt}=50\,V$, for normal operation, Curve *Lp*: maximum sustained touch voltage, $U_{Lt}=25\,V$, for particular situations; generally wet areas) (Based on Figure C2, IEC TR 61200-413, Copyright © 1996 IEC Geneva, Switzerland. www.iec.ch.

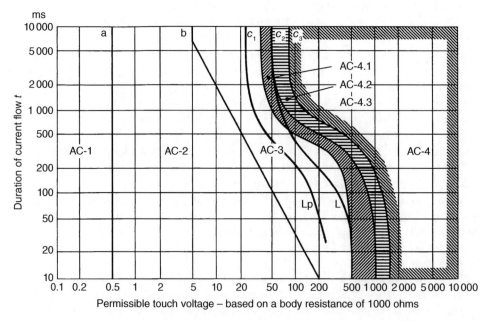

Figure 11.8 Curves Lp and L: maximum duration of permissible touch voltage. (Adapted from IEC/TS 60479-1 ed.4.0 Copyright © 2005 IEC Geneva, Switzerland. www.iec.ch).

Provided that protection devices are chosen whose clearance times are less than the maximum permitted duration, then the likelihood of severe injury is remote. The prospective touch voltage that occurs as a result of a phase-to-ground fault depends to a large extent upon the LV earthing system used.

Electric shocks result from either direct or indirect contact with an exposed source of current. Direct contact occurs when contact is made with a live phase conductor. This is generally prevented by the use of both insulation materials and physical barriers in electrical installations.

Indirect contact occurs through contact with metallic components that have been accidentally elevated above earth potential as the result of an insulation failure elsewhere in an installation. Thus from the viewpoint of electrical safety, we are primarily interested in the effects of phase-to-ground faults.

Protection against indirect contact is provided by the earthing system, which limits the touch voltage and by the protection system which clears the fault within an acceptably short period of time. The latter is accomplished through the use of fuses or magnetic circuit breakers, or in the case of small ground fault currents, through the application of residual current devices.

11.2 Residual Current Devices

Residual current devices (RCDs – also referred to as earth leakage circuit breakers or ground fault interrupters) provide an effective method of detecting and clearing earth fault currents, without requiring access to the earth conductor(s) concerned. They are based on the use of a differential transformer to measure the small *residual* currents that arise in such situations. RCDs have been in use since the 1920s when the technology was applied to HV transmission lines to detect residual currents of the order of 10% of the line rating. The need to protect machinery from the damaging effects of earth faults became evident in the 1930s, and the technology was improved to achieve trip currents as small as 500 mA.

During the 1960s both French and Austrian engineers developed single-phase earth leakage circuit breakers aimed at the protection of human life, operating at currents as low as 25–30 mA. In 1962, Charles Dalziel designed the first solid state RCD in the USA. This was capable of interrupting ground fault currents as small as 5 mA, for which he received a patent in 1965.

Today RCDs are an essential component in the interruption of small shock currents, and are found in virtually every home and industry. The IEC standard 61008 relates to RCDs without integral overcurrent protection, designated as RCCBs, while 61009 applies to RCDs with integral overcurrent protection; these are designated as RCBOs.

11.2.1 Operation of the RCD

The RCD operates on the principle of a differential transformer, as shown in the two-wire circuit of Figure 11.9. It consists of a magnetic circuit upon which there are three windings. Two of these carry load current, one from the phase terminal to the load, the other from the load back to the transformer neutral connection. These windings are wound anti-phase with respect to each other, so that when they carry the same current their ampere-turns cancel, resulting in zero flux developed in the core.

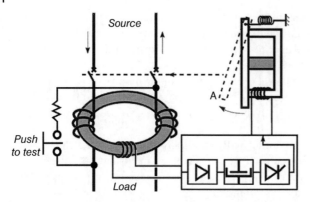

Figure 11.9 RCD operating principle. (Extracted and modified from Cahier Technique No 114, *Residual Current Devices in LV* with kind permission from Schneider Electric).

The third winding senses any current imbalance between the other two. Thus, in the event that leakage current does find its way to ground, the core imbalance will result in a residual flux, proportional to the residual current flowing in the earth conductor(s). The resulting sense voltage is rectified, filtered and applied to the trip coil of the isolation switch, which breaks both source conductors.

The isolation switch is held closed by a permanent magnet contained within its magnetic circuit. When the trip current threshold is exceeded, the DC flux in the circuit is nulled, and the isolator opens under the action of a spring. Once open, the air gap in the magnetic circuit is sufficient to prevent re-closure, and the isolator must therefore be reset manually.

The test button causes a small current to be bled from the outgoing active terminal to the neutral, without it passing through the neutral winding. This creates an imbalance in the core, and the resulting voltage in the sense coil trips the isolator. This facility enables users to periodically check that the RCD is functioning correctly.

In addition to the two-wire RCD described above, three- and four-wire circuits are also possible, so long as the protective earth conductor is not one of them. Under normal operation the vector sum of the currents flowing through the toroid adds to zero. Therefore any imbalance that does occur represents a genuine residual ground current, on which the unit will trip.

There are a couple of items relating to RCDs that should be stressed. Firstly, an RCD will not limit the magnitude of the ground fault current flowing; it will only limit the duration for which it flows. Secondly an RCD will not trip in the event of a phase-to-phase or a phase-to-neutral fault, since these do not affect the ampere-turn balance within the core.

11.2.2 RCD Sensitivities and Operating Times

The sensitivity of an RCD is the *rated residual operating current*, given the symbol $I_\Delta n$. Table 11.1 shows high, medium and low sensitivities commonly used in RCDs. Low and medium sensitivities are used for the protection of equipment and to assist in the prevention of fires ignited during low level insulation faults. Medium and low sensitivity RCDs are also used together to allow discrimination between circuits in order to isolate the faulty circuit.

Table 11.1 RCD sensitivities.

Sensitivity	I∆n (mA)
High	6–10 or 30
Medium	100–300–500 or 1000
Low	3000–10,000 or 30,000

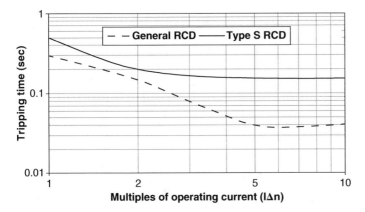

Figure 11.10 RCD trip time vs ground fault current in multiples of *I*∆*n*, for general and type S RCDs.

The high sensitivity RCDs are used for the purpose of shock prevention. In general residential applications, 30 mA is generally the trip threshold setting, but in hospitals and medical centres 6 or 10 mA may be used instead. In the USA, trip thresholds as low as 5 mA are also used.

The IEC standard 61008 defines two types of RCD based on tripping times. Type G (general) use RCDs will trip within 300 ms for ground fault currents less than $2I_{\Delta}n$ while type S (selective) devices are fitted with a short operating delay which can also be helpful in achieving discrimination between circuits. Figure 11.10 shows the tripping characteristic for both these types, as multiples of the RCD operating current.

Finally, Figure 11.11 shows a residential RCCB rated at 30 mA which can carry a through current of 40 A. The test button is located at the top of the device, and the isolation switch is below it.

11.2.3 Limitations of RCDs

RCDs are very good at doing what they're designed for, but there are several situations in which they will not operate, and these should be understood. Firstly, a three-phase

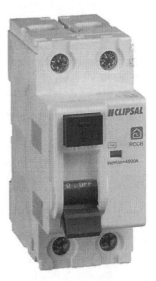

Figure 11.11 A 30 mA RCD used to protect socket outlet circuits. (Note the test button and the isolation switch) (Image kindly supplied by Schneider Electric).

RCD will not operate in the event of a phase-to-phase or a phase-to-neutral fault, regardless of its magnitude. This is because such faults do not generate residual currents that flow back to the star point, via one or more earth conductors. Secondly, a fault on the secondary side of a transformer will not cause an RCD on its primary winding to trip, for precisely the same reason.

There are also some situations in which a single-phase RCD may trip unexpectedly. These situations arise when a parallel path from the appliance neutral to the supply star point is inadvertently created. This can occur, for example, when using an oscilloscope, if the earth terminal of either channel is connected to the neutral conductor of an appliance protected by an RCD. By so doing, a parallel path is created back to the transformer's star point, via the oscilloscope's earth conductor. This path is likely to be slightly higher in impedance compared with the neutral conductor, but it may permit just enough neutral current to flow to unbalance and trip the RCD. Unexplained RCD trips should always be investigated.

11.3 LV Earthing Systems

The standard IEC60364 defines three earthing systems which are used in various forms worldwide. These are abbreviated by a two letter code: TT, TN and IT. The first letter applies to the connection of the LV distribution transformer's neutral terminal.

- T (terre or ground) implies that the transformer neutral is solidly connected to earth.
- I (isolated) implies that the transformer neutral is isolated, but in practice this may mean a high impedance earth connection.

The second letter relates to exposed conductive parts of the customer's premises, for example water or gas reticulation piping, metallic parts of buildings etc. These are generally bonded together to provide an equipotential installation in order to avoid the possibility of exposure to different potentials under fault conditions.

- T implies that they are connected to earth at the customer's premises.
- N implies that they are connected to the LV transformer's neutral terminal (via a distributed neutral conductor).

11.3.1 TT Earthing System

The TT earthing system sees both the transformer neutral and the customer's switchboard connected to earth, as shown in Figure 11.12. Therefore any phase-to-ground fault current is limited by the earth impedances both at the customer's premises and at the transformer. Assuming the fault impedance to be zero, the fault current, I_f is approximately given by:

$$I_f = \frac{V_{ph}}{Z_f} = \frac{V_{ph}}{Z_S + Z_L + Z_{PE} + Z_T} \approx \frac{V_{ph}}{Z_{PE} + Z_T}$$

where V_{ph} is the faulted phase potential, Z_f = the fault loop impedance = $(Z_S + Z_L + Z_{PE} + Z_T)$, Z_S is the source impedance (largely that of the local MV-LV

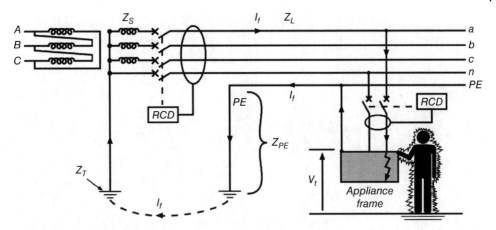

Figure 11.12 TT earthing system. (Adapted from Cahier Technique No 114, *Residual Current Devices in LV* with kind permission from Schneider Electric).

transformer), Z_L is the impedance of the active conductor, Z_{PE} is sum of the protective earth conductor impedance and the customer's earth impedance and Z_T is the transformer earth impedance.

Of these impedances, Z_{PE} and Z_T are generally the largest. Therefore the associated touch potential V_t is approximately:

$$V_t = I_f Z_{PE} \approx \frac{V_{ph} Z_{PE}}{Z_{PE} + Z_T}$$

Z_{PE} and Z_T are likely to have similar magnitudes and therefore $V \approx V_{ph}/2$. Since this is in all cases beyond the permissible continuous touch voltage of 50 volts (or in wet areas 25 volts), then the protection equipment must limit the exposure duration to a value less than that specified in Figure 11.7.

The fault current in a TT system, being limited by earth impedances, may not be sufficient to operate the installation's short-circuit protective devices (fuses or magnetic circuit breakers), which must be dimensioned according to the rated load current. Therefore residual current devices like those shown in Figure 11.13, must be used to detect and clear the fault, in which case we require:

$$V_{Lt} = 50 \geq I\Delta n Z_{PE}$$

where $I\Delta n$ is the operating current of the RCD and V_{Lt} is the maximum sustained touch voltage (also known as the *safety voltage*).

11.3.2 TN Earthing System

The TN system of LV earthing has the customer's switchboard and all other exposed metallic components solidly connected to the LV transformer neutral, which itself is earthed at the transformer. The phase-to-neutral fault impedance is therefore relatively low, and large fault currents can flow as a result. These are often well in excess of normal load currents and therefore the *short circuit protective devices* (SCPDs) will quickly clear such a fault. The TN earthing system is pictured in Figure 11.14.

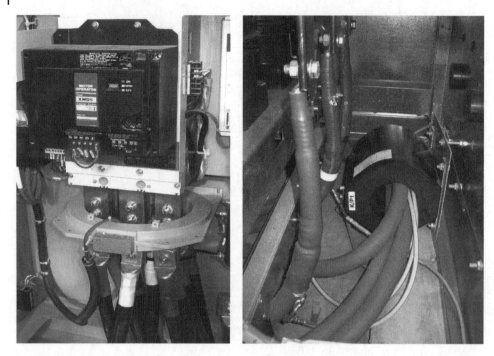

Figure 11.13 Left: An RCD current transformer on an LV circuit breaker; $I\Delta n = 300$ mA. Right: An RCD current transformer on an 11 kV feeder. (Note that since the 11 kV cable sheaths pass through the CT, then in order to remove the effect of any sheath current, the sheath earth conductors must pass back through the CT in the opposite direction).

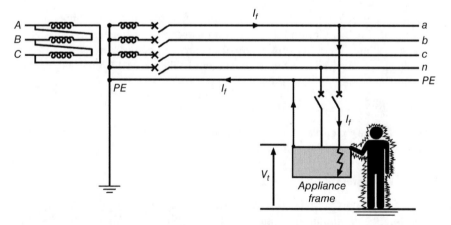

Figure 11.14 TN earthing system. (Adapted from Cahier Technique No 114, *Residual Current Devices in LV* with kind permission from Schneider Electric).

Because of the heavy fault currents flowing through the transformer and supply conductor impedances, the phase voltage V_{ph} seen by the customer will fall during the fault, typically by at least 20%.

The fault current for a phase to neutral fault is therefore given by:

$$I_f \approx \frac{0.8\,V_{ph}}{\left(Z_L + Z_{PE}\right)}$$

And the touch voltage V_t is equal to the drop in the protective earth conductor, thus:

$$V_t \approx I_f Z_{PE} = \frac{0.8\,V_{ph} Z_{PE}}{\left(Z_L + Z_{PE}\right)}$$

If these impedances are approximately equal then:

$$V_t \approx \frac{0.8\,V_{ph}}{2} = 0.4\,V_{ph}$$

Therefore in 230/400 volt systems the touch voltage will be about 90 volts, and since this is in excess of the sustained touch voltage limit of 50 volts, then according to Figure 11.7 the SCPD must clear the fault in less than 450 ms. In cases where the impedance of the supply conductors is high or unknown, a combined circuit breaker/RCD may be used, i.e. an RCBO.

The satisfactory operation of SCPDs requires a fault current of sufficient magnitude that will ensure operation of the fuse or magnetic circuit breaker within the required clearance time. The fault current in turn depends upon the impedance of the faulted circuit, known as the *fault loop impedance*. This impedance, defined in Figure 11.15, includes the source and active conductor impedances to the fault, as well as the return path via the protective earth and neutral (PEN), back to the neutral terminal of the transformer, i.e. path 1-2-3-4-5-6. The fault impedance is assumed to be zero.

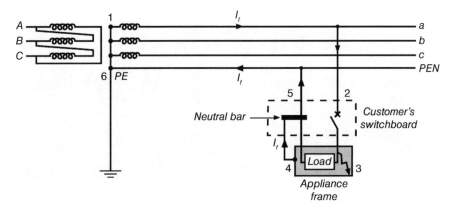

Figure 11.15 Earth fault loop definition.

The fault loop impedance, Z_f is sufficiently small if the following equation is met:

$$Z_f \leq \frac{V_{ph}}{I_a}$$

where Z_f is the fault loop impedance, V_{ph} is the faulted phase voltage and I_a is the current that will ensure operation of the SCPD (fuse or magnetic circuit breaker), within the required clearance time. (This is often taken to be the instantaneous trip current of a circuit breaker.)

Due to the relative size of the distribution conductors in the power network compared to those within the customer's premises, it is assumed that at least 80% of the faulted phase potential is dropped within the customer's premises. As a result, there exists a *maximum route length* for customer circuits, beyond which the SCPD will not operate within the required time. The maximum route length can be calculated using the following equation, which assumes negligible inductive conductor impedance (i.e. that the phase and PEN conductors run in close proximity to each other and are therefore predominantly resistive).

$$L_{max} = \frac{0.8\, V_{ph} S_{ph} S_{pen}}{I_a \rho \left(S_{ph} + S_{pen} \right)}$$

where L_{max} is the maximum route length in metres, ρ is the resistivity at the working temperature $\Omega\text{mm}^2/\text{m}$, $(22.5 \times 10^{-3}$ for Cu or 36×10^{-3} for Al), S_{ph} is the customer's phase conductor cross-sectional area in mm^2 and S_{pen} is the customer's PEN conductor cross-sectional area in mm^2.

Should a proposed circuit length exceed L_{max}, then the cross-section of the conductors must be increased or the circuit's current rating must be reduced. Larger conductors may also be demanded by the *minimum voltage drop requirements* for circuits approaching this length.

Where local regulations permit, an RCD may be used to protect long circuits. This makes the calculation of the fault loop impedance unnecessary. In such cases, the RCD threshold must be chosen so that:

$$I_{\Delta n} \leq \frac{0.8\, V_{ph}}{Z_L + Z_{PE}}$$

(This solution cannot be applied in the case of TN-C earthing, since here the neutral and the protective earth are the same conductor, see the following section.)

Variations on the TN Earthing System

Figure 11.16 shows several variations on the TN earthing system. These are designated TN-C, TN-S and TN-C-S. The TN-C system (Figure 11.16a) uses a common protective earth and neutral (PEN) conductor, whereas the TN-S system (Figure 11.16b) separates these conductors. While a combined neutral and PEN saves one conductor, it has the disadvantage that the return current flowing back to the neutral terminal of the transformer generates a small potential drop between the earth potential at the transformer

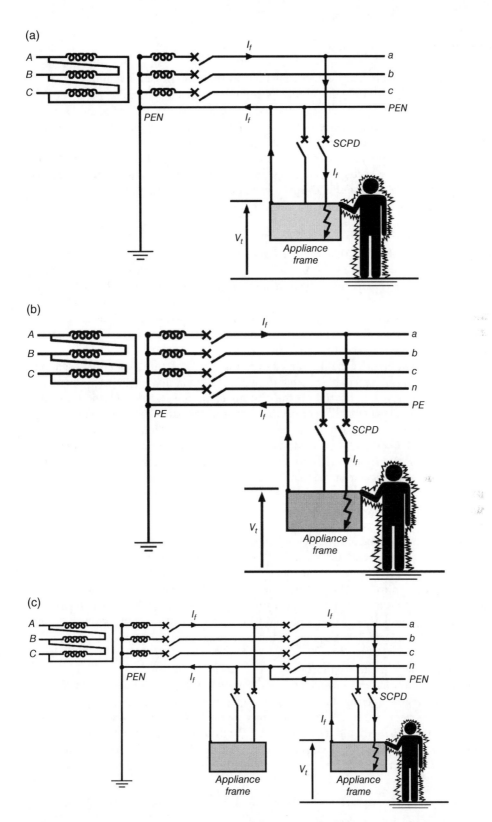

Figure 11.16 (a) TN-C earthing system (common PEN conductor) (b) TN-S earthing system (separate neutral and earth conductors) (c) TN-C-S earthing system (TN-C followed by TN-S). (Adapted from Cahier Technique No 178, *The IT Earthing System in LV* with kind permission from Schneider Electric).

and that at the customer's premises. Further, in the presence of a significant triplen harmonic component[1], the neutral may become overloaded, particularly if its dimensions are smaller than those of the associated phase conductors. In France, for example, the TN-C system is only permitted when the size of the PEN conductor exceeds $10\,mm^2$.

In the TN-S system the functions of the neutral conductor and the protective earth are separated by the provision of a distributed earth conductor, shown in Figure 11.16b and 11.16c. The TN-S system offers the opportunity of isolating the neutral conductor via the SCPD, while maintaining the protective earth connection. (It is not permitted to connect a TN-C system downstream of a TN-S.)

Earthing of the Neutral Conductor

Because of the possibility that the PEN may carry sizeable fault currents, in many countries it is common to earth this conductor at frequent intervals along its length, thereby providing a *global grounding system* in which the division of earth fault currents provides a reduction in the earth potential rise (EPR) that otherwise might occur. In the USA and the UK when a TN-C system is used, multiple earth connections are made at regular intervals to the PEN in the public network, while in Germany this conductor is earthed upstream of each customer connection.

In Australia and New Zealand a variation to the TN-C earthing system is used, one known as the *multiple earthed neutral* (MEN) system, in which each customer connection includes an earth rod on the customer's premises, to which the PEN conductor is connected via an MEN connection between the earth bar and the neutral bar in the customer's switchboard, as depicted in Figure 11.17 Thus the MEN system is effectively TN-C upstream of the customer's switchboard and TN-S downstream of it.

All exposed metal surfaces are connected to the switchboard earth conductor, including water and gas reticulation piping. The earthing of the neutral conductor at each customer's earth electrode ensures the existence of a global grounding system both in residential and city areas.

As with all TN systems, a phase-to-neutral fault in the MEN system will result in large fault currents that will be cleared by the associated protective device(s). Assuming that the phase and neutral conductors have the similar cross-sectional areas, the touch potential V_t will generally be about $0.4\,V_{ph}$, as derived earlier. Provided the fault is cleared quickly (typically within 400 ms), there should be minimal risk of fire or electrocution. RCDs can be successfully used within the customer's premises, since the neutral and the earth conductors are separate, and RCDs are now incorporated into all new Australian MEN installations as well as in older ones as they are upgraded.

Limitations of the MEN Earthing System

There are several situations in which the MEN system does not perform well, and while these do not arise frequently, when they do, the risks to personnel can be high. The first occurs in the event of an MV to LV contact or an MV earth fault within the local distribution transformer. This generates an EPR at the distribution transformer, which is impressed on the LV neutral. Because of the neutral-earth connection in the customer's switchboard, any such EPR is transferred to all earthed fixtures within the customer's

---

1 Triplen harmonics are multiples of three times the fundamental frequency and as a result they are co-phasal.

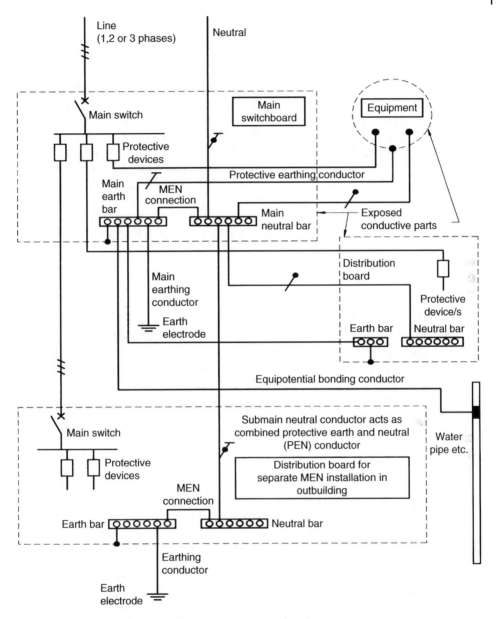

Figure 11.17 MEN switchboard arrangement (AS/NZS 3000:2007)). (© Standards Australia Limited. Copied by John Wiley & Sons with the permission of Standards Australia under Licence 1609-c119).

premises. This poses a significant safety risk to a person touching objects bonded to the earth bar, while standing on a surface at the local earth potential, for example when touching an outside tap.

The second situation arises in the event of a broken neutral conductor, upstream of the customer's switchboard. In the case of either a TT or a conventional TN earthing system, such a break would immediately interrupt the supply, thereby highlighting the fault.

However, under the MEN system, a broken neutral is partially masked by the neutral-earth link, which allows return current to flow back to the transformer's neutral via the customer's earth electrode. Depending on the electrode resistance and the current magnitude, this situation may lead to a local EPR being impressed upon all items connected to the earth bar. The use of RCDs on circuits within the installation will not detect this situation, as the phase current remains balanced by the neutral current flowing back towards the earth bar.

If the customer's earth electrode resistance is high, the effect of a broken neutral will be felt within the installation, usually in the form of dim lights or poorly operating appliances, but it may go undetected until a shock is received. (An example of such mal-operation relates to standard RCDs, which may not function correctly in the presence of a reduced phase to neutral voltage.)

Some distribution companies have been sufficiently concerned about this risk that they have issued devices capable of detecting a broken neutral to their customers. Others are questioning whether a change to the TT earthing system might be a viable alternative.

11.3.3 IT Earthing System

In the IT earthing system, the LV transformer neutral is not earthed, or alternatively it is earthed through a high impedance. This may be from a few ohms in the case of resistively earthed systems to several hundred kilo-ohms for fully floating systems, as illustrated in Figure 11.18. Exposed metal surfaces within the customer's premises are connected to ground. Despite the lack of a solid earth connection, the LV neutral point remains close to ground potential. This is due to the presence of distributed network capacitances between each phase and ground (Figure 11.18), which effectively tie the LV neutral-to-ground potential. These capacitances arise in LV cables between the

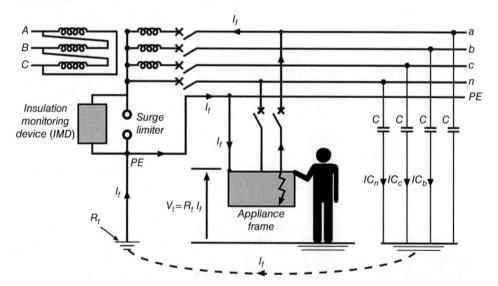

Figure 11.18 IT earthing system: first fault generates no hazardous touch voltages. (Adapted from Cahier Technique No 178, *The IT Earthing System in LV* with kind permission from Schneider Electric).

phase conductor and the earthed sheath, in transformers between phase windings and the earthed core, and to a lesser extent between overhead lines and ground. These capacitances are small and are approximately equal on each phase; they generally lie in the range 0.1–5 μF depending on the size of the LV network. As a result, under normal conditions the total positive sequence capacitor current flowing back to the transformer neutral is close to zero.

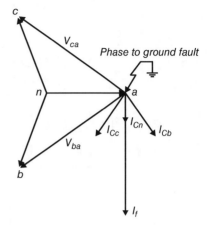

Figure 11.19 IT earthing system: phase-to-ground fault.

Because of the high impedance of both the earth and the distributed capacitances, the presence of a *single-phase-to-ground fault* will not generate significant fault current, and therefore the resulting touch potential will be very low. As a result, the system can remain operational in the presence of the first fault, but this must be located and rectified before a second fault develops, which may result in high fault currents and therefore high touch voltages as well. The ability of the LV system to remain operational in the presence of the first fault is a particular advantage of IT earthing; for this reason, it is often used in applications where the AC supply is critical, such as hospitals and medical centres.

The fault current that flows for the first phase-to-ground fault can be calculated with reference to Figure 11.18 as follows. Because the neutral normally lies very close to ground potential, the current I_f will also be close to zero. However, in the event of a phase-to-ground fault on phase a, as shown, the capacitances associated with phases b and c will now each see a line voltage. The phasor diagram in Figure 11.19 summarises this situation. Ic_b, Ic_c and Ic_n will flow through ground towards the transformer star point. The magnitudes of Ic_b and Ic_b will be given by:

$$|Ic_b| = |Ic_c| = \sqrt{3}\, V_{ph}\omega C$$

and their vector sum will equal:

$$Ic_b + Ic_c = 3\, V_{ph}\omega C$$

Assuming that the neutral-to-ground capacitance is of a similar size to the phase capacitances, then the current Ic_n will equal $V_{ph}\omega C$ and the total fault current will then be:

$$I_f = 4\, V_{ph}\omega C$$

For a 230/400 V, 50 Hz system, with $C = 5\,\mu F$ then $I_f = 1.4$ A and the resulting touch voltage V_t, will be given by:

$$V_t = I_f R_t = 14 \text{ V} \left(if\ R_t = 10\ \text{ohms.} \right)$$

Since V_t is less than the continuous touch voltage of 50 volts, even for relatively high earthing resistance and phase capacitances, the LV system can remain in operation

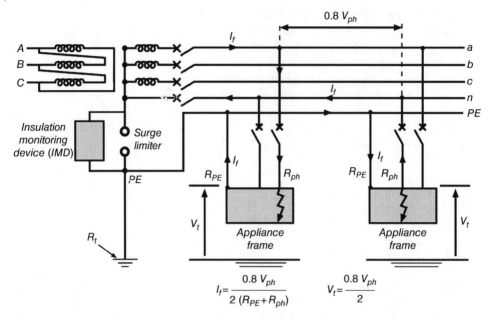

Figure 11.20 Second fault on an IT earthed LV network. (Adapted from Cahier Technique No 172, *System Earthing in LV* with kind permission from Schneider Electric).

while the fault is located and repaired. However, the occurrence of a second fault before the first has been repaired can lead to hazardous touch voltages being generated.

For example, consider a second fault, this time between neutral and ground, as shown in Figure 11.20. The second fault effectively generates a phase-to-neutral fault with fault current I_f flowing in the PE conductor. The associated voltage drops in the phase and PE conductors can create dangerous touch voltages of the order of $0.4\,V_{ph}$, which must be cleared quickly. The magnitude of the phase-to-neutral fault will generally be such that the associated SCPD will clear the fault, but if the distance between the two faults is sufficiently large, then the cabling impedances may reduce the fault current below the threshold of the SCPD and then RCDs may be required for fault clearance instead.

Where the second fault to ground occurs on another phase, the interrupting circuit breaker must now interrupt a phase-to-phase fault which must also be cleared quickly. It is therefore a necessary requirement that all circuit breakers used in IT earthing systems be capable of interrupting a single pole phase-to-phase fault.

A maximum route length can also be calculated for circuits within an IT earthing system, in a similar way to that already discussed for the TN system. In this case, the following equations apply:

1) For IT circuits where the neutral is distributed, the fault potential will be a phase voltage, and the maximum route length is given by:

$$L_{\max} = \frac{0.8\,V_{ph}S_{ph}S_{pen}}{2I_a\rho\left(S_{ph} + S_{pen}\right)}$$

2) For IT circuits where the neutral is not distributed the fault potential will be a line voltage, and the maximum route length is given by:

$$L_{max} = \frac{0.8\,V_{ph}S_{ph}S_{pen}\sqrt{3}}{2I_a\rho\left(S_{ph}+S_{pen}\right)}$$

(Note that a factor of 1/2 has been applied to both these equations since there is now twice the length of each phase and protective earth conductor within customer premises, as compared with the TN earthing system. This is shown in Figure 11.20.)

Locating the first fault in an IT earthing system is not easy, due to the small capacitive currents involved and the need to determine which LV feeder is responsible for the fault. One obvious method is to disconnect feeders in turn and note which carries the fault current, but this technique requires interruption to the supply, which defeats the main advantage of the IT system in being able to operate safely through the first fault. More recent approaches include the injection of a low frequency signal between the transformer neutral and ground. The resulting current can then be traced to the offending feeder with handheld or permanently installed detectors, tuned to the injected signal frequency. It is expected that an insulation monitoring device will be installed between the transformer neutral and ground in an IT system, to highlight the existence of the first fault.

11.4 LV Earthing Systems used Worldwide

Table 11.2 shows the LV earthing system used in a selection of countries. The most common earthing system used throughout Europe is TT. In the United States, Germany, Australia and New Zealand the TN system (or modifications of it, as described already) are used. In each case, the neutral is grounded at multiple points along its length thereby achieving a very low earth resistance connection. This arrangement significantly reduces the earth potential rise during fault conditions since the many parallel paths to ground share the fault current.

11.5 Medium Voltage Earthing Systems

The earthing of MV distribution networks is aimed at minimising both voltage and thermal stresses on equipment, while simultaneously providing an acceptable degree of personal safety. MV earthing systems tend to fall into two broad classifications, *directly earthed* or *impedance earthed*. Directly earthed systems have the MV neutral point solidly grounded and as a result ground fault currents tend to be large and require the immediate disconnection of the supply in order to prevent thermal damage to equipment. On the other hand (high) impedance earthed networks generate small ground fault currents and therefore present little danger of thermal damage to equipment. While they often may be left in service until the fault can be located, they suffer from over-voltage conditions on the un-faulted phases that are not present in directly earthed networks.

Table 11.2 LV earthing systems used worldwide.

Country	LV distribution voltage	Earthing system	Comments
Germany	230/400	TN-C TT	LV transformer earth <2 ohms, Neutral grounded upstream of each customer connection point for both TT and TN.
Belgium	230/400	TT	Customer's earth connection <100 ohms 30 mA RCD for socket outlets.
Spain	230/400	TT	Customer's earth connection <800 ohms with 30 mA RCD at supply point.
France	230/400	TT	Customers earth connection <100 ohms 30 mA RCD for socket outlets.
Great Britain	240/415	TN-C TT	TN-S & TN-C used in town areas, Earth connection of the neutral provided by distributor; resistance <10 ohms. TT used in rural areas.
Italy	230/400	TT	RCD threshold is a function of customer's earth resistance ($I\Delta n < 50\,V/R_{earth}$).
Japan	100/200	TT	Customer's earth connection <100 ohms 30 mA RCD.
Norway	230/400	IT	(High earth resistivity region) 30 mA RCD trips domestic supply on second fault.
Portugal	220/380	TT	Customers earth connection <100 ohms
USA	120/240	TN-C	Neutral conductor is earthed at customer's premises.
Australia	230/400	MEN (A special case of TN-C)	Customer's earth electrode is connected to the distributed neutral conductor via MEN link in switchboard. RCDs are incorporated in all new installations and as part of modifications.
New Zealand	230/400	MEN (A special case of TN-C)	Customer's earth electrode is connected to the distributed neutral conductor via MEN link in switchboard.

(Extracted from Cahier Technique No 173, *Earthing Systems Worldwide & Evolutions* with kind permission from Schneider Electric).

HV to MV transformers are generally either delta-star or star-star connected, in the latter case with the HV star point unearthed. These arrangements prevent zero-sequence current from flowing from the MV network back into the HV network, while providing a neutral terminal with which the MV potentials can be referenced to ground.

MV to LV three-phase transformers are generally either delta-star (Dyn11) or star-zigzag (Yzn11) connected. These vector groupings provide a low impedance path for LV zero-sequence currents (so that unbalanced LV loads can be supported), while simultaneously preventing them from entering the MV network. Despite this, the MV earthing

Figure 11.21 Transferred potential
from an MV fault to the LV network.

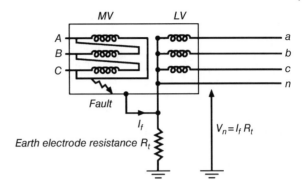

system can affect the operation of the LV network should a disruptive breakdown arise within the MV-LV transformer. This presents a danger to users of the LV network, and is usually manifested as a phase-to-tank fault on the MV side, resulting in significant zero-sequence current flowing to ground through the earth conductor attached to the transformer tank, as shown in Figure 11.21. This produces an EPR equal to the product of the fault current and the earth resistance at the transformer (R_t).

The problem arises when the LV network earth shares the same earth electrode as the transformer tank, in which case the LV neutral and phase windings are also exposed to the resulting EPR. This effect is known as a *transfer potential* from MV to LV. It is generally not a serious problem in an MV-LV substation, provided all exposed metallic surfaces are bonded to the substation earth grid, since the entire substation environment is raised by the same potential. However, it can become a serious problem in the wider LV distribution network, where a transient potential arises between the incoming neutral and the local earth, which may result in an LV insulation failure.

The IEC standard 60364-4-442 stipulates that the MV-LV substation earthing must be such that the LV network is *not* exposed to an earth voltage of (1) $V_{ph} + 250\,\text{V}$ for more than 5 seconds, (2) $V_{ph} + 1200\,\text{V}$ for less than 5 seconds. (In the case of IT earthing: V_{ph} becomes $\sqrt{3}\,V_{ph}$.)

This places a corresponding requirement on the rated withstand voltage between the phase and earth potential of LV customer equipment, which is typically around 1500 volts.

In order to obviate this risk, the MV earth fault current is often limited, by fitting a either neutral earthing resistor or a neutral earthing reactor between the MV neutral and ground at the HV-MV substation. This approach is used in France and frequently in the UK as well.

An alternative to impedance earthing is to use HV-MV transformers with a high zero-sequence impedance, which limits the maximum zero-sequence current that can flow in the event of a fault. Recall that for a single-phase-to-ground fault, the sequence networks are series connected, and therefore the fault current is given by:

$$\text{Single Phase Fault Level} = I_f = \frac{3\,V_{ph}}{\left(Z_1 + Z_2 + Z_0\right)}$$

where Z_1, Z_2 and Z_0 are the positive, negative and zero-sequence impedances respectively.

Table 11.3 Actual transmission substation fault levels in an Australian transmission substation.

Bus voltage (kV)	Three-phase fault level (kA)	Max single-phase fault level (kA) (based on 1.1 pu voltage)	Positive sequence impedance (pu)		Negative sequence impedance (pu)		Zero-sequence impedance (pu)	
			R_1	X_1	R_2	X_2	R_0	X_0
220	12.8	15.3	0.002	0.023	0.002	0.024	0.001	0.011
110	14.7	16.5	0.002	0.039	0.002	0.041	0.001	0.025
22	10.1	5.6	0.006	0.286	0.006	0.288	0.000	*0.961*

The positive sequence impedance of the HV-MV transformer is generally dimensioned to provide the three-phase fault level required at the MV bus, which is usually set to suit the fault capability of downstream switchgear. However, since this fault current is unaffected by the zero-sequence impedance, this can be dimensioned to reduce transformer's single-phase-to-ground fault level to an acceptable level.

As an example, a particular Australian transmission substation has both 220 kV and 110 kV busses, in addition to a 22 kV MV bus for local distribution. The published fault levels and sequence impedances for this substation, expressed in per unit on a 100 MVA base, are shown in Table 11.3.

The single-phase-to-ground fault levels on the 220 kV and 110 kV busses are determined largely by the positive and negative sequence reactances, since the zero-sequence impedance is relatively small by comparison. However, on the 22 kV bus this fault level is largely set by the high zero-sequence impedance of the 110 kV:22 kV transformer, in this case being 0.961 pu. As a result the 22 kV single-phase fault level is only about one-third of that of the 220 kV or 110 kV busses.

By limiting the single-phase fault current the transfer potential at the MV-LV transformer will also be limited; however, it is usually necessary to take additional steps to ensure an acceptably low transfer potential. This is frequently achieved by:

1) earthing the LV neutral conductor in multiple locations along its route, thereby effectively producing a global earthing system, achieving a very low overall earth resistance and providing multiple paths to ground for the fault current. This is particularly effective in TN systems.
2) separating the transformer tank earth from the LV neutral earth connection by a suitable distance, so as to minimise the transfer potential between the two. This is often achieved in aerial distribution systems using pole-mounted transformers, by earthing the transformer tank at the base of the transformer pole and the LV neutral one span away, on the next pole. This technique is effective in TT systems where the customer's earth connection is remote from that of the transformer. By making the LV neutral earth connection somewhat remote from the transformer too, the neutral-earth potential in the customer's premises can be kept low during disruptive breakdown faults. This approach is used in most countries where the TT system is used, as well as in Australia and New Zealand.

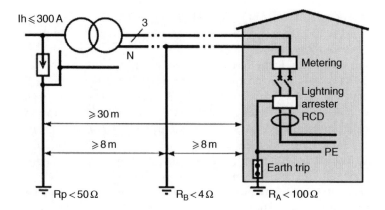

Figure 11.22 MV earthing requirements adopted in France (TT System). (Extracted from Cahier Technique No 172, *System Earthing in LV* with kind permission from Schneider Electric).

Figure 11.22 shows the MV earthing requirements used in France. Here the LV earth connection must be at least 8 m from the transformer tank earth and also from the nearest dwelling, and the latter must also be at least 30 m from the MV transformer.

In the USA the MV transformer neutral is frequently distributed along with the MV phase conductors. This permits single-phase MV-LV distribution transformers to operate from a phase voltage rather than a line voltage, as is the case elsewhere in the world. Where an MV neutral conductor is available, it is connected to the LV neutral at each distribution transformer. As a result any MV ground fault current generated within a distribution transformer has a metallic path back to the MV neutral terminal. This substantially reduces the ground fault current flowing and therefore the transfer potential as well.

There are two options for earthing an MV neutral conductor, *uni-grounded* and *multi-grounded*. In a uni-grounded system the distributed MV neutral conductor is earthed only at the MV transformer. As a result, the zero-sequence component of the load current returns via the neutral conductor, while any earth fault current flows via the ground to the transformer neutral terminal. This makes detection of ground fault currents considerably easier.

The MV earthing example above requires a multi-grounded neutral. While this ensures a low transfer potential, it does so at the expense of the ease of discrimination between an earth fault current and any zero-sequence load current, since they share the same path back to the transformer neutral terminal, which includes both the neutral conductor and the earth.

11.5.1 Un-Earthed MV Neutrals

In some countries, the MV neutral is unearthed, in a similar fashion to the IT system of LV earthing. In such cases the network capacitance between each phase and ground, ties the MV neutral terminal close to ground potential. As with the IT earthing system, the first phase-to-ground fault will generate a small residual capacitive fault current which, depending on the nature of the fault, may result in an arc. From the point of view of disruptive breakdown, such a fault will not generate large currents, and as a result the transfer potential seen in the LV network will generally be low. The MV supply can therefore remain in service until the fault is found.

Should an arc occur, however, significant over-voltages may be generated. This is as a result of the intermittent nature of the arc current, the stray inductance associated with the fault path and the capacitance associated with the faulted phase. The arc current flowing in the fault extinguishes as it passes through zero; this occurs at the instant when the voltage across the fault capacitance is at its maximum (since a capacitive current leads the voltage by 90°). Thus when the fault first occurs, the faulted phase is pulled to ground potential, and the neutral terminal therefore rises by an amount equal to the peak voltage of the faulted phase. The arc current extinguishes at this time and the associated capacitance retains this charge thereafter. The system of phase voltages now operates one peak phase voltage above ground potential. Then, 180° later, the faulted phase achieves a potential equal to twice the peak phase voltage above ground, which may be sufficient for the arc to restrike. Arc fault current now flows through the capacitor to ground via any stray inductance, included within the fault path. This LC circuit rings at its resonant frequency, and the potential on the capacitor becomes reversed. As before, the arc current extinguishes as it passes through zero, having existed for about a half cycle of the LC resonant frequency. The faulted phase now lies at a potential of −2 times the peak phase potential, and after a further 180° this phase voltage reaches −4 times this level. The arc will again likely reignite and flow in the LC circuit, once again reversing the potential on the capacitance. This process repeats, and thus the neutral to ground voltage alternates in sign and grows in magnitude until eventually an insulation failure occurs somewhere in the circuit. This may be in the transformer itself or in MV cabling associated with it. The insulation requirement of transformers and associated equipment that are to be used on ungrounded systems must therefore be considerably higher than those used in solidly grounded systems.

Figure 11.23 shows the capacitive current paths as a result of a fault on phase c of an MV transformer. Under this condition the distributed capacitances C_a and C_b are each excited by a line voltage, and the resulting capacitive currents I_{Ca} and I_{Cb} therefore flow in the fault.

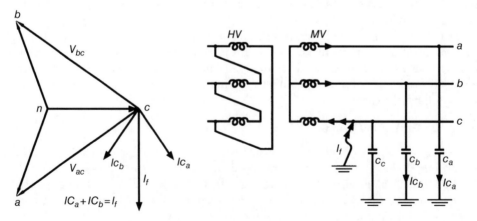

Figure 11.23 Capacitive fault currents in an unearthed MV transformer.

As for the IT LV earthing system, the capacitive current magnitudes are given by:

$$|Ic_a| = |Ic_b| = \sqrt{3} \, V_{ph}\omega C$$

and their vector sum equals:

$$Ic_a + Ic_b = 3 \, V_{ph}\omega C$$

Similarly with IT earthing networks, insulation monitoring is required between the neutral terminal and ground to detect the presence of the first fault, which may be located while the MV system is in operation.

Unearthed MV neutrals are sometimes used in Germany, Italy, Ireland and Japan, but in order to remove the risk of over-voltages generated through arcing faults, either *impedance earthing* or *resonant earthing* is frequently used instead.

11.5.2 Resonant Earthing (Compensated Earthing)

Resonant earthing is a novel system devised by Professor W Petersen in 1916. It uses a reactance inserted between the MV transformer neutral and ground (the Petersen coil) to resonate with the distributed network capacitance during phase-to-ground faults. Such an arrangement appears in Figure 11.24.

The capacitive current flowing to ground during a phase-*c*-to-ground fault is given by:

$$Ic_a + Ic_b = 3 \, V_{ph}\omega C$$

Also flowing in the fault is the inductive current, set up by the voltage on the faulted phase and the neutral reactance X_L. It is given by:

$$I_L = \frac{-V_{cn}}{X_L} = \frac{-V_{ph}}{X_L}$$

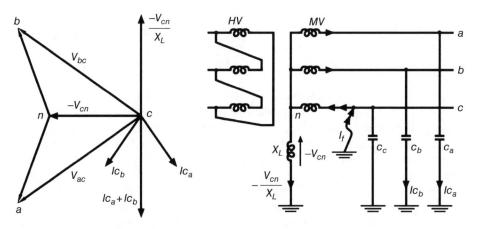

Figure 11.24 Resonant earthing system and phasor diagram.

By choosing the Petersen coil reactance appropriately, the total fault current can be reduced to near zero. Thus:

$$I_f = 3 V_{ph} \omega C - \frac{V_{ph}}{X_L} \approx 0$$

Therefore

$$3 \omega C \approx \frac{1}{\omega L}$$

For this to occur we require that:

$$\frac{1}{\sqrt{LC}} \approx \sqrt{3} \omega$$

This reduction in the magnitude of an earth fault current has the desirable characteristic of extinguishing most arc faults and thereby virtually eliminates the over-voltage phenomenon that often accompanies them. It also allows the MV distribution system to continue operating until the fault can be located and possibly even be repaired. For this reason the Petersen coil is also referred to as an *arc suppression coil* or a *ground fault neutraliser*.

To achieve complete compensation, the LC circuit must be resonant at $\sqrt{3}$ times the system frequency. Because the network's distributed capacitances vary with its configuration and with the connected load, it is necessary for the Petersen coil to be dynamically tuned to the network. This is achieved by altering a motor-controlled magnetic shunt within the reactor's magnetic circuit. Since the distributed capacitances are seldom exactly balanced, a small neutral current flows even during normal operation, creating a small zero-sequence voltage across the Petersen coil. This voltage is monitored, and the Petersen coil is tuned until it is minimized.

The cancellation of the residual capacitive current by the inductive current from the Petersen coil is seldom exact, and losses in the Petersen coil also result in an additional small current, in phase with the faulted winding potential flowing in the fault. Losses associated with the Petersen Coil can be modelled by the inclusion of a parallel connected resistance. This resistance is often augmented with an external resistance, and these collectively generate a resistive component of fault current which aid in fault detection. Since only the inductive component of the fault current flows in the Petersen Coil, the EPR is small, and therefore the potential transferred to the LV network will be as well.

Ground fault detection in resonant earthed systems can be difficult. It is relatively easy to detect the presence of a phase-to-ground fault by monitoring the zero-sequence voltage across the Petersen coil, but finding its exact location is not. One method is to progressively open each feeder in turn, until the fault condition disappears, but this is generally too disruptive and defeats the ability of a resonant earthed network to continue operating in the presence of the first fault. Detecting the small fault currents that occur in resonant systems may also be difficult, particularly if the compensation is nearly complete. One approach is to install a zero-sequence watt-metric relay on each feeder that originates from the transformer. (See problem 1, Chapter 12.)

11.5.3   Impedance Earthing

Impedance earthing of MV systems requires the inclusion of either a resistor or a reactor between the MV neutral terminal and ground. This earthing impedance can be used to limit the magnitude of the ground fault currents and hence the damage they do, and it is usually dimensioned so that the resulting fault currents are less than 1000 A. This limits the EPR in the case of faults in MV-LV transformers described above, and thus the potential transferred to the LV network.

Ground fault currents in impedance earthed systems are sufficiently large as to permit easy detection, by measuring either the zero-sequence current or voltage at the neutral terminal; however, these generally require the immediate disconnection of the supply. Resistive grounding impedances offer the advantage of damping over-voltages, but must be physically large to cater with the energy dissipated. On the other hand, inductive impedances dissipate very little energy but can create large over-voltages during fault clearance.

11.5.4   Over-Voltages

While the inclusion of an impedance between the neutral terminal and ground will reduce the phase-to-earth fault level, it does generate an over-voltage problem during a fault. This is because the voltage dropped across this impedance forces the neutral potential to rise. In the case of a solid phase-to-ground fault, the neutral potential will rise by an amount almost equal to the voltage of the faulted phase, and as a result the potential between healthy phases and ground becomes almost equal to the pre-fault line voltage. This over-voltage has implications for the winding insulation within the transformer, since it must now withstand an elevated potential during fault conditions.

Effectively Earthed Systems

IEC60044.2 (inductive voltage transformers), defines *earth fault factor* as the ratio between the maximum phase-to-ground voltage of a healthy phase during an earth fault, to the pre-fault phase-to-ground voltage. Distribution systems where the neutral point is solidly grounded are called *effectively earthed*, and in such systems the earth fault factor is less than 1.4. This is because in a solidly earthed system during a ground fault, the maximum voltage to ground of the healthy phases cannot exceed 80% of the phase to phase voltage, thus:

$$\frac{V_{healthy\ phase}}{V_{pre\ fault}} < \sqrt{3} \times 0.8 \approx 1.4$$

For an effectively earthed system we require that $X_o/X_1 < 3$ and $R_o/X_1 < 1$, where X_1 is the positive sequence impedance of the system at the fault and X_o and R_o are the zero-sequence reactance and resistance respectively. From the foregoing discussion ungrounded, resonant and impedance earthed systems are not effectively earthed.

Insulation Grading

Transformer manufacturers take advantage of the fact that when a star wound transformer is to be effectively earthed, it is possible to reduce the quantity of insulation at the neutral end of each phase winding. Since the star point will be firmly held at ground potential, there will be very little insulation stress between the neutral end of each phase winding and the earthed transformer core, even under fault conditions. Manufacturers use this fact by progressively increasing the winding insulation away from the neutral terminal, until at the phase terminal the winding is fully insulated. This technique is known as *graded insulation* or *non-uniform insulation*; however, it can only be applied to transformers whose neutral terminals will be solidly earthed. Transformers to be used in impedance earthed applications, including resonant earthing, must therefore be uniformly insulated, since considerable insulation stress occurs at the neutral end of the phase windings during phase-to-ground faults.

Table 11.4 provides a summary of the MV earthing systems used in various countries.

Table 11.4 MV earthing in some countries.

Country MV potentials	MV earthing system	Transformer frame to LV neutral connection	Insulation requirements
Germany 10 & 20 kV	Unearthed or resonant earthed $I_f < 60$ A	Connected if I_f. $R_t < 250$ V	Phase to phase (uniform)
Australia 11, 22 & 33 kV	Directly earthed $I_f =$ several kA	Separated, except if $R_t < 1\,\Omega$	Phase to earth (non-uniform)
Belgium 6.3 & 11 kV	Impedance earthed $I_f < 500$ A	Separated distance > 15 m	Phase to phase (uniform)
France 20 kV	Impedance Earthed Overhead: $I_f < 300$ A Underground: $I_f < 1000$ A	Separated, except if: $R_t < 3\,\Omega$ $R_t < 1\,\Omega$	Phase to phase (uniform)
Great Britain 11 kV	Directly earthed or Impedance earthed $I_f < 1000$ A	Separated, Except if $R_t < 1\,\Omega$	Phase to earth (Non-uniform) Phase to phase (uniform)
Italy 10–15 & 20 KV	Unearthed $I_f < 60$ A	Separated	Phase to phase (uniform)
Japan 6.6 kV	Unearthed $I_f < 20$ A	Connected if $R_t < 65\,\Omega$	Phase to phase (uniform)
Portugal 10 to 30 kV	Impedance Earthed Overhead: $I_f < 300$ A Underground: $I_f < 1000$ A	Separated, Except if $R_t < 1\,\Omega$	Phase to Phase (Uniform)
United States 4 to 25 kV	Directly Earthed or via a Low Impedance $I_f =$ several kA	Connected The MV neutral is usually distributed and it is also connected to the LV neutral, thus providing a metallic return path for MV fault currents.	Phase to earth (non-uniform)

($I_f =$ max earth fault current, $R_t =$ transformer grounding resistance) (Extracted from Cahier Technique No 173: *Earthing Systems Worldwide & Evolutions* with kind permission from Schneider Electric).

11.6   High Voltage Earthing

Because of the risks of over-voltages arising from impedance earthed transformers, and the expense of providing uniformly insulated windings at the potentials used in HV transmission, HV transformers are generally solidly earthed, and therefore can be fitted with non-uniform insulation. As a result HV earth fault currents are high, and immediate clearance is required on detection.

11.7   Exercise

Inspect some of the aerial MV/LV networks in your locality, particularly the location of MV/LV distribution transformers. Photograph the nameplate of any MV-LV transformers you find (a steady hand, a digital camera and a telephoto lens will be useful here), and in particular take note of the following:

1) Transformer voltage ratings, MV & LV
2) Overcurrent protection provided, MV & LV
3) Over-voltage protection provided, MV (if fitted)
4) Isolation provisions provided, MV & LV
5) Transformer vector grouping used
6) Earthing of pole fixtures and transformer tank and size of earth conductor
7) Earthing of the LV neutral conductor, particularly at or near the transformer as well as elsewhere on the LV feeder

As a guide, the photos below show typical MV-LV distribution transformers from several European countries, all having many items in common.

Transformer 1
Country: Australia
Rating: 22 kV:415 V, 100 kVA
Distribution: MV & LV *aerial*
Connection: Dyn11
Impedance: 4%
HV fuse protected (explosive drop out type)
LV fuse protected (withdrawable, mounted on LV cross-arm)
Pole fixtures and transformer tank earthed at pole LV neutral earthed one span away ($\approx$50 m)

Transformer 2
Country: Belgium
15 kV:410 V, 100 kVA
Distribution: MV *underground* (very low risk of MV to LV contact) LV *aerial* (bundled)
Connection: Yzn11
Impedance: 4%
HV fuse protected (HV isolator provided)
LV fuse protected (isolatable fuse links in enclosure at foot of pole)
Pole fixtures and transformer tank earthed at pole LV neutral earthed one span away (≈50 m)

Transformer 3
Country: France
20 kV:400 V, 100 kVA
Distribution: MV aerial, LV aerial (bundled)
Connection: Yzn11,
Impedance: 3.8%
HV fuse protected (Fixed type)
LV fuse protected (isolatable fuse links in enclosure at foot of pole)
Pole fixtures and transformer tank earthed at pole, LV neutral earthed one span away (≈50 m)
Earth distributed

Transformer 4
Country: Greece
15 kVkV:410 V, ≈ 250 kVA
Distribution: MV underground (very low risk of MV to LV contact) LV aerial (bundled)
Connection: Dyn11
Impedance 4.11%
HV fuse protected (fixed type, HV isolator provided on both incomers)
HV surge diverters fitted
LV fuse protected in LV enclosure at base of pole
LV distribution either bundled or aerial, 5 conductors
Pole fixtures and transformer tank earthed at pole LV neutral earthed remotely, earth distributed

Transformer 5
Country: Turkey
15 kV:400 V, 250 kVA
Distribution: MV: aerial, LV: cable to ground
mounted cubicle
Connection: Dyn11
Impedance: 4.11%
HV fuse protected (fixed type)
HV surge diverters fitted
LV fuse protected (LV circuit breaker in ground
cubicle?)
Pole fixtures and transformer tank earthed at
pole, LV neutral earthed remotely

11.8   Problems (*Earthing Grid Design*)

A substation earthing system provides a way of safely passing fault currents to ground without exceeding equipment limits or exposing persons in or near the substation to the risk of electric shock. Earth grid design can require considerable effort, both in field testing and in subsequent design analysis. The IEEE standard IEEE 80 *Guide for Safety in AC Substation Grounding* provides an excellent overview of this process, but there are many pitfalls, and earthing practitioners generally require considerable experience and the appropriate software in order to consistently produce compliant designs. Successful earth grid design requires an understanding of many parameters; some of the more important ones are discussed below.

1)   The *ground potential rise* GPR (also known as *earth potential rise*, EPR) is the maximum difference in voltage possible between the earth grid of a substation and a remote ground point (i.e. a reference earth), one that is assumed to represent a potential of zero volts. Under normal conditions the GPR will be very close to zero, but under fault conditions it increases to a potential equal to the product of the earth grid current and the earth grid resistance.

2)   The *earth grid current* I_G determines both the GPR and the step and touch voltages within the grid. It is equal to the earth fault current in situations where the ground circuit is the only return path to the source, or in cases where multiple return paths exist, that fraction of the fault current flowing into the earth grid.

3)   The *grid resistance* is the impedance presented by an earth grid to the flow of fault current; it is equal to the GPR divided by the earth grid current.

4)   *Step and touch voltages* arise within a substation due to the voltage gradients that occur between adjacent grid conductors and around the periphery of the earth grid. The step voltage is the maximum difference in potential likely between the feet of a person standing on the ground, their feet being separated by a distance of 1 metre. The touch voltage is the difference in potential between

the feet of a person standing on the ground and that of a grounded metallic object with which they are in hand contact. Figure 11.25 illustrates these potentials within a substation environment.

Figure 11.26 shows how hazardous voltages can arise outside a substation during a fault to ground, where the GPR or a fraction of it may be transferred to a member of the public.

5) The *fault clearance time* t_f and the *shock duration* t_s are particularly important parameters. The fault clearance time is determined by the speed of the upstream primary protection equipment and it effectively places an upper limit on the shock duration for a person in the substation. The shock duration inversely affects the fibrillation threshold current given by Equation (11.1), repeated here. In the following analysis we will assume that $t_s = t_f$.

$$I = \frac{K}{\sqrt{t_s}}$$

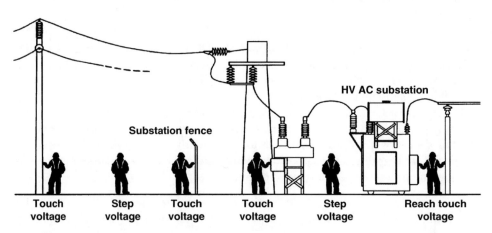

Figure 11.25 Step and touch potentials. (Courtesy of Energy Networks Association *Power System Earthing Guide Part 1: Management Principles*, 2010).

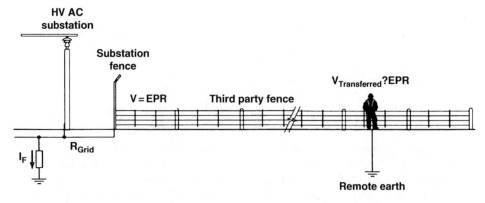

Figure 11.26 Transfer potential developed between a fence and a remote earth. (Courtesy of Energy Networks Association *Power System Earthing Guide Part 1: Management Principles*, 2010).

As t_s increases, the threshold current falls, and therefore so do the permissible step and touch potentials. This in turn places tighter requirements on the performance of the earth grid. In addition, longer clearance times also lead to an increase in the size of grid conductors, since they must carry grid current for a longer period during a fault. The fault clearing time must therefore be carefully calculated, usually with advice from protection engineers familiar with the upstream protection scheme.

6) The *soil resistivity* ρ_s is fundamental to the performance of the earth grid and directly affects the grid resistance. In regions where this is high the earth grid will require more conductors and/or ground rods in order to achieve a given grid resistance. Soil resistivity is a function of the soil model applicable to the site, which may be uniform or horizontally multi-layered, and in some cases vertically multi-layered. The soil resistivity profile is determined by a detailed measurement process across the proposed site by experienced engineers, followed by an analysis of the results, prior to the commencement of any construction work.

7) High resistivity *surface layers* such as crushed rock are often used throughout a substation to increase the apparent body resistance and to permit an increase the permissible step and touch voltages, thereby easing the earth grid performance requirements. In the following example such a layer will be necessary to achieve a compliant earthing design.

Earth Grid Design

We will evaluate the design of a simple earth grid suitable for a small ground-mounted distribution substation (like that in Figure 1.10), fed from the 22 kV aerial feeder shown in Figure 11.27. We will follow the iterative design process outlined in IEEE 80, whereby the design of the proposed earth grid is progressively refined until both the step and touch voltages fall below the permissible levels calculated for the substation. This may require reducing the spacing of ground conductors, the inclusion of deep driven ground rods or the provision of alternative current paths back to the source substation. The last may involve the provision of an overhead ground conductor, and improving the footing resistance of transmission towers in the vicinity, so that multiple earthing points may be established.

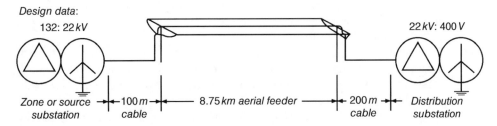

Figure 11.27 Network configuration.

Impedance Data

	132 kV:22 kV transformer (pu on 100 MVA base)	22 kV cables	Aerial feeder
Z_1	0.006 + j0.286	0.247 + j0.194 (Ω per km)	0.196 + j0.317 (Ω per km)
Z_2	0.006 + j0.286	0.247 + j0.194 (Ω per km)	0.196 + j0.317 (Ω per km)
Z_0	0.000 + j0.961	0.465 + j0.068 (Ω per km)	0.356 + j1.476 (Ω per km)

Primary protection fault clearance time: $t_f = 0.3$ sec ($= t_s$)
Soil resistivity: 40 ohm-m, uniform soil model
Crushed rock resistivity (wet): $\rho_s = 2500$ ohm-m
Thickness of crushed rock: 150 mm
Earth grid burial depth: 0.5 m (This is chosen to achieve an approximately constant moisture level around grid conductor, thereby minimising any seasonal variations.)
Land area available: 6 m × 6 m

IEEE 80 Design Process

The design process is outlined in the flowchart in Figure 11.28 and is summarised below:

Step 1: Determine the proposed site location and the area available for the installation of subsurface earth grid conductors. Obtain a soil model for the proposed location, based on a number of resistivity measurement traverses to sufficient depth to detail the soil layering across the site. (In this example, a uniform soil structure with resistivity of 40 ohm-m has been assumed for simplicity.)

Step 2: Determine the grid conductor size (Section 11.3 IEEE 80). This will require knowledge of the actual fault current and its duration, which will be determined in part by the grid resistance. In this example, we shall initially assume grid conductors 10 mm in diameter (80 mm²), chosen largely on the basis of physical strength. This choice can be refined once the grid current is known, should the conductor capacity prove insufficient.

Step 3: Evaluate the permissible step and touch voltages for the substation (Section 7, IEEE 80). These are based on Dalziel's equations and include the resistance of the human body, the fault clearance time, and the shoe resistance on the surface layer, which may be the underlying soil or an introduced layer of crushed rock.

Step and touch voltages for both surfaces are evaluated for 50 kg and 70 kg individuals. These limits are compared with the actual potentials occurring within the substation earth grid to determine design compliance. For controlled areas such as substations, step and touch potentials pertaining to 70 kg individuals are used to assess compliance, whereas in uncontrolled public spaces the 50 kg limits are used instead.

Step 4: Grid design: The earth grid design should include a buried conductor network arranged in a mesh covering the area to be grounded. Sufficient cross conductors must be provided to achieve an acceptably low earth grid resistance while providing sufficient control of ground surface potential gradients. Where space is limited or where an underlying low resistivity layer exists, ground rods may also be driven deep into the ground at regular intervals. IEEE 80 provides useful guidance regarding earth mesh layouts and earth rod placement in Section 9.

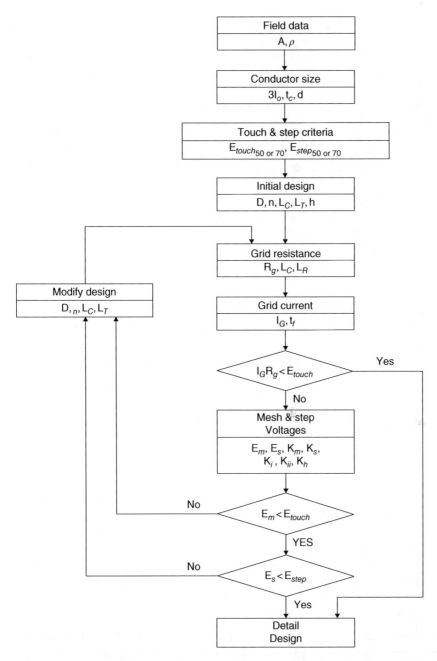

Figure 11.28 Earth grid design process. (Adapted and reprinted with permission from IEEE. Copyright IEEE 80 (2013) All rights reserved. Permission for further use of this material must be obtained from IEEE. Requests may be sent to stds-ipr@ieee.org).

Step 5: Calculate the resistance of the proposed earth grid according to the appropriate soil model. This may be either a uniform or multi-layer soil. In the case of multi-layer soils a detailed computer analysis is generally required due to the complex calculations involved.

Step 6: Evaluate the maximum fault current that will flow into the earth grid for an earth fault occurring in the substation; this may either be a phase-to-ground or a phase-to-phase-to-ground fault, depending on likelihood and severity. This current should be sufficiently inflated to allow for any future expansion of the source substation.

Since the earth grid current is not the same as the fault current in cases where an alternative return path exists, a *current division factor* generally must first be calculated. (In this example the only return path to the source substation is via the mass of the earth, therefore the division factor is unity.)

Step 7: Compute the *ground potential rise* (GPR) between the earth grid and a remote earth, from the product of the earth grid current and the earth grid resistance. If this is less than the calculated permissible touch voltage the earth grid design is compliant from the point of view of persons within the substation, and no further analysis is required. If not, the earth grid design may still be acceptable provided that the step and touch voltages within its boundary are less than the permissible values calculated in step 3. Additional measures may still be required to control the effects of the GPR on persons at a remote earth point, who may come into contact with conductors leaving the substation, metallic piping or fencing for example.

Step 8: Calculate the grid mesh voltage (which is defined as the maximum touch voltage that exists within the mesh of the earth grid) and the step voltage generated within the earth grid. A discussion of these potentials is given in Section 16.5 of IEEE 80.

Step 9: If both the earth grid step and touch voltages calculated in step 8 lie below the levels calculated in step 3 then the design is compliant. However, if this condition is not satisfied it must be revised in an effort to reduce both the grid resistance and the GPR. This may include reducing the earth grid electrode spacing, increasing its area, and driving additional earth rods around the perimeter.

Sample Grid Design Analysis

The analysis that follows includes many of the equations presented in IEEE 80. These frequently require the calculation of constants based on the geometry and parameters associated with the earthing installation, the calculation of which is quite detailed and yet it adds little to understanding the process. So as to highlight only the important aspects, the values of such constants will be provided where they arise. The interested reader will find an explanation of these in IEEE 80. For convenience we will adopt the symbols presented in IEEE 80.

Proceed as follows:

1) Calculate the maximum single-phase-to-ground fault level ($3I_0$) and the ground fault X/R ratio seen at the source substation. This occurs with an elevated bus voltage; assume a maximum of 1.1 pu. Answer: 5650 A, $X/R = 127$.

2) The feeder impedance substantially reduces the earth fault current seen at the distribution substation. Recalculate the above quantities, assuming a return ground path resistance of zero.

Answer: 1557 A, $X/R = 3.77$. (In this case because of the low X/R ratio and the relatively long clearance time, the effects of any transient DC component may be safely neglected.)

3) The equations for permissible step and touch voltages below are based on Equation (11.1) and the value of 1000 ohms as the resistance of the human body. They also include an allowance for the resistance provided by shoe contact with the ground surface, which can be made large if material such as crushed rock is spread at a reasonable depth over the entire area to be grounded.
Step voltage for 50 kg and 70 kg persons:

$$E_{step70} = (1000 + 6C_s\rho_s)\frac{0.157}{\sqrt{t_s}} \quad \text{and} \quad E_{step50} = (1000 + 6C_s\rho_s)\frac{0.116}{\sqrt{t_s}}$$

Touch voltage for 50 kg and 70 kg persons:

$$E_{touch70} = (1000 + 1.5C_s\rho_s)\frac{0.157}{\sqrt{t_s}} \quad \text{and} \quad E_{touch50} = (1000 + 1.5C_s\rho_s)\frac{0.116}{\sqrt{t_s}}$$

where C_s is the surface layer de-rating factor (Section 7.4 IEEE 80), ρ_s is the surface material resistivity (ohm-m) and t_s is the maximum shock duration (sec), i.e. the primary fault clearance time.

Since there will be no fence around this substation, we use the 50 kg step and touch potentials in assessing the performance of the earth grid. Evaluate these voltages assuming that: (a) no additional insulating material is placed on the ground surface (C_s = 1.0) (b) a 150 mm thick insulating layer of crushed rock is applied (C_s = 0.77). Answer: (a) E_{step50} = 263 volts, $E_{touch50}$ = 224 volts, (b) E_{step50} = 2660 volts, $E_{touch50}$ = 824 volts.

The difference between cases (a) and (b) is due to the insulating properties of the crushed rock layer, which may be used in some cases to achieve design compliance.

4) The initial design for the earth grid is shown in Figure 11.29. The grid resistance can be calculated from the equation:

$$R_g = \rho\left[\frac{1}{L_T} + \frac{1}{\sqrt{20A}}\left\{1 + \frac{1}{1 + h\sqrt{20/A}}\right\}\right]$$

where ρ is the soil resistivity (ohm-m), h is the depth at which the grid is buried (m), L_T is the total length of conductors and ground rods used in the grid (m) and A is the area of the grid (m^2).

Evaluate the grid resistance and express your result as a per unit quantity. Calculate the single-phase-to-earth fault current, assuming a bolted fault to the earth grid, including the calculated grid resistance, remembering that $3I_o$ flows into the grid resistance. Answer: R_g = 6.04 ohms or 1.25 pu, $I_F = I_G$ = 1160 A.

5) Calculate the ground potential rise (GPR) from the fault current and the grid resistance. Answer: GPR = 7000 V.

6) Since the ground potential rise exceeds the permissible touch potential, including the effects of a layer of crushed rock, we must now consider the step and touch potentials experienced within the grid. If these are less than the permissible values calculated above, the design is considered compliant, if not we must revise the grid design and repeat the process.

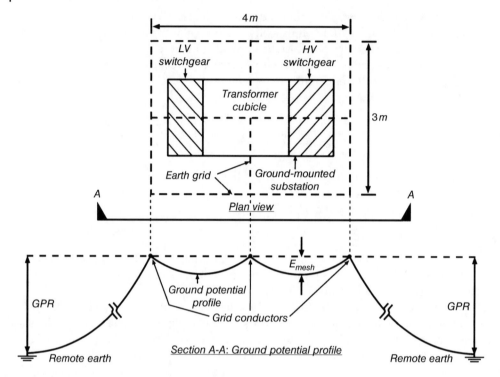

Figure 11.29 Initial grid plan and ground potential profile.

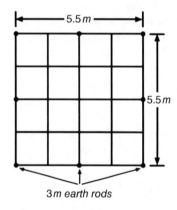

Figure 11.30 Revised earth grid design.

Using the equations below, compare the mesh voltage to the permitted touch voltage. Also compare the grid step potential with the permitted step potential calculated in step (3). Is the initial grid design compliant?

$$E_{mesh} = \frac{\rho K_m K_i I_G}{L_M} \text{ and } E_{step} = \frac{\rho K_i K_s I_G}{L_S}$$

where K_m is the spacing factor for mesh voltage, in this case $K_m = 0.696$, L_M is the total length of both grid conductors L_C and ground rods L_R, K_i is a correction factor for grid geometry, in this case $K_i = 1.09$, K_s is spacing factor for step voltage, in this case $K_S = 0.525$ and L_S is the effective ground conductor length for step potentials $= 0.75L_C + 0.85L_R$, where L_C is the total length of grid conductors and L_R is the total length of ground rods, in metres (of which there are presently none).

Answer: $E_{mesh} = 1676$ V, $E_{step} = 1686$ V. Although the step voltage experienced within the earth grid is less than the permitted value of 2660 volts, the initial grid design is *not* compliant because the mesh voltage exceeds the permissible touch voltage of 824 volts.

7) Figure 11.30 shows a revised grid design in which the area has been increased, additional conductors have been included and ground rods have been added around the periphery.

Calculate the new grid resistance R_g, the grid current I_G and the resulting GPR that will now occur. Answer: $R_g = 3.29$ ohms, $I_G = 1354\,A$ and therefore the GPR $= 4455\,V$.

8) Since the GPR still exceeds the permissible touch voltage, the grid compliance must again be assessed by evaluating the mesh and step potentials. Calculate these and compare them with the permissible limits and hence determine the grid compliance.

For the new grid design: $K_m = 0.593$, $K_i = 1.384$ and $K_s = 0.691$.

Answer: $E_{mesh} = 563$ volts, $E_{step} = 840$ volts. Since these potentials are less than the permitted touch and step voltages (824 and 2660 volts respectively), the revised design is now compliant in view of the use of a crushed rock layer.

Comments on Grid Design

IEEE 80 includes tables showing the current carrying capacity of copper grounding conductors as a function of fault current and duration. It also provides equations for the calculation of appropriate conductor sizes. In this example the grid current is relatively small and the conductors were chosen entirely on the basis of mechanical strength rather than current carrying capacity.

Where grid conductors cross, they must be welded together, the welds being of a similar cross-section to the conductors themselves, to enable any injected current to distribute itself as evenly as possible throughout the grid and into the earth. All electrical equipment in the substation must be bonded to the earth grid, and the bonding conductors used must also be adequately dimensioned.

On completion of a substation it is usual to perform commissioning tests to confirm that the performance of the earth grid is as expected. This may involve the injection of DC test currents between different parts of the earth grid and items of substation equipment to confirm low impedance connections, as well as the injection of off-frequency AC current into the earth grid from a remote earth, in order to measure the resulting GPR and mesh (touch) voltage. The values obtained can then be scaled according to the magnitude of the expected earth fault current, in order to determine compliance.

Transfer Potentials

An earth grid is primarily designed to preserve the safety of persons in a substation, and while it may be compliant in this respect, a dangerous transfer potential may still exist in cases where the GPR exceeds the touch voltage, as in this example. Transfer potentials arise when a person in a remote location touches a metallic object connected to the substation earth during fault conditions and thus becomes exposed to all or part of the GPR. Therefore depending on the location of the substation, further controls or risk reduction strategies may still be necessary to ensure the safety of general public.

There are numerous instances in which transfer potentials may arise, and we will consider three: communication circuits, low voltage neutral conductors and metallic pipelines and fences leaving or passing near to the substation. Metallic communications circuits leaving a substation can represent a major danger to personnel and communications equipment, depending on the amplitude of the possible GPR. In such cases it is usual to isolate metallic circuits by inserting an optical isolation device. This is often in the form of a short copper-to-fibre-to-copper connection, whereby any GPR is harmlessly dropped across the fibre linking the two copper circuits. In some cases, however, the requirements of the relevant communication standard may determine the allowable GPR at a proposed site. In any event the communication service provider must be aware of any hazards presented by the substation to both their assets and personnel.

Potentials transferred by an LV neutral conductor were partly discussed earlier. The provision of periodic earth connections along a radial neutral conductor assist in reducing the general transfer potential, at the risk of introducing local voltage gradients in the vicinity of each connection. Such risks can usually be managed by the provision of a ground mat or grounding loop. These earths also provide multiple return paths for fault current, thereby reducing both the grid current in the substation in question and the GPR. This is broadly the principle of the MEN system used in Australia and New Zealand.

Pole-top transformers provide a convenient means of separating HV and LV earth connections should this be necessary to control the transfer potential. By separating the HV and LV earths only a fraction of the GPR is transferred into the LV network. In such cases the LV neutral is often earthed at least one span away from its transformer.

Metallic piping and fences running from substation or passing through its zone of influence also need to be considered, with the relevant service provider notified of any hazards. In some cases it may be necessary to fit insulated pipe or fence sections at specific locations, of sufficient length across which part of the transfer voltage can be dropped.

Where the risk is considered excessive, several options are available. Firstly, in substations relatively close to the source, a ground wire can be run between the two earth grids. This provides a metallic return path directly to the source which reduces the grid current and thus the resulting GPR. Where distribution transformers are supplied by underground cables, the cable sheath is generally used for this purpose. However, the coupling of the two substations can have adverse consequences for the distribution substation which will be exposed to the effects of any earth fault in or near the source substation. This must be considered by the designer.

Secondly, where the substation in question is a reasonable distance from the source, the feeder impedance substantially reduces the fault current (as in this example). In such cases it may be feasible to further reduce the grid resistance through the inclusion of additional conductors or ground rods to reduce the GPR. This is only effective where the grid resistance is small with respect to that of the rest of the network; otherwise the resulting increase in grid current may offset any reduction in grid resistance.

Thirdly, the fault level in the parent substation itself may be reduced in magnitude by the inclusion of neutral earthing resistors or reactors. Petersen coils, for example, are particularly effective in reducing the GPR. Alternatively, in some cases it may be possible to decrease the fault clearance time t_s in the zone substation, allowing higher permissible step and touch voltages. While this approach can be effective, in most cases it will generally require a protection upgrade which may be impractical.

11.9   Sources

1 IEC International Standard IEC/TS 60479. *Effects of current on human beings and livestock Part 1 General aspects*, 2005, International Electrotechnical Commission, Geneva.

2 IEC International Standard IEC60364 *Protection for safety – protection against electric shock Part 4-41*, 2005, International Electrotechnical Commission, Geneva.

3 Schneider Electric. Cahier Technique No 173. *Earthing systems worldwide and evolutions*. 1995.

4 Schneider Electric. Cahier Technique No 172. *System earthing in LV*, 2004.

5 Schneider Electric. Cahier Technique No 114. *Residual current devices*, 2006.

6 Schneider Electric. Cahier Technique No 178. *The IT earthing system in LV*, 1999.

7 Parise G, *A new summary on the IEC protection against electric shock*, IEEE Transactions on Industry Applications Vol 49. No 2 March/April 2013.

8 Standards New Zealand LV Supply Earthing Systems Committee. *Low voltage supply earthing systems – is TT an acceptable alternative to MEN?* EEA Conference and Exhibition, Auckland, June 2013.

9 Schneider Electric. Electric Installation Guide: *Connection to the MV utility distribution network* Chapter B, 2008.

10 Standards Australia AS/NZ 3000. *Wiring rules*, 2007, Standards Australia, Sydney.

11 Jeff Roberts, Dr Hector J Altuve, Dr Daqing Hou. *Review of ground fault protection methods for grounded, ungrounded and compensated distribution systems*. Schweitzer Engineering Laboratories. Pullman, WA. USA.

12 Eaton White Paper. *Transient over-voltages on ungrounded systems from intermittent ground faults*, WP09-12. 2009.

13 IEEE Standard IEEE 80. *IEEE guide for safety in AC substation grounding*, 2013, IEEE Power Engineering Society, New York.

14 IEEE Standard IEEE 81, *IEEE guide for measuring earth resistivity, ground impedance and earth surface potentials of a grounding system*, 2012, IEEE Power Engineering Society, New York.

15 Energy Networks Association. *EG-0 Power system earthing guide, part 1 management principles*, 2010, ENA, Barton, ACT.

16 Energy Networks Association. *EG-1 Substation earthing guide*, 2006, ENA. Barton, ACT.

12

Introduction to Power System Protection

Power system protection is a relatively complex subject and requires a knowledge of the power system fundamentals presented in the preceding chapters. A thorough description of all the various protection schemes in use would fill an entire textbook. This chapter aims to provide the reader with an introduction to the underlying philosophy of power system protection and an exposure to some of the more common techniques in use. There are many excellent texts specifically on the subject of protection, several of which appear in the sources list at the end of this chapter.

We will principally confine our attention to the MV, HV and EHV networks, in which power system equipment is of a size and importance sufficient to justify the use of complex protection schemes. In contrast, the LV network generally comprises smaller and cheaper items of equipment, serving fewer customers and does not justify such protection. It is usually only *overcurrent* protected using fuses or magnetic circuit breakers, commensurate with the size and value of the assets.

12.1 Fundamental Principles of Protection

The primary function of the protection system is to rapidly disconnect any faulty piece of equipment from the power system in such a way that damage to both the equipment concerned and to the wider network is minimised, while causing the least disturbance possible to the connected load.

During fault conditions, network voltages near the fault become depressed, and for major faults this depression may extend over a considerable distance. Heavy fault currents also create large phase shifts throughout a network, and synchronism can be lost in parts of a network should transient stability limits be exceeded. If the network is already heavily loaded then its tolerance to additional fault currents is reduced, and unless faults are cleared quickly there is an increased risk that synchronism between different regions may be lost. This can lead to the *islanding* of parts of the network which may in turn lead to widespread blackouts. A fast-acting protection system is therefore important, not only from the point of view of the faulted equipment, but also for preserving the stability of the overall network.

Each MV or HV transformer, transmission line or generator is generally equipped with its own circuit breaker, capable of disconnecting it from the network in the event of a fault. Unlike magnetic circuit breakers used in LV applications, MV and HV circuit

AC Circuits and Power Systems in Practice, First Edition. Graeme Vertigan.
© 2018 John Wiley & Sons Ltd. Published 2018 by John Wiley & Sons Ltd.

breakers have no current sensing capability. Fault detection is achieved by supplying a *protection relay* with current and/or voltage information from the circuit being protected. The relay compares these parameters with pre-programmed internal settings, and issues a trip signal to the associated circuit breaker, should one or more of its settings be exceeded. Depending on the application, such a signal may be sent immediately (resulting in a rapid fault clearance), or alternatively after a short tripping delay, designed to permit a relay and circuit breaker nearer the fault to clear first.

12.2 Protection Relays

Protection relays were originally electromechanical devices which either operated on the principle of magnetic attraction of a movable armature to close a set of contacts, or from the rotation of an induction disc. The latter is seldom used now, but the former still finds application in many areas, not the least of which is in delivering the trip signal generated by an electronic protection relay, in the form of a voltage-free contact closure.

By way of an example, an electromechanical overcurrent relay built in the 1960s is shown in Figure 12.1. This is excited by the secondary current from a CT, and providing this exceeds the relay's *pick-up current*, the disc will begin to turn against the restraining torque of a hairspring. After a period of rotation (determined by the relay setting), the

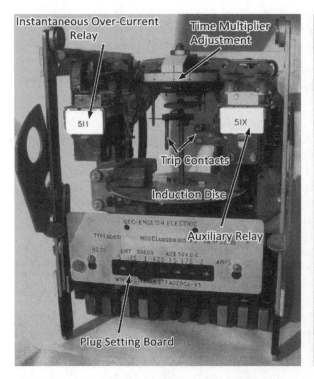
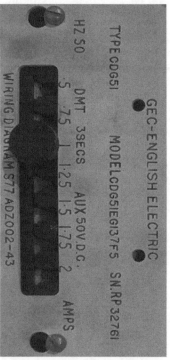

Figure 12.1 English Electric electromechanical IDMT overcurrent relay element and plug board from the 1960s.

trip contacts are closed by a pin turning with the disc, tripping the associated circuit breaker. The torque produced by the disc increases with the magnitude of the CT current and the number of turns on the exciting winding, the latter being determined by the *plug board setting*. The relay in Figure 12.1 is designed for a 1 amp CT and the current setting can be adjusted between 0.5 and 2 amps, by inserting the plug into the appropriate receptacle. Over-current relays of this kind therefore generate trip times that vary inversely with fault current.

Static Relays

Electromechanical relays were used successfully for many years, but these mechanical devices are subject to drift and fatigue, and since they lie inoperative for long periods, periodic testing – and in some cases cleaning – is necessary in order to be sure that a given relay will operate correctly when called upon to do so. In the 1960s and 1970s electromechanical relays began to be replaced with electronic relays using analogue circuits to synthesise the required protection characteristics. These were known as *static* relays since they had no moving parts. Whereas electromechanical relays were powered by the CTs and/or VTs in the circuits they protected, static relays require a secure and reliable source of power to operate. The substation DC tripping supply is used for this purpose. This potential is provided by a battery and it generally lies in the range between 24 and 250 volts DC, with 125 volts commonly used.

Digital and Numerical Relays

The advent of the microprocessor in the 1970s and 1980s saw analogue relays replaced by digital devices in which currents and voltages are represented digitally, and an algorithm is used to implement the protection function. Early digital relays generally performed the task of a single protection element and multiple devices were required to provide a range of protection functions. Advances in technology, the arrival of digital signal processors and parallel processing, have enabled many functions to now be performed by one device, known as a *numerical relay* or a *multi-function relay*. Numerical relays can therefore perform the tasks of multiple protection elements and may have spare processing capacity to enable other parameters to be displayed, including active and reactive power flows and a harmonic analysis of voltages and currents. Capture of current and voltage waveforms is available from some numerical relays, so that faults can be analysed after the event. In addition, self-diagnostic testing is generally provided, so that in the event of an internal fault an operator can be alerted.

12.3   Primary and Backup Protection (Duplicate Protection)

Primary protection is designed to trip only that equipment within a defined protection zone. It is therefore fast, and it isolates a minimum portion of the network. So as to ensure reliability, duplicate primary protection schemes are frequently employed, often referred to as *A* and *B* or *X* and *Y* protection. These schemes may be duplicates of one another or they may comprise different protection elements. They are designed to be entirely independent and therefore they use separate CTs and, where possible, different VT windings as well. Further redundancy is provided by using different relays for each

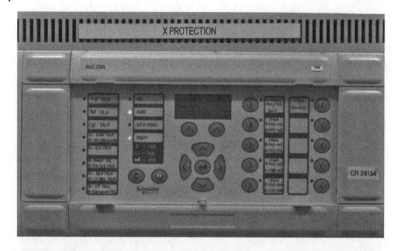

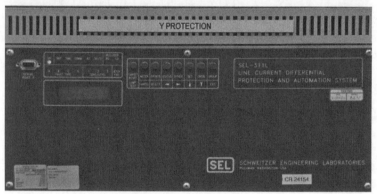

Figure 12.2 X and Y protection: Schneider P546 and Schweitzer 311 L numerical relays.

scheme, often from different manufacturers, like those in Figure 12.2. In major substations, tripping batteries are also duplicated, so that both primary protection schemes operate from separate DC supplies.

Redundant circuit breakers are rarely provided due to the associated expense, but duplicate trip coils in the same breaker are generally provided where duplicate protection is employed. In any event, the health of the trip coil is vital since failure to detect an open circuited coil could be disastrous. For this reason, *trip supervision* circuitry is used on critical circuit breakers. This generally involves continuously passing a small current through the trip coil and any other contacts in series with it to ensure that the circuit is continuous, but insufficient to initiate a trip or to overheat the winding. An alarm is generated if the circuit is found to be inoperative. Duplicate protection schemes generally operate entirely independently of each other, either being able to initiate a trip.

Check Relays

In addition to the main protection relay, high risk schemes such as bus-zone or sensitive earth fault protection, generally include a separate check relay in order to provide an independent confirmation of a fault or to provide discrimination. (For example, this may be an overcurrent relay looking for a residual current flowing towards a transformer neutral terminal.) In such cases both the main and the check relay must concur for a trip to be initiated. Operation of the check relay will never in itself cause a trip.

In modern numerical bus-zone protection, the main and check elements often occur within the same relay, although its architecture usually ensures that decisions are made by two independent algorithms. For schemes such as sensitive earth fault protection the check relay is a separate device, even in today's numerical environment.

Backup Protection

Backup protection is provided to clear a fault should the primary scheme(s) fail to operate. This may be through the operation of a nearby graded overcurrent relay or perhaps by a circuit breaker failure relay that has detected current flow after the breaker concerned has been sent a trip command.

Backup schemes may also be located in another substation, remote from the primary protection system, so that no item of equipment is common to both. (An example of this is the Zone 2 setting of a remote substation's distance relay.) Backup protection tends to be slower than the primary system it protects and it generally interrupts considerably more of the network.

Reliability

The reliability of a protection system is a function of its design, the equipment used and the care taken in its construction. The simpler and more robust the design the more likely it will be to operate correctly when required to do so. Modern numerical relays provide multiple protection functions and apply any necessary phase shifts digitally. As a result, they require relatively simple current and voltage connections, although the design of the relay itself is more complex. On the other hand complicated connection arrangements are frequently required with static and electromechanical relays to provide the correct phase relationships between the parameters being measured. This introduces the possibility of wiring errors which may not become apparent until the relay fails to operate correctly in service. The commissioning of protection schemes is therefore particularly important. The required protection functions are simulated during commissioning by injecting appropriate currents and voltages into each relay to ensure that it operates as expected in all scenarios.

12.4 Protection Zones

Protection schemes generally operate in defined zones, each being associated with one or more protection relays protecting the primary equipment within the zone. In order to ensure no portion of a network is omitted there is generally an overlap between adjacent zones, as shown in Figure 12.3. Here each transformer and outgoing feeder has its own protection zone, as does each half of the 33 kV bus. The boundary of each zone is determined by the location of the associated current transformers, and in this case

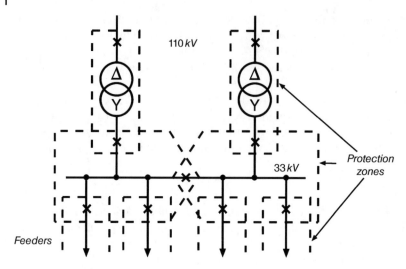

Figure 12.3 Zones of protection.

CTs are required on both sides of each circuit breaker. This is not always practical how-ever and when CTs must be located on one side only (as is frequently the case in metal clad switchgear), blind spots can occur. In such cases in order to completely isolate a fault it will be necessary to trip a breaker in the adjacent zone as well.

Each of the relays associated with Figure 12.3 should be sufficiently sensitive to operate on the smallest fault likely to occur, and it must only operate for faults within its zone of protection, i.e. it should be able to discriminate against out of zone faults. Discrimination (or selectivity) in radial distribution networks is often achieved using time/current grading, where all relays operate under an inverse time-current characteristic in which each relay operates faster for larger fault currents. By providing an appropriate time grading interval between adjacent zones, the relay closest to the fault can be made to operate first, thereby tripping a minimum of equipment in response to the fault. Should the closest circuit breaker fail to operate, then that in the adjacent zone will clear once the grading interval has expired. This will however be at the expense of a longer clearance time and generally a greater disturbance to the network.

Unit Protection

Time/current grading is inherently slow in its operation since relays closer to the source of supply are delayed in their operation in order to permit those closer to the fault to operate. Generators, transmission lines and busbars all require fast protection for the reasons already described. In such cases unit protection is employed to provide a fast fault clearance in a defined zone of protection. Current transformers are required at each zone boundary, from which the associated protection relay compares the current entering with that leaving. Should an in-zone fault occur, the substantial residual cur-rent created will be seen by the relay, resulting in the tripping of all associated circuit breakers. An important requirement of a unit protection scheme is its ability not to operate in the event of a fault external to the zone of protection. The scheme must remain stable for external faults (often called through faults).

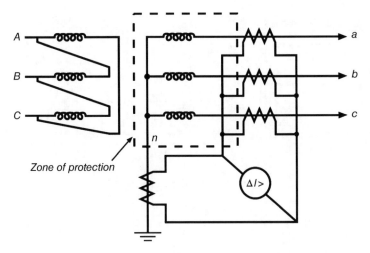

Figure 12.4 Restricted earth fault (REF) unit protection scheme.

Figure 12.4 shows an example of unit protection in the form of *restricted earth fault* (REF) scheme surrounding the phase windings of a star connected transformer. The REF relay is essentially an overcurrent relay operating on the residual current arising from the CT connection shown. This will only occur in the event of a ground fault within the protected zone and because unit protection is not required to grade with any other, it can be made fast acting. (See Section 12.6.2 for more detail.).

12.5 Overcurrent Protection

Overcurrent protection was one of the earliest protection schemes to evolve and it is one of the simplest and most widely used, particularly in radial MV distribution networks. Overcurrent relays are provided with a single set of CTs, and they monitor phase and/or earth fault currents at a particular location. They are frequently placed at multiple locations throughout a radial network and graded in such a way that the relay nearest the fault clears before those further upstream can operate. In this way, a minimum network disruption is achieved.

There are two principal methods by which overcurrent relays can discriminate a fault: time and time and current. In the case of smaller networks in which the fault level changes little, time-based discrimination must be used. All relays are set to pick up at the same fault current, simultaneously starting their tripping delay timers. Those further downstream have progressively shorter delays, and the relay closest to the fault trips its circuit breaker first; once the fault is cleared, all other relays reset. This is known as a *definite time overcurrent scheme,* but it has the disadvantage that relays nearer the source (where the fault level is higher) require longer operating times.

In networks where the fault level decreases substantially with the distance from the source (due to long feeders or the presence of intermediate transformers) a combination of time and current can be used in fault discrimination. The relay in Figure 12.1 is designed for such an application. It has a *standard inverse* time-current characteristic,

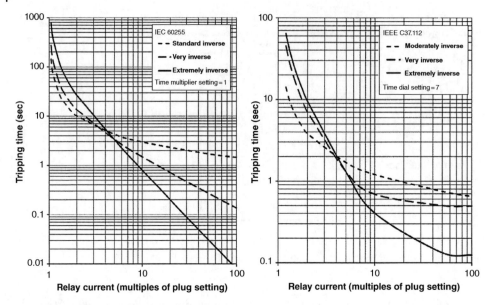

Figure 12.5 Relay time-current characteristics.

like that shown in Figure 12.5. This relates the relay's operating time to the current flowing in its excitation winding, expressed in multiples of the plug setting value. Higher currents generate larger torques, resulting in faster rotation of the disc and therefore faster relay operation. The relay in Figure 12.1 is designed to be connected to a 1 A CT and has plug settings ranging from 0.5 to 2 A, in 0.25 A steps. It is shown with a plug setting of 1 A and therefore according to this time-current characteristic, with 10 A flowing it will trip in 3 seconds.

Figure 12.5 shows three inverse time-current curves defined in the European standard IEC 60255 *Measuring Relays and Protection Equipment: Functional Requirements for Overcurrent Protection* and three curves from the North American standard IEEE C37.112 *Standard Inverse Time Characteristic Equations for Overcurrent Relays*. Equations describing these curves appear in Table 12.1. The standard inverse curve is used in most applications, whereas the *very inverse* characteristic is useful when the fault level decreases rapidly, its steep slope allowing a rapid reduction in tripping times for larger fault currents. The *extremely inverse* characteristic is used where grading against downstream fuses. Because of these inverse-time characteristics and the fact that for large currents these relays operate in a definite minimum time, they are collectively referred to as *inverse definite minimum time* (IDMT) relays.

An additional relay adjustment is provided by the *time multiplier setting* (TMS) as defined in IEC 60255, referred to as the *time dial* (TD) setting in IEEE C37.112. Both these adjustments alter the operating time of a relay by changing the angular distance through which the disc must turn to effect a trip. A family of standard inverse tripping characteristics is shown in Figure 12.6, where the time multiplier settings range from 0.1 to 1.0. Time multiplier settings are infinitely adjustable in electromechanical and static relays and nearly so in digital relays. The TMS dial can be seen in Figure 12.1.

Table 12.1 Inverse time-current characteristic equations.

IEC 60255		IEEE C37.112	
Standard inverse (SI)	$t = TMS\left[\dfrac{0.14}{M^{0.02}-1}\right]$	Moderately inverse	$t = \dfrac{TD}{7}\left\{\left[\dfrac{0.0515}{M^{0.02}-1}\right]+0.114\right\}$
Very inverse (VI)	$t = TMS\left[\dfrac{13.5}{M-1}\right]$	Very inverse	$t = \dfrac{TD}{7}\left\{\left[\dfrac{19.61}{M^2-1}\right]+0.491\right\}$
Extremely inverse (EI)	$t = TMS\left[\dfrac{80}{M^2-1}\right]$	Extremely inverse	$t = \dfrac{TD}{7}\left\{\left[\dfrac{28.2}{M^2-1}\right]+0.1217\right\}$

t = tripping time
M = plug setting multiple
TMS = time multiplier setting $(0 \le TMS \le 1)$

t = tripping time
M = plug setting multiple
TD = time dial setting $(1 \le TD \le 15)$

(Adapted and reprinted with permission from IEEE. Copyright IEEE C37.131 (2012) All rights reserved. Permission for further use of this material must be obtained from IEEE. Requests may be sent to stds-ipr@ieee.org) (IEC 60255-151 ed.1.0 Copyright © 2009 IEC Geneva, Switzerland. www.iec.ch)

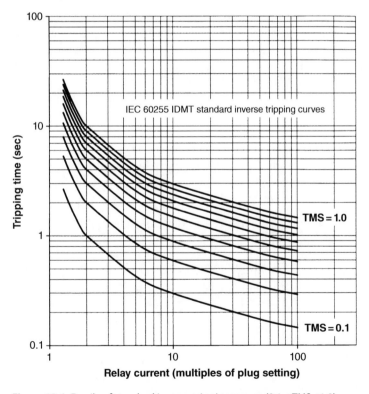

Figure 12.6 Family of standard inverse tripping curves $(0.1 \le TMS \le 1.0)$.

Pick-up/Reset Ratio

The pick-up current is the minimum current that will cause the disc to turn. A relay will not pick up for secondary currents equal to the plug setting, but will do so for currents between 1.05 and 1.3 times this value. Should the load current just exceed the pick-up current, the disc will slowly begin to turn, or in the case of a digital relay, the tripping delay timer will commence. If the current subsequently falls below the pick-up value the relay should reset, allowing the disc to return to its rest position. There is a little hysteresis between the pick-up and the reset currents, defined by the pick-up/reset ratio which typically lies in the range 0.9–0.95. Therefore the current setting must lie sufficiently above the maximum load current to enable the relay to correctly reset in such circumstances, avoiding a nuisance trip, i.e.

$$Current\ Setting \geq \frac{Max\ Load\ Current}{Pickup/reset\ ratio} \tag{12.1}$$

Grading of Overcurrent Relays

In the case of IDMT relays a *grading margin* is provided between relays in adjacent zones through the choice of appropriate time multiplier settings so that there is little chance of any relay operating before that immediately downstream has had an opportunity to clear the fault. Grading margins include allowance for relay timing errors and overshoot, CT composite error and circuit breaker operating time. A typical grading margin for an electromechanical relay is about 0.4 seconds, or 0.3 seconds for a digital relay. It is the additional time required before the upstream relay operates, should the downstream one fail. The grading margin is applied at the maximum fault current seen by each relay. Provided that all relays use the same tripping curve, this will ensure that satisfactory discrimination will also exist for lower fault currents as well.

The TMS or TD can be estimated from the tripping curves, or obtained more precisely from the equations:

$$TMS = \frac{required\ operation\ time}{operation\ time\ for\ TMS = 1}\ or\ TD = \frac{required\ operation\ time}{operation\ time\ for\ TD = 7} \tag{12.2}$$

Example: Determine the trip time of a standard inverse (SI) overcurrent relay with a plug setting of 1.5 A and TMS of 0.4 when connected to a 1000:1 CT with a primary fault current of 10,000 A.

Solution: The CT secondary current will equal 10 A which equates to 6.67 multiples of the plug setting (i.e. $M = 6.67$). The trip time can be read from Figure 12.6, or it may be calculated using the SI equation. By either method the trip time $t \approx 1.45$ seconds.

Overcurrent Relay Connections

Overcurrent relays may be installed on three-wire circuits where, under normal conditions $I_A + I_B + I_C = 0$, as depicted in Figure 12.7, allowing both phase and earth faults to be detected. Traditionally the arrangement of Figure 12.7a would have been adopted using three relays of the type shown in Figure 12.1. Two of these would be used to detect phase overcurrents, one in the A phase and one in the C phase, while

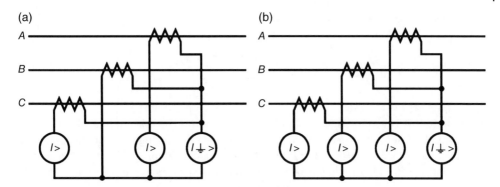

Figure 12.7 (a) Two-phase overcurrent and earth fault connection (b) Three-phase overcurrent and earth fault connection.

the third would see the residual current and detect earth faults. Since the third element only sees zero-sequence current, it can be provided with quite a low current setting (20–30% of the phase current setting is typical), necessary to limit damage from earth faults or to operate on the small earth fault currents that flow in impedance-earthed networks.

A phase-to-phase fault will result in overcurrents on at least two phases, and these will be seen by at least one phase overcurrent element. However, in the special case of a delta-star transformer protected by an overcurrent relay on the HV side, a phase-to-phase fault on the LV will result in twice the fault current appearing on one primary phase, which may be the one without an overcurrent element. While such a fault will be cleared, this may not occur as fast as if the heavily faulted phase had its own overcurrent element. These days all phases are provided with a dedicated overcurrent element from either a digital or a numerical relay in the arrangement shown in Figure 12.7b.

12.5.1 Instantaneous (High-Set) Overcurrent Elements

Overcurrent relays are often provided with an instantaneous or high-set element that trips without an appreciable delay should the high-set current threshold be exceeded. The current in the faulted circuit however will be interrupted a few cycles later, once the breaker has opened.

Instantaneous elements permit an upstream relay to be graded at the instantaneous setting of the relay immediately downstream and not the maximum fault current at that location, therefore enabling faster fault clearance to be achieved. They are usually applied in locations where a substantial change of impedance – and therefore of fault level – occurs between zones. They should be set so that they do not see faults in the next zone of protection, otherwise discrimination may be lost. For example, an instantaneous overcurrent element on the HV side of a transformer must be set so that it only sees faults at the HV terminals; it must never overreach into the LV protection zone if discrimination is to be preserved.

Transient Over-Reach

The transient DC offset early in a fault current waveform will be seen by an instantaneous relay and may cause it to operate prematurely and overreach into the next zone.

The *transient overreach percentage* is defined as:

$$Transient\,Overreach = \frac{\left(I_{RMS\,pick\,up} - I_{offset\,pick-up}\right)}{I_{offset\,pick-up}} \times 100\% \tag{12.3}$$

where $I_{RMS\,pick\,up}$ is the relay's steady state RMS pick-up current and $I_{offset\,pick-up}$ is the steady state RMS current which, when fully offset, causes the relay to pick up.

To avoid transient overreach the instantaneous current setting must be increased by this percentage, which in the case of high X/R ratios, can be large. This may result in a setting that is unusable should it exceed the maximum steady state in zone fault current, or result in the relay overreaching into the adjacent zone.

Overcurrent Protection: Current Transformer Requirements

Because IDMT overcurrent protection is inherently slow the transient performance of the CTs used is generally not critical, and some CT saturation in the first few cycles can be tolerated since the relay will usually operate well, after the transient component has died away. Class P transformers are generally used in such applications, but the accuracy limit factor chosen must be sufficient to permit the steady state fault current to flow in the connected burden. For example a 500:1, 15 VA, 10P20 core will deliver a maximum of 20 A to a 15 Ω burden with a composite error less than or equal to 10%; it will therefore be suitable for steady state primary fault currents of up to 10,000 A at rated burden. Where the total connected burden is less than the rated burden, the accuracy limit factor increases according to:

$$ALF = ALF_{Rated} \frac{\left(R_{ct} + R_{bRated}\right)}{\left(R_{ct} + R_{bApplied}\right)} \tag{12.4}$$

and the maximum primary fault current increases in proportion. Since the burden presented by modern numerical relays tends to be very small, the cabling burden is usually a significant component of the total burden seen by the CT, which may be considerably less than the rated burden.

In the case of instantaneous protection however, where fast operation is required, the transient performance of the CTs concerned is important. Due allowance must therefore be made for any DC transient component of the fault current. This will require the CT to generate larger CT secondary voltages, increased by the *transient factor* $(1 + X/R)$. In such cases IEC class PX or TP or IEEE class C CTs should therefore be used.

12.5.2 Directional Overcurrent Relays

There are numerous applications in which the correct operation of an overcurrent relay depends on detecting the direction of a fault current. For example, a fault on one of two parallel connected feeders, like those shown in Figure 12.8, will result in fault currents flowing from both busses towards the fault. As a result, all relays will see the fault and if each has a simple overcurrent element, it is possible that all breakers will trip, resulting in complete loss of supply to the load. Instead, in such cases, it is usual to fit directional overcurrent relays at the load, operating whenever the fault current exceeds the relay setting and is flowing away from the bus into the protected feeder. Such an arrangement

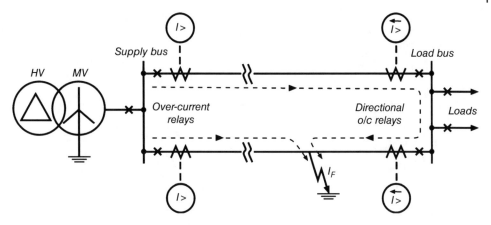

Figure 12.8 Parallel connected feeders.

will ensure that the faulted feeder is tripped while the healthy one remains in service. The directional elements can be provided with low time multiplier settings since they do not need to be graded with the upstream relays.

The fault current direction is defined as that in which power flows towards the fault. In general this will be a reactive power flow, since the network impedance limiting the fault current is itself largely reactive. The directional element is therefore based on a VAr-meter, configured to measure the VArs flowing towards the fault. It must correctly discriminate the direction of this power flow, and allow a trip only when VArs flow in the reverse direction.

In Chapter 10, we saw that reactive power could be measured using a wattmeter supplied with a voltage in quadrature with the phase concerned. Therefore in order to measure the A phase VArs, we supply the A phase wattmeter element with a scaled version of the BC line voltage together with the A phase current; it therefore measures $\left(1/\sqrt{3}\right)V_{BC}I_A\cos\left(\phi_A-90°\right)=V_AI_A\sin\left(\phi_A\right)$. Such an instrument is therefore most sensitive to currents which are either in phase or 180° out of phase with the BC line voltage. Similarly this VAr element is least sensitive to currents lagging or leading the BC line voltage by 90°.

Electromechanical instruments must produce a torque in order to turn the disc, and exhibit a *maximum torque angle* (MTA), which is defined as the angle between the applied voltage and current at which the relay produces maximum torque, i.e. the angle at which the relay is most sensitive. In the example above, the MTA is 0°, as depicted in Figure 12.9. In the case of static and digital relays, this angle is referred to as the *relay characteristic angle* (RCA). For currents that lie 90° away from the MTA or RCA, the relay has zero sensitivity. In terms of electromechanical relays this is referred to as the zero torque line.

A directional relay must determine whether the reactive power flow is towards the faulted feeder, in which case a trip signal must be generated, subject to a sufficiently large fault current flow. Alternatively, if the reactive flow is away from the feeder, then relay operation must be restrained, regardless of the magnitude of the fault current seen. Directional overcurrent relays therefore require two elements, a directional element and an overcurrent element, both of which must concur in order for a trip signal to be generated.

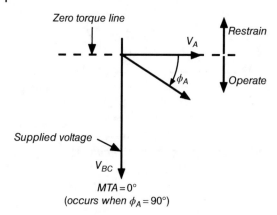

Figure 12.9 Maximum torque angle.

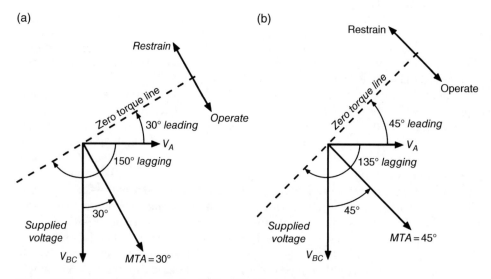

Figure 12.10 (a) The 90-30° connection (b) The 90–45° connection.

In protection applications the quadrature voltage is known as a *polarising voltage*. There are several different voltages that could be used for polarising a relay, but in practice relatively few are used. Directional earth fault relays use the zero-sequence voltage, V_o (or in practice the residual voltage, $3V_o$) for relay polarisation, since this is directly related to the zero-sequence earth fault current.

Directional phase overcurrent relays generally use a quadrature voltage connection as described above, although the applied voltage is usually internally phase shifted to produce an MTA or RCA other than zero degrees. Figure 12.10a shows such an arrangement. As above, the *BC* line voltage is supplied to the *A* phase element, to which an additional 30° anticlockwise phase shift is applied internally, yielding an MTA of 30°. The relay therefore has a maximum sensitivity to currents lagging the *A* phase voltage by 60°. It exhibits 50% of this sensitivity to currents in phase with the *A* phase voltage, and 86% to quadrature currents. By shifting the MTA in this way, the relay provides a useful response for currents leading the *A* phase voltage by 30° to lagging currents of about 150°. This arrangement is summarised as a 90-30° connection.

Figure 12.10b depicts the 90-45° connection in which the supplied voltage has been internally phase shifted towards the *A* phase voltage by 45°. This provides an MTA of 45° and a useful response for currents leading the *A* phase voltage by 45° and lagging it by up to 135°. The relay must operate for power flows from the bus towards the faulted circuit and restrain for flows away from it.

12.6 Differential Protection

Differential protection is a form of unit protection frequently used in the protection of transformers, transmission lines and busbars. It is based on Kirchhoff's current law, and operates by comparing the phase currents flowing on either side of a transformer, or between the ends of a transmission line, as depicted in the single-phase circuit of Figure 12.11. An out-of-zone fault generates a through-fault current in each CT which circulates through the secondary windings, bypassing the relay. On the other hand, an in-zone fault may be seen by one CT, or alternatively it may be fed from both ends of the line. In either event, the secondary current(s) will flow through the relay resulting in a rapid trip.

12.6.1 Transformer Differential Protection

Differential protection frequently forms part of transformer protection schemes. Depending on a transformer's vector grouping, there is often a phase shift between primary and secondary voltages and currents. Such a shift must be allowed for in a differential protection scheme so that the primary and secondary CT currents are compared in-phase, otherwise a spill current will flow in the relay, possibly causing an unwanted trip. Electromechanical and static relays derive a complementary phase shift from the CT connection itself. This concept is illustrated in Figure 12.12 which shows differential protection applied to a delta-star Dyn11 transformer, in which the LV line currents lead those on the HV side by 30°. A complementary 30° lag in the measured LV currents is achieved by connecting the CTs on the delta side of the transformer in a star configuration, and those on the star side in delta. (Note that the delta connected CT windings on the LV side introduce a factor of √3 between the current flowing in each CT and those seen by the relay. A ratio correction will also be necessary to accommodate the difference in magnitude between HV and LV currents.) In the case of electromechanical and static relays this is achieved through the use of interposing current transformers in the

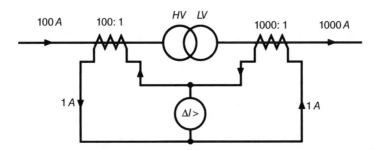

Figure 12.11 Basic differential protection.

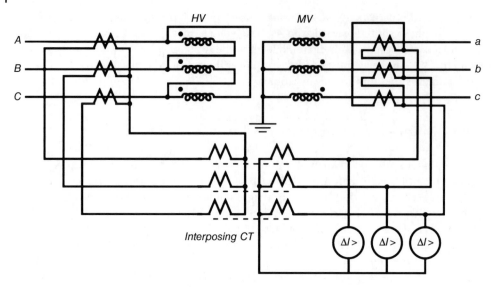

Figure 12.12 Transformer differential protection.

CT secondary circuit as shown in Figure 12.12, but in digital and numerical relays both phase and ratio corrections are implemented in software, permitting all CTs to be star connected, regardless of the transformer's configuration.

(A dot convention is commonly used in transformer schematics like Figure 12.12 to identify winding polarities. The end of each winding marked with a dot corresponds to the same polarity.)

Transformer Magnetising Current
The transformer's magnetising current must also be catered for in a differential scheme. In the steady state, this is generally only a few per cent of the transformer's rated current, but as discussed in Section 7.5.4, at switch-on the magnetic inrush current may become very large indeed, momentarily exceeding the rated current. Such an inrush current will appear as an internal fault, and the relay must be restrained from operating while the steady state flux is established. The DC component present during magnetic inrush (see Figure 7.18) means that the current waveform does not possess half wave symmetry, and therefore contains even harmonics (see Section 13.2.2). Of these the second harmonic is usually the largest, and its presence is frequently used to temporarily restrain the relay's operation.

Figure 7.18 also shows that the inrush current is close to zero for a considerable fraction of each cycle. This feature can also be used to inhibit relay operation until a steady state flux is established in the core.

Zero-Sequence Current
An earthed star or zigzag connected winding on one side of the transformer can deliver zero-sequence currents to an out-of-zone earth fault which will be misinterpreted by a differential relay as an in-zone fault. It is therefore necessary to prevent the passage of zero-sequence currents in the associated CT secondary circuit. In the case of electromagnetic or

static relays this is achieved by including a delta connection, either in the CT's windings themselves or in those of an associated interposing CT. Digital and numeric relays have the software capability of filtering zero-sequence currents in addition to making phase and ratio corrections. Figure 12.12 illustrates this for a Dyn11 transformer where HV zero-sequence line currents are blocked by the transformer's delta winding. Zero-sequence currents can flow in the star connected LV windings but are excluded from the secondary circuit by the delta connected CTs, which are also required for the phase correction.

Tap-Changers

Where a tap-changer is present, CT ratios are usually set to suit the principal tap (i.e. that for which the nominal primary voltage will generate nominal secondary voltage, on an unloaded transformer). A tap-changer dynamically alters the transformer's effective turns ratio, partially upsetting the primary–secondary current balance, which can lead to a spill current flowing in the relay. However, correct operation is ensured through the use of a biased differential relay in which the spill current required to operate the relay increases with the magnitude of the current flowing.

Biased Differential Relays

Magnetising current, the presence of a tap-changer or CT ratio errors can all contribute to an erroneous spill current flowing in the relay's operating winding. These effects could be negated by setting the relay's differential operating current above the maximum spill current likely to arise from such sources; however, this would unnecessarily limit the sensitivity of the relay. Instead, the relay operating current is made to increase in approximate proportion to the magnitude of the fault current. This is achieved through the use of bias windings to restrain the relay's operation, as shown in Figure 12.13. This significantly improves the low fault sensitivity by permitting a smaller differential current to trip the relay under light load conditions.

The bias characteristics for an Areva KBCH transformer differential protection relay appear in Figure 12.14. The bias current (sometimes called the restraint current) is equal to the average of the magnitudes of the CT secondary currents I_1 and I_2, while the minimum differential tripping current, $(I_1 - I_2)$ can be set in the range between $0.1I_n$ and $0.5I_n$, where I_n is the rated CT secondary current. The tripping characteristics lie

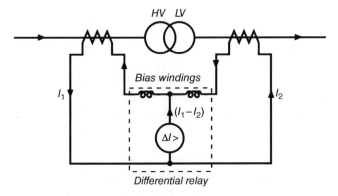

Figure 12.13 Biased differential protection.

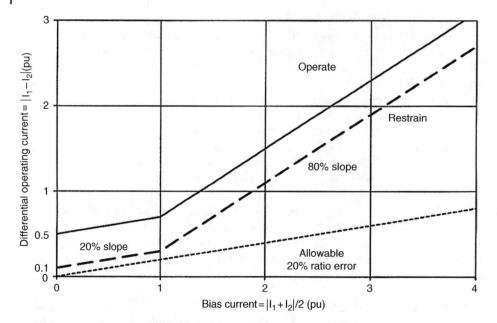

Figure 12.14 Bias characteristic for an AREVA KBCH transformer differential protection relay.

above the likely magnetising current of the transformer concerned and have a slope of 20% up to rated current, to allow for operation throughout the transformer's tapping range and any CT mismatch. Beyond I_n, the slope rises to 80% to allow for the effects of CT saturation at high through currents occurring due to the transient DC fault current component.

Differential Protection: CT Requirements

IEC class *PX*, *TP* or ANSI class *C* current transformers are generally used in transformer differential protection applications, and it is important that they do not heavily saturate during through-fault conditions. The biased relay characteristic will accommodate some saturation, but if this is to be avoided, the knee point voltage must be chosen to accommodate the transient DC component of the fault current. The knee point voltage must exceed the maximum secondary winding voltage. A safety factor of 2–4 is often applied, therefore:

$$E_k \geq 2I_{s\max}\left(R_{ct} + R_b\right)\left(1 + X / R\right) \approx 2I_{s\max}\left(R_{ct} + R_b\right)K_{td} \tag{12.5}$$

where E_k is the knee point voltage, $I_{s\max}$ is the maximum symmetrical secondary current under fault conditions, R_{ct} and R_b are the CT winding and the burden resistances respectively, including the cabling resistance, X and R are the primary network reactance and resistance up to the point of the fault and K_{td} is the transient dimensioning factor (typically 10–25).

In the case of IEC class *PX* or *TP* transformers the knee point voltage forms part of the specification and appears on the nameplate, while with ANSI class *C* CTs it is related to the rated secondary terminal voltage, and can be easily found from the corresponding magnetising characteristic.

12.6.2   Restricted Earth Fault Protection (REF)

Detecting faults to ground within a star connected LV transformer winding, especially those close to the star point can be particularly difficult from the HV side of a transformer. The LV fault current depends on the fault location, the leakage inductance associated with the faulted portion of the winding and the earthing system employed. In the case where impedance earthing is used, the fault current varies almost linearly with fault position, from very low values for faults close to the neutral terminal to the single-phase-to-ground fault level for faults at the phase terminal. This current is reflected into the HV winding according to the effective turns ratio, which itself is also proportional to the faulted fraction of the winding. The fault current observed on the HV side therefore depends on the square of faulted fraction of the winding, and as shown in Figure 12.15, this can be difficult to detect for faults in the lower third of the winding.

In the case of a solidly earthed neutral ground fault, currents become considerably larger, depending on the fault potential and the leakage impedance of the LV winding, which varies non-linearly with fault position. In such cases there is also a discontinuity between HV and LV fault currents, making fault detection difficult.

A form of unit protection known as *restricted earth fault protection* is frequently applied to such windings and is illustrated in Figure 12.4. This differential scheme balances the phase currents against the neutral current and therefore ignores through faults, only operating in the event of an in-zone earth fault. The direct measurement of the fault current permits the use of sensitive settings and fast relays, substantially increasing the fraction of each winding protected.

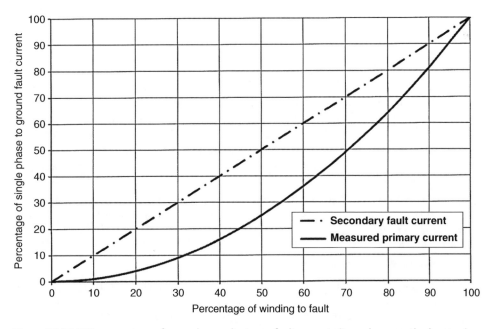

Figure 12.15 REF comparison of secondary and primary fault currents (impedance earthed system).

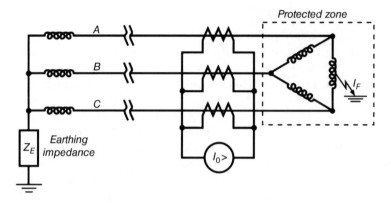

Figure 12.16 Delta winding REF protection.

While all the phase CTs in such a scheme are notionally identical, the effect of the transient DC component of a through-fault current can result in the saturation of one or more, particularly if all are not equally burdened. This can result in spill current flowing in the relay and with it the possibility of mal-operation. A stabilising resistor is usually inserted in series with the relay, effectively making the combination voltage sensitive. This forces the spill current to flow through the very low magnetising impedance of the saturated CT, thereby avoiding relay mal-operation. In this configuration the scheme is referred to as high impedance REF protection. The value of the stabilising resistor depends on the setting current of the relay and the CT knee point voltage of the CTs; its selection is detailed in the next section.

When an in-zone fault occurs, one CT delivers current to the high impedance relay, and providing that its setting voltage is comfortably less than the CT's knee point voltage, the relay will operate correctly. As mentioned above, a knee point voltage between 2 and 4 times the required setting voltage is usually specified.

Delta connected windings may also be protected using a REF scheme, as depicted in Figure 12.16. In this case since the phase currents normally sum to zero, any residual current is an indication of an earth fault. The fault current magnitude will depend on the earthing system used and the position of the fault on the winding. The minimum potential to ground on a delta winding occurs at its midpoint, and its magnitude is equal to half the corresponding phase potential. The winding impedance will be a sizeable fraction of the transformer's impedance and in cases where the earthing impedance is moderate, the earth fault current may be of the order of 1 pu or perhaps less, and since this current will be delivered by two phases, it may be difficult to detect. Subject to this limitation, both sides of a transformer may include a restricted earth fault protection scheme.

12.6.3 Busbar (or Bus-Zone) Protection

Busbars are generally very reliable and on the rare occasions when faults do occur, they are often the result of operator error or very occasionally by the failure of voltage or current transformers. Because of the critical nature of busbars and the loads they serve, busbar protection must be fast and reliable; it must operate correctly when required,

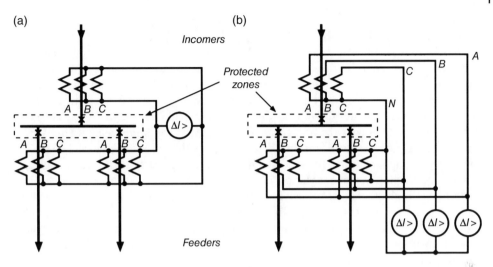

Figure 12.17 (a) Bus-zone earth fault protection (b) Bus-zone phase and earth fault protection.

tripping only the faulted bus section, while minimising any damage resulting from the fault. Through faults are much more common, and it is critical that busbar protection remains stable when the bus supplies an external fault.

In order to clear a fault, all the circuit breakers supplying the faulted bus must be tripped, which will usually result in a considerable loss of load. The consequences of an incorrect bus protection operation can therefore be severe, and the possibility of this is one reason for not providing busbar protection, particularly in small MV substations. In larger substations however, where the fault level is higher and the load larger, an uncleared bus fault can result in considerable damage, possibly resulting in the loss of the entire bus. In such cases busbar protection is usually provided to mitigate such risks. To ensure correct operation a bus fault must generally be seen by at least two independent protection elements before any breakers are tripped. Where a bus is sectionalised through the use of bus coupler breakers, only the faulted zone need be tripped and the protection scheme must be designed accordingly.

Bus-zone protection usually relies on differential unit protection, whereby currents incoming to the zone are balanced against those leaving it (see Figure 12.17), through the use of either low impedance biased relays or high impedance unbiased ones. Such schemes can be made to operate very fast on relatively low differential currents.

Differential High Impedance Bus-Zone Protection
Figure 12.17 shows two arrangements for differential bus-zone protection. In Figure 12.17a circulating CT currents balance on an instantaneous basis, and since the single overcurrent relay only sees the zero-sequence current, it will only operate for ground faults located between the incoming and outgoing CTs.

The three-element scheme in Figure 12.17b operates in a similar fashion, but since a relay is provided for each phase, it has the advantage of also being able to detect phase-to-phase faults as well as ground faults. Modern relays provide overcurrent elements for all phases and therefore this arrangement is now preferred.

Through-Fault Stability

During out-of-zone faults, the fault current will be supplied by all the bus incomers (current sources), delivering it to the faulted outgoing feeder. Consequently one current transformer may see the entire fault current, the transient component of which may be sufficient to cause it to saturate, while the others carrying considerably less current, remain unsaturated. The result of this will be to generate a spill current in the relay, which may result in an unwanted trip. One way to avoid this is through the use of a high impedance relay. This may be a low impedance current calibrated relay with a series connected stabilising resistor, making it behave as a voltage sensitive device, or a high impedance voltage calibrated relay, in which the stabilising resistance is fitted internally.

Figure 12.18 shows such a situation in which a ground fault occurs on an outgoing feeder and the fault current is delivered by two incomers; the faulted feeder CT is assumed to fully saturate as a result. In this state, this CT sees no flux change throughout much of each cycle, and it consumes most of its current internally as magnetising current, therefore as a first approximation it can be represented as a short circuit, as shown in the figure.

The reflected fault current in the secondary circuit, $I'_F = (I'_A + I'_B)$ divides between the relay, including its stabilising resistance R_R, and the wiring resistance R_{wC} in series with saturated CT winding resistance R_{ctC}. The currents flowing into the CT (I'_C) and the relay (I_R) therefore become:

$$I'_C = (I'_A + I'_B)\frac{R_R}{(R_R + R_{ctC} + R_{wC})} \approx (I'_A + I'_B) \ if \ R_R \gg (R_{ctC} + R_{wC})$$

$$I_R = (I'_A + I'_B)\frac{(R_{ctC} + R_{wC})}{(R_R + R_{ctC} + R_{wC})} \approx 0 \ if \ R_R \gg (R_{ctC} + R_{wC})$$

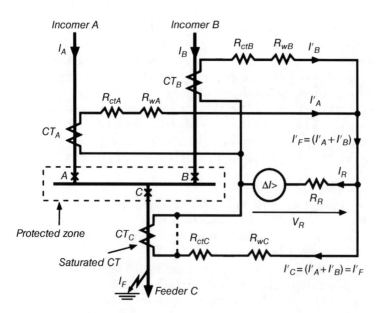

Figure 12.18 Through-fault operation of high impedance bus-zone protection.

Therefore by choosing R_R appropriately, the spill current can be made as small as necessary to preserve the through-fault stability of the scheme. The voltage developed across the relay and its stabilising resistance (V_R) is therefore determined by the maximum fault current and the maximum wiring and winding resistances of any saturated CT. This is referred to as the stability voltage and is given by:

$$V_R = I'_F\left(R_{ct} + R_w\right) \tag{12.6}$$

Where a voltage calibrated high impedance relay, like the MFAC14 is used (one including an internal stabilising resistor), the setting voltage V_S, must be greater than V_R. On the other hand, if a current calibrated low impedance relay such as the MCAG14 is used instead, together with an external stabilising resistor, the setting current $I_S\left(=V_S/R_R\right)$, must be chosen so that:

$$I_S > I'_F\left(R_{ct} + R_w\right)/R_R \tag{12.7}$$

Finally, the assertion that a fully saturated CT can be replaced by a short circuit is in practice not quite correct. Every half cycle the flux density must swing from its saturation value in one direction to that in the other, creating a rapid change of $2B_{sat}$. During this interval the CT produces a short pulse of current, one that is rich in harmonics. Many circulating current relays tune the relay winding to the fundamental frequency, making it insensitive to the harmonics created by saturated CTs. The inclusion of a capacitor in the relay circuit also makes the relay less sensitive to the transient DC component.

In the case of in-zone faults, all CTs carrying a fraction of the fault current drive the relay and its stabilising resistance. Since the latter has been chosen to preserve through-fault stability, it is quite possible that with the fault current from several CTs flowing, the burden presented may cause an excessively high voltage to develop, one which may threaten the CT winding insulation. In such circumstances while it is likely that some saturation will occur, it is usual to fit a non-linear zinc oxide resistor (known as a Metrosil), in parallel with the relay and its stabilising resistor to limit the amplitude of any voltage developed. This device consumes very little current when the relay voltage is low, and effectively clamps the peak voltage by consuming large amounts when the voltage rises, by which time the relay will have operated.

Effective Primary Current Setting

During fault conditions all parallel connected CTs are excited to the setting voltage (or possibly beyond), and therefore the effective setting current should also include the magnetising current consumed within each CT which, depending on the magnitude of the setting voltage, may not be negligible. The effective primary current required to operate the scheme is increased in proportion, and is given by:

$$I_P = N\left(I_S + nI_{eS}\right) \tag{12.8}$$

where I_P is the primary tripping current, I_S is the setting current, N is the CT turns ratio, n is the number of parallel connected CTs in the scheme and I_{eS} is the excitation current at the relay setting voltage.

In order to obtain a fast fault clearance time the effective primary tripping current should be less than 30% of the minimum phase-to-earth fault current. In large switchboards, with many parallel connected CTs, the cumulative effect of the magnetising currents may lead to an excessive tripping current, which is another reason why the three-element arrangement of Figure 12.17b is preferred.

High Impedance Bus-Zone Applications

High impedance busbar protection requires the secondary windings of all current transformers to be connected in parallel, together with the relay and its stabilising resistor. These CTs must be virtually identical and dedicated to the scheme; they cannot be shared with other protection functions. The relay effectively becomes an over-voltage device, sending trip signals to all breakers within its protection zone when its setting voltage is exceeded. The wiring for such a scheme is straightforward, and its cost is relatively low. These features make it attractive for simple bus arrangements like the sectionalised bus in Figure 14.7 which can be divided into two zones, one on either side of the bus coupler. For more complex arrangements where feeders can be switched from one bus to another, a secondary replica of the primary feeder arrangement is required and in such cases low impedance bus-zone protection is preferred.

Low Impedance Bus-Zone Protection

Low impedance differential relays can also be used in bus-zone protection schemes and are arranged in a similar manner to the transformer differential protection discussed earlier. Stability is similarly assisted through the use of bias windings or through the use of a software implemented bias characteristic generated within a numerical relay.

The choice between high impedance and low impedance bus-zone protection depends to a large extent on the complexity of the particular bus application. In the case of complex reconfigurable busbars, where incomers and outgoing circuits can be switched from one bus to another, low impedance bus-zone protection is generally preferred. The reason for this choice is partly due to the differences between these protection schemes and partly due to the flexibility of modern numerical relays.

Consider applying high impedance bus-zone protection to the double bus arrangement in Figure 14.8. In this configuration, incoming and outgoing circuits can be easily moved from one bus to another, by placing the associated circuit breakers into the appropriate cubicles, or by closing the appropriate isolator. Since separate zones of protection are required for each bus, the CT secondary currents must be reconfigurable to exactly mirror the primary circuit arrangement. This effectively demands that the CT currents be automatically switched whenever the bus is reconfigured. While it is possible to achieve this using the auxiliary switches within the circuit breaker cubicles (activated when the breaker is inserted) the switching of CT currents is fraught with danger. The risk of a switch mal-operation may be relatively low, but its consequences for the protection scheme are high. An open circuited CT will generate dangerously high voltages and its absence from the scheme will result in a spill current likely to trip the relay.

Low impedance schemes can be configured in a way that does not require the switching of CT currents. While these appear to operate on a similar principle to high impedance schemes, the significant difference between them is that a low impedance relay is sensitive to current and not voltage. While fault current can be detected by conventional low impedance relays like those in Figure 12.17, it need not be. So long as the vector sum of the CT secondary currents can be calculated, the presence of an in-zone fault can be determined with certainty. This fact and the flexibility of numerical relays has enabled low impedance bus-zone protection to be applied to reconfigurable busbar arrangements without the need to switch CT currents.

While low impedance schemes do require an image of the primary circuit to be generated (often called a *dynamic bus replica*), this is not in the form of switched CT currents. Instead it is sufficient to supply a numerical relay with a digital indication as to which bus a given breaker is connected, from which it can build a software replica of the primary circuit. Position information is obtained from auxiliary switches associated with the isolator connecting the breaker to the bus, either from within the circuit breaker cubicle in the case of metal clad switchgear, or directly from within the isolator, in the case of an aerial switchyard.

The secondary current from each CT is also supplied to the relay as shown in Figure 12.19, and from the dynamic bus replica the relay vectorially sums all the CT currents appropriate to each zone. When a circuit breaker is relocated to another zone its CT current will appear in the associated summation, therefore any residual current arising indicates the presence of an in-zone fault, for which all the breakers in that zone must be tripped.

Should an auxiliary switch mal-operate and provide the relay with incorrect position information, a false trip could result. This contingency is mitigated by the use of both normally open and normally closed position switches, the logical states of which must agree if a valid breaker position is to be inferred. When these disagree, the relay alarms, and switching operations must cease until the problem has been rectified. As mentioned earlier, bus-zone protection schemes also generally include a check relay to confirm the existence of a fault; this will prevent a protection mal-operation in the presence of a position error.

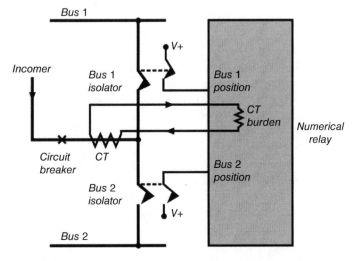

Figure 12.19 Current and position information supplied to a numerical relay.

Because a low impedance relay has access to the current from each CT, it can implement any necessary bias characteristic, generating a bias (or restraining) current from the algebraic sum of the CT currents, and the operating (or differential) current, from their vector addition. This information is used to determine a dynamic trip threshold for the relay of the kind shown in Figure 12.14.

Other functions such as *breaker fail* protection can also be provided by a low impedance numerical relay, where if a post-trip current is detected in any circuit breaker, the tripping of all breakers supplying the failed device is triggered. Numerical relays also permit the use of current transformers having different ratios and because each CT is only burdened by the relay, it can if necessary be used for other functions as well.

Current Transformer Requirements

High impedance schemes require a greater degree of CT performance and in order to preserve the through-fault stability, the setting voltage for a high impedance scheme V_S must exceed the stability voltage V_R. As a result, the CT knee point voltage (E_k) must also exceed the setting voltage. Relay manufacturers generally require a safety margin of at least a factor of two, yielding:

$$E_k > 2I'_F \left(R_{ct} + R_w \right) \tag{12.9}$$

IEC class *PX* or *TP* current transformers are suitable for use in both high and low busbar protection applications as are ANSI class *C* transformers, since the knee point voltage and the secondary winding resistance are readily available.

12.7 Frame Leakage and Arc Flash Busbar Protection

Bus-zone protection in MV metal clad switchgear can also be achieved using a *frame leakage* technique whereby the switchgear frame is grounded through a current transformer connected to an overcurrent relay. Since most switchgear faults involve phase-to-ground faults, any ground fault within the switchgear will be seen by this CT, allowing a rapid trip to be initiated by the overcurrent relay. Frame leakage protection requires that the switchboard be supplied from a grounded source so that a substantial fault current will flow in the event of a ground fault. The switchboard structure must also be insulated from ground, including the isolation of all cable sheaths, so there is no possibility of earth fault current flowing through this CT from an external source. In the case of sectionalised busbars, each bus section (or zone) is insulated from the next and equipped with its own frame leakage CT and dedicated overcurrent relay. This arrangement allows only the faulted section to be tripped.

Figures 12.20a and 12.21 show a three-section 11 kV busbar with frame leakage protection. Here each section is insulated from the concrete floor as well as from its neighbours by sheet rubber and the frame must also be independent of the reinforcement within the concrete slab. The centre section includes a bus coupler breaker, while those on either side each have a transformer incomer and outgoing distribution feeders. Figure 12.20b shows the frame leakage CTs on a similar switchboard. Each bus section has its own frame leakage CT while the bus coupler section has two, one of which supplies current to the bus *A* frame leakage relay and the other to the bus *B* relay, as

(a)
(b)

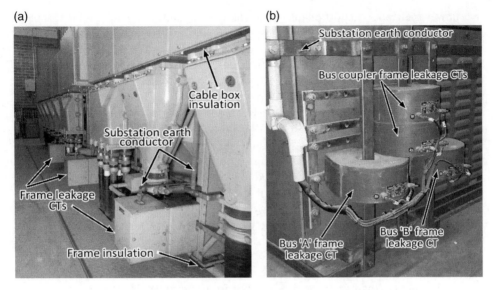

Figure 12.20 (a) Sectionalised frame leakage protected bus (b) Frame leakage CTs.

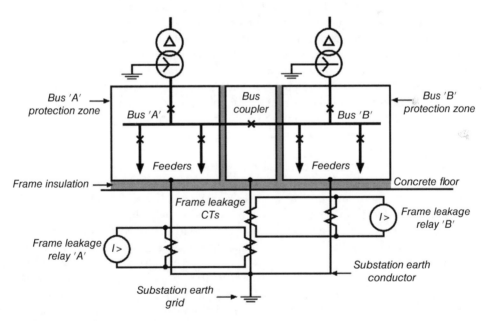

Figure 12.21 Typical two bus frame leakage arrangement.

depicted in Figure 12.21. A fault in either section will result in tripping all the associated breakers including the bus coupler, while a fault in the bus coupler section will generally result in the loss of the entire switchboard.

In practice a switchboard frame will generally see a non-infinite leakage impedance to ground in parallel with that of the local earth grid. So long as the former is large compared with the latter, the frame leakage scheme will operate correctly. The leakage impedance may however introduce a portion of any external ground fault current to the switchboard frame, which will then flow to the station earth grid through the frame leakage CT(s). Should this current become excessive the scheme may mal-operate, and for this reason any leakage impedance should be large with respect to the frame to earth impedance, and the overcurrent relays should be set accordingly.

A zero-sequence check relay is often used in conjunction with frame leakage protection schemes, one based on the presence of a neutral current or on the existence of a residual current or voltage within the MV winding supplying the bus. The check relay will generally have an instantaneous setting, and bus-zone trips will be contingent on the operation of both the frame leakage relay and the check relay.

Frame leakage schemes are seldom used in modern switchboards due to the need to preserve the frame insulation throughout the life of the switchboard, while ensuring that accidental foreign earth connections do not introduce earth currents that may be interpreted as fault currents. There have been many nuisance trips of frame leakage busbar protection schemes for this reason.

Arc Flash Busbar Protection

The requirement to insulate the entire switchboard frame and cable glands from ground is expensive and must be maintained throughout the life of the switchboard. This has led to a decline in the use of frame leakage schemes. In recent years fibre optic sensors capable of detecting the light from an arcing fault have become available. Together with arc flash monitor relays, these provide a simple alternative to frame leakage and circulating current based bus protection systems. They can be retro-fitted to existing metal clad switchgear or fitted to new switchgear during manufacture and can detect both phase-to-ground and phase-to-phase faults and provide very fast fault clearance times.

Arc flash sensors fall into two categories: the first is a naked optical fibre, capable of collecting light from an arc flash anywhere along its length and delivering it to an arc flash relay. This type of sensor can detect a fault anywhere throughout the bus chamber of a switchboard. The second type of sensor detects light from an arc flash through a lens at the end of an optical fibre. These are used to detect arc flash energy in specific locations such as in circuit breaker chambers or in cable termination enclosures.

Arc fault detection protection schemes also require some form of check protection to prevent nuisance trips. The check function is usually provided by a neutral earth fault or phase overcurrent relay or an overcurrent element within the arc monitor relay itself. This element is not required to time out or to generate a trip, but simply to pick up, confirming the existence of an internal fault. The arc monitor relay will issue a trip command when an arc has been detected and a fault current has been detected within the switchboard. In a similar fashion to frame leakage and high/low impedance bus protection schemes, arc fault protection can also be applied to sectionalised busbars, so that the minimum equipment is tripped in the event of a fault.

12.8 Distance Protection (Impedance Protection)

The protection schemes discussed in the preceding sections each depend solely on current information being supplied to a protection relay. Distance protection (also called *impedance* protection) is used to protect HV and EHV transmission lines and uses both current and voltage information in order to determine the impedance of the line between the relay and a fault. Because the impedance of a transmission line is proportional to its length, an impedance relay can be made capable of detecting a fault anywhere along the length of the line being protected. The maximum distance at which an impedance relay is capable of clearing a fault is known as its *reach*, and is determined by pre-programmed impedance settings representative of the line to be protected.

An impedance relay can be thought of as a voltage controlled overcurrent relay, where the voltage seen by the relay acts as a restraining influence. For faults occurring far from the relay, the line voltage it sees is relatively large, being given by the product of the fault current and the line impedance to the fault. In this situation, the relay requires a relatively large fault current in order to operate. On the other hand for close-in faults the line voltage seen by the relay will be small and the relay will not be so restrained, and will therefore trip on a smaller fault current. Unlike overcurrent relays whose reach is dependent on the local fault level which may vary with the state of the network, the reach of an impedance relay remains substantially constant.

The protection elements within distance relays compare the voltage seen at the relay with that generated across a replica of the line impedance Z_L, carrying the observed fault current. A trip command is issued when the magnitude of the observed voltage falls below that seen across Z_L. An impedance relay with this characteristic is non-directional and will trip whenever the apparent line impedance falls inside the impedance characteristic, regardless of the direction of the fault current flowing.

The impedance characteristics of a distance relay are summarised in a resistance-reactance (R-X) diagram, like that shown in Figure 12.22a, which represents the impedance values for which the relay will operate. The line AB in Figure 12.22a represents the positive sequence impedance of the line to be protected, and is characterised by a dominant inductive component. The relay is assumed to lie at the origin and it will operate when the apparent fault impedance falls within the circle, i.e. within the relay's reach. A fault at point F on the line, for example, will cause the relay to operate.

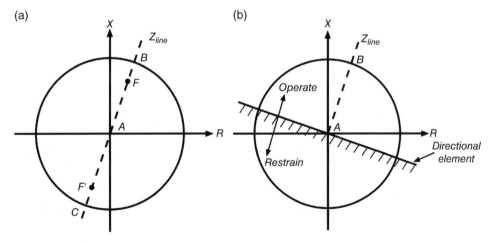

Figure 12.22 (a) Impedance circle (b) Characteristic of a directional impedance relay.

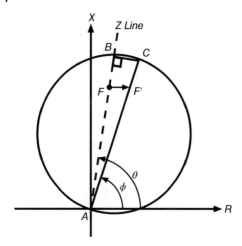

Figure 12.23 The 'mho' characteristic.

In this case, because the R-X characteristic is centred on the origin, the relay is said to be non-directional, since it will trip for faults in either direction so long as the apparent impedance seen by the relay falls within its impedance circle. A fault at point F', behind the relay, will therefore cause this relay to mal-operate.

In order to ensure the correct discrimination is achieved, a directional element is fitted to the relay, as shown in Figure 12.22b. This element appears as a straight line on the impedance diagram and it will only allow the relay to respond to faults in front of the relay, i.e. along the line being protected.

Alternatively, the *mho* characteristic shown in Figure 12.23 is more commonly used. Since this characteristic passes through the origin, it is directional in nature having maximum reach for impedances lying along the diameter *AC*. The angle ϕ, between the R axis and the line *AC* is the relay characteristic angle (also called the maximum torque angle), while angle θ is the angle of the line impedance *AB*.

A bolted fault at point F on the line will result in an apparent impedance seen by the relay equal to *AF*. However, in cases where the fault resistance is non-zero, the apparent impedance seen by the relay becomes slightly more resistive. For example, if *FF'* represents the arc resistance associated with a fault, then the apparent impedance becomes *AF'*. The relay is said to under-reach slightly since the apparent impedance is greater than the line impedance to the fault. By choosing the relay characteristic angle equal to θ rather than the angle ϕ, an allowance can be made for any fault resistance that may arise. Since *ABC* is a right-angle triangle, the reach of the relay (*AC*) is related to the fraction of the line to be protected *(AB)* according to:

$$AB = AC\cos(\theta - \phi) \tag{12.10}$$

The relay's setting must therefore be chosen to achieve the desired reach along the line to be protected.

Co-ordination of Distance Relays

The actual reach achieved by a distance relay depends upon the accuracy of the information supplied to it. Errors in current and voltage information, the impedance values programmed into the relay and even those in the relay itself, all contribute to the tolerance on the reach achieved. Satisfactory coordination of distance relays is achieved by employing a multi-zone approach to reach and trip delay setting. Three zones of protection are usually employed, although modern digital and numerical relays offer the possibility of more.

Zone 1 is typically set at 70–80% of the length of the line being protected. This is to ensure that the relay cannot see beyond the end of its line. Zone 1 protection is directional, responding only to faults along the line, allowing the Zone 1 element to trip instantaneously.

Zone 2 is also directional and covers the remainder of the line. It is typically set to 120% of the line impedance and therefore it also sees part of the next line. It must

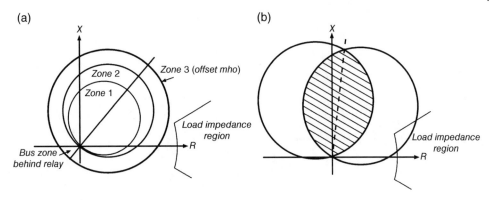

Figure 12.24 (a) Zones of protection (b) Lenticular mho characteristic.

therefore be time graded with any other relays that lie within its reach and provides backup protection for the remote busbars.

Zone 3 is often set to reach beyond Zone 2 with correspondingly longer delay times. The Zone 3 impedance circle can also be offset to include the origin as shown in Figure 12.24a. This will permit the relay to provide backup protection for the local busbars immediately behind it.

Lenticular Impedance Characteristic

Figure 12.24a also includes a typical load impedance characteristic. In cases where the Zone 3 reach is large, there is a risk that the load may encroach on the relay characteristic, possibly causing nuisance trips, particularly under emergency conditions when the line may be heavily loaded. To avoid this, a lens-shaped (lenticular) characteristic may be used instead, as shown in Figure 12.24b. This can be obtained from the intersection of two mho characteristics with one displaced towards the X axis, thereby providing an improved immunity against load encroachment.

Quadrilateral Impedance Characteristic

In addition to the circular impedance characteristics discussed, a quadrilateral characteristic similar to that in Figure 12.25 is also available. This is useful for the earth fault protection of MV and HV lines where vegetation can give rise to highly resistive faults. The relay's forward reach along the line is determined by the line impedance at point *B* which can be set independently from its resistive reach at *C*. Care must be taken in the choice of this parameter so as to avoid problems of load encroachment.

Distance Relay Impedance Measurements

Phase voltage and line current information is supplied to an impedance relay which sees an apparent secondary impedance given by:

$$Z_{sec} = \frac{V_{sec}}{I_{sec}} = \frac{V_{pri}/(VT\ Ratio)}{I_{pri}/(CT\ Ratio)} = Z_{pri}\left[\frac{CT\ Ratio}{VT\ Ratio}\right] \tag{12.11}$$

The relay's reach must therefore be specified in terms of the equivalent secondary impedance of the portion of the line to be protected.

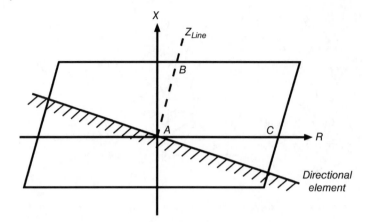

Figure 12.25 The quadrilateral impedance characteristic.

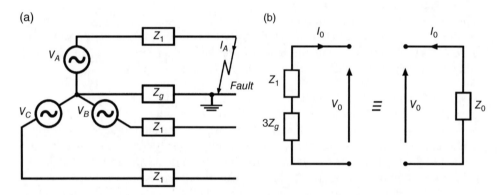

Figure 12.26 (a) Phase-to-ground fault (b) Comparison of zero-sequence networks.

Distance relays evaluate the positive sequence impedance Z_1, between the relay and the fault. In the case of three-phase, phase-to-phase and phase-to-phase-to-ground faults the relays measure this impedance by evaluating the ratio of the line-to-line voltage and the current flowing between the faulted phases. For faults involving the A and B phases for example, Z_1 is calculated from:

$$Z_1 = \frac{(V_A - V_B)}{(I_A - I_B)} \tag{12.12}$$

In the case of phase-to-ground faults the fault current must return to the neutral terminal via the ground impedance Z_g (which includes any neutral earthing impedance that may be fitted), as shown in Figure 12.26a. In this case the ratio of the phase-to-ground voltage and the phase current will not yield the positive sequence impedance of the line, since $I_A = V_A/(Z_1 + Z_g)$. However, modifying the impedance equation as shown below will result in the correct evaluation of the positive sequence impedance.

$$Z_1 = \frac{V_A}{(I_A + 3I_0 Z_g / Z_1)} = \frac{V_A}{(I_A + 3I_0 K_n)} \tag{12.13}$$

where K_n is called the *neutral compensation factor* $= Z_g/Z_1$, which is often approximated as a real number, and $3I_0$ is the residual current seen by the relay.

The replica phase voltage V_A is generated in the relay by the sum of two voltage drops: I_A through the positive impedance Z_1 and the residual current $3I_0$ through Z_g. The latter is adjusted so that for an earth fault at the maximum reach of the relay, V_A equals the voltage measured at the relay, thus $V_A = I_A Z_1 + 3I_0 Z_g$.

By comparing the zero-sequence networks in Figure 12.26b and remembering that $3I_0$ flows in Z_g, we obtain the relationship $Z_g = (Z_0 - Z_1)/3$ and therefore $K_n = (Z_0 - Z_1)/3Z_1$. A distance relay requires three impedance comparators for phase-to-phase faults and three for phase-to-ground faults, for each of its zones of protection.

Voltage and Current Transformers

As mentioned, voltage transformers provide a distance relay with phase-to-ground voltage information. The sudden loss of potential from a voltage transformer may suggest a close-in fault sufficient to instantly trip the relay, or alternatively it may simply mean the failure of either a VT primary or secondary fuse. Voltage supervision circuitry within a distance relay is necessary to prevent spurious trips in the latter case, while ensuring that the relay operates correctly in the former.

Impedance relays evaluate the line's positive sequence impedance from voltages and currents at the system frequency only, and therefore any transient DC current present will be ignored. However, it is important that the DC current does not saturate the CTs and thereby corrupt the underlying symmetrical fault current information. The knee point voltage E_k should therefore be chosen according to:

$$E_k \geq I_{s\max}\left(R_{ct} + R_b\right)\left(1 + X/R\right) \tag{12.14}$$

IEC class *PX* or *PXR* CTs or IEEE class *C* CTs would be suitable for distance relay applications (see Chapter 9).

12.9 Problems

1 *Resonant earth fault detection* (Refer to Section 11.5.2)

Figure 12.27 shows an MV substation fitted with resonant earthing, from which three aerial feeders originate. A residual current transformer is fitted to each one, for the purpose of fault location when phase-to-ground faults occur. Under normal conditions the residual current flowing to the Petersen coil is near zero, since the capacitive currents flowing from each phase sum to approximately zero. However, when a phase-to-ground fault occurs the balance is upset and a residual current exists in each CT. The purpose of the protection scheme is to detect that corresponding to the faulted feeder, so the fault can be quickly repaired.

A A phase-to-ground fault exists on the *c* phase of feeder 3. Use a phasor diagram to show that in this condition the residual current on any feeder leads the residual voltage $(3V_o)$ by 90° and therefore that the residual current is driven by the residual voltage. Finally show that if the inductance is tuned correctly, no fault current will flow.

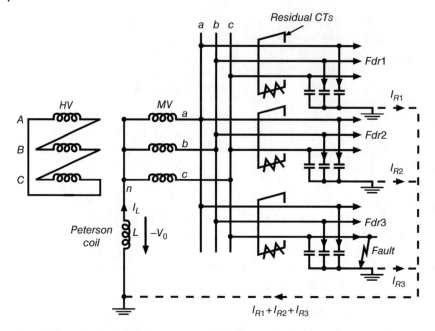

Figure 12.27 Faulted cable in a resonant earthed system.

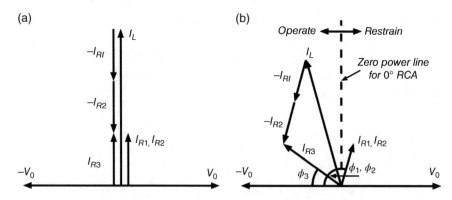

Figure 12.28 Discrimination between faulted and un-faulted feeders.

B Show that $-V_o$ drives the inductor current, and that the inductor current is equal to $-V_o/X_L$.

C Show that the residual current in the faulty feeder is given by: $I_{R3} = I_L - I_{R1} - I_{R2}$. Use this equation to obtain the phasor diagram in Figure 12.28a assuming that both the Petersen coil and the feeders are lossless.

D The protection scheme must discriminate between the residual current on the un-faulted feeders and that on the faulty one. Figure 12.28a suggests that this is difficult, since all residual currents lie in phase and are approximately the same magnitude. However, the situation changes slightly when the resistance in both the Petersen coil and the feeders is taken into account. Show that the phasor diagram in Figure 12.28b results when both these elements are slightly resistive.

One way of achieving this discrimination is to provide a wattmetric protection element for each feeder. This is essentially a zero-sequence wattmeter with the current from the feeder's residual CT supplied to its current terminals, and the residual voltage $(-3V_o)$, applied to its voltage terminals (known as the *polarising voltage*). The wattmeter on the faulted feeder will indicate the forward zero-sequence power flow $V_o I_{R3} \cos(\phi_3)$, while those on the un-faulted feeders 1 and 2, will indicate the reverse power flows $V_o I_{R1} \cos(\phi_1)$ and $V_o I_{R2} \cos(\phi_2)$ respectively, because these phase angles are in excess of 90°. By polarising the relay in this way, effective discrimination can be achieved between the faulty and the healthy feeders.

An additional resistance is often inserted in parallel with the Petersen coil to increase the relay discrimination. This occurs at the expense of a small resistive current flowing in the faulted phase. In the case of a resonant earthed system it is generally not necessary to immediately trip the faulted feeder, so long as the fault can be quickly located and repaired.

2   The simple radial network shown in Figure 12.29 is protected by an overcurrent relay on the each side of the transformer. The three-phase fault level on the HV side is 10 kA. The relays use the standard inverse curves shown in Figure 12.6 and have a pick-up/reset ratio of 0.9, and plug settings from 0.5 to 2.0 in 0.25 A steps.

  A   Determine the fault level on the LV side of the transformer.

  B   Assuming that the relay on the LV side need not grade with any relays further downstream, determine a suitable plug setting, the TMS required and the time taken for this relay to trip on an LV fault.

  C   Choose a suitable current setting and TMS for the HV relay to achieve a grading margin of 0.4 seconds between the two relays.

  D   Calculate the trip time for a fault occurring on the HV side of the transformer.

  E   What grading margin will exist between these relays for a fault current of 5000 A on the 22 kV side?

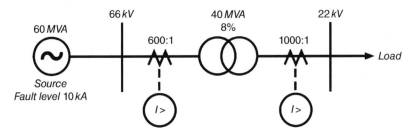

Figure 12.29 One line diagram.

Solution:

  a)   If the fault level on the 66 kV bus is 10 kA and the rated current there is 525 A, the per-unit impedance there is $525/10,000 = 0.0525$ pu which scales as $0.0525 \times 40/60 = 0.035$ pu on a 40 MVA base. The fault level on the 22 kV side of the transformer is therefore $1/(0.035 + 0.08) = 8.69$ pu, and since the 22 kV rated current is 1050 A, the fault level is 9130 A.

b) The plug setting for the LV relay should ideally be greater than $1050/0.9 = 1167$ A so a current setting of 1.25 A will suffice. Since there is no need to grade with any downstream relays the TMS setting can be set to the minimum of 0.1. A fault on the 22 kV bus will cause this relay to see $9130/1000 = 9.13$ A ($M = 9.13/1.25 = 7.3$), causing it to trip in 0.34 seconds.

c) The rated 66 kV current is 525 A and the relay setting should exceed $525/0.9 = 583$ A, so a relay current setting of 1 A will suffice. A 9130 A fault on the LV side is reflected into the HV side as 3043 A, which will appear as 5.07 A in the relay, i.e. $M = 5.07$. This must result in a trip time of $(0.34 + 0.4) = 0.73$ seconds. With a TMS of 1.0 this relay will trip in 4.24 seconds, and therefore the required TMS is $(0.73/4.24) = 0.172$.

d) A three-phase fault on the HV bus will result in 16.67 A flowing in the relay, i.e. $M = 16.67$. With a TMS of 0.18 the relay will trip in 0.43 seconds.

e) For a 5000 A fault on the 22 kV side, the LV relay will see 5 A thus $M = 4$, and from the equation the trip time will be 0.5 seconds. This fault current will be reflected as 1667 A on the HV side, and therefore the HV relay will see $1667/600 = 2.77$ A, thus $M = 2.77$ and the HV trip time will be 1.17 seconds. The grading margin between the two relays will be $1.17 - 0.5 = 0.67$ seconds. Therefore providing discrimination is achieved at the maximum fault current it will be maintained at lower currents as well.

3 The CT connections for a transformer differential protection scheme are shown in Figure 12.30. Based on the CT connections, what is the vector grouping of the transformer?

(Answer: Yd11.)

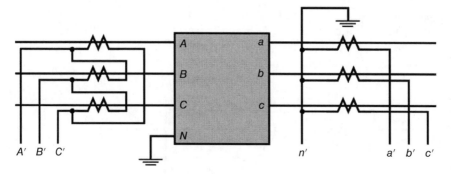

Figure 12.30 Primary and secondary CT connections.

4 A current calibrated MCAG14, 1 A attracted armature circulating current relay is to be used in a restricted earth fault protection scheme on the LV side of a solidly earthed 30 MVA, 110 kV:11 kV Dyn11 transformer, in which 1600:1 CTs are to be used. The relay has setting currents of 0.2, 0.3, 0.4, 0.5, 0.6, 0.7 and 0.8 A and the tripping curve is shown in Figure 12.31. The transformer impedance data appears in Table 12.2, and each CT has a winding resistance of 9 Ω with a worst case lead resistance of 1 Ω.

A Determine the phase-to-earth fault level assuming the upstream impedance is negligible.

(Answer: $I_{Fault} = 5431$ A.)

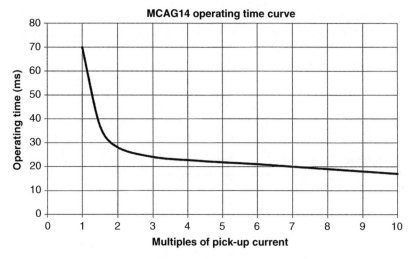

Figure 12.31 MCAG14 operating characteristic.

Table 12.2 Transformer impedances.

Transformer impedances (30 MVA base)	
$Z_1 = Z_2$	Z_0
$0.007 + j0.21$ pu	$0.0 + j0.45$ pu

B Calculate the stability voltage V_R, choose a suitable setting voltage V_S as well as the knee point voltage of the CTs.

(Answer: $V_R = 34.5$ volts, a suitable setting voltage therefore might be 40 V, thus $E_K \approx 80$ V.)

C Assuming a relay setting current of 0.4 A is chosen, calculate the stability resistance R_R assuming that the relay resistance is negligible, and determine the voltage that would be seen across the relay and stabilising resistance for in-zone faults in the absence of CT saturation.

(Answer: $R_R = 100\,\Omega$, 345 V.)

D Calculate the effective primary earth fault current required to operate the relay. Assume that the exciting current of each CT at its knee point voltage is 50 mA. It is usually acceptable to scale this current linearly to the setting voltage.

(Answer: 800 A.)

E Estimate the relay operating time corresponding to the fault current calculated in (a).

(Answer: Operating time ≈ 20 msec.)

F Repeat parts (c) through (e) for a setting current of 0.8 A.

(Answer: $R_R = 100\,\Omega$, 345 V, Ipri = 1440 A, operating time ≈ 23 msec.)

5 Figure 12.26a shows a circuit with a phase-to-ground fault. The source and line impedance is represented by Z_1, the positive sequence impedance, while the ground path impedance is represented by Z_g.

A Show through the inclusion the neutral compensation factor K_n in the equation below, that the positive sequence impedance results, i.e.:

$$Z_1 = \frac{V_A}{\left(I_A + 3K_n I_0\right)}$$

B Show by comparing the earth fault current calculated from Figure 12.26a and that predicted by symmetrical component theory, that:

$$Z_g = \frac{\left(Z_0 - Z_1\right)}{3}$$

C Why does this approach not work for a phase-to-phase-to-ground fault?

6 Show that for the case of a phase-to-phase-to-ground fault, the ratio $(V_A - V_B)/(I_A - I_B)$ is equal to the positive sequence line impedance.

7 A 132 kV transmission line has the following per unit impedances on a 100 MVA base. $Z_1 = Z_2 = 0.05 + j0.18$ pu and $Z_0 = 0.15 + j0.55$ pu. Zone 1 is set to 80% of the line length with the RCA equal to the line angle.

A Calculate the neutral compensation factor, K_n.
(Answer: $K_n \approx 0.68\angle 0°$.)

B Determine the secondary impedance setting required for Zone 1 if the VT ratio is 132 kV:110 V and the CT ratio is 1000:1.
(Answer: $Z_{sec} = 21.7$ ohms)

C A phase-to-earth fault on the line generates a phase current of $2.8\angle -80°$ pu, and a zero-sequence current of $0.4\angle -70°$ pu. Will the relay operate if it sees a voltage of $0.607\angle -3.3°$ pu on the faulted phase? Justify your answer.
(Answer: No)

D What is the maximum voltage the relay would require under these circumstances in order for it to operate?
(Answer: 0.539 pu.)

8 A three-phase fault somewhere on the line described in question 7 generates a current of $4.91\angle -79.2°$ pu. The corresponding voltage seen by the relay was $0.458\angle -4.7°$ pu.

A What secondary impedance was seen by the relay?
(Answer: 13.5 secondary ohms)

B Did the Zone 1 comparator generate a trip?
(Answer: Yes)

C At what point on the line did the fault occur?
(Answer: 62% of the line's length.)

9 If the ground impedance in Figure 12.26a ($Z_g = R_g + jX_g$) has the maximum impedance according to the requirements for effective earthing in Section 11.5.4, determine the earth fault factor. You may assume that the positive and negative sequence impedances are equal and wholly reactive. Is this as you expect? (Answer: 1.35)

12.10   Sources

1 GEC Alsthom, *Protective relays application guide*, 1990, GEC Alsthom Protection and Control Limited, Stafford UK.

2 Areva T&D, *Network Protection and Automation Guide*, 2005, Areva, Paris.

3 IEEE Std C37.112 *IEEE Standard inverse-time characteristic equations for overcurrent relays*, 1996, IEEE Power Engineering Society, New York.

4 IEC 60255-151 *Measuring relays and protection equipment* (2009) International Electrotechnical Commission, Geneva.

5 IEEE Std C37.110 *Guide for the application of current transformers used for relaying purposes*, 2007, IEEE Power Engineering Society, New York.

6 Sethuraman Ganesan, *Selection of current transformers & wire sizing in substations*, ABB Inc, Allentown, PA.

7 Alstom *MCAG14/34, MFAC14/34 Application guide, high stability circulating current relay*, R6136D, 2013, Alstom.

8 Blackburn Jl, Domin TJ, *Protective relaying, principles and applications*, 2007, CRC Press, Boca Raton, FL.

9 Behrendt K, Costello D, Zocholl S E, *Considerations for using high impedance or low impedance relays for bus differential protection*, 49th Annual Industrial & Commercial Power Systems Technical Conference, Schweitzer Engineering Laboratories, April 2013, revised February 2016.

10 Kasztenny B, Brunello G, *Modern cost-efficient digital busbar protection solutions* 29th Annual Western Protective Relay Conference, GE Power Management, Ontario, October 2002.

11 Ekanayake J, Karunanayake J, and Terzija V, *Modern power system protection*, 2017, Wiley Publishing.

12 Horowitz SH, Phadke AG, *Power system relaying*, 4th Edition, 2014, Wiley Publishing.

13 Anderson P M, *Power system protection*, 1998, Wiley – IEEE Press.

13

Harmonics in Power Systems

Ideally the currents flowing throughout the power system should be sinusoidal, since the voltages exciting them are sinusoidal, or nearly so. However, when a customer's load contains non-linear (or distorting) elements then non-linear currents will occur as a result. Such currents can be resolved into a component at the system frequency, henceforth referred to as the fundamental frequency (f_1), and harmonic components occurring at integral multiples of the fundamental (nf_1). Therefore we might expect 2nd, 3rd, 4th and 5th harmonic currents, for example, to appear as a result of distorting loads connected to the network; however, we will see that under most conditions even harmonics are generally absent.

Harmonics are not new; as long ago as 1916 the third harmonic, caused by the saturation of machines and power transformers, was noticeable. Improvements in transformer design reduced this problem substantially until the 1930s when the introduction of large rectifier equipment for rail transport again introduced significant harmonics into the power system. Interestingly, the main problem presented by harmonics was interference with the open wire telephony circuits of the day.

Prior to 1960 one of the main sources of harmonic distortion was television receivers with power supplied from a half wave rectifier. As a result each receiver introduced a small DC current into the LV network, as well as even and odd harmonics. The DC currents in the LV network approximately cancelled due to the fact that the AC connector could be reversed, but any residual current created a DC flux in distribution transformers, increasing their magnetising losses slightly. During the period 1960–75 the availability of thyristors led to the introduction of phase-controlled dimmers for domestic and industrial lighting and heating applications. These produced considerable harmonic distortion, as well as high frequency radio emissions, and led to the publication of the first standards restricting harmonics in power systems.

The 1980s and early 1990s saw a rapid expansion in the use of on line switch mode power supplies in television receivers, personal computers and numerous other consumer appliances. These employed single-phase full wave rectifiers and capacitive filters which consumed short duration, high amplitude current pulses, close to the peak of the AC cycle. This resulted in a substantial flattening of the peak of the AC waveform, as well as the generation of many low-order harmonics, particularly the third, fifth and seventh. This period saw the development of European standards on consumer products, and by 1995 IEC 61000-3-2 defined requirements and harmonic current limits for four classes of consumer equipment, Classes A–D.

AC Circuits and Power Systems in Practice, First Edition. Graeme Vertigan.
© 2018 John Wiley & Sons Ltd. Published 2018 by John Wiley & Sons Ltd.

Today, in addition to a vast array of consumer electronics, there are also many large non-linear devices connected to the power system. With the advent of high current thyristors and insulated gate bipolar junction transistors (IGBJTs), AC to AC and AC to DC converter circuits frequently appear in industrial loads. Since this equipment is inherently non-linear, the current it consumes is often rich in harmonics.

The presence of harmonics in the power system leads to many adverse effects, including increased losses in transmission lines, transformers and rotating machines, in addition to the possible mal-operation of some items of customer equipment. We will show that the presence of harmonic currents requires that the power factor definition be amended to include their effects, and that harmonic currents always act to reduce the total power factor of a load.

Harmonics generated by one customer may interfere with the load of another, and to avoid this, the background level of individual harmonics in the system must be carefully managed. This task falls broadly into two areas. Firstly, it is the responsibility of customers to ensure that any harmonic distortion their load creates does not exceed the *emission limit* defined in the relevant standard or in the supply agreement with the network service provider. This may require selection of equipment designed to avoid excessive harmonic generation, or the installation of harmonic filters to ameliorate their effects. The connection agreement either places limits on the customer's contribution to the harmonic voltage distortion or to the harmonic currents that can be injected into the network at the customer's connection point, or alternatively at the point of common coupling between the customer's load and other loads on the network. (A limit on the harmonic currents that may be injected is relatively easy to enforce, since the harmonic current spectrum can be measured with an energy meter having power quality capabilities. On the other hand, determining the component of harmonic voltage distortion introduced by a particular customer is not so straightforward, unless the harmonic impedances are known at the point of common coupling.)

Secondly, the network service provider also has a responsibility to ensure that the level of harmonic distortion *allocated* to each new customer does not result in an overall harmonic level sufficient to cause disturbances to any customer equipment. This usually requires an investigation of a customer's proposed load prior to connection, in addition to a knowledge of the background level of harmonic distortion pre-existing in the network. (It is much easier to determine the background voltage distortion levels prior to connection of the proposed load than post connection.)

Harmonic Propagation

Harmonic sources can usually be modelled by a *harmonic current source* connected in parallel with a distorting load. As a result, harmonic currents are shared between those linear components of the load and the upstream network where they generate small harmonic voltages across the local network impedance. This produces a degree of *harmonic voltage distortion* which is propagated throughout the network. Thus the effect of one customer's distorting load is frequently felt by others.

Where harmonic disturbances must be suppressed, it is common to install harmonic filters within or close to the offending load. Large variable speed drives, for example, are often designed to minimise the export of harmonic currents to the upstream network. Alternatively, harmonic filters can be installed in parallel with a load. These are circuits designed to provide an alternative low impedance path for selected harmonics, which may then flow harmlessly to ground rather than into the local network. In this way, harmonic voltage distortion can be minimised, together with the harmonic current the load injects into the network.

13.1 Measures of Harmonic Distortion

There are several simple numerical measures of the harmonic content of either a voltage or a current. We will consider the *total harmonic distortion* (THD), the *total demand distortion* (TDD) and the *crest factor* (CF). However, we initially consider the RMS current arising from a fundamental current together with an assortment of its harmonics, which can be expressed in the form:

$$I_{RMS} = \sqrt{I_1^2 + I_2^2 + I_3^2 + I_4^2 + \ldots + I_h^2 + \ldots} = \sqrt{\sum_{h=1}^{h_{max}} I_h^2} \qquad (13.1)$$

A similar result applies for voltage, thus:

$$V_{RMS} = \sqrt{\sum_{h=1}^{h_{max}} V_h^2} \qquad (13.2)$$

Where I_h and V_h are the current and voltage magnitudes of the harmonic h respectively and h_{max} is the maximum harmonic order present.

The *total harmonic current distortion* (THD$_i$) is the ratio of the RMS harmonic current present to the fundamental current, expressed as a percentage:

$$THD_i = \frac{\sqrt{\sum_{h=2}^{h_{max}} I_h^2}}{I_1} \times 100\% \qquad (13.3)$$

(Note that the RMS summation in Equation (13.3) excludes the fundamental component.)

Similarly the *total harmonic voltage distortion* (THD$_v$):

$$THD_v = \frac{\sqrt{\sum_{h=2}^{h_{max}} V_h^2}}{V_1} \times 100\% \qquad (13.4)$$

Handheld instruments are capable of determining the *THD* of voltage and current waveforms, and in some cases they are also able to resolve the levels of the individual harmonics present.

It is possible for a load to exhibit quite high THD$_i$ at times when the fundamental current demand is low, without imposing an unacceptable burden on the network, and thus the THD$_i$ can provide a slightly misleading estimate of a load's actual harmonic content. The *total demand distortion* (TDD) is a way of quantifying the harmonic content in terms of the fundamental current, corresponding to the maximum VA demand. The TDD is the ratio of the (worst case) RMS harmonic current to the fundamental component at maximum demand. The expression for TDD is therefore similar to that for THD_i, except that the RMS harmonic current is referenced to the maximum demand load current I_L, rather than the fundamental current recorded at the time of the measurement, I_1. The TDD will therefore be lower that the THD_i and is used to define the harmonic current limits in IEEE Std 519-2014, *Recommended Practice and Requirements for Harmonic Control in Electric Power Systems*.

$$TDD = \frac{\sqrt{\sum_{h=2}^{h_{max}} I_h^2}}{I_L} \times 100\% \qquad (13.3a)$$

As well as THD and TDD, the crest factor of a waveform (usually a current) is also used as a measure of its harmonic content. This is defined as follows:

$$Crest\ Factor = \frac{Peak\ Value}{RMS\ Value}$$

The crest factor provides a measure of the 'peakiness' of a waveform. So, for a current or voltage without significant harmonic content, the crest factor will be close to $\sqrt{2}$. Crest factors higher than this correspond to peaky waveforms, while values less than $\sqrt{2}$ relate to flat topped waveforms, both of which imply that harmonics are present. However, a heavy harmonic content is generally characterised by large crest factors. Such loads impose a severe burden on the source of power supplying them, since unless the peak current can be delivered, the supply voltage at the load will become distorted. Well-designed loads will therefore have crest factors approaching $\sqrt{2}$.

While all these indices provide a useful quantitative measure of harmonic distortion, they provide no information as to which of the harmonic(s) present have the largest amplitude and therefore should be targeted in the first instance. We will see that harmonic components roughly diminish as $1/h$ (or sometimes as $1/h^2$), and therefore it is frequently the lower-order harmonics that are most problematic.

13.2 Resolving a Non-linear Current or Voltage into its Harmonic Components (Fourier Series)

Sinusoidal waveforms, be they voltage or current, have the unique property that they only contain energy at one frequency. It was shown in 1822 by Joseph Fourier, in his solution to the heat equation, that any other periodic wave shape can be resolved into a series of sinusoidal components at different frequencies; specifically the fundamental frequency and a series of harmonics. The fundamental frequency is the reciprocal of the period T of the waveform and harmonics are integer multiples of this frequency.

In the steady state analysis of AC networks, voltages and currents are represented by periodic functions. A function $f(t)$ is said to be *periodic* with period T, if for any t we can write:

$$f(t) = f(t + T)$$

Any periodic function may be decomposed into a sum of its Fourier components, consisting of a DC term (or average value), a component at the fundamental frequency, plus a series of harmonic components, represented by sine and cosine functions at integer multiples of the fundamental frequency. The Fourier series of a periodic current $i(t)$ can be expressed as follows:

$$i(t) = i_{DC} + \sum_{k=1}^{\infty} \left[a_k \cos(k\omega t) + b_k \sin(k\omega t) \right] \qquad (13.5)$$

where $\omega \, (= 2\pi/T)$ is the angular frequency of the fundamental component and i_{DC} is the average value of the waveform $i(t)$, i.e.

$$i_{DC} = \frac{1}{T} \int_0^T i(t) \, dt$$

The weightings a_k and b_k are known as the Fourier coefficients, and are given by:

$$a_k = \frac{2}{T} \int_0^T i(t) \cos(kwt) \, dt \quad \text{and} \quad b_k = \frac{2}{T} \int_0^T i(t) \sin(kwt) \, dt$$

Thus any function (except for purely sinusoidal functions, where a_k and $b_k = 0$ for all $k \neq 1$), can be expressed as a DC term (should one exist), a component at the fundamental frequency, plus a sum of harmonic components.

By way of example, consider the current waveform shown in Figure 13.1. This is typical of the current drawn by a power supply consisting of a full wave bridge rectifier with a simple capacitive filter, as found in many personal computers and other consumer equipment. The waveform consists of a current pulse near the crest of the voltage; its harmonic spectrum shows that it contains no even harmonics, and the magnitudes of the third and fifth harmonics are comparable to that of the fundamental. As a result, the total harmonic current distortion of this waveform is in excess of 100% and its crest factor is about 3.5. In recent years, manufacturers of computing equipment have begun to move away from this type of power source, constructing instead *power factor corrected* supplies. These present an almost resistive load, and do not consume significant harmonic currents.

13.2.1   Fourier Simplifications Due to Waveform Symmetry

Not all waveforms contain both sine and cosine terms. For example, if a waveform has odd symmetry there are no cosine terms. Odd symmetry exists if $i(t) = -i(-t)$, after any DC term has been removed, therefore it can be shown that $a_k = 0$. Similarly the sine terms are absent if the signal has even symmetry whereby $i(t) = i(-t)$, after any DC terms have been removed, thus $b_k = 0$.

Of more importance from the point of view of power system harmonics is the concept of *half wave symmetry*, which exists if the positive and negative halves of a waveform are identical in shape but opposite in sign, i.e. $i(t) = -i(t + T/2)$. When a waveform has half wave symmetry it contains no even harmonics. Half wave symmetry is common in power systems and thus most current or voltage waveforms contain only odd harmonics. The current waveform in Figure 13.1 is a case in point. Note that the presence of a DC

(a)

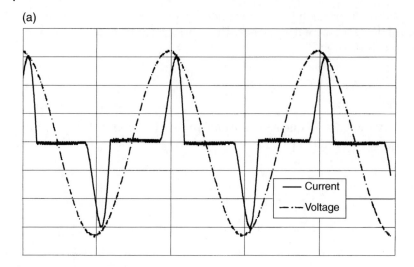

(b)

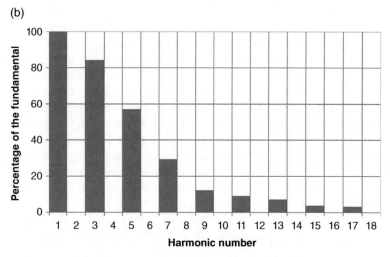

Figure 13.1 (a) Personal computer current and voltage waveforms (b) Current harmonic spectrum.

term precludes half wave symmetry. As a result of half wave symmetry, the Fourier coefficients can be expressed in the form:

$$a_k = \frac{4}{T} \int_0^{T/2} i(t)\cos(kwt)\,dt \text{ and } b_k = \frac{4}{T} \int_0^{T/2} i(t)\sin(kwt)\,dt$$

Examples: Consider the square waveform shown in Figure 13.2. By inspection it has both odd symmetry and half wave symmetry, and thus it contains no cosine terms and no even harmonics.

In this case, the only non-zero Fourier coefficients are b_k when k is odd. Thus:

$$b_k = \frac{4}{T} \int_0^{T/2} V \sin(k\omega t)\,dt = \frac{4V}{k\pi}\left(\text{for k odd}\right).$$

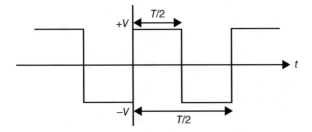

Figure 13.2 Square waveform.

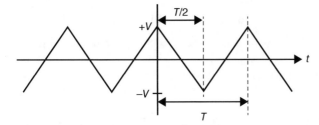

Figure 13.3 Triangular waveform.

This result is interesting, since the odd harmonics present decrease as $1/k$, thus higher-order harmonics tend to have smaller amplitudes. This fact is significant in reducing the impact of large non-linear loads on power systems, as will be shown later.

Consider the triangular waveform in Figure 13.3. Here even symmetry and half wave symmetry exist, so the triangular waveform will have no sine terms and no even harmonics. Accordingly a_k is given by:

$$a_k = \frac{4}{T} \int_0^{T/2} V\left(1 - \frac{4t}{T}\right)\cos(k\omega t)\,dt = \frac{8V}{k^2\pi^2}.$$

And the harmonics decrease as $1/k^2$.

Finally, consider the half wave rectified current waveform in Figure 13.4. This has a DC term and therefore has no half wave symmetry, although it does have even symmetry, thus $b_k = 0$.

$$\text{So: } a_k = \frac{4}{T} \int_0^{T/4} I\cos(\omega t)\cos(k\omega t)\,dt = \frac{(-1)^{(k/2)+1}\,I}{k^2-1}\left(\text{for k even}\right).$$

In this case, there are no odd harmonics, and this waveform contains harmonic components whose amplitude decrease approximately as $1/k^2$. The DC component of this current is given by:

$$I_{DC} = \frac{1}{T} \int_{-T/4}^{T/4} I\cos(\omega t)\,dt = \frac{I}{\pi}$$

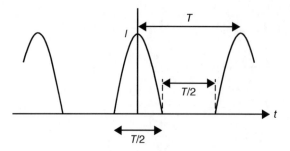

Figure 13.4 Half wave rectified current.

13.2.2 Transformer Inrush Current (Second Harmonic Restraint)

Since half wave symmetry occurs frequently, odd harmonics generally occur much more often than even ones. There is one particular instance, however, where a transient DC term occurs and with it the brief existence of even harmonics. The situation in question relates to the inrush currents that flow when a transformer is energised. Depending on the instant of switching, the magnetising current may take on values very much higher than normal. This phenomenon was discussed in Section 7.5.4, and Figure 7.18 illustrates typical magnetising currents that arise immediately after switch on. Until the B–H loop re-centres itself on the origin, the magnetising current possesses a DC component in addition to even harmonics.

This fact is used to advantage by protection engineers who must protect transformers from faults at switch-on, as well as during normal operation. The large inrush current can cause false trips in transformer differential protection relays (which compare primary and secondary currents on an instantaneous basis). Since the magnetising current is consumed within the transformer, the large inrush appears as an internal fault to a differential relay, which therefore must be restrained for a short period following switch-on.

Inrush current can be distinguished from fault current by detecting the presence of even harmonics, notably the second harmonic, since it is generally the largest. In this instance, transformer protection relays are usually restrained from operating for a period, so long as there is a sufficiently large second harmonic component present. This restraint may mean that trips are inhibited entirely until the steady state has been achieved (which can take tens of seconds) or alternatively, a group of higher trip settings are enabled instead, so that should a fault also occur at switch on, the transformer will still be protected. Once the steady state has been reached and the even harmonics have disappeared, the normal protection settings are restored.

13.3 Harmonic Phase Sequences

Since large non-linear industrial loads such as variable speed drives are generally well balanced, so are the harmonics that they create. In other words currents or voltages of a particular harmonic tend to exist as balanced three-phase systems in their own right. This enables them to assume either positive, negative or zero-sequence characteristics.

Consider the case of a balanced three-phase system of currents from a distorting load, including harmonics. The fundamental phase currents may be expressed as:

$$I_{A1} = I\cos(\omega t)$$
$$I_{B1} = I\cos(\omega t + 240°)$$
$$I_{C1} = I\cos(\omega t + 120°)$$

In general, the *hth* order harmonic takes the form:

$$I_{Ah} = F_h I\cos(h\omega t)$$
$$I_{Bh} = F_h I\cos(h\omega t + h240°)$$
$$I_{Ch} = F_h I\cos(h\omega t + h120°)$$

where F_h is a *harmonic scaling factor*, and depends on the shape of the current waveform in question. Generally F_h will diminish as h increases, possibly as $1/h$ or $1/h^2$.

Of significance here is the fact that not only is the frequency of the fundamental scaled by the factor h, but also its phase as well. These equations demonstrate that when the fundamental currents are balanced then so are their harmonics. However, this does not necessarily mean that the phase sequence of these harmonics is the same as that of the fundamental. Consider, for example, the case of the third harmonic:

$$I_{A3} = F_3 I\cos(3\omega t)$$
$$I_{B3} = F_3 I\cos(3\omega t + 720°) = F_3 I\cos(3\omega t)$$
$$I_{C3} = F_3 I\cos(3\omega t + 360°) = F_3 I\cos(3\omega t)$$

Since the third harmonic phase shifts are three times as large as those of the fundamental, the resulting third harmonic currents are co-phasal. In other words, the third harmonic is a zero-sequence harmonic. This property is also true of any harmonic that is a multiple of three, i.e. the 3rd, 6th, 9th etc. These harmonics are often called *triplen harmonics*.

Non-triplen harmonics may have either a positive or a negative phase sequence. In the case of the 5th harmonic we find:

$$I_{A5} = F_5 I\cos(5\omega t)$$
$$I_{B5} = F_5 I\cos(5\omega t + 1200°) = F_5 I\cos(5\omega t + 120°)$$
$$I_{C5} = F_5 I\cos(5\omega t + 600°) = F_5 I\cos(5\omega t + 240°)$$

Comparing these equations with those of the fundamental, we see that the B and C phasors have swapped positions, and therefore the phase sequence of the 5th harmonic is *ACB*; it is therefore a negative sequence harmonic. In contrast the 7th is a positive sequence harmonic, as shown below.

$$I_{A7} = F_7 I\cos(7\omega t)$$
$$I_{B7} = F_7 I\cos(7\omega t + 1680°) = F_7 I\cos(7\omega t + 240°)$$
$$I_{C7} = F_7 I\cos(7\omega t + 840°) = F_7 I\cos(7\omega t + 120°)$$

Harmonic phase sequences are summarised in Table 13.1.

Table 13.1 Harmonic phase sequences.

Positive sequence harmonics (+)	Negative sequence harmonics (−)	Zero-sequence harmonics
1st	2nd	3rd
4th	5th	6th
7th	8th	9th
10th	11th	12th
13th	14th	15th

13.3.1 Voltage Harmonic Distortion

Even harmonics are rare in power systems, so the odd harmonics tend to be the most troublesome, and since these generally diminish with the harmonic number, it is the first few odd harmonics (3rd, 5th, 7th, 11th and 13th) that can be problematic. Longer duration harmonic currents (existing for 10 minutes or more), cause additional losses in rotating machines, transformers and transmission lines, while in the very short term (less than 3 seconds), elevated levels of harmonic voltage distortion may lead to the mal-operation of some items of customer equipment.

In order to avoid the propagation of harmonic currents throughout the entire network, harmonic standards require that network operators maintain relatively low levels of harmonic voltage distortion, particularly at transmission potentials. Because most distorting loads exist in the LV network (and to a lesser extent in the MV), it is to be expected that voltage distortion will be highest at LV potentials, becoming progressively less so at MV and HV. Some typical *harmonic planning levels* used by network owners at various network voltages are shown in Table 13.2. These can be considered a quality objective to be achieved in a particular locality where various distorting loads are connected. They are used in the establishment of the *harmonic emission limits* assigned to individual customers, taking into account all distorting loads connected to the local network, with the aim of sharing the capacity of the network to absorb harmonic currents between all customers.

Table 13.2 shows that harmonic planning levels are considerably reduced at HV potentials as compared to those at MV and LV. In addition, as the harmonic number

Table 13.2 Typical harmonic planning levels for low-order, non-triplen harmonics.

Harmonic	Typical voltage distortion planning limits (% of Fundamental)				
	EHV	33–69 kV	11–22 kV	6.6 kV	400 V
5th	2.00	3.1	5.1	5.3	5.5
7th	2.00	2.7	4.2	4.3	4.5
11th	1.5	1.9	3.0	3.1	3.3
13th	1.5	1.8	2.5	2.6	2.8

increases, the planning level reduces. This is largely due to the natural tendency for the harmonic amplitude to reduce with increasing harmonic number. More will be said about planning levels and emission limits in Section 13.8.

Finally, in the unusual case where the distorting load is particularly unbalanced, then so will be the harmonics it creates. In this situation, all harmonics behave in a similar fashion to unbalanced fundamental currents, each having positive, negative and zero-sequence components.

13.4 Triplen Harmonic Currents

Because of their zero-sequence characteristics triplen harmonics can be the source of localised overheating problems, particularly in LV neutral conductors where the residual current flows towards the star point of the supply transformer. Since network owners attempt to balance LV loads evenly across all three phases, the neutral conductor is generally not expected to carry a significant current. However, this situation can change dramatically in the presence of triplen harmonic currents (notably the third), since these accumulate in the neutral and may become sufficiently large to overload this conductor.

Consider the case of a distorting LV load including a 'modest' 40% third harmonic component, in addition to the fundamental. The resulting RMS phase current will be:

$$I_{RMS} = \sqrt{I_1^2 + I_3^2} = \sqrt{I_1^2 + 0.4I_1^2} = 1.08I_1$$

This current is little different from that which might be expected to flow in each phase in the absence of harmonics, but the amplitude of the neutral current will be an unexpected $3 \times 0.4I_1 = 1.2I_1$. This can overload the neutral conductor which may be of a smaller size than the associated phase conductors. One method often used in larger LV networks to reduce this effect is to install a grounded zigzag transformer on the LV bus, close to the distorting load. Operating in a similar way to an earthing transformer, this arrangement provides a low zero-sequence ground impedance, diverting much of the 3rd harmonic current away from the supply transformer and out of the neutral conductor(s), in addition to reducing the transformer's triplen harmonic loss. To assist this, a small inductance is often inserted in the neutral connection of the distribution transformer, to increase its zero-sequence impedance and therefore further reduce the triplen harmonic current flowing.

An alternative way of avoiding triplen neutral currents is to operate a star connected load without a neutral connection, although clearly this method is unsuitable for LV loads. Without a neutral connection, triplen currents cannot flow and, as a result, a triplen oscillation occurs at the star point, in a similar fashion to that described in Section 7.5.1. This method is frequently used in voltage support capacitor banks and harmonic filters, where it is usually desirable to avoid triplen currents.

13.5 Harmonic Losses in Transformers

Winding temperatures in all transformers are exacerbated by the presence of harmonic currents, particularly as the harmonic number increases, due to the effects of the additional eddy current losses occurring within the conductors themselves. These losses arise as a result of the electromagnetic field surrounding a current carrying conductor, produced by the current itself. When the associated flux passes through the face of a conductor, a small potential is generated which causes eddy currents to circulate around the conductor periphery, in much the same way as eddy currents occur within magnetic laminations. The associated losses increase the temperature of the winding and any other metallic components in which they occur. They are proportional to the square of the electromagnetic field strength (and thus the square of the current producing it), as well as to the square of the frequency of the field.

Eddy current losses are frequently concentrated towards the ends of the winding closest to the core (usually the LV winding) since the flux there has a tendency to fringe towards the core, passing through the face of conductors as it does so. Because these losses vary as the square of both the current and frequency, high-order harmonics can have a disproportionate influence, despite their having generally lower amplitudes.

Some small distribution transformers are built without oil immersion cooling. This saves considerably on size, weight and expense, but these transformers are totally reliant on natural airflow for their cooling. They are known as *dry transformers*, and as shown in Figure 7.1b, their windings are cast in epoxy resin. Because they are not flooded with oil, the heat generated must be dissipated by conduction through the windings themselves and convection and radiation from their surfaces. This can lead to high hot spot temperatures occurring, particularly in the presence of a harmonic rich load. If the hot spot temperature exceeds the insulation's limiting temperature, it will become degraded, reducing the expected life of the transformer.

The cooling of all transformers becomes considerably more difficult in the presence of harmonics, to the extent that harmonic rich loads may demand the de-rating of the transformer concerned.

13.5.1 Harmonic Loss Factor

The IEEE standard C57.110 *Recommended Practice for Establishing Liquid Filled and Dry Type Power and Distribution Transformer Capability when Supplying Non-Sinusoidal Load Currents*, considers the effect of the harmonic content of the load current in establishing the effective capacity of a transformer when supplying non-sinusoidal load currents. A transformer's load losses (those associated with its load current as opposed to magnetising losses) can be expressed, in terms of the symbolism of C57.110, as:

$$P_{LL} = I^2R + P_{EC} + P_{OSL} \tag{13.6}$$

where P_{LL} is the total load loss (watts), I^2R is the resistive portion of the load loss, and I is the RMS load current flowing, P_{EC} is the winding eddy current loss and P_{OSL} is the other stray losses.

An allowance is made for the eddy current loss (P_{EC}) occurring at the fundamental frequency in the design of a transformer. However, this loss increases substantially in the presence of harmonic load currents and so the rated capacity of the transformer must often be reduced when a significant harmonic content is present.

The other stray losses include eddy current losses in the tank walls, clamping structures and internal busbars etc., and since these also increase the cooling load on liquid-filled transformers, they must also be included in the overall transformer loss. Dry transformers on the other hand, are not so restricted, since losses in clamping structures do not generally affect the temperature of the windings, as both reject heat independently into the atmosphere.

Since the eddy current loss is proportional to the square of the harmonic frequency, the total eddy current loss, including contributions at both the fundamental and harmonic frequencies, is given by:

$$P_{EC} = P_{EC-O} \frac{\sum\limits_{h=1}^{h=h\max} I_h^2 h^2}{I^2} = P_{EC-R} \frac{\sum\limits_{h=1}^{h=h\max} I_h^2 h^2}{I_R^2} = P_{EC-R} \sum\limits_{h=1}^{h=h\max} I_{h(pu)}^2 h^2$$

where P_{EC-O} is the measured eddy current loss at the fundamental frequency, P_{EC-R} is the rated eddy current loss at the fundamental frequency, I_h is the RMS current of order h, I is the measured RMS transformer secondary current $= \sqrt{\sum\limits_{h=1}^{h=h\max} I_h^2}$, I_R is the rated fundamental current of the transformer and $I_{h(pu)}$ is the per-unit harmonic current (relative to the rated fundamental current).

IEEE C57.110 defines the *harmonic loss factor* (F_{HL}) for a transformer as:

$$F_{HL} = \frac{P_{EC}}{P_{EC-O}} = \frac{\sum\limits_{h=1}^{h=h\max} I_h^2 h^2}{I^2} = \frac{\sum\limits_{h=1}^{h=h\max} I_h^2 h^2}{\sum\limits_{h=1}^{h=h\max} I_h^2} = \frac{\sum\limits_{h=1}^{h=h\max} (I_h/I_1)^2 h^2}{\sum\limits_{h=1}^{h=h\max} (I_h/I_1)^2} \tag{13.7}$$

where I_1 is the RMS fundamental load current.

F_{HL} quantifies the increase in the eddy current loss occurring in the presence of harmonic currents. As suggested by Equation (13.7), it can be calculated from per-unit measurements of the harmonics present as assessed using a handheld harmonic analyser. F_{HL} can therefore be readily determined for any particular load spectrum. It is used in determining by how much a transformer must be de-rated when supplying a particular spectrum of harmonic currents, as shown in the following example for a dry type transformer.

13.5.2 Harmonic Capacity Constraints in Dry-Type Transformers

Consider the de-rating necessary for a dry transformer when supplying a harmonic rich load. Since the stray loss component P_{OSL} does not substantially affect the winding temperature of a dry transformer, it can be ignored. Equation (13.6) can then be written in per-unit form, where the rated resistive loss $I_R^2 R$ is taken as the base:

$$P_{LL(pu)} = 1_{pu} + P_{EC(pu)}$$

However, since in the presence of harmonics the measured load current may exceed the rated current, each term is scaled according to the square of the measured per-unit load current $I_{(pu)}^2$ to yield the actual per-unit load loss $P_{LLact(pu)}$, dissipated by the transformer.

$$P_{LLact(pu)} = I_{(pu)}^2 + I_{(pu)}^2 P_{EC(pu)} = I_{(pu)}^2 + I_{(pu)}^2 P_{EC-O(pu)} F_{HL}$$

When the transformer carries its rated fundamental current the measured eddy current loss $P_{EC-O(pu)}$, becomes equal to the rated eddy current loss $P_{EC-R(pu)}$, thus we may write:

$$P_{LLact(pu)} = I_{(pu)}^2 + I_{(pu)}^2 P_{EC-R(pu)} F_{HL} \tag{13.8}$$

If the transformer is not to be overheated by the additional eddy current losses, then the actual load loss $P_{LLact(pu)}$ must remain less than or equal to the rated load loss, $P_{LL-R(pu)}$ applicable to the transformer, i.e.:

$$P_{LL-R(pu)} \geq I_{(pu)}^2 + I_{(pu)}^2 P_{EC-R(pu)} F_{HL} \tag{13.9}$$

We may rearrange Equation (13.9) to obtain the maximum per-unit load current that can be permitted to flow in the transformer in the presence of harmonics, subject to preserving its rated load loss, thus:

$$I_{max(pu)} = \sqrt{\frac{P_{LL-R(pu)}}{1 + F_{HL} P_{EC-R}}} \tag{13.10}$$

Example: Consider a transformer in which the load current has the harmonic spectrum shown in Table 13.3. Assume that the magnitude of the rated eddy current loss $P_{EC-R(pu)}$ occurring at the fundamental frequency is 15% of the rated $I^2 R$ load loss. Therefore the total rated load loss $\left(P_{LL-R(pu)}\right)$ is 1.15 pu.

From the harmonic spectrum the RMS load current I can be calculated, since $I = \sqrt{\sum_{h=1}^{19} I_h^2} = 1.037(pu)$. Next, the harmonic loss factor F_{HL} can be evaluated, as shown in Table 13.4.

Table 13.3 Harmonic Spectrum.

h	1	5	7	11	13	17	19
I_h/I_1	1.00	0.24	0.12	0.048	0.029	0.015	0.009

Table 13.4 Harmonic calculations.

h	(I_h/I_1)	$(I_h/I_1)^2 h^2$
1	1.00	1.00
5	0.24	1.44
7	0.12	0.705
11	0.048	0.279
13	0.029	0.142
17	0.015	0.065
19	0.009	0.029
		$\Sigma = 3.66$

Thus:

$$F_{HL} = \frac{\sum\limits_{h=1}^{h=19}(I_h/I_1)^2 h^2}{\sum\limits_{h=1}^{19}(I_h/I_1)^2} = \frac{3.66}{1.037} = 3.40$$

We can now apply Equation (13.10) to obtain the per-unit maximum non-sinusoidal load current which the transformer can safely support:

$$I_{\max(pu)} = \sqrt{\frac{P_{LL-R(pu)}}{1+F_{HL}P_{EC-R}}} = \sqrt{\frac{1.15}{1+3.4\times0.15}} = 0.87\,\text{pu}$$

Therefore as a result of the harmonic content in its load, this transformer must be de-rated by 13%. A similar approach is used in evaluating the maximum load current for oil-filled transformers, except that in this case the other stray losses (P_{OSL}) must also be taken into account, since they also contribute to the transformer top oil temperature and therefore to the winding temperature as well. Examples of such calculations are provided in IEEE C57.110.

13.5.3 K-Factors

The concept of a K-factor was introduced by the Underwriter's Laboratory in the USA in the early 1990s. This index is an indication of a transformer's ability to supply a harmonic rich load while operating within the temperature limitations of its insulation. The K-factor is defined as:

$$K = \sum_{h=1}^{h=h\max}\left(\frac{I_h}{I_R}\right)^2 h^2 = \frac{1}{I_R^2}\sum_{h=1}^{h=h\max} I^2 h^2 \tag{13.11}$$

where I_h is the RMS value of harmonic current h and I_R is the rated transformer current.

The K-factor is clearly similar to the harmonic loss factor, except that it is based on the rated transformer current I_R rather that the actual load current flowing I. As a result, the K-factor for a given load varies inversely with the size of the transformer.

Manufacturers build *K-rated* transformers capable of supplying load currents of a particular K-factor. These oversized transformers require no further de-rating, provided the K-factor of the connected load is less than or equal to that for which the transformer has been built. Designers take the following steps to reduce the effects of eddy current losses in a K-rated transformer:

1) Design the core to operate at a lower flux density than might normally be chosen, so as to reduce the magnetising losses.
2) Provide oversized windings comprising several parallel connected conductors to mitigate the increased resistance due to the skin effect at harmonic frequencies. This technique allows harmonic currents to be carried by several conductors. By so doing, the total resistance at each harmonic is reduced, since a greater portion of each conductor is used to carry current, reducing the total harmonic loss in the winding.

3) Provide increased clearances between the core and other metallic components such as the windings and the tank. The use of non-magnetic materials (where possible) and the avoidance of closed circulating current paths within clamping structures, as well as the use of magnetic shielding materials to avoid inducing eddy currents within metallic components, all contribute to a reduction in these losses.
4) Provide an oversized neutral conductor to avoid the excessive heating effects of triplen harmonics.

Table 13.5 shows some typical transformer K-ratings and their likely applications.

Table 13.5 K-factor transformer ratings and applications.

K-factor rating	Harmonic content	Likely application
1	Nil	Sinusoidal loads, resistive heating, incandescent lighting, motors (without VSDs)
4	16% 3rd, 10% 5th 7% 7th, 5.5% 9th	Welders, induction heaters, high intensity discharge lighting, fluorescent lighting
9	150% of the harmonic loading of a K4 transformer	Healthcare facilities, schools, office buildings
13	200% of the harmonic loading of a K4 transformer	UPS & VSD systems, healthcare facilities, schools
20	Very large	Critical care areas in hospitals, operating facilities, data processing equipment, computer installations

13.6 Power Factor in the Presence of Harmonics

Just as harmonics increase the losses within individual transformers, they also increase the losses in the distribution system since harmonic currents increase the reactive power consumed through the presence of harmonic VArs. The presence of harmonics requires that the power factor definition be amended accordingly. In the presence of harmonic distortion, we may use Equations (13.1) and (13.3) to express the RMS load current as follows:

$$I_{RMS} = I_1 \sqrt{1 + (THD_i/100)^2}$$

Similarly, from Equations (13.2) and (13.3), we may write:

$$V_{RMS} = V_1 \sqrt{1 + (THD_v/100)^2}$$

The power delivered to a load will generally include a major component at the fundamental frequency, as well as minor components at each harmonic, provided that both harmonic voltages and harmonic currents exist. The total power can therefore be expressed as:

$$P_{Total} = \sum_{h=1}^{\infty} V_h I_h \cos(\phi_h) = P_1 + P_2 + \ldots + P_h$$

Where P_h is the power associated with harmonic h. The power factor in the presence of harmonics is still defined by the ratio of the power consumed to the volt-amps supplied; however, the latter now includes a harmonic component.

$$PF_{total} = \frac{P_{Total}}{V_{RMS}I_{RMS}} = \frac{\sum_{n=1}^{\infty} V_n I_n \cos(\phi_n)}{V_{RMS}I_{RMS}} = \frac{\sum_{n=1}^{\infty} V_n I_n \cos(\phi_n)}{V_1\sqrt{1+(THD_v/100)^2}\, I_1\sqrt{1+(THD_i/100)^2}}$$

(13.12)

Equation (13.12) can usually be simplified. Since the total harmonic voltage distortion is generally small with respect to the fundamental, the vast majority of the power is supplied by the fundamental voltage and current, i.e. $P_{Total} \approx V_1 I_1 \cos(\varphi_1)$. This assumption is quite reasonable because THD_v is likely to be less than 5%, and consequently $V_{RMS} \approx V_1$. (On the other hand, THD_i may be very high, often in excess of 100%). Therefore Equation (13.12) can be rewritten in the form:

$$PF_{total} \approx \frac{P_1}{V_{RMS}I_{RMS}} \approx \frac{V_1 I_1 \cos(\phi_1)}{V_1 I_1 \sqrt{1+(THD_i/100)^2}} = \frac{\cos(\varphi_1)}{\sqrt{1+(THD_i/100)^2}}$$

(13.12a)

$$= PF_{displacemnt} \cdot PF_{distortion}$$

In this case the displacement power factor, $\cos(\phi_1)$ is that previously defined for the sinusoidal case. When harmonics are present, a new term known as the *distortion power factor* also arises, which further reduces the total power factor of the load. In order to increase the displacement power factor it is necessary to increase $\cos(\phi_1)$ through the addition of power factor correction capacitors. The distortion power factor on the other hand can only be increased by including harmonic filtering near the offending load, so that the harmonic currents it produces do not flow back into the upstream network. Interestingly, passive harmonic filters also improve the displacement power factor in addition to reducing harmonic levels, since they appear capacitive at the fundamental frequency.

Equation (13.12a) can be summarised by the diagrams in Figure 13.5, where the total power factor may be approximated by:

$$PF_{Total} \approx \frac{P_1}{V_{RMS}I_{RMS}} = \cos(\phi_1)\cos(\gamma)$$

where $\cos(\gamma) = 1/\sqrt{1+(THD_i/100)^2}$.

We may simplify the expression for the total volt-amps delivered to the circuit, by defining the harmonic VArs as:

$$S_T^2 = (V_{RMS}I_{RMS})^2 \approx V_1^2 I_1^2 + V_1^2 \sum_{h=2}^{\infty} I_h^2$$

$$S_T^2 \approx S_1^2 + V_1^2 \sum_{h=2}^{\infty} I_h^2 = (P_1^2 + Q_1^2) + Q_h^2$$

(a)

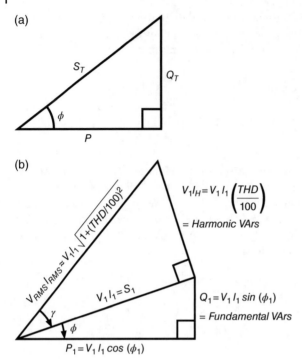

Figure 13.5 (a) Simplified power triangle in the presence of harmonics (b) Modified power triangle.

(b)

where the harmonic VArs are defined as $Q_h^2 = V_1^2 \sum_{h=2}^{\infty} I_h^2$. This quantity is sometimes called the *distortion power*, and given the symbol D. If we define the *total reactive power* flowing as Q_T, where $Q_T^2 = Q_1^2 + Q_h^2$, then we may write $S_T^2 = P_1^2 + Q_T^2$, suggesting the simplified power triangle shown in Figure 13.5a, where the angle Φ is defined by:

$$\cos(\Phi) = \cos(\phi)\cos(\gamma) = PF_{Total}$$

This equation suggests that the power triangle can be redrawn to include the effect of the harmonic VArs, as shown in Figure 13.5b. The distortion power factor further reduces the total power factor since it increases the RMS current yet does not contribute to the power delivered to the load; harmonic currents only increase the network losses.

Note that Figures 13.5a and 13.5b can no longer be related to a phasor diagram, since the harmonic VArs rotate at angular frequencies different from that of the fundamental, and the angle Φ no longer represents the phase shift between the fundamental voltage and current. This diagram does however illustrate the Pythagorean relationship between the quantities concerned, and it is useful from this point of view. Figure 13.6 shows the variation in distortion power factor as a function of THD_i; when this reaches 100% $\cos(\gamma) = 0.707$.

Example: Let us consider a numerical example to illustrate these concepts. The table below shows data collected from measurements on a small inverter-driven compact fluorescent lamp, the current of which is rich in harmonics.

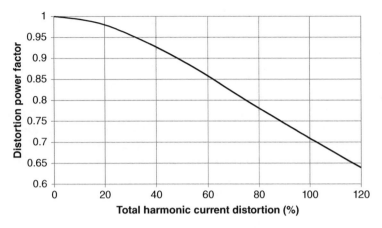

Figure 13.6 Distortion power factor variation with current THD.

Compact fluorescent lamp measured data

V_{RMS} (V)	I_{RMS} (mA)	Power (W)	Current THD (%)
241	140	20.2	112.6

Using the equations above and this small amount of data, the information shown in the following table can be obtained. From the THD_i value and the RMS current, the fundamental current I_1 can be found. This leads to a quite reasonable displacement power factor of 0.9. However, when the reactive harmonic power is taken into account, the total power factor falls to 0.6. This poor result is largely due to the high harmonic component in the supply current and the harmonic VArs that the lamp thus consumes. The above calculations were possible from just four pieces of information, each available from a handheld power analyser.

Power factor analysis

Fundamental current	Fundamental reactive power (Q_1)	Fundamental power factor $\cos(\phi)$	Total reactive power (Q_T)	Total power factor $\cos(\Phi)=\cos(\phi)\cos(\gamma)$
92.6 mA	9.67 VArs	0.9	27.02 VArs	0.6

Non-linear devices exhibiting such poor power factors consume a disproportionately large portion of the network capacity in comparison to the power they consume. Unfortunately some manufacturers occasionally state the displacement power factor, but omit to include the distortion power factor, making their product specifications appear better than they really are.

13.7 Management of Harmonics

The impact that a given distorting load will have on a network depends upon the capacity of the network at the point of connection, which is frequently expressed in terms of the three-phase *short circuit current* I_{SC}. Small loads may be defined as those whose maximum demand current is less than 1 per cent of I_{SC}. Many 'small' customers are unaware that the harmonic content of their load needlessly increases network losses, but since their contribution is small this is generally of little consequence.

'Large' loads, on the other hand, demand currents that represent a considerably larger fraction of I_{SC}, 5 or 10% of I_{SC} for example. 'Large' customers must actively manage the effects that their distorting load has on the network, and harmonic emission limits are frequently included in connection agreements in order to enforce this. Harmonic management is often achieved by installing harmonic filters in parallel with the offending load, or by using *harmonic cancellation techniques*. These involve arranging the load in such a way that current flows smoothly over as much of the AC cycle as possible. Important examples of this appear in the design of *electrowinning* rectification equipment and variable speed drives (VSDs – also known as *adjustable frequency drives*, *ASDs*) where rectifier circuits are often arranged to achieve a high pulse number.

13.7.1 Harmonic Cancellation Techniques (Rectifier Pulse Number)

It is often necessary to rectify the three-phase AC supply in order to produce a source of DC potential, perhaps for use in a variable frequency drive, a converter circuit or for the electrolytic deposition of metals such as zinc or tin. In all these applications a three-phase bridge rectifier circuit is commonly used to produce a relatively smooth DC potential.

The circuit depicted in Figure 13.7 is a *naturally commutated six-pulse* rectifier, so called because of the use of uncontrolled rectifier diodes to produce a DC waveform containing six ripple pulses per cycle of the fundamental. These switch in a sequence that presents the load with whichever line voltage is instantaneously the most positive; the DC voltage generated being equal to 1.35 times the magnitude of the RMS AC line voltage.

Figure 13.8 shows the diode and line currents for a rectifier supplied from a set of star connected windings, each conducting for 120° every half cycle. When the rectifier is supplied from a delta connected set, the line current adopts the stepped waveform shown in Figure 13.9 and flows continuously throughout each cycle.

The AC line current of either rectifier configuration contains harmonics of order $np \pm 1$, where p is the pulse number (in this case six), and n is a positive integer. Thus a six-pulse rectifier's line current includes 5th and 7th harmonics as well as 11th and 13th, 17th and 19th and so on, as shown in Figure 13.8. As the harmonic number h increases, so the harmonic current amplitude decreases, in roughly inverse proportion.

Although the amplitude spectrum is the same for each winding configuration, the phase spectrum is not, and when these line currents are added, a substantial reduction

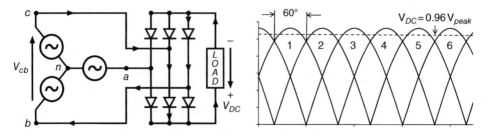

Figure 13.7 Three-phase bridge rectifier and voltage waveforms.

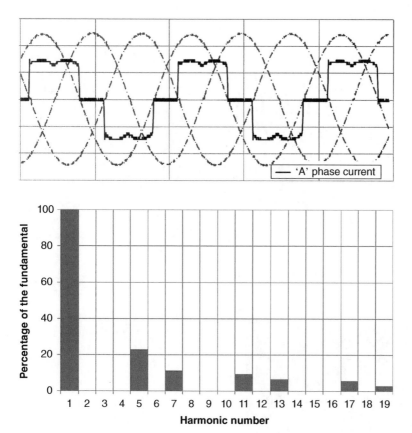

— 'A' phase current

Figure 13.8 Star connected three-phase rectifier line voltages, line current and harmonic spectrum.

in the lowest harmonics results occurs due to harmonic cancellation. Because of this there is a significant benefit to be gained from using a rectifier with a high pulse number, since this means that the first harmonic present will have frequency of $p-1$ times the fundamental, and if p is sufficiently large, its amplitude will be relatively small. Twelve-pulse rectifiers are commonly used on this account, and can be built by introducing a 30° phase shift between two six-pulse rectifiers, as shown in Figure 13.9.

A 30° phase shift is usually achieved by supplying one six-pulse rectifier from a delta connected transformer winding and the other from a star connected one. The traditional method for combining the outputs of two such rectifiers is to use an inter-phase transformer (shown in Figures 13.9 and 13.10), which supports the difference between their output voltages while summing their currents. This transformer usually has a single turn winding on each side, made by passing the DC busbars through a transformer core, in much the same way that an AC busbar passes through a current transformer. The busbars are arranged so that the DC currents from the rectifiers flow in opposite directions through the core, avoiding the creation of a DC flux, while supporting the cumulative ripple voltage that exists between them. In this way, each rectifier continues to operate as an independent six-pulse unit, while the output voltage and current of the combination appear as twelve-pulse.

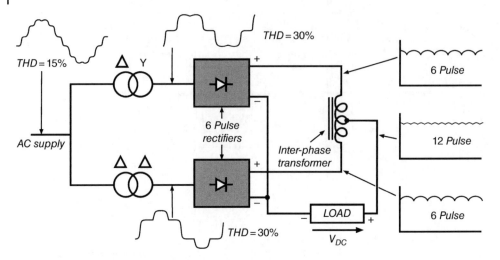

Figure 13.9 Typical construction of a twelve-pulse rectifier.

Figure 13.10 A 20 kA inter-phase transformer used in a small electrolytic zinc plant.

Modern electrolytic rectifiers generally do not employ inter-phase transformers; they rely instead on the reactance of the outgoing DC busbars across which to drop the difference in ripple voltages. For this approach to be successful, these busbars must be of sufficient length and each must only carry the current from one rectifier bridge.

Figure 13.11 shows the effect of adding the winding currents from each six-pulse bridge, the combination of which provides a considerably smoother line current. The twelve-pulse harmonic spectrum begins at the 11th harmonic and, as mentioned, contains components at $np \pm 1$ times the fundamental frequency. However, as shown in the figure, the degree of cancellation of the 5th and 7th harmonics depends on the precision of the matching between the component rectifiers. When the star and delta bridges are not precisely matched in terms of DC voltage and current, residual

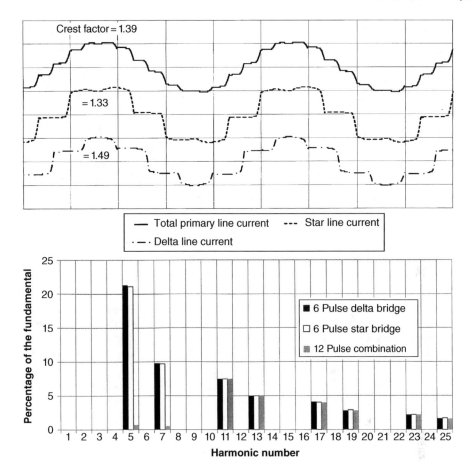

Figure 13.11 Harmonic cancellation in a twelve pulse rectifier.

5th and 7th harmonic currents may remain, although these will generally be much lower in amplitude than those of the 11th and 13th.

Similar arrangements are used in the design of variable speed drives (VSDs), where a variable frequency three-phase system is synthesised from a DC voltage produced by rectifying the incoming AC supply. Low voltage drives often use 12- or 18-pulse rectifiers instead of a 6-pulse bridge, while medium voltage drives frequently use 24-pulse rectifiers in order to reduce the supply current distortion to acceptably low levels.

However, this harmonic cancellation concept can be extended much further. Aluminium and zinc producers, for example, frequently pass DC currents in excess of 200 kA through their electrolytic cells. Currents of this magnitude are traditionally obtained from a fleet of naturally commutated rectifiers, each one phase shifted from the others by a small angle, so that when seen from the supply side, a very high pulse number results. For example, when a fleet of six 12-pulse rectifiers is employed, each phase shifted from its neighbours by 5°, a pulse number of 72 results. The first significant harmonics are the 71st and the 73rd, and as a result the resulting line currents will be almost sinusoidal, since the amplitudes of these harmonics will be very low.

13.7.2 Harmonic Filters

In the past three decades there has been a move away from naturally commutated rectifiers for electrolytic plants towards thyristor controlled rectifiers. Thyristors provide the ability to adjust the DC output voltage by delaying the point on the line voltage waveform at which each thyristor fires (i.e. by controlling its firing angle). This avoids the expense and complexity of an on-load tap-changer within the transformer tank.

Because both the amplitude and phase of current harmonics in controlled rectifiers are strongly influenced by the firing angle, the phase shifting techniques described above are generally not successful, since small differences in firing angles will result in incomplete harmonic cancellation. As such, achieving a high pulse number will not provide a reliable means of harmonic reduction. Instead it is usual to install a harmonic filter on the supply bus, close to the rectifier, to provide a path to ground for harmonic currents that would otherwise flow into the upstream network.

There are also many other large industrial devices connected to the power system today, including inverters, static VAr compensators (SVCs), cycloconverters, DC Motor controllers and high voltage DC (HVDC) transmission equipment, all of which employ thyristors or insulated gate bipolar transistors (IGBTs). This equipment also consumes currents rich in harmonics and it is often necessary to install harmonic filtering equipment to suppress the harmonic currents that would otherwise flow into the local network. There are various harmonic filter circuits used for this purpose, but we will consider one of the more common, the series tuned LC circuit, which is designed to provide a current sink at a discrete harmonic frequency.

Figure 13.12a shows a typical single stage *series resonant* harmonic filter used to sink non-triplen harmonic currents. The series connected tuning inductances supply two star connected sets of capacitors, collectively resonant just below the target harmonic. This circuit presents a low impedance at the target frequency and thus provides a local sink for harmonic currents. The filter's single-phase equivalent circuit appears in Figure 13.12b, the resistive component of which (R_f) represents the losses associated with the tuning reactors.

The capacitors are frequently split into two star connected groups as shown, with a balance current transformer inserted between the star points. So long as both star networks remain balanced, there will be no current flowing in this CT. However, if a partial capacitor failure occurs in one or more phases (usually as a result of the open-circuiting of one or more of the internal capacitive elements), then a proportional unbalance current will flow in the CT. Small currents will not warrant tripping the filter and may simply raise an alarm, but a complete capacitor failure will result in a large current flow, requiring the immediate tripping of the filter.

Series and Parallel Resonance

A typical impedance plot for a 50 Hz high voltage 5th harmonic filter appears in Figure 13.13. The series resonant frequency (or notch frequency) is determined solely by the filter components and is given by:

$$f_{series} = 1/2\pi\sqrt{L_f C_f} \tag{13.13a}$$

In this example f_{series} is 240 Hz.

The local source inductance L_s appears in parallel with the filter elements, and it therefore creates a parallel tuned circuit, resonant just a little below the

(a)

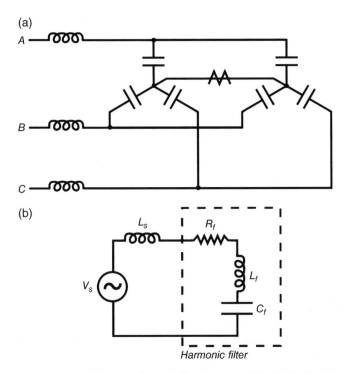

(b)

Figure 13.12 (a) Typical schematic for a harmonic filter (b) Single-phase equivalent circuit.

target frequency. The parallel resonant frequency is given by Equation (13.13b), occurring in this case at about 208 Hz.

$$f_{parallel} = 1/\left(2\pi\sqrt{\left(L_s + L_f\right)C_f}\right)$$ (13.13b)

While externally a parallel resonant circuit appears as an open circuit, internally there can be very large currents exchanged between the inductance and capacitance, particularly if the Q of the circuit is high. The danger of parallel resonance occurring at or near an active harmonic must be taken into account when installing harmonic filters or power factor correction capacitors. This is because large and destructive currents can be exchanged between the filter capacitance and the source should a parallel resonance be excited by harmonic currents already present in the network. Should such a situation occur, the capacitors will usually experience a substantial over voltage, usually followed by rapid failure.

The series resonant circuit is generally tuned a little below the target harmonic, since this will permit a partial capacitor or inductor failure without permitting the circuit's resonant frequency to rise significantly above the target harmonic. Were this to occur, then the target harmonic may fall close to the parallel resonant frequency, presenting the possibility that the harmonic will be amplified as a result.

Away from these resonant frequencies, the filter's impedance rises significantly with respect to that of the source, thus in these regions the network impedance is dominated by the inductive component of the source, scaled according to frequency. This effect is

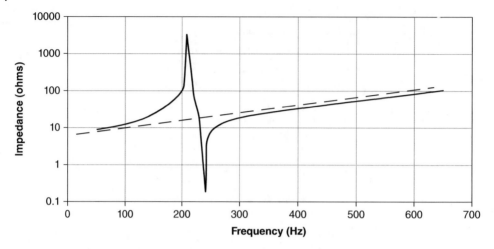

Figure 13.13 Impedance plot of a 50 Hz single stage HV 5th harmonic filter.

shown by the dotted line in Figure 13.13, which represents the system impedance in the absence of the filter.

Distorting loads frequently generate current at several harmonic frequencies, and therefore require the use of multi-stage harmonic filters. Figure 13.14 shows the frequency response of a medium voltage, four-stage 50 Hz series tuned LC filter, including the connected load.

In this example, the stages are tuned to the 5th, 7th, 11th and 13th harmonics, each corresponding to a notch in the response. Between adjacent notch frequencies lies a parallel resonance. In this case, most are not particularly sharp, due to the damping effect of the load, but each could give rise to unwanted harmonic amplification. Fortunately, the parallel resonant frequencies all lie close to even harmonics where little or no harmonic energy is expected.

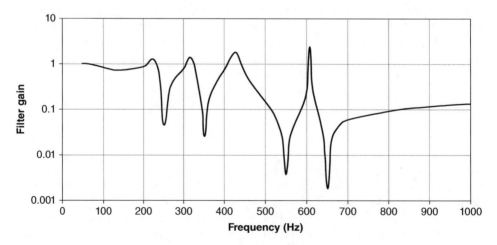

Figure 13.14 Frequency response of a 50 Hz four-stage MV harmonic filter and load.

Figure 13.14 illustrates an important point: whenever it is necessary to introduce a notch frequency into the spectrum, an associated parallel resonance will always occur a little below it, one that must not coincide with an active harmonic. For example, if the 7th harmonic current is to be targeted with a series tuned filter, it is usual to also install a 5th harmonic filter as well, so as to force a low system impedance at the 5th harmonic, and hence avoid any parallel resonance that may otherwise occur there.

The installation of power factor correction and voltage support capacitors also gives rise to parallel resonant frequencies. It is usual therefore to include a *de-tuning reactance* in such circuits to ensure that the parallel resonant frequency does not lie near an active harmonic. This reactance is chosen so that the series resonant frequency generally lies well below the 5th, thereby preventing 5th harmonic currents from overloading the capacitors. Such circuits are said to be de-tuned. The de-tuning reactance is also useful in limiting the transient inrush current that flows into the capacitors at switch on. Its impedance is usually expressed in terms of the de-tuning ratio p, which is given by:

$$p = 100 \left(\frac{X_L}{X_C} \right) \%$$

where X_L is the reactance of the inductor at the fundamental and X_C is the reactance of the capacitance at the fundamental.

Thus if $p = 7\%$ the reactor impedance is 7% of that of the capacitor at the fundamental frequency. The series resonant frequency is given by:

$$f_{resonant} = \frac{1}{2\pi\sqrt{LC}} = \frac{10 f_1}{\sqrt{p}} \text{Hz} \qquad (13.14)$$

where L is the reactor inductance, C is the effective capacitance, f_1 is the fundamental frequency and p is the de-tuning ratio.

Table 13.6 shows some commonly used series resonant frequencies in the application of power factor correction and voltage support capacitances.

Triplen harmonics are eliminated from harmonic filters and PFC equipment by connecting the capacitors in an ungrounded star configuration, as is usually the case in MV distribution networks, or by using a delta connection, as is generally found in LV applications.

Table 13.6 Common series resonant frequencies used for voltage support and PFC capacitors.

Series resonant frequency (50 Hz systems)	Series resonant frequency (60 Hz systems)	De-tuning ratio p (% of Xc at the fundamental)
134 Hz	160 Hz	14
189 Hz	227 Hz	7
204 Hz	245 Hz	6
210 Hz	252 Hz	5.67

Active Harmonic Filters

While passive harmonic filters are robust and require little maintenance, they modify the system impedance spectrum and are generally only able to target a discrete range of harmonic frequencies. Active filters on the other hand can target all harmonics up to about the 50th without introducing changes to the impedance spectrum. They are essentially semiconductor converter circuits connected in parallel with an offending load, as shown in Figure 13.15. They measure and invert the load's harmonic content, and by injecting it back onto the bus they cancel the harmonic currents flowing from the load, leaving only the fundamental component in the supply.

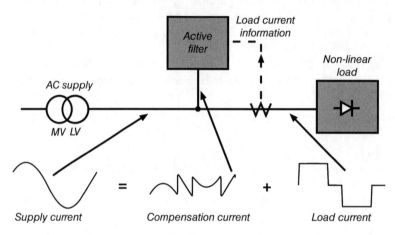

Figure 13.15 Active harmonic filter operation.

Because the converter circuit appears as a high impedance current source, the active filter does not modify the system impedance, and therefore does not introduce unwanted resonances. Active filters have a finite current capacity, and therefore must be carefully chosen to suit the harmonic spectrum of the intended load. They are particularly useful for providing harmonic cancellation for a collection of VSDs. They can also be used to provide reactive support at the fundamental, but in this respect they are an expensive way to provide power factor correction.

13.8　Harmonic Standards

The International Electrotechnical Commission publishes the IEC 61000 series of standards on harmonics and power quality, including the following:

- IEC 61000-3-2 *Limits for harmonic current emissions (equipment input ≤ 16 A per phase)*
- IEC 61000-3-4 *Limits for harmonic current emissions in LV power supply systems for equipment input > 16 A per phase*
- IEC 61000-3-12 *Limits for harmonic equipment currents produced by equipment connected to the public LV system with input current ≥ 16A*
- Technical Report IEC 61000.3.6 (2012), *Assessment of emission limits for the connection of distorting loads to MV, HV and EHV power systems.*

The last document provides an excellent introduction to electromagnetic compatibility and harmonic assessment, and in many countries it carries the weight of a standard, while in others it has been formally adopted as such. In the USA the IEEE Std 519 (2014), *IEEE Recommended practice and requirements for harmonic control in electric power systems* prescribes the limits for harmonic emissions at all voltage levels.

A detailed analysis of all these documents is beyond the scope of this book; however, we will consider the Technical Report IEC 61000-3-6, which explains the broader issues that form the basis for allocating the harmonic absorption capacity of the power system. We will also discuss the current and voltage distortion limits prescribed in IEEE 519, and the conditions under which they apply.

13.8.1 Electromagnetic Compatibility, Planning and Allocation Levels

The electromagnetic compatibility is the ability of all equipment connected to the power system to operate satisfactorily in the presence of disturbances, which may include harmonics and voltage fluctuations, such as voltage notching and flicker. For electromagnetic compatibility to be maintained, the immunity level of the equipment concerned must be greater than the emission levels of the devices creating the disturbance. For the purposes of the present discussion we shall assume that disturbances relate to the effects of harmonic voltage distortion.

The power system generally has a low impedance and as a consequence is relatively well regulated. It is therefore reasonably tolerant of the harmonic currents that are injected by non-linear loads. Some of these currents add destructively while others accumulate; however, the resulting voltage distortion occurring throughout the network can be kept acceptably low provided customers take steps to ensure that their loads do not exceed the prescribed harmonic allocation limits.

Harmonic currents are in most cases independent of the system impedance and are relatively easily to measure. Harmonic voltage distortion on the other hand contains a background component at each harmonic, as a result of the injection of harmonic currents elsewhere in the network, plus a component produced by currents flowing from each customer's distorting load through the local network impedance. It is therefore difficult to accurately determine what proportion of the voltage distortion is attributable to any particular customer, unless the network harmonic impedance is known at the connection point. Harmonic impedances can be difficult to predict, particularly when there are capacitors within the local network that may introduce resonances.

The primary intent of harmonic standards is to restrict the harmonic voltage distortion to an acceptable level and to share the harmonic capacity of the network between all present and future customers, roughly in proportion to the size of their connected loads. Rather than allocating a proportion of the permitted harmonic voltage distortion to individual customers, harmonic standards generally limit the magnitude of harmonic currents that may be injected instead, thus avoiding the need to determine the harmonic impedance associated with each customer connection.

One consequence of an excessive level of harmonic voltage distortion is the mal-operation of some items of customer equipment, in either the LV or MV networks. Because harmonic voltage distortion is a continuously varying quantity, a statistical approach is adopted in its measurement. This is expressed in terms of the *probability density* curves shown in Figure 13.16a, which depict the relationship between the local equipment

(a)

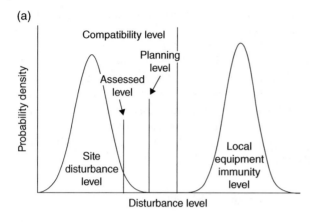

(b)

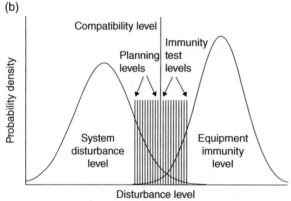

Figure 13.16 (a) Site disturbance levels (b) Global disturbance levels (IEC/TR 61000-3-6 ed.2.0 Copyright © 2008 IEC Geneva, Switzerland. www.iec.ch).

immunity levels and the site disturbance level produced by non-linear loads in a particular locality. The area beneath the immunity probability density curve to the left of a particular disturbance level represents the probability that the equipment will be disrupted by that disturbance. In this case, the disturbance level can be thought of as the magnitude of a given harmonic, expressed as percentage of the fundamental.

In the context of a particular location in a network we expect that the likelihood of overlap between the disturbance and immunity distributions will be small or non-existent, and that a suitable planning level can easily be chosen for any particular harmonic that will not result in local equipment malfunction. This situation is depicted in Figure 13.16a.

On the other hand when viewed globally, the entire network may experience a small overlap between disturbance and immunity levels, as shown in Figure 13.16b. This is in recognition of the fact that the network owner cannot control the harmonic levels at all locations throughout the network, all of the time. The immunity to a particular disturbance of a piece of equipment is a function of its design, and immunity test levels are usually agreed upon between manufacturers and users, to ensure acceptable product performance.

The compatibility level is used by the network owner to determine the electromagnetic compatibility of the entire network, and may be thought of as an upper bound to the planning level. It is based on the 95% disturbance level of the network, meaning that for 95% of the time the disturbance level will be less than the measured value.

The planning levels for harmonic voltages or currents are usually set by the network owner, and may vary slightly from one location to another, although in some countries they are prescribed in national standards. They can be considered as internal quality objectives for the network and vary with the harmonic number and the voltage level concerned. They will always be less than or equal to the associated compatibility level. The compatibility and indicative planning levels for individual harmonic voltages appear in Tables 13.7 and 13.8.

For very short disturbances of 3 seconds or less, both the compatibility and the indicative planning levels may be increased by a factor k_{hvs}, where:

$$k_{hvs} = 1.3 + \frac{0.7}{45}(h-5)$$

These tables are divided into three classes of harmonic: odd non-triplen harmonics, odd triplen harmonics and even harmonics. Even harmonics will not usually be present in the network due to the preponderance of half wave symmetry, and as a consequence the compatibility levels are set somewhat lower than for the other classes. Odd non-triplen harmonics generally represent the majority of system disturbances, particularly in the lower orders, and therefore the levels specified are considerably higher than for even harmonics. Note that as the harmonic order increases the compatibility level decreases, in a similar way to which harmonic amplitudes themselves tend to decrease. Odd triplen harmonics are in a class of their own. Of these the third is the most important since it can exist in LV and to a lesser extent in some MV networks. At higher orders, odd triplen harmonics generally exist only at small amplitudes, hence the reduction in level beyond the 9th.

The emission limit (sometimes called the allocation level) is the maximum permitted disturbance level imposed on an individual customer's load, usually specified as

Table 13.7 LV and MV compatibility limits.

Odd, non-triplen harmonics		Triplen harmonics		Even harmonics	
Harmonic order (h)	Harmonic voltage (%)	Harmonic order (h)	Harmonic voltage (%)	Harmonic order (h)	Harmonic voltage (%)
5	6	3	5	2	2
7	5	9	1.5	4	1
11	3.5	15	0.4	6	0.5
13	3	21	0.3	8	0.5
$17 \leq h \leq 49$	$2.27\left(\dfrac{17}{h}\right)-0.27$	$21 \leq h \leq 45$	0.2	$10 \leq h \leq 50$	$0.25\left(\dfrac{10}{h}\right)+0.25$

The compatibility level for the total harmonic distortion (THD) is 8%

(IEC/TR 61000-3-6 ed.2.0 Copyright © 2008 IEC Geneva, Switzerland. www.iec.ch)

Table 13.8 MV, HV and EHV indicative planning limits.

Odd, non-triplen harmonics			Triplen harmonics			Even harmonics		
	Harmonic voltage (%)			Harmonic voltage (%)			Harmonic voltage (%)	
Harmonic order (h)	MV	HV & EHV	Harmonic order (h)	MV	HV & EHV	Harmonic order (h)	MV	HV & EHV
5	5	2	3	4	2	2	1.8	1.4
7	4	2	9	1.2	1	4	1	0.8
11	3	1.5	15	0.3	0.3	6	0.5	0.4
13	2.5	1.5	21	0.2	0.2	8	0.5	0.4
$17 \leq h \leq 49$	$1.9\left(\dfrac{17}{h}\right)-0.2$	$1.2\left(\dfrac{17}{h}\right)$	$21 \leq h \leq 45$	0.2	0.2	$10 \leq h \leq 50$	$0.25\left(\dfrac{10}{h}\right)+0.25$	$0.19\left(\dfrac{10}{h}\right)+0.16$

Indicative planning levels: $THD_{MV} = 6.5\%$ and $THD_{EHV} = 3\%$

(IEC/TR 61000-3-6 ed.2.0 Copyright © 2008 IEC Geneva, Switzerland. www.iec.ch)

part of the network connection agreement. It is generally a fraction of the local planning level, and is chosen in proportion to network capacity consumed by the load. The allocation level is effectively an upper bound to the emissions permitted from the load in question.

13.8.2 IEEE Std 519 IEEE Recommended Practice and Requirements for Harmonic Control in Electric Power Systems

IEEE Std 519 is principally concerned with ensuring that the network voltage distortion limits are maintained. It specifies acceptable voltage distortion limits (THD_v) for all network phase voltages at the point of common coupling (PCC) between the customer concerned and the local network; these are shown in Table 13.9.

The point of common coupling is the closest point in the network to the customer connection at which other customer(s) are or could be connected. For many customers this will generally be the low voltage bus from which the supply originates and from which other customers are supplied, but for LV customers supplied from a dedicated MV transformer, the PCC lies at the point of connection to the MV network. This distinction is important since the standard must only be complied with at the PCC, and not necessarily throughout the customer's plant, where LV harmonic levels are likely to be considerably higher.

Provided that the network owner ensures that all customers adhere to these limits, then the absorption capacity of the network should be such that the voltage limits will be met. IEEE Std 519 permits the limits in Table 13.9 to be exceeded in the very short term (3 seconds). Voltage harmonics exceeding the daily 99th percentile (i.e. those exceeded only 1% of the time) very short time values must be less than 150% of those given in the table. The weekly 95th percentile short time (10 minutes) values, however, must be less than those in the table.

As previously mentioned, it is difficult to accurately apportion voltage harmonic distortion to any particular customer. IEEE Std 519 recognises this by defining limits to the harmonic currents that a given load can inject into the network. The current distortion limits for system voltages between 120 V and 69 kV are given in Table 13.10. These are defined in terms of the *short circuit ratio* which is defined as the ratio of the short circuit current at the PCC to the maximum demand current of the load concerned: I_{SC}/I_L. The standard defines I_L as the sum of the maximum demand current measured in each of

Table 13.9 IEEE 519 Voltage harmonic distortion limits.

Bus voltage at the PCC	Individual harmonic (% of fundamental)	THD$_v$ (%)
$V \le 1\,kV$	5.0	8.0
$1\,kV < V \le 69\,kV$	3.0	5.0
$69\,kV < V \le 161\,kV$	1.5	2.5
$V > 161\,kV$	1.0	1.5

Table 13.10 IEEE 519 current distortion limits for system voltages between 120V and 69 kV.

Maximum harmonic current distortion (% I_L) (Individual odd order harmonics)						
I_{SC}/I_L	$3 \leq h < 11$	$11 \leq h < 17$	$17 \leq h < 23$	$23 \leq h < 35$	$35 \leq h \leq 50$	TDD
$I_{SC}/I_L < 20$	4.0	2.0	1.5	0.6	0.3	5.0
$20 < I_{SC}/I_L < 50$	7.0	3.5	2.5	1.0	0.5	8.0
$50 < I_{SC}/I_L < 100$	10.0	4.5	4.0	1.5	0.7	12.0
$100 < I_{SC}/I_L < 1000$	12.0	5.5	5.0	2.0	1.0	15.0
$I_{SC}/I_L > 1000$	15.0	7.0	7.0	2.5	1.4	20.0
Even harmonics are limited to 25% of the values listed above						

the proceeding twelve months, divided by 12. Table 13.10 also imposes additional restrictions on the harmonic spectrum of the load by limiting the maximum total demand distortion.

Small loads for which the short circuit ratio is large consume relatively little network capacity and therefore can be permitted to inject a larger relative proportion of harmonic current. On the other hand, large loads with small short circuit ratios consume much more network capacity and are therefore considerably more restricted in terms of the harmonic currents they can inject.

The daily 99th percentile very short time harmonic currents must be less than twice the levels given in Table 13.10, while the weekly 99th percentile short time currents must be less than 150% of these limits. Finally, the weekly 95th percentile short time currents must be less than 100% of the limits in the Table 13.10.

13.8.3   Assessment of MV Customer Emission Levels

IEC 61000.3.6 defines three stages of assessment, which are used in sequence to decide if a particular distorting load may be connected to the MV network without exceeding the local planning levels. Small distorting loads may well pass stages 1 or 2, and thus may be connected with relatively little detailed analysis. However, larger loads often require detailed system studies to determine the pre-existing harmonic levels at the PCC and suitable emission limits and the conditions under which the connection may be permitted.

The Australian publication, Handbook 264 *Power quality recommendations for the application of AS/NZ 61000.3.6 and AS/NZ 61000.3.7* provides more information as well as numerical examples for stage 1, 2 and 3 assessments. This is an excellent source of information to which the interested reader is referred. The main requirements of each stage are described below.

Stage 1 Assessment

The stage 1 assessment is similar to the short circuit ratio approach employed in IEEE 519. It is based on the customer's agreed power consumption S_i (which will usually be

Table 13.11 Harmonic weighting factors for some common MV loads.

Equipment	Typical THD (%)	Weighting factor (W_j)
6-pulse converter with capacitive filtering	80	2.0
6-pulse converter with capacitive and partial inductive filtering	40	1.0
6-pulse converter with heavily inductive load	28	0.8
12-pulse converter	15	0.5

(IEC/TR 61000-3-6 ed.2.0 Copyright © 2008 IEC Geneva, Switzerland. www.iec.ch)

defined in terms of the contract maximum demand), relative to the system short circuit fault level at the point of common coupling, S_{sc}. If $S_i / S_{sc} < 0.2\%$ then the distorting load may be connected without further investigation. If this inequality is not satisfied then a weighted summation of the customer's distorting loads may be computed, to produce an effective distorting power, S_{Dwi} where:

$$S_{Dwi} = \sum_j S_{Dj} W_j$$

where S_{Dj} is the power of distorting equipment j of customer i, and W_j is a load dependent weighting factor, some of which appear in Table 13.11.

If the ratio of the customer's weighted distorting power to the system short circuit power is less than 0.2%, then connection of the load is permitted without further investigation.

Stage 2 Assessment

Stage 2 assessments allow for higher emissions than stage 1, based on the capacity of the wider MV network to absorb harmonic currents through diversity and harmonic cancellation. The limits allocated to each customer are commensurate with the *fraction* of the total available power from the PCC.

Stage 2 offers two approaches to harmonic assessment. The first applies to customers with an agreed power S_i less than 1 MVA, on condition that S_i/S_{sc} is less than 1%, and the background harmonic levels are acceptably low. It assigns limits to the harmonic currents that the customer may inject in a similar fashion to those imposed by IEEE 519. Table 13.12 shows some typical limits for odd harmonics as a function of the current corresponding to the customer's agreed power, I_L.

Table 13.12 Indicative emission limits for stage 2 assessments.

Harmonic order h	5	7	11	13	>13
Emission limit I_h/I_L (%)	5	5	3	3	$500/h^2$

(IEC/TR 61000-3-6 ed.2.0 Copyright © 2008 IEC Geneva, Switzerland. www.iec.ch)

The second approach that may be adopted under stage 2 is based on the general summation law defined by Equation (13.15). We are already familiar with this concept from Equation (13.1) in the calculation of the RMS value of a current (or voltage) comprising a series of harmonics. The general summation law is used for the aggregation of harmonic emissions (either voltages or currents) of order h and can be written as:

$$U_h = \sqrt[\alpha]{\sum_i U_{hi}^{\alpha}}$$ (13.15)

where U_{hi} is the magnitude of component i of the harmonic voltage (or current) of order h, U_h is the probabilistic magnitude of the aggregated harmonic voltage (or current) order h and α is the summation exponent given in Table 13.13, and depends on variation in the phase and magnitude of the individual harmonic.

Low-order odd harmonics have phase angles that vary little across the power system and therefore the chance of cancellation is limited, and as a result α is chosen conservatively for these frequencies. On the other hand higher order harmonics generally vary considerably in both magnitude and phase, offering more scope for cancellation, and therefore are assigned less conservative exponents.

While Equation (13.15) applies to harmonic emissions corresponding to a particular voltage level, it is also useful in combining quantities from different levels. For example, it can be used to aggregate the LV emissions transferred into the MV network.

The second approach defined under stage 2 apportions each customer a fraction of the *global voltage emission limit* (G_{hMV+LV}) of the total MV and LV load that can be supplied from the MV bus in question, according to:

$$E_{Uhi} = \alpha \sqrt{\frac{S_i}{S_t}} G_{hMV+LV}$$ (13.16)

where E_{Uhi} is the *harmonic voltage emission limit* of order h for distorting MV load i, S_i is the agreed power the distorting load, S_t is the total MV supply capacity, including any associated LV loads and α is the harmonic exponent, enabling smaller loads to receive a larger relative allocation.

The *MV planning level* for harmonic h (L_{hMV}) is expressed using the general summation law as the vector summation of the MV harmonic voltage distortion plus that contribution transferred from the upstream HV network (which is assumed to have already reached its planning level), thus:

$$L_{hMV} = \sqrt[\alpha]{G_{hMV+LV}^{\alpha} + \left(T_{hUM} L_{hUS} \right)^{\alpha}}$$

Table 13.13 Harmonic summation exponents.

Harmonic order h	α
$h < 5$	1
$5 \leq h \leq 10$	1.4
$h > 10$	2

(IEC/TR 61000-3-6 ed.2.0 Copyright © 2008 IEC Geneva, Switzerland. www.iec.ch)

where T_{hUM} is the transfer coefficient of harmonic voltage distortion from the upstream network to the MV network. This typically has a value around unity, although it may be as high as 3 in the presence of harmonic resonances.

From this equation we obtain the global contribution G_{hMV+LV} for the combined MV and LV load:

$$G_{hMV+LV} = \sqrt[\alpha]{L_{hMV}{}^\alpha - \left(T_{hUM} L_{hUS}\right)^\alpha}$$

Provided the load concerned can meet the limits defined by Equation (13.16) it may be connected without further investigation. Equation (13.16) provides an emission limit in terms of the voltage distortion at the PCC for harmonic h, rather than in terms of the current that can be injected at each significant harmonic, which is generally much more useful to the customer. If the harmonic impedance at the PCC is known, the corresponding current emission limit E_{Ihi} can be found since $E_{Ihi} = E_{Uhi}/Z_{hi}$, where Z_{hi} is the system impedance at harmonic h of customer i.

As mentioned earlier, harmonic impedances can be difficult to predict, especially when resonances exist in a network. At locations where there are no connected capacitances, low-order harmonics Z_{hi} can often be approximated by hx_1, where x_1 is the fundamental short circuit reactance at the PCC. This parameter will vary with changes in network configuration and it is best to err on the high side if a conservative approach is to be adopted. IEC 61000.3.6 suggests that for $h < 8$ an upper bound to Z_{hi} is $Z_{hi.\max} = 2hZ_1$.

Loads connected close to the HV-MV interface will generally see relatively low harmonic impedances, and can be allocated emission levels according to Equation (13.16). However, loads connected to long MV feeders remote from the MV bus, will experience considerably higher source impedances and proportionally lower short circuit currents. For similar agreed power levels these customers will therefore receive lower harmonic current allocations than those nearer the sending end.

Alternatively, the allocation of equal harmonic currents to all such customers would require low emissions limits in order to contain the voltage distortion at the far end of the line, in which case the network's capacity to absorb harmonic currents would be under-utilised. IEC 61000.3.6 suggests that, in these situations, an allocation based on harmonic VA be used instead; the current emission limit being given by:

$$E_{Ihi} = \frac{A_{hMV} S_i{}^{1/\alpha}}{\sqrt{X_{hi}}}$$

where A_{hMV} is the allocation constant, chosen for a given network, so that the planning level on the *weakest* feeder is preserved.

In this case, the harmonic current allocation is inversely proportional to $\sqrt{X_{hi}}$. The interested reader is referred to IEC 61000.3.6, Annex B, which includes examples of this approach.

Stage 3 Assessment

Equation (13.16) suggests that if all customers contribute no more than their share of the planning level at each harmonic, then the planning levels will not be exceeded. However, it is known that some customers generate very few harmonics, as a result of largely linear loads. Because of this, the strict use of Equation (13.16) will generally lead to voltage harmonic levels significantly below the planning levels.

The network owner therefore generally has some discretion to permit new customers to exceed the limits that would normally be imposed, and in so doing use some of the excess capacity within the network. Stage 3 assessments require a detailed study of the network background levels at the PCC as well as of the expected contribution of the new load, under various network configurations. Connection to a network under a stage 3 assessment may be on a conditional basis, thus postponing harmonic mitigation investment by the customer or possibly restricting the operation of distorting loads under certain network configurations.

13.9 Measurement of Harmonics

The assessment of the background harmonic levels present on a bus is always advisable prior to the connection of a large distorting load, i.e. one for which $I_{SC}/I_L < 100$. Such measurements can avoid subsequent arguments as to whether the contribution to the bus distortion by the new load is excessive. Standards IEEE 519 and IEC61000.3.6 both specify limits up to the 50th harmonic; however, accurate measurement of these may not always be possible with conventional voltage transformers. Capacitive VTs (CVTs) for example are tuned to the fundamental frequency and are therefore only accurate over a very narrow range of frequencies on either side of it. Inductive VTs may also exhibit resonances as low as 500 Hz, due to parasitic winding capacitances and, depending on the applied burden, are also likely to misreport harmonic levels.

Unless the actual frequency response is known and can be corrected for, the installed VTs generally should not be relied upon when making harmonic measurements, and non-conventional voltage transformers should be considered instead. These could be *RC* based voltage divider instruments or perhaps an optical VT, and may need to be installed for the measurement period. It is also possible to measure and integrate the current flowing in the capacitive divider of a CVT, in order to produce a faithful replica of the voltage spectrum, including at least the first 50 harmonics. Proprietary equipment is available for this purpose, which can be retrospectively fitted to most capacitive voltage transformers.

Where injection limits are based on current harmonic content, soon after a load is connected its harmonic content should be measured to ensure compliance with the prescribed limits (such as those in Table 13.10). In this regard, conventional inductive current transformers will usually be quite adequate for harmonic frequencies up to the 50th. If required, higher-order harmonics should perhaps be measured with a non-conventional current transformer such as a Rogowski coil, which will provide a considerably broader frequency response.

Measurement Periods and Target Frequencies

The standards require that harmonic measurements be made over very short intervals of 3 seconds and short intervals of 10 minutes. These should be accumulated over a period of one day and one week respectively. The 99th percentile of the very short time values and the 95th percentile of the short time values are then compared with the harmonic voltage or current limits, as described in Section 13.8.2.

Measurements should be made at times when the maximum harmonic disturbance is expected, i.e. when all distorting loads are in operation. The harmonic frequencies measured shall include, at a minimum, the 3rd, 5th and 7th, in addition to any harmonic frequencies specific to the equipment concerned. The latter, for example, will include the resonant frequencies of any harmonic filters present, or in the case of high pulse number converters, frequencies equal to $(np \pm 1)$, where p is the pulse number and n is an integer.

Where the harmonic limits are based on the contribution to the harmonic voltage distortion at the PCC, the local network configuration should also be considered when harmonic measurements are to be made. Equipment outages, for example, may temporarily leave a network in a state of reduced fault level, yielding an apparently higher system impedance. This will tend to accentuate the harmonic voltage distortion created by the load beyond the level that might normally be expected.

Very Short Time Measurements (3 Seconds)
Voltage or current measurements are generally made using a fast Fourier transform (FFT) technique. Instruments should be synchronised to the fundamental frequency and should sample the respective waveform over a period of about 200 ms, recording the value of each harmonic. The RMS averaging process shown in Equation (13.17) is then used over 15 consecutive intervals to obtain a 3-second very short time harmonic voltage value.

$$V_{n,vs} = \sqrt{\frac{1}{15} \sum_{i=1}^{15} V_{n,i}^2} \qquad (13.17)$$

where $V_{n,vs}$ is the very short (vs), 3-second RMS value of the nth harmonic voltage. A similar equation applies for harmonic currents.

Short Time Measurements (10 Minutes)
A 10-minute harmonic voltage value can be obtained by similarly averaging 200 consecutive 3-second values as shown in Equation (13.18).

$$V_{n,sh} = \sqrt{\frac{1}{200} \sum_{i=1}^{200} V_{(n,vs),i}^2} \qquad (13.18)$$

where $V_{n,sh}$ is the short (sh), 10-minute RMS value of the nth harmonic voltage. A similar equation applies for harmonic currents.

13.10   Problems

1   Derive Equation (13.1). Does this equation hold when the currents are not harmonically related but are arbitrary sinusoids instead?

2   **A** Show that where half wave symmetry exists then $a_k = b_k = 0$ for k even, and
    for k odd that $a_k = \frac{4}{T} \int_0^{T/2} i(t)\cos(kwt)dt$ and $b_k = \frac{4}{T} \int_0^{T/2} i(t)\sin(kwt)dt$.
    B Show that when $i(t)$ is an even function, the b_k terms become zero.
    C Show that when $i(t)$ is an odd function, the a_k terms become zero.

3 In this question, we consider the cancellation of the 5th and 7th harmonics in the 12-pulse rectifier configuration shown in Figure 13.9. The 30° phase shift required between the two rectifier bridges is achieved by using Dd0 and Dy11 connected transformers. The Dd0 connection introduces no phase shift between HV and LV windings, but the Dy11 connection introduces a 30° leading phase shift, as seen by the LV windings. Each 6-pulse rectifier functions entirely independently, emitting harmonic currents of order $6n \pm 1$. Therefore we may notionally express the AC current flowing in the LV delta winding in the form:

$$I_\Delta = I_1 \sin(\omega_1 t) + I_5 \sin(5\omega_1 t) + I_7 \sin(7\omega_1 t) + I_{11} \sin(11\omega_1 t) + I_{13} \sin(13\omega_1 t)$$
$$+ \ldots + I_{6n-1} \sin((6n-1)\omega_1 t) + I_{6n+1} \sin((6n+1)\omega_1 t)$$

where I_i is the amplitude of harmonic i (and may be negative).

A Write a similar expression for the LV star winding current, taking into account the 30° phase shift introduced by the Dy11 connection.

B These currents are reflected into the HV windings of their respective transformers, and collectively make up the supply line current. The Dd0 transformer presents a scaled version of I_Δ, but the Dy11 transformer introduces a 30° phase shift, the sign of which is dependent on the phase sequence of the harmonic concerned. It can be shown that the Dy11 connection introduces a 30° leading phase shift for positive sequence voltages from HV to LV, and a 30° lagging phase shift for negative sequence voltages, a characteristic possessed by half of the harmonics present. Demonstrate this using phasor diagrams similar to those in Figure 7.7.

C Therefore the summation of the HV currents depends upon the sequence of the harmonics concerned, and can be expressed as follows:

$I_{HV} = I_\Delta + I_Y \angle -30°$ for positive sequence harmonics (1st, 4th, 7th, 10th, 13th...) and $I_{HV} = I_\Delta + I_Y \angle +30°$ for negative sequence harmonics (2nd, 5th, 8th, 11th, 14th...)

Substitute the expressions for I_Δ and I_Y into these equations and show that the 5th and 7th harmonic currents cancel in the HV supply, while all other harmonics accumulate, yielding the spectrum shown in Figure 13.11.

4 A 6.6 kV pump motor is supplied from a dedicated 22 kV:6.6 kV transformer, shown in Figure 13.17. The motor is supplied from a 6-pulse VSD demanding 1500 kVA and having a displacement power factor of 0.85. The fault level at the 22 kV bus is

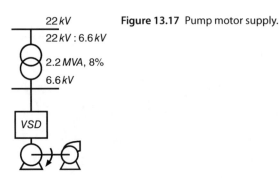

Figure 13.17 Pump motor supply.

22 kV
22 kV : 6.6 kV
2.2 MVA, 8%
6.6 kV

VSD

50 MVA and there are no nearby capacitors. A slightly pessimistic estimate for the harmonic emissions from the VSD suggests that $I_h = I_1 / h$.

A Determine the total power factor of the load seen at the 6.6 kV bus. Consider harmonics up to and including the 13th.

(Answer: 0.82)

B Find the source impedance seen at the 6.6 kV bus in ohms. (Assume for simplicity that all impedances are reactive.)

(Answer: 2.45 ohms)

5 We now consider the steps necessary to design a 50 Hz, 5th harmonic filter to suit the distorting load in question 4. The filter will be star connected and configured like that in Figure 13.12a and will provide power factor correction for the load while reducing the 5th harmonic emissions to within the limits prescribed by IEEE 519. It is to be tuned to the 4.8th harmonic (i.e. to a frequency 4.8 times the fundamental, just below the 5th harmonic).

A The filter's reactive power rating is usually chosen so that the load/filter combination achieves an acceptable power factor. Determine the fundamental reactive power required from the filter to increase the combined load's displacement power factor to 0.95, and hence find the filter's impedance X_F at the fundamental frequency.

(Answer: $Q_F = 371$ kVar, $X_F = 117.4$ ohms)

B Show that the capacitive reactance of the filter at the fundamental, X_C is related to X_F according to:

$$X_C = \left[\frac{h^2}{h^2 - 1} \right] X_F, \text{ and hence find } X_C.$$

(Answer: 122.7 ohms)

C Show also that the filter inductor's fundamental reactance X_L, is given by $X_L = X_C / h^2$, where h is the harmonic order of the filter (in this case 4.8), and thus obtain X_L.

(Answer: $X_L = 5.32$ ohms)

Figure 13.18 shows the equivalent circuit with the filter installed. The harmonic currents divide between the filter impedance and that of the source. As expected, the largest harmonic current absorbed by the filter will be the 5th; however, it will also absorb other harmonic components in proportion to the filter and source impedances.

D Assuming that the harmonic spectrum remains as specified in question 4, calculate the magnitude of each harmonic current (up to and including the 13th) flowing into the filter as well as those flowing into the upstream supply. Find the RMS current supplying the combined load.

(Answer: $I_{RMS} = 118.4$ A)

Figure 13.18 Division of harmonic currents.

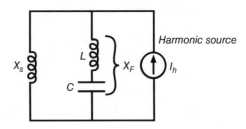

E The capacitor's voltage rating is not assessed according to the RMS summation of the fundamental plus the harmonic voltage components, as one might expect. Instead the arithmetic sum of these components is used to approximate the peak voltage that the capacitor must sustain and thus to provide a degree of redundancy in the selection of the capacitor's voltage rating. Therefore to obtain the capacitor voltage rating we use the equation:

$$V_{rated} = \sum_h I_h \left(X_C / h \right)$$

The fundamental voltage will influence the capacitor rating to a large extent, and therefore the maximum likely fundamental voltage should be used in this assessment. Find the largest fundamental current flowing through the filter, occurring when the local bus voltage is at a maximum of 1.1 pu, and hence determine the largest fundamental voltage appearing across the capacitor. (Why is this larger than the supplied phase voltage?)

(Answer: $I_{1max} = 35.7$ A, $V_{phase} = 4380$ V)

F From the amplitudes of the harmonic currents evaluated in part (d), obtain the capacitor voltage rating for this filter. Determine also the capacitance value required for each capacitor can.

(Answer: $V_{rated} = 5165$ volts; $C = 13 \mu$F (50 Hz), $C = 10.8 \mu$F (60 Hz))

G For this application the reactors will probably be individually constructed on gapped iron cores, and at the 5th harmonic they can be expected to have a Q value of about 50. In addition to the inductance value, the RMS winding voltage rating must also be specified. This is calculated using a similar equation to that used for the capacitor rating:

$$V_{rated} = \sum_h I_h X_L h$$

Calculate the inductance and value required and the winding voltage rating.

(Answer: $L = 16.9$ mH (50 Hz), 14.1 mH (60 Hz); $V_{rated} = 1590$ V)

H Show that the filter reduces the level of the 5th harmonic to within the limits imposed by IEEE 519.

6 If the transformer in question 4 is a dry type, calculate the harmonic loss factor (F_{HL}) and the de-rating factor necessary as a result of the load's harmonic content:

A for the load without the filter

B for the combined load.

Assume that the magnitude of the rated eddy current loss at the fundamental frequency is 12% of the rated I^2R loss.

(Answer: no filter: $F_{HL} = 3.72$, de-rating factor 0.88; combined load: $F_{HL} = 1.4$, de-rating factor $= 0.98$)

7 Demonstrate that as a result of the filter in question 4, the 5th harmonic current emission as seen at the PCC lies within the limits prescribed by IEEE 519. Is the installation similarly compliant on the 6.6 kV bus?

13.11   Sources

1 IEC *TR IEC 61000-3-6 Assessment of emission limits for the connection of distorting loads to MV, HV and EHV power systems*, 2012, International Electrotechnical Commission, Geneva.

2 IEEE Power Engineering Society, Std 519 *IEEE Recommended practice and requirements for harmonic control in electric power systems*, 2014, IEEE Power Engineering Society, New York.

3 Standards Australia / Standards New Zealand, *Power quality recommendations for the application of AS/NZ 61000.3.6 and AS61000.3.7 (HB 264)*, 2012, Standards Australia, Sydney.

4 Standards Australia / Standards New Zealand, *TR 61000.1.4 General historical rationale for the limitation of power frequency conducted harmonic current emissions from equipment in the range up to 2 kHz.*, 2012, Standards Australia, Sydney.

5 Gosbell VJ, Perera S, Browne, T, *Harmonic allocation using IEC/TR 61000.3.6 at the distribution/transmission interface*, University of Wollongong, 13th International Conference on Harmonics & Quality of Power, 2008.

6 Browne TJ, Gosbell VJ, Perera S, Falla LM, Windle PJ and Perera ACD, *Experience in the application of IEC/TR 61000.3.6 to harmonic allocation in transmission systems*, 41st International Conference on Large High Voltage Electric Systems, Paris, 2006.

7 Gosbell VJ, Robinson D. *Allocating harmonic emission to MV customers in long feeder systems*, 2008, Integral Energy Power Quality Centre, Brisbane.

8 Blooming TM, Carnovale DJ. *Application of IEEE Std 519-1992 harmonic limits*. IEEE IAS Pulp and Paper Industry Conference, Appleton WI, Eaton Electrical, 2005.

9 Blooming TM, Carnovale DJ, Dionise PE, *Price and performance considerations for harmonic solutions*. Eaton/Cutler-Hammer.

10 IEEE Power Engineering Society, *Std C57.110 IEEE recommended practice for establishing liquid filled and dry type power and distribution transformer capacity when supplying non-sinusoidal load currents* 2008, IEEE Power Engineering Society, New York.

11 IEEE Power Engineering Society, *Std 1531 IEEE guide for application and specification of harmonic filters*, 2003, IEEE Power Engineering Society, New York.

12 Wakileh G J, *Power system harmonics* 2001, 1st Edition, Springer.

13 Energy Networks Association, *Guideline for power quality: harmonics. Recommendations for the application of the joint Australian/New Zealand technical report TR IEC 61000.3.6: 2012* (Revised 2013), SAI Global, Sydney.

14 Dugan RC, McGranaghan MS, Santoso S, Beaty HW, *Electrical power systems quality* 2012, 3rd Edition, McGraw Hill, New York.

14

Operational Aspects of Power Engineering

In this chapter we examine some of the operational issues that practising engineers encounter every day, with a view to providing the new engineer with an introduction to some of the more important items. The material presented includes a discussion of the different bus arrangements and switchgear used in AC networks, electrical safety from the point of view of both electric shock and arc flash injury, limits of approach as well as isolation and switching procedures. Familiarity with these will assist in a more rapid integration into the profession. We begin with a discussion of device numbers and a description of the one line diagram.

14.1 Device Numbers

In chapter 12 symbols were used in some illustrations to identify protection relays. In addition to these symbols (defined in IEC 60617- Part 7), device numbers and acronyms are also defined in IEEE Standard C37.2 where they provide a shorthand notation for the function of devices installed in electrical installations. While most relate to electrical functions, some mechanical functions such as flow and pressure switches are also included.

Historically each device number represented an individual item of equipment, such as protection relays. Today they also apply to particular *protection functions* generated by the hardware or software contained in a multi-function relay. Device numbers are often preceded by a numerical prefix which may indicate a particular unit in a multiple unit installation and/or the potential of the bus concerned. For example, device number *52* applies to circuit breakers, and in a substation with three identical transformers the associated breakers may be identified by an alphabetic prefix *A*, *B* or *C*. In addition, circuit breakers on the HV side of a transformer frequently carry a *1* prefix and therefore are labelled *A152*, *B152* and *C152*. Those on the LV side are often prefixed by *2* and appear as *A252*, *B252* and *C252*. Prefixes may extend to other potentials as well; therefore *352* or *452* will signify circuit breakers operating at progressively lower potentials. Any device number may include prefixes.

Similarly, a suffix can also be added to a device number to further define its application. An inverse time overcurrent relay, for example, has the device number *51*. A suffix *G (51G)* implies that it is connected to a current transformer in the ground circuit of the device it is protecting; it therefore provides an inverse ground fault overcurrent function. (An alternate suffix *N* (neutral) is used in cases where the relay is connected to a residual source of current or voltage for the same purpose.) Suffixes are also frequently used to identify multiple

AC Circuits and Power Systems in Practice, First Edition. Graeme Vertigan.
© 2018 John Wiley & Sons Ltd. Published 2018 by John Wiley & Sons Ltd.

devices operating at the same potential. In some countries, current transformers use the device number *96*, so multiple current transformers all associated with the HV windings of transformer *A* in the example above would be identified as *A196A, A196B* and *A196C*.

Device numbers and also acronyms are used in drawings and technical documents. They also frequently appear in the labelling used to identify the equipment itself. Tables 14.1 and 14.2 include some of the more common device numbers and acronyms defined in IEEE C37.2, together with the equivalent symbols defined in the IEC standard IEC 60617.

Table 14.1 Common IEEE C37.2 device numbers and their IEC 60617 equivalent symbols.

Device number (IEEE C37.2)	Symbol (IEC 60617 – Part 7)	Description
21	Z <	Impedance relay (distance relay)
24	U/f >	Over excitation relay (over fluxing, *volts/Hz*)
27	U <	Under voltage relay
29	—	Disconnector or isolation device
40	—	Loss of excitation
46	$I_2 >$	Negative sequence current relay
47	U2 >	Negative sequence voltage relay
50	I ≫	Instantaneous overcurrent relay
50N	$I_0 ≫$	Instantaneous ground fault relay
51	I >	Inverse time overcurrent relay
51G	$I_0 >$	Inverse time ground fault relay
52	—	AC circuit breaker
152	—	Primary side circuit breaker
252	—	Secondary side circuit breaker
55	Cos ϕ >	Power factor relay
59	U >	Over voltage relay
64	$I_0 >$	Ground fault relay
67	I >	Directional overcurrent fault relay
67N	$I_0 >$	Directional ground fault relay
68	$I_2 >$	Transformer inrush blocking relay
78	ϕ >	Phase angle relay
81R	df/dt	Rate of change of frequency
81U	f <	Under-frequency relay
81O	f >	Over-frequency relay
87	ΔI >	Differential relay
96	—	Current transformer (some countries only)
97	—	Voltage transformer (some countries only)
99	—	Multi function relay (some countries only)

Table 14.2 Common acronyms defined in IEEE C37.2.

Acronym	Description
AFD	Arc flash detector
CLK	GPS clock
DDR	Dynamic disturbance recorder
DFR	Digital fault recorder
ENV	Environmental data recorder (wind, temp, icing, etc.)
HIZ	High impedance ground fault detector
HST	Historian (data gathering equip, for trend analysis)
LGC	Logic scheme (programmed protection, remedial action etc.)
MET	Substation metering equipment (W, VAr, VA, PF, energy)
PMC	Phasor data concentrator
PMU	Phasor measurement unit
PQM	Power quality measurement unit (harmonics and disturbances)
RIO	Remote input/output device
RTU	Remote terminal unit / data concentrator
SER	Sequence of events recorder (SOE) (fault data recording)
TCM	Trip circuit monitor (trip circuit supervision, TCS)

(Adapted and reprinted with permission from IEEE. Copyright IEEE C57.2 (2008). All rights reserved. Permission for further use of this material must be obtained from IEEE. Requests may be sent to stds-ipr@ieee.org)

14.2 One Line Diagram (OLD)

The one line diagram (also called a single line diagram SLD) is a convenient way of providing a concise overview of part or all of a power system; indeed, we have already seen simple one line diagrams in many of the proceeding chapters. They show only one phase and are usually drawn so that the highest potential appears at the top, while lower potentials and individual loads appear towards the bottom of the drawing. The one line diagram contains a representation of all the *primary equipment* such as transformers, circuit breakers, disconnectors, busbars, voltage and current transformers, as well as the *secondary equipment* comprising associated protection relays, energy metering, monitoring, control and communications equipment. It does not show the detailed wiring between these elements, but it does indicate the general interconnections between primary and secondary equipment, for example, which CTs and VTs supply particular protection relays or energy meters. Protection functions are represented in terms of the relevant device number, symbol or acronym.

The one line diagram also generally includes the voltage, current and VA ratings of all items of primary equipment and thus it provides a concise graphical description of the electrical installation at a particular site. Because of this, it is also used as a convenient data input method by network analysis programs for fault level and load flow studies. It is also one of the first documents used in planning maintenance activities and in fault finding, and should be kept up to date as system changes occur.

By way of example, Figure 14.1 shows the one line diagram of a small hydro generation station. The generator operates at 3.3 kV, delivering up to 5.1 MW into the local 33 kV network via a star-delta step-up transformer and a dedicated aerial feeder. The drawing shows all the primary and secondary equipment as well as the associated protection relays and energy metering.

The generator is a star connected machine with its neutral terminal cable connected to a 380 ohm *neutral earthing resistor* (NER), dimensioned to avoid serious winding damage in the event of a stator earth fault. The delta star transformer T1 takes the machine potential to 33 kV for connection to the local distribution network. Machine breaker A52 is located on the transformer's HV side, the HV star point being grounded through circuit breaker B52 only when the machine operates in island mode supplying the local area while isolated from the remainder of the grid. At all other times the neutral ground reference is provided from the remote 33 kV substation. Isolator A29 permits the machine to be removed from the grid for maintenance purposes while the station continues to be supplied via the station service transformer.

Duplicate *A* & *B* primary protection is provided for the machine and the machine/transformer combination by two numerical Schweitzer SEL-700G generator protection relays. The *A* protection relay A99A, provides fast differential protection across the machine windings using CTs A96A and A96D, and impedance and overcurrent protection using A96A. It also provides over-excitation, under/over voltage and frequency, reverse power and loss of excitation protection from voltage information provided by voltage transformer A97B. This relay includes an automatic synchronisation package that compares the machine voltage and frequency to that of the network, sending signals to both the *auto voltage regulator* (AVR) and the governor to permit the machine to be synchronised without operator intervention.

The duplicate protection relay A99B also provides these protection functions from the 33 kV bus, while providing differential machine and transformer protection by comparing the machine winding current with that delivered by the outgoing aerial feeder. The relay applies the necessary current scaling and phase shifting through software, avoiding the need for complex CT connections. Duplicate transformer HV earth fault protection (50G) is provided by CTs A196D and A196E. Both relays also include stator fault protection by detecting the potential developed across the NER via VT A97A. Each also records watt, VAr, voltage and current information for *system control and data acquisition* (SCADA) purposes.

The outgoing 33 kV aerial feeder is separately protected by a Schweitzer SEL311L feeder protection relay, which provides differential, distance and overcurrent protection. In this case an optical communications link provides current comparison between the relays at each end of the feeder. Finally current and voltage transformers A196C and A197B provide an energy meter with information relating to the net energy delivered to the feeder. All the above information is readily available from the one line diagram, demonstrating its usefulness in summarising the essential characteristics of a network.

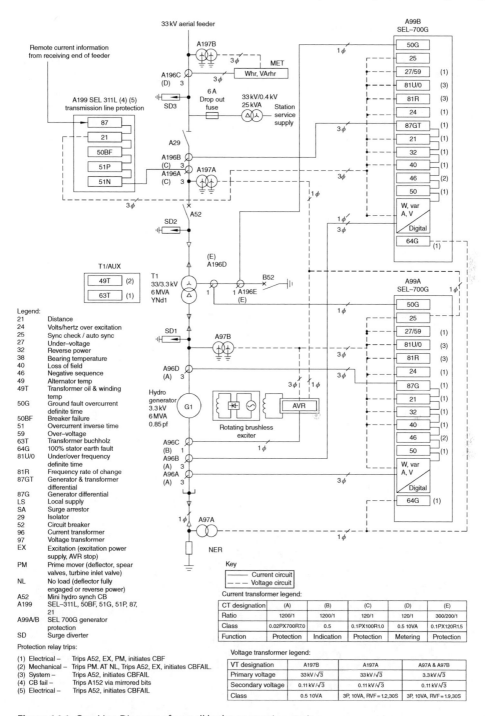

Figure 14.1 One Line Diagram of a small hydro generating station.

14.3   Switchgear Topologies

We shall investigate several of the more common MV and HV bus arrangements, but first we begin with a brief discussion of the insulating media used in MV and HV switchgear. Traditionally, air has been used to provide dielectric insulation between busbars and conductors in HV and EHV outdoor *aerial switchyards*. However, the large clearances required demand a sizeable footprint, which can present a problem where space is limited. Figure 14.2 shows part of a large urban substation occupying several hectares. Such a substation should ideally be located very close to the load centre it serves, but the land requirement frequently means that a compromise must be made, often leading to longer than desirable distribution feeders.

Circuit breakers used in aerial switchyards are generally classified as *dead tank* or *live tank*. As the name suggests, the interruption chambers of a dead tank breaker are at ground potential, with each current interrupter being contained within its own 'tank', as shown in Figure 14.3a. The HV terminations are made via bushings around which are positioned the necessary current transformers. Live tank breakers on the other hand, use bushing-like interruption chambers and operate at phase potential with respect to ground. They do not have provision for current transformers, which must be provided separately. Because both these circuit breakers periodically require maintenance, it is customary to install an isolation switch (or disconnector) on the supply side, in order to permit the breaker to be earthed while maintenance activities are carried out. An isolation switch also provides a visual indication of disconnection.

Figure 14.2 Part of an urban aerial switchyard.

(a)

(b)

Figure 14.3 (a) 220 kV Dead tank circuit breaker (b) Live-tank circuit breakers (left) and associated current transformers (right).

In the 1970s *gas insulated switchgear* (GIS) became commercially available as an alternative to aerial switchyards, and while it is considerably more compact, it is also more expensive. This mainly indoor technology uses sulphur-hexafluoride gas (SF_6) to insulate the busbars and to provide the necessary arc quenching in circuit breakers. Because of its extremely good insulating properties, the phase-to-phase and phase-to-ground clearances are very much smaller than those required in air. This provides for very compact equipment, in which both the busbars and the associated switchgear are enclosed in a heavy metallic casing, sufficient to contain the gas pressure and, more importantly, the effects of a fault. Figure 14.4 shows a suite of 145 kV double bus GIS switchgear in which the busbars are contained within the circular chambers running the length of the suite.

GIS technology finds application in city and urban areas where space is at a premium. It can also be used to advantage in the upgrading of older aerial switchyards, where the limited space makes it difficult to build a new switchyard within an operating old one. A suite of GIS switchgear can often be installed in a vacant section of a yard, allowing circuits to be progressively transferred from old to new, with minimal load disruption.

SF_6 is a particularly potent greenhouse gas, and manufacturers are now mixing it with other gasses to reduce the possible impact on the environment should it escape. As environmental concerns about the ongoing use of SF_6 grow, manufacturers are also turning to other gases for arc quenching and dielectric insulation duties. Carbon dioxide is finding application in MV and HV circuit breakers as a quenching medium, and although it too is a greenhouse gas, it is much less environmentally damaging than SF_6. As part of this trend, vacuum circuit breaker technology, which has been widely used up to 52 kV, is now being investigated for use at higher potentials.

Figure 14.4 Double bus 145 kV GIS switchgear.

Indoor MV *metal enclosed* switchgear (also called *metal clad switchgear*) uses air to insulate the internal busbars. It consists of a series of abutting cubicles, one for each of the incoming and outgoing circuits, through which pass the AC busbars (see Figure 14.8b). Connection is made between the busbars and an outgoing circuit when the associated breaker is racked onto the busbars and closed. The circuit breaker contacts mate with the busbar stubs, connecting the circuit side to the bus through the breaker. Isolation is achieved when the circuit breaker is opened and racked off the busbars. As a result, metal clad switchgear does not require the isolation switches necessary in aerial switchyards.

Figure 14.5 shows an 11 kV vacuum switch temporarily mounted on a trolley enabling it to be withdrawn from its cubicle. The right-hand photo shows the circuit breaker spring-loaded 'fingers' that make electrical contact with the busbar stubs.

Last century, transformer oil was commonly used as an insulating medium in MV circuit breakers due to its excellent arc quenching properties, and many oil insulated switches are still providing good service. However, over time, oil becomes contaminated with carbon and combustible gasses such as acetylene and ethylene, and it must be periodically replaced, imposing an ongoing maintenance requirement. Because of this and the risk of fire, oil is no longer preferred. Modern MV circuit breakers generally use either a vacuum or SF_6 gas, while HV breakers generally use SF_6 as the quenching medium.

14.3.1 Common Busbar Configurations

There are numerous ways in which incoming feeders and the loads they supply can be configured on a busbar. Some arrangements, while more expensive, offer more load security and operational flexibility than others. It is generally the importance of the load that dictates the choice of bus configuration.

Figure 14.5 Left: A single bus 11 kV vacuum circuit breaker, withdrawn from its cubicle. Right: Circuit breaker: busbar contacts (foreground), vacuum interruption canisters (background).

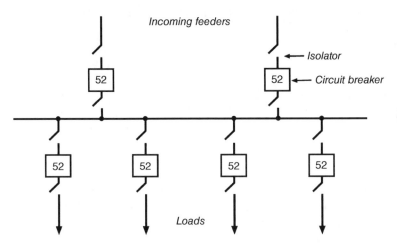

Figure 14.6 Single bus arrangement.

Non-critical loads, for example, may be supplied from a simple *single bus* arrangement, like that shown in Figure 14.6. This provides very little operational flexibility and in the unlikely event of a bus fault all load will be lost. Critical loads on the other hand, are usually provided with a degree of supply redundancy. Typically this may include duplicate incomers from separate transformers, feeding the load from two independent bus sections, so that a fault on one will not result in a loss of load from the other.

Operational flexibility should also be considered. For example, is it possible to remove circuit breakers or protection relays for maintenance, without disturbing the load? Load segregation is also important in the case of critical loads, which are often fed from a different bus section to non-critical load. This removes the possibility of critical load loss due to a fault in a piece of non-critical equipment. All these items should be considered when choosing a particular bus configuration.

Single Bus Configuration

The simplest configuration is the single bus arrangement shown in Figure 14.6, drawn to represent an HV aerial bus where each circuit breaker is provided with the isolation switches necessary to permit circuit breaker maintenance. While this is the least expensive arrangement, as mentioned above, it suffers from the distinct disadvantage that a bus fault will lead to the loss of the entire bus. Individual circuits must also be removed from service in order to maintain the associated circuit breaker or the protection relay. On the other hand, this configuration is easy to operate and expand, while bus zone and feeder protection is straightforward to implement.

Sectionalised Bus

Figure 14.7 shows a modification to the HV single bus arrangement with the inclusion of a *bus-coupler circuit breaker* and its associated isolation switches. A sectionalised bus can preserve half the load in the event of a bus fault, since only the faulty portion need be shed. Alternatively, if the bus coupler is normally run open, then in the event that one incomer feeder trips, the coupler can be closed, allowing the entire load to run from the remaining feeder. This arrangement is frequently used in small substations supplied from two distribution transformers (each able to carry the full load), where the combined fault level is beyond the interruption capacity of the station's switchgear, thus providing $(n-1)$ redundancy.

Double Bus Configuration

A much more flexible double bus configuration appears in Figure 14.8a. This schematic is appropriate to medium voltage metal clad switchgear, where each breaker can be withdrawn from its cubicle to complete the isolation, thus removing the need for isolation switches. Two independent buses are provided; these run through separate chambers in each cubicle. Each feeder can be connected to either bus, depending on where the circuit breaker is inserted into its cubicle. The arrangement shown in Figure 14.8b has top and bottom buses arranged in separate chambers, with the outgoing circuit termination located in between. This may be coupled to the top bus by inserting the breaker in the upper position, thereby connecting the circuit side to the top bus, or to the bottom bus by inserting it in the lower position and connecting the circuit side to

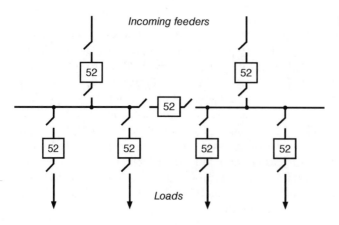

Figure 14.7 Sectionalised bus.

(a)

(b)

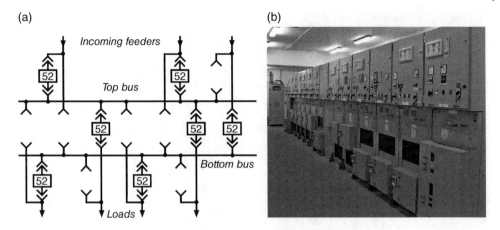

Figure 14.8 (a) MV double bus configuration (b) 11 kV double bus switchboard.

the lower bus. Figure 14.8b shows such an arrangement in which some loads are connected to the top bus with others on the bottom. In this case the circuit breakers can be racked up and down on a permanent carriage inserted into the cubicle. Other configurations use circuit breakers similar to those shown in Figure 14.5 and are placed on a temporary carriage for connection to either the top or bottom bus.

The double bus arrangement permits critical loads to be quarantined from non-critical, thereby preventing a fault in a non-critical piece of equipment from disturbing a critical load. Frequently, a critical bus is provided with a degree of feeder redundancy so that in the event that one incomer trips, no load will be lost. In addition, a sectionalised double bus provides the advantages described above, so that in the unlikely event of a bus fault, the bus zone protection will only trip that portion of the load on the faulted bus. This configuration also permits the load to be transferred to one bus, should the other require maintenance or extension. MV busbars are segregated from one another in separate chambers, so that access to one is not compromised by the presence of the other.

Figure 14.9 shows the side view of a suite of double bus GIS switchgear. The circular bus chambers are located one above the other at the front, and run the length of the suite. The switchgear is divided into individually pressurised bays, one for each incoming or outgoing circuit. These are connected via HV cables which enter from the basement, at the rear of each bay. Each circuit is supplied with a voltage transformer in addition to a chamber containing the necessary protection and metering current transformers. Each incoming or outgoing circuit passes through a circuit breaker before being routed to two isolators, one for the top bus and one for the bottom. Sectionalising breakers are inserted at intervals along each bus in order to minimise the load loss in the event of a fault.

This particular equipment has a rated SF_6 pressure of 6.3 bar, although it can safely operate at pressures as low as 5.5 bar (by which time low gas pressure alarms will certainly have been generated). The SF_6 gas provides dielectric insulation for the busbars and associated switchgear, while acting as the arc quenching medium for the circuit breakers as well.

Figure 14.9 Double bus 145 kV GIS switchgear.

Figure 14.10 shows a sectional view of bays *A* and *B* in a double bus aerial switchyard together with the corresponding portion of the power circuit one line diagram, which shows only the primary devices in the circuit. Not shown in this figure are six additional bays that accommodate two local generators, two transformers, two additional transmission lines, and two bus section circuit breakers and a bus coupler breaker.

Bay *A* terminates an incoming transmission line via a dead tank breaker A52, and a combination isolation/earth switch, A29. It can be coupled to either bus via bus-selection isolators A29, AB29 and B29. Then A29 and B29 are pantograph isolators, which operate vertically on a scissor principle and can be conveniently located beneath a busbar when space is limited; AB29 is a conventional double break isolator. This bay is also equipped with three single-phase capacitive voltage transformers (A97) for synchronising, protection and metering purposes. Bay *B* supplies a 30 MVA, 110 kV:22 kV distribution transformer, via dead tank circuit breaker B52.

The protection and metering current transformers required for each bay reside within the associated dead tank circuit breaker, while the CTs associated with the transformer's low voltage winding are contained within the transformer tank.

A variation of the double bus arrangement is sometimes used in HV aerial switchyards where outgoing circuits are supplied by two circuit breakers, one from each bus, as shown in Figure 14.11. While this provides a very flexible and secure supply (since any bus fault can be isolated without loss of load), it is particularly expensive since two circuit breakers are required per circuit.

Breaker-and-a-Half Configuration

Figure 14.12 shows the breaker-and-a-half configuration, so called because three circuit breakers are required for every two incoming or outgoing circuits. It uses two normally energised busses, and allows any circuit breaker to be removed for maintenance

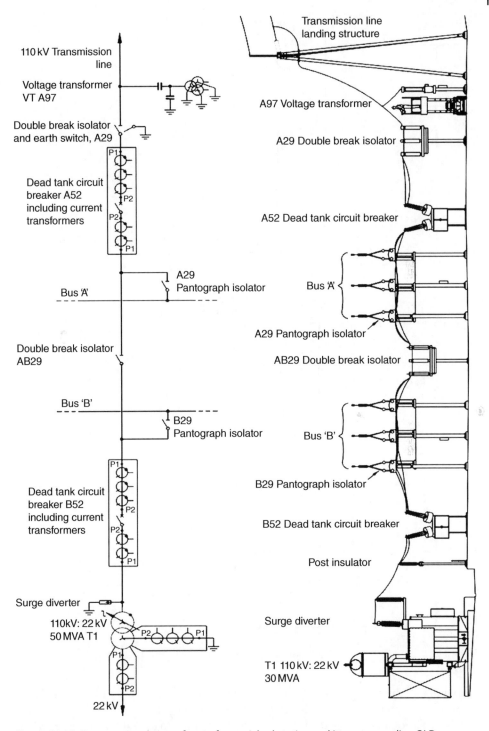

Figure 14.10 Cross-sectional view of part of an aerial substation and its corresponding OLD.

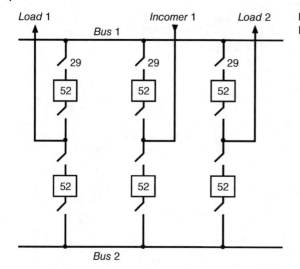

Figure 14.11 Aerial switchyard double bus arrangement.

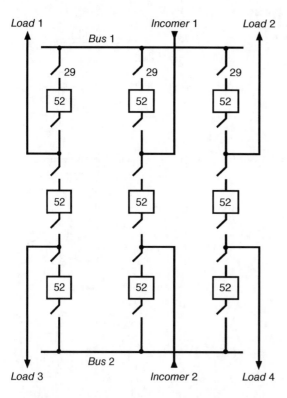

Figure 14.12 Breaker-and-a-half bus configuration.

without disturbing the associated loads. It provides a highly flexible and secure arrangement and allows bus faults to be isolated without loss of load.

In the event of a circuit breaker failure, however, some load will be lost. Should the centre breaker fail then depending on the position of the associated isolators, both circuits may be lost, although each can be readily restored. On the other hand, if either of

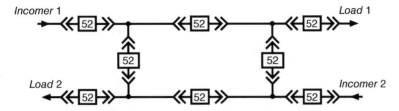

Figure 14.13 MV ring bus.

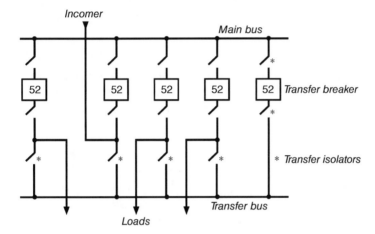

Figure 14.14 Main and transfer bus configuration.

the outside breakers fails only one circuit will be lost. The protection arrangements are somewhat more complex, however: each circuit requires its own voltage transformer as well as duplicate current transformer(s) in each adjacent circuit breaker.

Ring Bus Configuration

The ring bus, shown in Figure 14.13, is often used in MV cable networks where supply must be distributed between several small substations, for example around a large hospital or throughout a university campus. It operates as a closed loop with two or more incoming feeders and multiple out-goers, each of which effectively has two sources of supply. A fault in any part of the ring can be isolated with minimal load disruption, and circuit breakers can be removed at any time for maintenance, without loss of load. The ring bus protection requirements are similar to those for the breaker and a half scheme, and in the case of aerial substations, it can be readily adapted to a breaker and a half configuration.

Main and Transfer Bus Configuration

In the case where the circuit breaker maintenance must be performed without a loss of load the main and transfer bus arrangement can be used, as depicted in Figure 14.14. Under normal operation the main bus supplies the load and the transfer bus is de-energised. When a circuit breaker must be removed for maintenance the transfer breaker supplies the load from the transfer bus, without loss of load. As with the single bus configuration, a main bus fault will result in the loss of the entire load.

Since the transfer circuit breaker is capable of supplying any incoming or outgoing circuit, its protection relay must first be programmed with the appropriate settings before any switching can take place.

14.3.2 Circuit Arrangements on a Busbar

It is tempting to assume that the potentials at all points along a busbar are virtually the same, and therefore both the incoming feeders and outgoing loads supplied can be arranged without restriction. While this may be approximately so for active power flows, reactive flows require more careful planning. Further, since a busbar has a finite current rating, feeders and loads must be interspersed so that no part of the bus becomes overloaded during normal or abnormal operations.

Reactive Voltage Drop

The distribution of any reactive power sources (capacitors) or heavily reactive loads on a busbar should be carefully considered. Any busbar has a small resistive and inductive impedance, the inductive component of which is generally dominant and becomes particularly important when considering the reactive power flowing along the bus. The inductive impedance of medium voltage busbars is typically around 0.15 milliohms/m and the small voltage drops that arise when loads are adversely arranged can easily upset the current balance between incoming feeders. This is illustrated in Figure 14.15a, where three inductive loads are supplied from an MV bus, in which the inductive reactance is shown. In order to improve the power factor and suppress harmonics, two harmonic filters have also been installed at the opposite end of the bus. The reactive power flowing from these to each load generates small reactive drops (δV) along the bus, in such a way that the voltage at point A is slightly higher than that at point B. This means that there is less voltage dropped across incomer 2 and, as a result, it supplies less current than incomer 1. This arrangement means that it is not possible to fully load feeder 2 without overloading feeder 1. A better configuration is shown in Figure 14.15b, where by interspersing the filters and loads there is little net reactive drop between the incomers, and therefore both share the current equally.

For loads without capacitive support, the reactive demand of each will still generate reactive voltage drops, and there are still advantages in arranging them symmetrically with respect to the incomers. If load 2 has the lowest power factor, for example, it is best installed at the centre of the bus, as shown in Figure 14.15b.

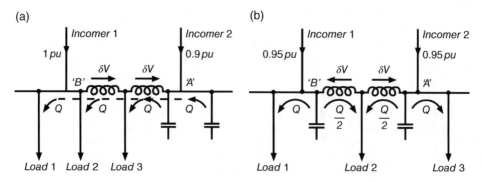

Figure 14.15 (a) Sub-optimal load configuration (b) Improved load configuration.

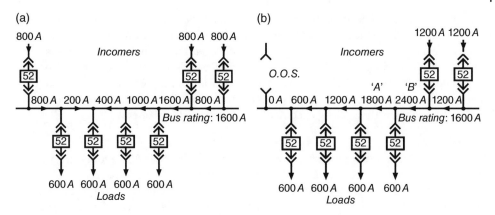

Figure 14.16 (a) Normal bus operation (b) LH feeder OOS, bus sections A and B overloaded.

Bus Overload

Critical loads are frequently provided with supply redundancy, so that should an incoming feeder trip unexpectedly (an $n-1$ situation) those remaining can carry the load. Consideration must be given to the resulting current distribution in such a situation, to ensure that sections of the bus do not become overloaded. Figure 14.16a depicts an arrangement in which the 1600 A bus only operates within its current rating when all feeders are connected. Should the left-hand feeder trip or be taken out of service (OOS), then although those remaining are capable of supporting the load, the bus will become overloaded at points A and B, as shown in Figure 14.16b. A better arrangement would see an incomer positioned between each pair of outgoing circuits.

14.4   Switching Plans, Equipment Isolation and Permit to Work Procedures

Switching Plans

From time to time all electrical installations must be reconfigured in order to cater for changing load requirements or perhaps to provide access to part of the network for maintenance or augmentation purposes. Before any switching operation can be commenced a switching plan must be written. This document is prepared with reference to the one line diagram and identifies the reason for the switching, the equipment concerned, together with all isolation and earthing points. It includes detailed and numbered step-by-step instructions for the switching process, in the exact sequence in which they will be carried out.

The aim of any switching plan is to safely achieve the desired outcome without unexpected loss of load or overloading any particular items of equipment. This frequently involves altering the normal bus arrangement to accommodate the required change before the subject equipment is switched or removed from service. For example, it may be necessary to rearrange the loads connected to a double bus when an incoming feeder is to be taken out of service. As a result, the switching process may become quite complex and the sequence in which the steps must occur often becomes critical.

A switching plan will be prepared by staff familiar with the plant concerned, to the extent that all eventualities can be foreseen in advance and accommodated as required. It will generally be written by one person and checked by another (often the switching operators who will execute the plan). Switching plans generally include provision for the operator to record the time that each step was completed. Where a particular switching operation occurs frequently, it is common for organisations to prepare standard switching plans that can be executed by suitably qualified operators.

Isolation Procedures

In order to safely isolate an item of equipment, fundamental switching practices must be used. These are generally quite simple and logical, and will form part of the switching plan. Any piece of equipment that is in service must first be de-energised. This requires switching off the main circuit breaker as well as every other associated energy source. This may include sources of mechanical energy; for example, sources of compressed air, pressurised hydraulic oil or water must also be isolated, so that no part of the equipment can operate unexpectedly.

The steps that must be followed for the complete electrical isolation of a piece of equipment can be summarised as follows:

a) *Correctly identify* the circuit to be isolated, as outlined in the switching plan.
b) *De-energise each energy source* associated with this equipment. This step requires opening the main circuit breaker as well as any auxiliary sources of electrical or mechanical energy.
c) *Isolate each energy source* so that it cannot easily be unexpectedly brought back into service. This requires removing MV circuit breakers from the busbars and opening the isolators associated with HV circuit breakers. The tripping and closing supply fuses associated with the circuit breaker is also generally removed. Some LV circuit breakers can also be withdrawn from the busbars into an isolated position, as shown in Figure 14.17. In the case of LV circuit breakers which cannot always be removed from the bus, it is usual to withdraw the isolation fuses provided specifically for this purpose.

Figure 14.17 A 3500 A LV circuit breaker, withdrawn from its cubicle in the isolated position.

d) *Prove dead.* In the case of HV and MV circuits this is generally achieved with a *test stick* or a similar device. The MV test stick shown in Figure 17.18a is essentially a resistive voltage divider incorporating an analogue indicator that is used to prove an MV circuit dead. This device requires physical contact with the bus conductor concerned, usually achieved by lifting the circuit side shutters in the cubicle, to expose the busbar stubs. The length of the stick keeps the operator at a safe distance from a live busbar.

HV aerial circuits can be proven dead by bringing a voltage detection device like that in Figure 14.18b (fixed to a long fibreglass pole) into close proximity with the bus conductors. This device emits an audible tone when close to any live conductor. (Its operation can therefore be checked by bringing it very close to any LV circuit). LV circuits are usually proven dead with a multimeter.

All circuits should be treated as live until proven dead, and therefore the necessary personal protective equipment must be worn during all such tests.

In circuits where substantial capacitance exists, such as harmonic filters or power factor correction equipment, a suitable period of time must be allowed to elapse for the capacitors to discharge, otherwise hazardous DC potentials may still exist in the circuit. Internal discharge resistors are usually built in to both MV and LV capacitors to absorb stored energy after switch-off, a process that usually requires about 10 minutes. It will be necessary to test the capacitor potentials at the end of this period to be sure that discharge has indeed taken place.

In the case of LV circuits, care should be taken to ensure that the circuit isolated is in fact the one to be worked on. While this may sound obvious, the main circuit breaker is frequently located in a switchboard remote from the equipment itself, and therefore if the wrong circuit breaker is opened in error, the subject equipment will still be alive. The need to prove dead at the equipment itself is therefore paramount. To obviate this problem, industrial LV equipment is frequently equipped with a *motor safety station* (MSS). This consists of an isolation switch and a set of line fuses, adjacent to the equipment concerned. Thus proving dead can be performed at the worksite where there is minimal risk of misidentification.

In the case of mechanical systems, gauges should be checked to ensure that pressure vessels have discharged to zero. Similarly, any mechanism capable of storing energy must be operated and subsequently tested to ensure that all stored energy has been dissipated.

(a)

(b)

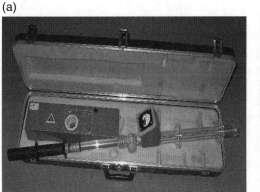
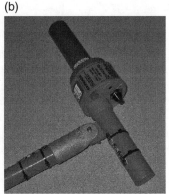

Figure 14.18 (a) Resistive 11/22 kV test stick (b) 'Modiework' voltage detector.

e) *Earthing.* In the case of HV and MV apparatus, an earth must be applied at the point of supply, so that staff working on it, do so downstream of the earthed point. This will generally be on the circuit side (load side) of the circuit breaker.

Where possible two earths should be applied, a *functional earth* at the supply circuit breaker itself, and an *operational earth* near the actual work site, so that maintenance staff work between earths. The operational earths frequently consist of a set of trailing earth conductors, connected at one end to the line conductors and to an earth point provided for this task at the other. Dual earthing minimises the risk of induced voltages in earthed transmission lines and switchyard circuits.

LV systems generally need not be earthed as part of the isolation process. This is because, due to the relatively low potentials and the small circuit lengths involved, it is not possible for de-energised conductors to have dangerous potentials induced within them from adjacent live conductors (either through capacitive coupling or magnetic induction). So long as an LV circuit has been isolated, proven dead and tagged, work can generally commence.

f) *Tagging each isolation point.* So that others can see that the equipment is out of service, an *out of service tag* is placed at each isolation point. This identifies the equipment concerned; it states the reason that it has been taken out of service and by whom, as well as the time and date of the isolation.

Finally, the isolation boundary around the equipment is usually delineated with rope or barrier tape so that maintenance staff do not accidentally stray into live areas. While isolation procedures appear straightforward, they are often fraught with dangers, particularly when a group of people are involved with a maintenance task on equipment that has been removed from service. Problems occasionally arise either through lack of communication or miscommunication within a work group, and they usually occur when the equipment is initially taken out of service or when it is returned to service. For example, people may attempt to commence work before it is safe to do so, or through work group confusion the equipment may be returned to service before all maintenance tasks have been completed. Either event can prove disastrous for the people concerned. To avoid such risks *permit to work* procedures have been devised that ensure that no work commences prior to the isolation being completed, and preventing the equipment from being returned to service until everyone has left the worksite.

Permit to Work

In order to prevent accidents from occurring permit to work procedures are used throughout all industries, the details of which tend to vary slightly from industry to industry and frequently from site to site. Accordingly everyone working on a particular site must be trained in the permit to work procedure used at that site. The following is a general discussion only and does not reflect the particular requirements that may be applied at different sites.

Permit to work procedures are used any time maintenance or construction activities are undertaken and when a work crew must be given access to a piece of equipment or to a portion of a site. The permit to work document applies to the crew conducting the work and it is used to ensure the safe isolation and de-isolation of equipment.

Generally all isolations will be performed by one or perhaps two people, as specified in the switching plan. They must be familiar with the site and appropriately trained to perform the required isolation(s). In general, they will not be part of the maintenance work crew. Once the isolation procedure is complete, the isolation officer will demonstrate to the work crew's representative that each piece of equipment has been correctly isolated, and when satisfied they will sign the isolation sheet, formally accepting that the worksite is safe and ready for work to commence. That person then effectively becomes in charge of the maintenance task and is responsible for communicating with the entire work crew.

Personal Danger Tags

Before work can commence, each member of the work group must sign on to the job attendance register and place their token at every isolation point in order to prevent the equipment from being re-energised before the work has been completed. Some industries use a personal danger tag as this token. In such cases, each isolation point contains a personal danger tag from everyone working on the job. The equipment cannot be returned to service until everyone in the work crew has signed off the attendance register and removed their danger tag from each and every isolation point.

The personal danger tag includes the name and in some cases a photograph of its owner, and effectively states that this person is still working on the equipment, and until this tag has been removed by its owner, the equipment cannot be returned to service. Should members of the work group leave the site without removing their personal danger tag(s) they must return and do so before the equipment can be returned to service; it is usually a dismissible offence to remove someone else's personal danger tag.

Personal danger tags can be a little inconvenient when there are many isolation points. A more recent alternative is the concept of a personal safety lock. This system is a little simpler and does not require each member of the work group to visit every isolation point.

The safety lock concept requires that the isolation officer places a safety lock at each isolation point, in such a way that the equipment concerned cannot be returned to service without removing the lock. The keys to each lock are then placed in a lock box similar to that in Figure 14.19. The isolation officer places his lock on the tag, together with one from every member of the workgroup concerned. In this way, de-isolation is physically not possible until every lock has been removed by its owner.

Figure 14.19 Lock box secured with two personal safety locks.

14.5   Electrical Safety

The risk of electric shock was considered at length in Chapter 11, including the various earthing schemes used worldwide in order to prevent it. While most electrical safety codes require maintenance on electrical equipment be carried out in a de-energised state, provision usually exists to conduct live electrical work in circumstances where this is not possible, or where a greater risk may exist if the work is conducted de-energised. For example, this may include electrical fault finding or maintenance on networks involving life support equipment, as well as in proving dead equipment that has just been isolated. Live electrical work requires a detailed job hazard analysis, a special permit to work, the identification of all risks and, where possible, their elimination. Above all it must be carried out by suitably trained and supervised personnel.

During such activities it is necessary for electrical practitioners to come into relatively close proximity to live conductors, which raises the question of personal safety. In addition to the risk of electric shock arising from contact with a live conductor, an arc flash injury simply requires that the victim be in the vicinity of an arc fault. Unfortunately, arcing faults can occasionally be triggered as a result of live electrical work.

The consequences of arc flash injury were poorly understood for many years, probably because arc flash incidents occur very infrequently, and although the probability of an arc flash incident is quite low, its consequences can be extreme. In recent decades, the risks associated with arc flash injury have become better appreciated and specific arc flash personal protective equipment (PPE) has been developed. Many instrumentalities now require arc flash rated PPE to be worn whenever any live electrical work is undertaken. The following sections discuss both these risks and the PPE required to mitigate them.

14.5.1   Electric Shock Injury

Provided practitioners maintain a safe distance from live conductors the risk of shock should remain acceptably low. Many national safety standards provide guidance with respect to electrical shocks, including the *National Fire Protection Association*[1] publication *NFPA 70E®*, *Electrical Safety in the Workplace®*, published in the USA. This document defines two shock protection boundaries with respect to any live conductor; the limited approach boundary and the restricted approach boundary, as outlined in Figure 14.20.

Limits of Approach

The limited approach boundary represents the limit of approach for an unqualified person. Should there be a need for such a person to cross this boundary, they must be accompanied by a qualified person who will advise of the hazards specific to the site.

The restricted approach boundary may only be crossed by qualified personnel, provided that steps are taken to mitigate the shock risk. This includes the wearing of full body cover, closed footwear and insulated gloves. It may also require that the energised electrical conductors be insulated from the qualified person, as well as from any other conductive object. Finally, the qualified person shall be insulated from any conductive object at a different potential.

---

1 Electrical Safety in the Workplace® and NFPA 70E® are registered trademarks of the National Fire Protection Association, Quincy, MA, USA.

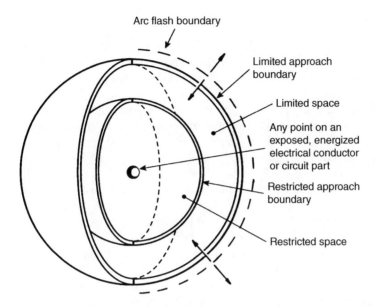

Arc flash boundary

Limited approach boundary

Limited space

Any point on an exposed, energized electrical conductor or circuit part

Restricted approach boundary

Restricted space

Figure 14.20 Limits of approach boundaries (NFPA 70E®). (Reproduced with permission from NFPA 70E®, Electrical Safety in the Workplace, Copyright©2014, National Fire Protection Association. This reprinted material is not the complete and official position of the NFPA on the referenced subject, which is represented only by the standard in its entirety).

Table 14.3 defines generic limited and restricted approach boundaries for a range of AC potentials. While these boundaries may change with jurisdictional requirements, they are broadly representative of those used worldwide.

Table 14.3 Typical Australian limits of approach boundaries for AC shock prevention.

Nominal system voltage (phase to phase)	Limited approach boundary (mm) (un-qualified persons)	Restricted approach boundary (mm) (qualified persons)
<1000 V	1000	Not specified
11 kV	2000	700
22 kV	2000	700
33 kV	2000	700
50 kV	3000	750
66 kV	3000	1000
110 kV	3000	1000
132 kV	3000	1200
220 kV	4500	1800
275 kV	5000	2300
330 kV	6000	3000
400 kV	6000	3300
500 kV	6000	3900

14.5.2 Arc Flash Injury

Electric shock is not the only risk to which electrical practitioners are exposed. The arc generated when a major electrical fault occurs also presents a substantial burn risk to electrical personnel. The first significant work on the risks associated with arc flash injury appeared in 1982 in the form of a paper published by American engineer Ralph H Lee, entitled *The other electrical hazard: electric arc burns.* His was the first work to quantify the power available from an arc as a function of the local fault level, and he also quantified the threshold energy density for a curable burn as being 1.2 calories/cm^2 or 5 joules/cm^2. This figure is widely recognised today and it defines the *arc flash boundary*, being the distance from an arc at which unprotected skin will receive a second-degree burn. Lee also derived an equation for the distance to the arc flash boundary from a three-phase arc in terms of the local fault level, and realised that the level of PPE required to protect against arc flash injury far exceeded that in use at the time for electric shock prevention. He also made suggestions as to the type of protective equipment that should be worn by practitioners operating within the arc flash boundary.

An arc is formed when the dielectric strength of the air between two electrodes is exceeded, and the resulting ionised gas creates a small current path. If sufficient current is permitted to flow, the surrounding temperature will increase very rapidly, explosively vaporising the electrode material, creating a metal plasma which further contributes to the current path. Provided sufficient conductive vapour is available the arc will continue to grow, generally requiring about 40 V to sustain each centimetre of length, almost independent of the current flowing. In low voltage systems this drop can absorb a sizeable portion of the available potential and as the electrodes are consumed, the arc length may become sufficiently long for it to self-extinguish. Arcs in high voltage systems can become considerably longer and in such cases the local short-circuit impedance limits the magnitude of the arc current, which will usually be sustained until the upstream protection operates. Temperatures in excess of 20,000°C can be generated in the vicinity of an arc, and the pressure wave created by the vaporisation process, showers everything in the area with hot gasses, metallic particles and metal plasma. Arc flash injuries can be severe and are often fatal.

As an example, Figure 14.21 shows the arc damage sustained to a busbar in an electrowinning rectifier, in which the failure of a paper board insulating panel triggered an arc between the busbar and the grounded structure beneath. The fault current existed for about 600 ms, reaching in excess of 30 kA as it consumed the lower part of a 6 × 2 inch busbar, until the arc eventually became so long that it self-extinguished. The resulting pressure wave was sufficient to blow the doors off the thyristor cubicles and the ensuing fire engulfed the rectifier, causing in excess of $1 M in damage. Fortunately, no one was injured.

Lee showed that the arc energy density decreases as the square of the distance from the arc itself, and he described the conditions under which maximum arc power is generated, which can be demonstrated using Figure 14.22. A simplified equivalent circuit of a single-phase arc is shown in Figure 14.22a, in which the short-circuit impedance is assumed to be wholly inductive. The arc current is limited by the system reactance X, and flows through a plasma in which the fundamental components of the current and the voltage lie in phase. However, since the voltage necessary to sustain an arc is largely independent of the current flowing, the arc 'resistance' is non-linear. The arc current

Figure 14.21 Arc damage, 6″ × 2″ busbar. Fault current: 30 kA for 600 ms.

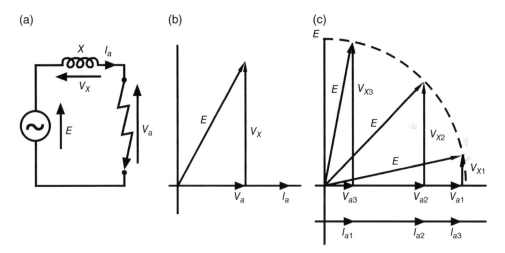

Figure 14.22 (a) Equivalent circuit (b) Phasor diagram (c) Variations in arc voltage and current.

also generally contains harmonics, but unless there are corresponding harmonic voltages present in the supply, the majority of the power in the arc will be derived from the fundamental.

Figure 14.22b shows the phasor relationship between the arc voltage V_a, the drop across the source reactance V_x, the open circuit voltage E and the arc current I_a. Three possible arc scenarios are illustrated in Figure 14.22c. Scenario 1 corresponds to a long arc in which most of the available potential is dropped across the arc path, leaving little to be dropped across the source reactance; as a result relatively little current flows. Scenario 3 corresponds to a short arc, where most of the available potential appears across the source reactance. As a result the current flow is large, almost equal to the single-phase fault current available at the site. However, since the arc power is given by the $V_a I_a$ product, in both of these scenarios this power is relatively low. This is not to say that these conditions are not dangerous – they certainly can be, but scenario 2 represents the maximum power condition which occurs when $V_a = V_x$. Because these voltages are in quadrature, each will equal 70.7% of the open circuit voltage and the $V_a I_a$ product will assume its maximum value.

Since it is impossible to predict which scenario will occur in any given situation, from a safety point of view, scenario 2 must be assumed, as it will produce the largest arc flash boundary. Further, since all three-phase conductors are usually co-located, a single-phase-to-ground fault will rapidly become a three-phase fault once the metal vapour expands to include the other two phases. The maximum possible three-phase arc power therefore becomes:

$$Arc\,Power\,(\mathrm{MW}) = \frac{\sqrt{3}}{2} V_{line}\, I_{bolted\,fault} \times 10^{-6} = \frac{MVA_{bolted\,fault}}{2} \tag{14.1}$$

Lee also derived an equation for the distance D in feet to the arc flash boundary for an unconstrained arc:

$$D = \sqrt{2.65\,MVA_{bolted\,fault}\;t} \tag{14.2}$$

Here t is the clearance time of the associated protective device in seconds. In the case of a fuse, t can be as little as 5 ms, whereas for a protection relay and circuit breaker, it is generally closer to 100 ms, and this value is frequently used in arc flash calculations. Clearly the longer an arc exists the larger will be the arc flash boundary and the higher the risk to anyone working within it. For a typical distribution transformer with an impedance of 5%, Equation (14.1) can be rewritten in terms of the rated transformer capacity as:

$$D = \sqrt{53\,MVA_{rated}\;t} \tag{14.3}$$

Figure 14.20 includes an indicative arc flash boundary, in addition to the limits of approach based on the avoidance of electric shock. Whereas the latter depend only on the system voltage, the arc flash boundary depends on the system fault level. Where this is high, the distance to the arc flash boundary may far exceed the limits of approach, and unless additional arc flash PPE is worn, working within the limited access boundary may be unsafe from an arc flash point of view.

Since it is usually necessary for practitioners to work within the arc flash boundary, additional arc flash rated PPE must also be worn. In such cases, the incident energy density E (cal/cm^2) to which a person may be exposed must be known if the appropriate PPE is to be chosen. This can be calculated in terms of the person's distance from a prospective arc as follows:

$$E = \frac{5.5 V_{line}\, I_{bolted\,fault}\,t}{D^2} \tag{14.4}$$

Once the prospective energy density has been calculated, suitably rated PPE can be chosen. Table 130.7(C)(16) in NFPA 70E® defines four categories of arc flash rated PPE based on the prospective energy density. As the latter increases, the degree of protection required increases in proportion. Category 1 requires PPE with an energy density withstand of $4\,cal/cm^2$, while category 4 clothing must collectively withstand $40\,cal/cm^2$.

(The complete version of NFPA 70E®, including this table and its associated explanatory notes can be viewed at no cost by visiting www.nfpa.org/freeaccess.)

NFPA 70E® also provides a table of common equipment found in industry, together with the corresponding arc flash category for PPE that must be used when live work is to be performed on each one. This table may be used, where appropriate, in choosing the required PPE; however, users should be aware that the energy density at the arc flash boundary is a function of the actual fault current, the voltage and importantly the actual fault clearing time. In situations where these parameters are different from those assumed in the table, it may be preferable to perform the calculations described in NFPA 70E® and choose the appropriate PPE accordingly.

For low voltage systems the assumption that arc current will equal the bolted fault current is frequently conservative, and for this reason the Lee equations may overstate the arc boundary or the energy density at a given distance from a low voltage arc. NFPA 70E® also includes similar equations derived by other authors, including those published in IEEE Standard 1584, *IEEE Guide for Performing Arc Flash Hazard Calculations*. This standard presents equations derived from data obtained from laboratory tests. Its results are more complex than those presented by Lee and include a wider range of voltages and fault currents in addition to providing results applicable to arcs formed within a switchboard cubicle (arcs in a box).

The Energy Networks Association in Australia has published ENA NENS 09-2014, *National Guideline for the Selection, Use and Maintenance of Personal Protective Equipment for Electrical Arc Hazards*. This document presents alternative equations to those in IEEE 1584 for the incident energy, following extensive testing in which it was discovered that considerably more plasma energy is directed from the end of three-phase busbars than from the sides during an arc fault.

Arc Rated PPE

The energy density rating of each arc flash PPE category, also called *arc thermal performance value* (ATPV), is the incident energy in cal/cm^2 that will result in a heat transfer through the material to the wearer of $1.2\,cal/cm^2$, sufficient to cause a second degree (curable) burn within about 100 ms. Arc rated clothing is assigned an ATPV, based on the results of extensive fabric testing. The ATPV of a garment increases with the weight of the fabric (g/m^2) and the number of layers used in its manufacture.

Arc rated fabrics are proven to be flame resistant and will self-extinguish once the initial heat source has been removed. They may be either produced from man-made synthetic materials engineered in such a way to have flame resistant properties, or from natural fibres such as cotton or wool, chemically treated to achieve the required flame resistant characteristics.

Two Category Approach to Arc Flash PPE

NFPA 70E® includes a provision for a simplified two category approach to arc flash protection. For routine work and any tasks that require category 1 and 2 arc flash protection, technical staff wear a long-sleeved shirt and trousers having an arc flash rating of at least $8\,cal/cm^2$, often in the form of a corporate uniform. When

Figure 14.23 Category 4 arc flash PPE.

hazardous tasks must be performed on live equipment requiring arc flash category 3 or 4 protection, the collective arc rating of the PPE worn must be equivalent to category 4, having an arc flash rating of at least $40\,\text{cal/cm}^2$. Such tasks include switching operations on some types of older metal clad switchgear, inspections of and measurements in live low voltage cubicles such as metering chambers and motor safety stations, and in proving dead LV and MV circuits. The arc flash suit shown in Figure 14.23 provides head-to-toe protection and is worn over the everyday PPE described above. The collective arc flash rating of all this PPE is equivalent to category 4.

Arc Flash in Low Voltage Cubicles

Low voltage cubicles deserve special mention with respect to arc flash injury. Because the low voltage restricted approach boundary lies within arm's reach, it has become accepted practice to conduct measurements, fault finding and inspection activities on live LV cubicles. Since LV arc flash injuries are relatively rare, the general impression is that live LV work is quite safe, and provided the appropriate precautions are taken and the correct PPE is used, this may be so. However, when something goes wrong and an arc is created in the confines of an LV cubicle (usually as the result of the work in progress), there is enormous scope for severe arc flash and/or pressure wave injury. Practitioners must accept that live LV work is potentially very dangerous and therefore the appropriate PPE (like that in Figure 14.23), must be worn, even for the smallest of tasks.

Manufacturers are now designing safer LV switchboards, by covering busbars with coloured insulation and shrouding them behind clear plastic panels. The LV cubicle in Figure 14.24 is an example of a well-designed metering low voltage CT chamber. Metering personnel frequently work in such cubicles, measuring the connected burden at the CT secondary terminals and occasionally comparing the current in the primary busbar with that flowing in the CT secondary. The risk to an unprotected person conducting such activities is extreme, despite the relatively low probability of an arc fault occurring.

A careful examination of Figure 14.24 will reveal that a Perspex panel has been close fitted over the front of the entire cubicle. The only items protruding through this are the potential fuses, the CT secondary terminal covers and the neutral terminal cover. These items can be accessed live without the risk of arc flash. Unfortunately, shrouded construction of this type is rare; many LV cubicles and switchboards in service contain exposed busbars or phase terminals, and collectively they represent some of the most hazardous electrical items. This situation is made worse because, as mentioned above, electrical practitioners have for years routinely carried out live work in such locations with minimal ill effects. Few appreciate that a dropped screwdriver or sealing wire, falling across exposed busbars has the potential to initiate a powerful arc with devastating results, particularly in a chamber enclosed on three sides.

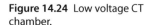

Figure 14.24 Low voltage CT chamber.

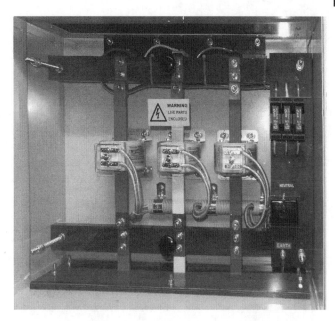

The further from the supply transformer the switchboard is located, the lower will be the fault current, but there will usually be sufficient energy in both the arc and the accompanying pressure wave, to cause severe injury to an unprotected person standing in front of an open LV cubicle.

In contrast, at the low voltage appliance level the potential fault currents are much lower, reduced by the impedance of building wiring and that of the appliance flex itself. But perhaps the most significant reduction in sustained fault level is due to the presence of fault current limiting devices such as fuses or circuit breakers. These will clear a fault very rapidly, limiting the duration of an arc fault, should one occur.

14.6 Measurements with an Incorrectly Configured Multimeter

Perhaps the one of the most common causes of arc flash injuries in LV switchboards is the measurement of current with a handheld multimeter. While this sounds innocuous enough, it can be particularly dangerous if the meter is configured incorrectly.

An ammeter is a low impedance device. Ideally it will present zero impedance to the circuit to which it is to be connected. In practice, however, it introduces a tiny resistance into the circuit (called a shunt) through which the current to be measured must flow. The small voltage developed across this shunt is interpreted by the meter and is displayed as a current. A voltmeter on the other hand is a high impedance device, ideally presenting an infinite impedance to the circuit to which it is connected. In practice a digital voltmeter will present between one and ten million ohms to the circuit to which it is connected. Such a resistance will not disturb the circuitry under test.

(a) (b) (c)

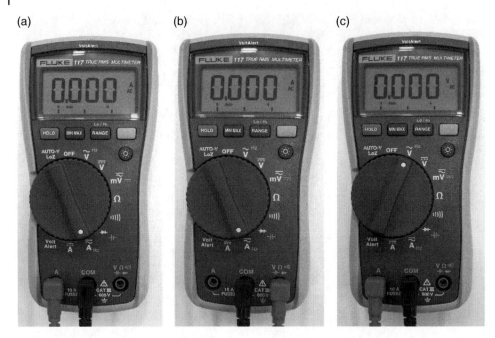

Figure 14.25 Three multimeter configurations (a) correct (b) incorrect (c) potentially dangerous.

Where a single instrument is required to perform both these functions, it must be configured differently for each purpose. To measure current, the low resistance shunt must be connected between the meter probes, while to measure voltage the shunt must be removed and substituted with the high resistance measuring circuit. This change in configuration cannot be accomplished using the range selection switch, since it is unable to carry the currents that the meter must be able to measure. Multimeter manufacturers therefore require the user to reconfigure the meter when changing from a voltage to a current measurement. This is achieved by inserting one of the two probes into a different receptacle on the face of the meter. Handheld multimeters generally have three receptacles into which the probes can be inserted. One is common to all measurements and it is labelled 'COM'. Of the other two, the one is for current measurements and is labelled 'A', while the other is for all other measurements and is labelled 'V Ω' etc., as shown in Figure 14.25.

Figure 14.25a shows a multimeter correctly configured for AC current measurement. The probes are inserted into the 'COM' and the 'A' receptacles, and the selector switch is set to amps AC (or DC) as required.

Figure 14.25b shows the same meter incorrectly configured for current measurement. The selector switch is in the correct position, but the probes are inserted as required for voltage measurement. This error is not usually dangerous[2], since the high impedance presented, prevents any current from flowing.

---

2 In the special case of a current transformer, introducing an high impedance into its secondary circuit, may lead to a large flux density in the CT core and possibly dangerous voltages being generated across the meter as a result.

Figure 14.25c depicts what is probably the most dangerous meter configuration error. Here the meter has its range selector switch configured for voltage measurement, but the probes are inserted into the current receptacles. This means that a very low impedance is inserted between them. When the meter is applied to a substantial potential difference in this configuration a very large current will flow, often sufficient to destroy the meter. This action can also be sufficient to trigger an arc fault if the arc struck as the operator withdraws the probes is able to link the terminals being measured.

Meter manufacturers attempt to limit the damage that may occur in this situation by including a fuse in series with the current receptacle, one that will rapidly clear when currents in excess of about 20 amps flow and one capable of interrupting a large current flow. This does not always protect the meter however, and depending on the magnitude of the fault current it may still be destroyed. The inclusion of a fuse will lessen the likelihood of operator injury; however, the use of arc flash PPE for all live measurements is essential. Extreme care should be exercised when using an instrument without a fuse.

14.7   Sources

1   IEC 60617–7 *Graphical symbols for diagrams – part 7 switchgear, control gear and protection devices*, 1996, International Electrotechnical Commission, Geneva.

2   IEEE Std C37.2 *IEEE Standard for electric power system device function numbers, acronyms and contact designations*, 2008, IEEE Power Engineering Society, New York.

3   NFPA 70E®, *Standard for electrical safety in the workplace*, 2015 Edition, National Fire Prevention Association, Quincy, MA.

4   Nack D, *Reliability of substation configurations* 2005, Iowa State University.

5   Lee RH, *The other electrical hazard: electric arc blast burns*, Paper IPSD 81–55, IEEE Industry Application Society Annual Meeting, Philadelphia. PA, October 5–9, 1981.

6   Lee RH, *Pressures developed by arcs*, IEEE Trans. Ind. Appl., vol IA-23, No. 4 pp. 760–764, July/Aug 1987.

7   Ammerman RF, Sen PK, Nelson JP. *Electrical arcing phenomena, a historical perspective and comparative study of the standards IEEE 1584 and NFPA 70E®*, Petroleum and Chemical Industry Conference, 2007, Calgary Alberta, Canada, 17–19 Sept 2007.

8   Prasad S. *Arc flash hazards the burning question*. IDC Electrical Arc Flash Forum, Melbourne, Australia 14th – 15th April 2010.

9   Electricity Networks Association, *NENS 09 – 2014 National guideline for the selection, use and maintenance of personal protective equipment for electrical arc hazards*, 2014, SAI Global, Sydney.

Index

AC Circuits and Power Systems in Practice, First Edition. Graeme Vertigan.
© 2018 John Wiley & Sons Ltd. Published 2018 by John Wiley & Sons Ltd.

Printed and bound by CPI Group (UK) Ltd, Croydon, CR0 4YY

17/04/2025

14658872-0001